Swarm Intelligence for Electric and Electronic Engineering

Girolamo Fornarelli
Politecnico di Bari, Italy

Luciano Mescia
Politecnico di Bari, Italy

T0336686

ENGINEERING
SCIENCE REFERENCE

Managing Director:	Lindsay Johnston
Editorial Director:	Joel Gamon
Book Production Manager:	Jennifer Romanchak
Publishing Systems Analyst:	Adrienne Freeland
Development Editor:	Myla Merkel
Assistant Acquisitions Editor:	Kayla Wolfe
Typesetter:	Christina Henning
Cover Design:	Nick Newcomer

Published in the United States of America by
Engineering Science Reference (an imprint of IGI Global)
701 E. Chocolate Avenue
Hershey PA 17033
Tel: 717-533-8845
Fax: 717-533-8661
E-mail: cust@igi-global.com
Web site: http://www.igi-global.com

Library of Congress Cataloging-in-Publication Data

Swarm intelligence for electric and electronic energy / Girolamo Fornarelli and Luciano Mescia, editors.
     p. cm.
  Includes bibliographical references and index.
  Summary: "This book provides an exchange of knowledge on the advances, discoveries and improvements of swarm intelligence in electric and electronic engineering, bringing together new swarm-based algorithms and their effect on complex problems and various real-world applications"--Provided by publisher.
  ISBN 978-1-4666-2666-9 (hardcover : alk. paper) -- ISBN 978-1-4666-2697-3 (ebook : alk. paper) -- ISBN 978-1-4666-2728-4 (pr : alk. paper) 1. Electrical engineering--Mathematics. 2. Electronics--Mathematics. 3. Swarm intelligence. I. Fornarelli, Girolamo, 1970- II. Mescia, Luciano, 1973-
  TK153.S953 2013
  621.30285'63--dc23
                                    2012029074

British Cataloguing in Publication Data
A Cataloguing in Publication record for this book is available from the British Library.

All work contributed to this book is new, previously-unpublished material. The views expressed in this book are those of the authors, but not necessarily of the publisher.

The authors would like to dedicate this work to their wives,
since so much time preparing this publication has been taken from them.

Table of Contents

Section 1
Circuit Design

Section 4
Scheduling and Diagnosis

Detailed Table of Contents

Section 1
Circuit Design

Design methodologies and tools for analog, mixed-signal and radiofrequency circuits are not automated like the digital ones due to the complexity of the issue. To this purpose, this section collects some research activities regarding the application of PSO approach to automated synthesis and optimization of electronic circuits and hardware. In particular, a PSO-based CAD tools able to develop microwave circuits and devices automatically, a PSO model to calculate small-signal model parameters of a transistor, and a multi-objective PSO for the optimal design of a versatile CMOS current conveyor transconductance amplifier will be illustrated.

Chapter 1

Massimo Donelli, University of Trento, Italy

In this chapter, a methodology for the unsupervised design of microwave devices, circuits, and systems is considered. More specifically, the application of the Particle Swarm Optimizer and its integration with electromagnetic simulators is discussed in the framework of the microwave circuits and devices design and optimization. The idea is to automatically modify the characteristics of the device in an unsupervised way, with the goal of improve the device performances. Such kind of CAD tool could be the solution to reduce the time to market and keep the commercial predominance, since they do not require expert microwave engineers and it can reduce the computational time typical of the standard design methodologies. To assess the potentialities of the proposed method, a selected set of examples concerning the design of microwave planar devices such as filters, splitters, and other microwave devices under various operative conditions and frequency bands are reported and discussed. The chapter also includes a brief discussion concerning different strategies, such as parallel computation, to reduce the computational burden and the elaboration time. The obtained results seem to confirm the capabilities of the proposed method as effectiveness microwave CAD tool for the unsupervised design of microwave devices, circuits, and systems. The chapter ends with some conclusions and considerations related to ideas for future works.

Gabriel Cormier, Université de Moncton, Canada
Tyler Ross, Carleton University, Canada

Circuit models play an important role in the design and optimization of microwave circuits (circuits in the GHz frequency range). These circuit models contain many parameters, including parasitic elements, necessary to correctly model the behavior of transistors at high frequencies. These models are often designed based on a series of measurements. Because of its ability to efficiently locate the global optimum of an objective function, particle swarm optimization (PSO) can be a useful tool when matching a model to its measurements. In this chapter, PSO will be used to calculate a transistor's small-signal model parameters, determine the noise parameters of the transistor, and design a microwave mixer. The mixer is designed at 39.25 GHz, and a comparison between measurements and simulation results shows good agreement.

Mourad Fakhfakh, University of Sfax, Tunisia
Patrick Siarry, University of Paris-Est Créteil, France

The authors present the use of swarm intelligence in the analog design field. MO-TRIBES, which is an adaptive user-parameter less version of the multi-objective particle swarm optimization technique, is applied for the optimal design of a versatile building block, namely a CMOS current conveyor transconductance amplifier (CCTA). The optimized CCTA is used for the design of a universal filter. Good reached results are highlighted via SPICE simulations, and are compared to the theoretical ones. The use of such adaptive optimization algorithm is of great interest in the analog circuit design, as it is highlighted in this chapter.

Jai Narayan Tripathi, IIT Bombay, India
Jayanta Mukherjee, IIT Bombay, India
Prakash R. Apte, IIT Bombay, India

This chapter is an overview of the applications of particle swarm optimization for circuits and systems. The chapter is targeted for the Analog/RF circuits and systems designers. Design automation, modeling, optimization and testing of analog/RF circuits using particle swarm optimization is presented. Various applications of particle swarm optimization for circuits and systems are explained by examples.

Section 2
Antenna and Optical Devices

In last years, PSO techniques received increasing attention in the electromagnetic community since they are promising algorithms for handling optimization problems in the area of the design of antennas and optical devices. To this purpose, this section illustrates recent results regarding PSO developments aimed at the efficient design of reconfigurable antennas, microwave absorbers, erbium doped fibre amplifiers and Raman amplifiers. Moreover, the proposed contributions show applications of PSO to the synthesis of linear antennas array, the estimation of the permittivity of multilayer structures, the recovering of parameters describing the energy transfer phenomena among the erbium ions.

Chapter 5

Arezoo Modiri, University of Texas at Dallas, USA
Kamran Kiasaleh, University of Texas at Dallas, USA

This chapter is intended to describe the vast intrinsic potential of the swarm-intelligence-based algorithms in solving complicated electromagnetic problems. This task is accomplished through addressing the design and analysis challenges of some key real-world problems, ranging from the design of wearable radiators to tumor detection tools. Some of these problems have already been tackled by solution techniques other than particle swarm optimization (PSO) algorithm, the results of which can be found in the literature. However, due to the relatively high level of complexity and randomness inherent to these problems, one has to resort to oversimplification in order to arrive at reasonable solutions utilizing analytical techniques. In this chapter, the authors discuss some recent studies that utilize PSO algorithm particularly in two emerging areas; namely, efficient design of reconfigurable radiators and permittivity estimation of multilayer structures. These problems, although unique, represent a broader range of problems in practice which employ microwave techniques for antenna design and microwave imaging.

Chapter 6

Sotirios K. Goudos, Aristotle University of Thessaloniki, Greece
Zaharias D. Zaharis, Aristotle University of Thessaloniki, Greece
Konstantinos B. Baltzis, Aristotle University of Thessaloniki, Greece

Particle Swarm Optimization (PSO) is an evolutionary optimization algorithm inspired by the social behavior of birds flocking and fish schooling. Numerous PSO variants have been proposed in the literature for addressing different problem types. In this chapter, the authors apply different PSO variants to common antenna and microwave design problems. The Inertia Weight PSO (IWPSO), the Constriction Factor PSO (CFPSO), and the Comprehensive Learning Particle Swarm Optimization (CLPSO) algorithms are applied to real-valued optimization problems. Correspondingly, discrete PSO optimizers such as the binary PSO (binPSO) and the Boolean PSO with velocity mutation (BPSO-vm) are used to solve discrete-valued optimization problems. In case of a multi-objective optimization problem, the authors apply two multi-objective PSO variants. Namely, these are the Multi-Objective PSO (MOPSO) and the Multi-Objective PSO with Fitness Sharing (MOPSO-fs) algorithms. The design examples presented here include microwave absorber design, linear array synthesis, patch antenna design, and dual-band base station antenna optimization. The conclusion and a discussion on future trends complete the chapter.

    Girolamo Fornarelli, Politecnico di Bari, Italy

    Antonio Giaquinto, Politecnico di Bari, Italy

    Luciano Mescia, Politecnico di Bari, Italy

The rapid increasing of internet services requires communication capacity of optical fibre networks. Such a task can be carried out by Er3+-doped fibre amplifiers, which allow to overcome limits of unrelayed communication distances. The development of efficient numerical codes provides an accurate understanding of the optical amplifier behaviour and reliable qualitative and quantitative predictions of the amplifier performance in a large variety of configurations. Therefore, the design and optimization of the optical fibre can benefit of this important tool. This chapter proposes an approach based on the Particle Swarm Optimization (PSO) for the optimal design and the characterization of a photonic crystal fibre amplifier. Such approach is employed to find the optimal parameters maximizing the gain of the amplifier. The comparison with respect to a conventional algorithm shows that the proposed solution provides accurate results. Subsequently, the presented method is used to study the amplifier behaviour by evaluating the curves of optimal fibre length, erbium concentration, gain, and pumping configuration. Finally, the PSO based algorithm is exploited to determine the upconversion parameters corresponding to a desired value of gain. This application is particularly intriguing since it allows recovery of the values of parameters of the optical amplifier, which cannot be directly measured.

    Alireza Mowla, K.N. Toosi University of Technology, Iran

    Nosrat Granpayeh, K.N. Toosi University of Technology, Iran

    Azadeh Rastegari Hormozi, K.N. Toosi University of Technology, Iran

In this chapter, the authors introduce the hybrid erbium doped fiber amplifier (EDFA)/fiber Raman amplifier (FRA) and its optimization procedure by particle swarm optimization (PSO). EDFAs, FRAs, and their combinations, which have the advantages of both, are the most important optical fiber amplifiers that overcome the signal power attenuations during the long-haul communication. After choosing a proper configuration for a hybrid EDFA/FRA, users have to choose its numerous parameters such as the lengths, pump powers, number and wavelengths of pumps, number of signal channels and their wavelengths, the signal input powers, the kind of the fibers and their characteristics such as the radius of the core, numerical apertures, and the density of Er3+ ions in the EDFA. As can be seen, there are many parameters that need to be chosen properly. Here, efficient heuristic optimization method of PSO is used to solve this problem.

Section 3
Control Optimization

PSO algorithms showed their full potentialities in terms of flexibility, robustness and reliability in modern control theory. In particular, they have been proposed to employ new optimal control techniques in complex systems consisting of a large number of autonomous agents. In this section, the first chapter illustrates the application of swarm intelligence in the emerging field of the swarm robotics. In particular, the distributed bees algorithm is proposed as a solution to distributed multi-robot task allocation. In the second chapter, the use of PSO is proposed to perform the optimal design of fractional order controllers. In the third chapter, a multi-objective particle swarm optimization algorithm is employed to find the optimal structure of a parallel kinematic manipulator-based machining robotic cell, minimizing the power consumed by the manipulator during the machining process in a robotic cell. Finally, the fourth chapter reports a generalized PSO, which is inspired by linear control theory, and its practical engineering applications in two fields, like the fault detection and classification of electrical machines, and the optimal control of water distribution systems.

Chapter 9

Aleksandar Jevtić, Robosoft, France

Diego Andina, E.T.S.I.T.-Universidad Politécnica de Madrid, Ciudad Universitaria, Spain

Mo Jamshidi, University of Texas, USA

This chapter introduces a swarm intelligence-inspired approach for target allocation in large teams of autonomous robots. For this purpose, the Distributed Bees Algorithm (DBA) was proposed and developed by the authors. The algorithm allows decentralized decision-making by the robots based on the locally available information, which is an inherent feature of animal swarms in nature. The algorithm's performance was validated on physical robots. Moreover, a swarm simulator was developed to test the scalability of larger swarms in terms of number of robots and number of targets in the robot arena. Finally, improved target allocation in terms of deployment cost efficiency, measured as the average distance traveled by the robots, was achieved through optimization of the DBA's control parameters by means of a genetic algorithm.

Chapter 10

Guido Maione, Politecnico di Bari, Italy

Antonio Punzi, Politecnico di Bari, Italy

Kang Li, Queen's University of Belfast, UK

This chapter applies Particle Swarm Optimization (PSO) to rational approximation of fractional order differential or integral operators. These operators are the building blocks of Fractional Order Controllers, that often can improve performance and robustness of control loops. However, the implementation of fractional order operators requires a rational approximation specified by a transfer function, i.e. by a set of zeros and poles. Since the quality of the approximation in the frequency domain can be measured by the linearity of the Bode magnitude plot and by the "flatness" of the Bode phase plot in a given frequency range, the zeros and poles must be properly set. Namely, they must guarantee stability and minimum-phase properties, while enforcing zero-pole interlacing. Hence, the PSO must satisfy these requirements in optimizing the zero-pole location. Finally, to enlighten the crucial role of the zero-pole distribution, the outputs of the PSO optimization are compared with the results of classical schemes. The comparison shows that the PSO algorithm improves the quality of the approximation, especially in the Bode phase plot.

Most machining tasks require high accuracy and are carried out by dedicated machine-tools. On the other hand, traditional robots are flexible and easy to program, but they are rather inaccurate for certain tasks. Parallel kinematic robots could combine the accuracy and flexibility that are usually needed in machining operations. Achieving this goal requires proper design of the parallel robot. In this chapter, a multi-objective particle swarm optimization algorithm is used to optimize the structure of a parallel robot according to specific criteria. Afterwards, for a chosen optimal structure, the best location of the workpiece with respect to the robot, in a machining robotic cell, is analyzed based on the power consumed by the manipulator during the machining process.

A generalization of the popular and widely used Particle Swarm Optimization (PSO) algorithm is presented in this chapter. This novel optimizer, named Generalized PSO (GPSO), is inspired by linear control theory. It enables direct control over the key aspects of particle dynamics during the optimization process, overcoming some typical flaws of classical PSO. The basic idea of this algorithm with its detailed theoretical and empirical analysis is presented, and parameter-tuning schemes are proposed. GPSO is also compared to the classical PSO and Genetic Algorithm (GA) on a set of benchmark problems. The results clearly demonstrate the effectiveness of the proposed algorithm. Finally, two practical engineering applications of the GPSO algorithm are described, in the area of electrical machines fault detection and classification, and in optimal control of water distribution systems.

Section 4
Scheduling and Diagnosis

Nowadays, many scientific experiments are conducted through complex and distributed scientific computations that are represented and structured as scientific workflows. Moreover, nondestructive testing and imaging techniques are widely used in many industrial and research applications. This section reports some recent studies of PSO to solve hard problem of scheduling, task allocation and diagnosis. In detail, the first chapter illustrates the implementation of PSO to determine all the changes needed in the electric transmission system infrastructure allowing the balance between the projected demand and the power supply, at minimum investment and operational costs. The second chapter proposes the application of PSO to solve the problem of short-term hydrothermal generation scheduling problem which consists of minimizing the fuel cost for thermal plants under the constraints of the water available for hydro generation in a given time period. The third chapter illustrates the ant colony optimization method in the framework of the nondestructive analysis of dielectric targets by using electromagnetic approaches based on inverse scattering. Finally, the forth chapter provides a survey of some common evolutionary algorithms used in electroencephalogram studies.

Chapter 13

Santiago P. Torres, University of Campinas (UNICAMP), Brazil
Carlos A. Castro, University of Campinas (UNICAMP), Brazil
Marcos J. Rider, São Paulo State University (UNESP), Brazil

The Transmission Expansion Planning (TEP) entails to determine all the changes needed in the electric transmission system infrastructure in order to allow the balance between the projected demand and the power supply, at minimum investment and operational costs. In some type of TEP studies, the DC model is used for the medium and long term time frame, while the AC model is used for the short term. This chapter proposes a load shedding based TEP formulation using the DC and AC model, and four Particle Swarm Optimization (PSO) based algorithms applied to the TEP problem: Global PSO, Local PSO, Evolutionary PSO, and Adaptive PSO. Comparisons among these PSO variants in terms of robustness, quality of the solution, and number of function evaluations are carried out. Tests, detailed analysis, guidelines, and particularities are shown in order to apply the PSO techniques for realistic systems.

Chapter 14

Víctor Hugo Hinojosa Mateus, Universidad Técnica Federico Santa María, Chile
Cristhoper Leyton Rojas, Universidad Técnica Federico Santa María, Chile

In this chapter, a particle swarm optimizer is applied to solve the problem of short-term Hydrothermal Generation Scheduling Problem – one day to one week in advance. The optimization problems have been formulated taking into account binary and real variables (water discharge rates and thermal states of the units). This proposal is based on a strategy to generate and keep the decision variables on feasible space through the correction operators, which were applied to each constraint. Such operators not only improve the quality of the final solutions, but also significantly improve the convergence of the search process due to the use of feasible solutions. The results and effectiveness of the proposed technique are compared to those previously discussed in the literature such as PSO, GA, and DP, among others.

Chapter 15

Matteo Pastorino, University of Genoa, Italy

Andrea Randazzo, University of Genoa, Italy

Electromagnetic approaches based on inverse scattering are very important in the field of nondestructive analysis of dielectric targets. In most cases, the inverse scattering problem related to the reconstruction of the dielectric properties of unknown targets starting from measured field values can be recast as an optimization problem. Due to the ill-posedness of this inverse problem, the application of global optimization techniques seems to be a very suitable choice. In this chapter, the authors review the use of the Ant Colony Optimization method, which is a stochastic optimization algorithm that has been found to provide very good results in a plethora of applications in the area of electromagnetics as well as in other fields of electrical engineering.

Chapter 16

Adham Atyabi, Flinders University, Australia

Martin Luerssen, Flinders University, Australia

Sean P. Fitzgibbon, Flinders University, Australia

David M. W. Powers, Flinders University, Australia

Electroencephalogram (EEG) based Brain Computer Interface (BCI) is a system that uses human brainwaves recorded from the scalp as a means for providing a new communication channel by which people with limited physical communication capability can effect control over devices such as moving a mouse and typing characters. Evolutionary approaches have the potential to improve the performance of such system through providing a better sub-set of electrodes or features, reducing the required training time of the classifiers, reducing the noise to signal ratio, and so on. This chapter provides a survey on some of the commonly used EA methods in EEG study.

Foreword

There has been an unprecedented recent growth of highly sophisticated methodologies, algorithms, and successful practices of optimization technologies applied to a spectrum of challenging real-world engineering problems.

This volume is positioned in a timely, rapidly expanding, and highly relevant area of system optimization. There is an ongoing timely and indisputably challenging quest to develop the comprehensive methodologies and solid algorithmic frameworks of population-based optimization augmented endowed with a variety of powerful metaheuristics. Within the realm of optimization tools, the techniques of particle swarm optimization (PSO) along with their numerous generalizations play a pivotal role. Here, some of them are worth highlighting: constriction factor PSO, comprehensive learning PSO, PSO with velocity mutation, and multi-objective PSO with fitness sharing.

Interestingly, the results of Google Scholar returned to the query "particle swarm optimization" show some interesting tendency: in 2000 there were 396 hits, while in 2005 this number went up to 3,140 hits. The year 2012 saw 4,010 hits, while a total number of hits over 2000-2013 is 25,300. All of these numbers speak loudly to the high relevance, potential, and a genuine practical importance of the discipline.

The book brings a series of interesting papers and forms a significant contribution to the body knowledge in the area of system optimization. The material is organized in several main sections concentrated on various application areas including electromagnetic fields, design and optimization of electronic devices, and automation and robotics, scheduling, diagnostics and classification with a plethora of convincing studies on the design of microwave circuits, antennas, analog filters, RF circuits, task allocation in swarms of robots, transmission expansion planning, short-term generation scheduling, to name only some of them.

Undoubtedly, the editors have done a superb job by bringing timely and important concepts and practice of the contemporary technology of PSO presented by experts in the field. The authors have offered an overall exposure of the key issues in a convincing way that, undeniably, will appeal to the broad spectrum of the readership.

Overall, the volume is a very much welcome, extremely timely, and very much needed publication, which elaborates on essential conceptual developments, fosters further endeavors in advanced methods of particle swarm optimization, and offers practitioners a wealth of valuable insights of practical relevance.

Witold Pedrycz
Department of Electrical and Computer Engineering, University of Alberta, Edmonton, Canada

Witold Pedrycz *(IEEE- M'88, SM'90, F'99) is a Professor and Canada Research Chair (CRC - Computational Intelligence) in the Department of Electrical and Computer Engineering, University of Alberta, Edmonton, Canada. He is also with the Systems Research Institute of the Polish Academy of Sciences, Warsaw, Poland. He also holds an appointment of special professorship in the School of Computer Science, University of Nottingham, UK. In 2009 Dr. Pedrycz was elected a foreign member of the Polish Academy of Sciences. His main research directions involve computational intelligence, fuzzy modeling and granular computing, knowledge discovery and data mining, fuzzy control, pattern recognition, knowledge-based neural networks, relational computing, and software engineering. He has published numerous papers in this area. He is also an author of 14 research monographs covering various aspects of computational intelligence and software engineering. Witold Pedrycz has been a member of numerous program committees of IEEE conferences in the area of fuzzy sets and neurocomputing. Dr. Pedrycz is intensively involved in editorial activities. He is an Editor-in-Chief of Information Sciences and Editor-in-Chief of IEEE Transactions on Systems, Man, and Cybernetics - part A. He currently serves as an Associate Editor of IEEE Transactions on Fuzzy Systems and is a member of a number of editorial boards of other international journals. He has edited a number of volumes; the most recent one is entitled "Handbook of Granular Computing" (J. Wiley, 2008). In 2007 he received a prestigious Norbert Wiener award from the IEEE Systems, Man, and Cybernetics Council. He is a recipient of the IEEE Canada Computer Engineering Medal 2008. In 2009 he received a Cajastur Prize for Soft Computing from the European Centre for Soft Computing for "pioneering and multifaceted contributions to granular computing."*

Preface

Bio-inspired algorithms make use of criteria occurring in natural phenomena, which they are inspired to, with the aim to find the solution of a specific problem. In particular, solutions can be searched by mimicking the behaviour of animal social groups. This kind of approach is called Swarm Intelligence and plays an important role in the field of such algorithms.

It is worthwhile to observe that the imitation does not allow an exact reproduction. Nevertheless, similarly to natural cases, this kind of algorithms requires examples to evaluate the possible solution by a measure of its quality. This evaluation requires the knowledge of the relation between the available examples and the measure of the solution.

In this sense, swarm intelligence based methods belong to the set of meta-heuristic ones; in fact they, need examples which are significant for the cases to be reproduced. In this field, Particle Swarm Optimization and Ant Colony have constituted methods of great interest since their introduction in the middle of the 90s, and as an increasing attention of the scientific community is focusing on Bees Algorithm.

After their first and necessary development phase these methods reached the maturity, as shown by their application to several fields as an alternative to traditional algorithms. This is due to many different aspects, but probably the most relevant one is that those algorithms imply the advantages of a simple implementation and, as a consequence, of a less complex application.

Such qualities affect the convergence properties of these methods in terms of both robustness and computational time. It followed that researchers developed a great number of methods aiming at solving real world engineering problems.

Therefore, now these methods represent an important alternative to improve past solutions especially in the two fields of design and optimization. In fact, in the former case they constitute an effective tool to find a solution, proving to be essential in complex situations in which traditional methods failed. In the latter case they can result efficient to identify the values of parameters that have to be refined in order to satisfy particular problem constraints or to obtain specific performance.

Electric and electronic engineering is not an exception in this context. In fact, procedures of design, optimization, and characterization of devices can benefit of the solutions of meta-heuristic methods, since the search of the solution is removed from experts' work and demanded to the automatic evolution of the design algorithm.

This book is aimed at showing this aspect of problem solution in the fields of electric and electronic engineering. In other words, the book reports neither all the possible solutions of a specific engineering problem nor the most recent solutions of each considered problem evaluating the multitude of applications of these algorithms to the different problems in electric and electronic engineering. Therefore, the main goal is collecting examples of the employment of these methods in problems which are fundamental in engineering from a conceptual point of view, providing the main ideas of their application to practical cases.

On the basis of these considerations, the proposed contributions face the problems of the design of circuits, devices and analog filters, the optimization of the synthesis of antennas, and systems based on optical fibres. It is shown that the concepts are applicable to cases characterized by circuit architectures requiring particular performance and devices whose behaviour is described by complex mathematical models.

The optimization and the design procedures could require that the swarm based algorithm interfaces to a different solver which evaluates the performance quality. This implies that such kind of methods allows automatic design and optimization guaranteeing also the application to the characterization of devices. As a consequence, those parameters, which cannot be derived from the system under analysis by means of the only observable variables, can be found indirectly.

These concepts can be extended to the employment of these algorithms in system control. Such control can be intended as the capability to define the parameters used by the system to control another one and as the identification of values characterizing its best architecture.

In a similar way, it is possible to argue that the capabilities of identifying the control parameters of a system can be exploited to arrange the best allocation of its resources and establish the optimal parameters during its working. This enables to fix the steps that have to be executed to assure the optimal working of the system, like the scheduling of the generation of electric energy or the planning of its distribution. Finally, it should be noted that these forecasting properties can be interpreted in terms of diagnosis of systems in which the parameter values point out behaviour anomalies.

All these problems can be solved by simple algorithms, which are based on the original versions of the proposed paradigms, or by deriving techniques which yield improved performance in their application field.

This scheme reproduces the organization of the book chapters, whose considered topics are affine. All the chapters have a practical spin-off in terms of real world application, otherwise all the problems could constitute a theoretic exercise which would not be significant in engineering.

The first part of the book reports applications of swarm intelligence to the circuit design and their optimization. In particular, in chapter 1, the problem of the design and optimization of microwave circuits are dealt with. In this chapter a methodology for the unsupervised design of microwave devices, circuits and systems is considered. In detail, the application of the Particle Swarm Optimizer and its integration with electromagnetic simulators is discussed in the framework of the microwave circuits and devices design and optimization. The idea is to automatically modify the characteristics of the device in an unsupervised way to improve the device performance. Such kind of CAD tool could be the solution to reduce the time to market and keep the commercial predominance. In fact, it can reduce the computational time which is typical of the standard design methodologies, avoiding the requirement of expert microwave engineers. In order to assess the potentialities of the proposed method, a selected set of examples, concerning the design of microwave planar devices such as filters, splitters, and other microwave components under various operative conditions and frequency bands are reported and discussed. The chapter also includes a brief discussion concerning different strategies, such as parallel computation, to reduce the computational burden and the elaboration time. Nowadays, the development of microwave devices and systems requires complex design techniques, high level of expertise, and a final tuning phase that could dramatically increase the costs and the time to market of the devices. In this framework, the application of evolutionary algorithms for the development of microwave CAD tools offer a possible solution to reduce the time to market and the devices cost. The proposed approach can be useful for microwave engineer and company involved in the design of microwave devices and systems, since it permits to design and modify a given device in an unsupervised manner without requesting an experienced engineer to operate.

As well as the design phase, device modelling can benefit of the advantages of particle swarm optimization. This is shown and discussed in chapter 2, in which the modelling of microwave transistors and the design of microwave circuits is performed by making use of PSO. In order to explore the topic of microwave transistor modelling, a commercial GaAs FET is used as an example in the first half of the chapter. The extraction process using particle swarm optimization is described in detail, and a small-signal transistor model and its noise parameters are successfully obtained. To illustrate the use of particle swarm optimization in a design problem, the second half of the chapter describes the design and optimization of a 39.25 GHz half-Gilbert cell mixer by a particle swarm algorithm. Transistor dimensions, transmission line lengths and impedances are all optimized to yield a mixer design with the optimal impedance matching, gain and stability. Since modern monolithic microwave integrated circuits (MMICs) are very costing to manufacture in terms of turnaround times, it is highly desirable to reduce the number of design-fabricate-test cycles needed until design specifications are achieved. Ideally, the first iteration would be successful. To achieve first-pass design success two things are required: accurate models and robust circuit designs. Swarm optimization is very useful for developing both high-quality models and optimal circuit designs, and both the extraction of a microwave transistor model and the optimization of a microwave mixer are discussed in the chapter. If implemented in the design process, these techniques have the potentiality to reduce both the time required to design a circuit and the costs associated with fabricating prototypes.

The problem of circuit design is also faced in chapter 3, where the optimal design of analog versatile building blocks, namely the CMOS current conveyor transconductance amplifier (CCTA), and its use for the design of universal active filters is presented. In detail, the chapter focuses on the AMS/RF optimization problem statement, giving details about a PSO based technique and its use to solve NP-hard problems. To this aim, an application to the optimal sizing of a current conveyor based current mode building block and its use for the design of a multifunction filter is discussed. Analog circuit design and synthesis have not been automated to a great extent so far. This is mainly due to their towering complexity. Optimizing the automatic sizes of the analog components is an important issue towards ability of rapidly designing true high performance circuits. For this reason, the Chapter is intended to skilled designers and to semaphores, as well. It details and highlights the use and the adaptation of a PSO algorithm to the optimal design of analog circuits.

If working at higher frequencies, circuits can present complications like parasitic effects, skin effects, coupling etc., which make their intuitive design complex. Thus, Design Automation is needed for making product design cycle easier. There are various methods for design automation. The most popular among them is to associate an algorithm with the circuit/system simulator, such that it can achieve the design goals. These algorithms can also be used for modelling and testing. Chapter 4 discusses the applications of Particle Swarm Optimization to Analog/RF circuits and systems. These applications can also be used for development of new EDA tools, therefore the chapter is intended for high frequency circuit designers and high speed system engineers.

The second part of the book shows how swarm intelligence can be applied to the design and the characterization of electronic devices in different fields, like antennas and optical fibres.

The intrinsic potentialities of the swarm-intelligence-based algorithms in solving complicated electromagnetic problems is discussed and proved in chapter 5, where some recent studies that utilize some variants of particle swarm optimization are discussed in two emerging areas, like efficient design of reconfigurable antennas and permittivity estimation of multilayer structures. The problem dependency of PSO algorithms is highlighted by the analysis of different benchmark problems. Comparisons in terms of final result accuracy and optimization cost are also presented. Subsequently, the authors illustrate how PSO is customized and utilized in challenging scenarios of designing reconfigurable and wearable antennas. In the last part of the chapter, the usefulness of PSO-based solutions in an even more challenging and attractive scenario is discussed. In that scenario, the algorithm is used to estimate the electrical properties of a lossy, multilayer structure as a model of human body tissue, as shown by the application of this study to tumor detection. Although unique, the considered problems represent a broader range of problems in practice employing microwave techniques for antenna design and microwave imaging. Therefore, this chapter could be very interesting for readers who are interested in creating customized PSO variants for their desired design and estimation problems. In particular, the reader is expected to get a good understanding of how he/she can take advantage of the flexibility of PSO algorithm and customize it for his/her particular problem.

The application of different PSO variants to common antenna and microwave design problems is also investigated in chapter 6. PSO is applied to both real and discrete-value optimization problems. These optimization problems can be solved by using single or multi-objective PSO algorithms. The book chapter is focused on major issues and challenges of PSO application to the design of devices which can be considered an important part of every modern system in wireless communications. Therefore, its intended target audience is comprised of a wide spectrum of professionals and researchers with a particular interest in the area of designing new algorithms as it will provide a set of different real world problems.

Chapter 7 proposes a PSO based approach for the optimal design and the characterization of a photonic crystal fibre amplifiers. A comprehensive explanation of the PSO algorithm and the basic phenomena involved in the amplification process is provided. The obtained numerical results show that the proposed solution can be considered as an useful tool to solve the optimization and characterization problem regarding several configurations of fibre amplifiers. The intended audience of this chapter is constituted by students, scientists and communication system engineers who might use or be involved with the considered devices. In particular, the reader can take advantage of the flexibility of PSO algorithm as very useful tool for understanding the behaviour of the erbium doped fibre amplifier, to provide accurate predictions for the design and optimization of the amplifier performance, to offer a novel and efficient characterization procedure allowing the recovering of most relevant spectroscopic parameters of rare earth ions. Moreover, the chapter topic fits in new emerging area as soliton transmission, bioscience, industrial, medical and surgical applications.

The optimization of optical fibre amplifiers constitutes the topic of the subsequent chapter 8. In particular the hybrid Erbium doped fibre amplifier (EDFA)/fibre Raman amplifier (FRA) and its optimization procedure by particle swarm optimization are introduced. Optical fibre amplifiers are the key elements in fibre optic communication systems that enable to overcome the signal power attenuations during the thousands of kilometres of the long-haul communication. EDFAs and FRAs are the most important optical fibre amplifiers and hybrid EDFA/FRA is a combined configuration that uses the advantages of the both mentioned amplifiers. After choosing a proper combination for a hybrid EDFA/FRA, a designer has to evaluate its numerous parameter values such as the fibre lengths, pump powers, number of pumps, wavelengths of pumps, number of signal channels and their wavelengths, the signal input powers for EDFA and FRA, the kind of the fibres used as the gain media of EDFA and FRA and their characteristics such as the radius of fibre core, numerical apertures, and the density of Er3+ ions in the EDFA. In this chapter, particle swarm optimization and its privileges over other heuristic optimization methods, in the optimization of optical fibre amplifiers are discussed. To achieve the optimized hybrid EDFA/FRA, the hybrid EDFA/FRA is simulated and the simulation is set as a computer program that gets the parameters which are to be optimized as the input variables and returns the amplification characteristics. Then, the simulation program is set as a sub-program of the more general program of the particle swarm optimization one providing the optimized values of the input parameters. It is straightforward to observe that the researchers and designers working in the specific considered field can benefit of the proposals of this chapter. Moreover, engineers and scientists who want to become familiar with the optical amplifiers and the methods of their optimization are advised to study this chapter.

The third part of the book analyzes the employment of swarm intelligence in the field of control optimization. In detail, chapter 9 introduces a swarm intelligence-inspired approach for target allocation in large teams of autonomous robots. For this purpose, the Distributed Bees Algorithm (DBA) is proposed and developed by the authors. The algorithm allows decentralized decision-making by the robots based on the locally available information, which is an inherent feature of animal swarms in nature. The algorithm performance are validated by physical robots. Moreover, a swarm simulator is developed to test the scalability of larger swarms in terms of number of robots and number of targets in the robot arena. Finally, improved target allocation in terms of deployment cost efficiency, measured as the average distance traveled by the robots, is achieved through optimization of the DBA's control parameters by means of a genetic algorithm. On the basis of the abovementioned topics, the chapter could be considered very interesting by scientists and engineers working in the field of the optimization of autonomous robot control.

Chapter 10 illustrates how particle swarm optimization can be used to develop efficient approximations of irrational Laplace-domain differential or integral operators. These operators are the main unit in transfer functions representing fractional-order controllers (FOC), that are an advanced and extended PID-type of controllers based on integral and derivative actions of non-integer (fractional) order. FOC can benefit of control loops by improving closed-loop performance, by obtaining more robustness to parameter variations, and by giving more design degrees of freedom. However, the realization of FOC requires approximation by rational transfer functions that exhibit stability and minimum-phaseness properties for control applications. Therefore, PSO is profitably used in the chapter to improve approximations that were developed in past research works. The topic of the chapter resides in the emerging field of fractional calculus and fractional-order control. It fits to the ever increasing need of high performance and robust devices and software for control systems engineering, for applications both in the analog and digital domains. In particular, industrial applications may benefit of the cited improvements because most of feedback loops are based on PID-controllers and fractional-order controllers extend and improve PID. Moreover, many applications ranging from bioengineering, biomechanical devices, communication, energy management, and environmental protection can benefit from the fractional calculus approach to modelling, simulation, and control. Scientists and researchers working in the field of signals, systems and control engineering constitute the target audience of this chapter.

The control of robots is the topic of chapter 11, which proposes a multiobjective PSO algorithm to optimize the structure of a parallel robot according to specific criteria. Moreover, the best location of the workpiece with respect to the robot for a chosen optimal structure is analyzed in terms of power consumed by the manipulator during the machining process in a robotic cell. The design of accurate machining task with parallel robots overcomes the limitation of using dedicated machine-tools. This is relevant for many industrial applications.

Finally, chapter 12 proposes ideas, analyses and engineering applications of a novel global optimization technique, named Generalized Particle Swarm Optimization (GPSO), which is a generalization of the popular and widely used Particle Swarm Optimization algorithm. This optimizer is inspired to linear control theory and enables direct control over the key aspects of particle dynamics during the optimization process, overcoming some typical flaws of classical PSO. Two practical engineering applications of the GPSO algorithm are described in the area of electrical machines fault detection and classification, and in optimal control of water distribution systems. The topic of the chapter could be of interest for all researchers dealing with global optimization techniques. Furthermore, the described algorithm represents an efficient tool for all those dealing with practical engineering problems involving optimization of functions with multiple local optima.

The last part of the book reports researches dealing with the application of PSO to scheduling and industrial diagnostic. To this purpose, chapter 13 describes a work on Electric Transmission Expansion Planning using DC and AC network models. Four particle swarm variations are proposed to solve the involved optimization problem. Several tests are performed using test and realistic power systems. Important conclusions are obtained about the use of DC and AC models, and PSO in Transmission Expansion Planning. This work can be particularly interesting for power system planners, researchers, consultants, and anyone involved in Electric Power Systems, especially in Transmission Expansion Planning.

In chapter 14 a Particle Swarm Optimizer is exploited to solve a short-term Generation Scheduling. The methodology can be applied by Independent Electricity System Operators (IESO), and by researchers in the field of Operation and Planning of Electrical Power Systems using Artificial Intelligence techniques.

Chapter 15 proposes the use of Ant Colony Optimization to realize the nondestructive analysis of dielectric bodies. In this contribution, the authors review the use of the Ant Colony Optimization method, which is a stochastic optimization algorithm that has been found to provide very good results in a plethora of applications in the area of electromagnetics as well as in other fields of electrical engineering. Electromagnetic approaches based on inverse scattering are very important in the field of nondestructive analysis of dielectric targets. In most cases, the inverse scattering problem related to the reconstruction of the dielectric properties of unknown targets starting from measured field values can be recast as an optimization problem. Due to the ill-posedness of this inverse problem, the application of global optimization techniques seems to be a very suitable choice. It follows that researchers and scientists in the field of electromagnetics could be very interested in the reading of this chapter.

Chapter 16 deals with topics in the field of biomedical engineering. This chapter presents some existing studies in the field of EEG based Brain Computer Interface (BCI) emphasizing on the impact of Evolutionary based approaches. On-line EEG based BCI systems include steps such as signal acquisition, pre-processing, feature extraction, dimension reduction, and classification. The chapter offers a survey of applied EA based approaches and discusses the achieved performance with special focus on dimension reduction, classifier training, and ensemble learning. EEG based BCI systems are ideal for patient with partial or complete paralysis due to the ability to provide communication channels that are independent from the brain's normal output pathways of peripheral nerves and muscles. Advantages such as providing high temporal resolution, being cost efficient (compare to similar imaging and recording devices), and non-invasive and risk free operant nature of the EEG recording procedure attract scientists to use this type of signal in a wide range of studies such as seizure, anaesthesia, sleep disorder, brain stork, Alzheimer, investigating brain tumour, brain wave abnormality, epilepsy, spike detection, consciousness, etc.

The previously reported information related to the contents of each chapter shows the growing and recent interest in bio-inspired algorithms. In particular, the attention has been focused on PSO based algorithms modelling swarm behaviours without neglecting the ant colony optimization and bees algorithm. These algorithms have been employed to solve a wide range of actual problems in the field of electric and electronic engineering. In particular, approaches based on these algorithms have often proved to be suited to face applicative and technological challenges. Therefore, this book should be intended to be a space in which to find knowledge about theoretical and applicative advances, new experimental discoveries and novel technological improvements of swarm intelligence in electric and electronic engineering. It is particularly interesting to offer a collection reporting the concepts related to the fundamental problems and the open challenges in the abovementioned engineering areas, showing how the swarm-based algorithms can be employed efficiently. In this sense, the target is to ensemble new swarm-based algorithms and challenge complex problems. At the same time, the original versions of the algorithms are reported in order to show how the basics of swarm intelligence can be useful to solve specific problems in different real-world applications. The resulting work constitutes a collection of high quality chapters which are representative of the existing research trends.

The target audience of the book is composed of professionals and researchers working in the field of electric and electronic engineering, e.g. computer science, power electronics, optimization and system control, and design of electric and electronic devices. These researchers can find the methods and the most prominent concepts of the applications of the swarm intelligence based algorithm to the fundamental problems of electric and electronic engineering in a unique book. Moreover, the volume can provide insights and support executives concerned with the management of expertise, knowledge, information, and development in different types of work communities and environments.

Girolamo Fornarelli
Politecnico di Bari, Italy

Luciano Mescia
Politecnico di Bari, Italy

Section 1
Circuit Design

Design methodologies and tools for analog, mixed-signal and radiofrequency circuits are not automated like the digital ones due to the complexity of the issue. To this purpose, this section collects some research activities regarding the application of PSO approach to automated synthesis and optimization of electronic circuits and hardware. In particular, a PSO-based CAD tools able to develop microwave circuits and devices automatically, a PSO model to calculate small-signal model parameters of a transistor, and a multi-objective PSO for the optimal design of a versatile CMOS current conveyor transconductance amplifier will be illustrated.

Chapter 1
Design and Optimization of Microwave Circuits and Devices with the Particle Swarm Optimizer

Massimo Donelli
University of Trento, Italy.

ABSTRACT

In this chapter, a methodology for the unsupervised design of microwave devices, circuits, and systems is considered. More specifically, the application of the Particle Swarm Optimizer and its integration with electromagnetic simulators is discussed in the framework of the microwave circuits and devices design and optimization. The idea is to automatically modify the characteristics of the device in an unsupervised way, with the goal of improve the device performances. Such kind of CAD tool could be the solution to reduce the time to market and keep the commercial predominance, since they do not require expert microwave engineers and it can reduce the computational time typical of the standard design methodologies. To assess the potentialities of the proposed method, a selected set of examples concerning the design of microwave planar devices such as filters, splitters, and other microwave devices under various operative conditions and frequency bands are reported and discussed. The chapter also includes a brief discussion concerning different strategies, such as parallel computation, to reduce the computational burden and the elaboration time. The obtained results seem to confirm the capabilities of the proposed method as effectiveness microwave CAD tool for the unsupervised design of microwave devices, circuits, and systems. The chapter ends with some conclusions and considerations related to ideas for future works.

INTRODUCTION

The Particle Swarm Optimizer (Kennedy, Eberhart, & Shi, 2001; Robinson & Rahmat-Samii, 2004; Clerc, & Kennedy, 2002) has been successfully adopted as a powerful optimization tools in several areas of applied electromagnetic (Grimaccia, Mussetta, & Zich, 2007; Mikki, & Kishk, 2006; Robinson & Rahmat-Samii, 2004; Mussetta, M., Selleri, S., Pirinoli, P., Zich, R.E., & Matekovits, L. 2008; Nanbo Jin, & Rahmat-Samii, Y. 2010; Yilmaz, A.E. 2010) such as microwave

DOI: 10.4018/978-1-4666-2666-9.ch001

imaging (Donelli & Massa, 2005; Caorsi, Donelli, Lommi, & Massa, 2004; Huang, & Mohan, 2007; Huang, C.-H., Chen, C.-H., Chiu, C.-C., & Li C. L. 2010; Genovesi, S., Salerno, E., Monorchio, & A., Manara, G. 2009), antenna design (Robinson, Sinton, & Rahmat-Samii, 2002; Jin, & Rahmat-Samii, 2005; Migliore, Pinchera, & Schettino, 2005, Azaro, De Natale, Donelli, Massa, & Zeni, 2006a; Azaro, De Natale, Donelli, & Massa, 2006b; Azaro, Boato, Donelli, Massa, & Zeni, 2006c; Azaro, Donelli, Franceschini, Zeni, & Massa, 2006c; Fimognari, Donelli, Massa, & Azaro, 2007; Jin, & Rahmat-Samii, 2007; Donelli, Martini, & Massa, 2009; Chamaani, S., Mirtaheri, S.A., & Abrishamian, M.S. 2011; Ismail, T.H., & Hamici, Z.M. 2010; Bevelacqua, P.J., & Balanis, C.A., 2009; Lin, C., Zhang, F.-S., Zhao, G., Zhang, F., Jiao, Y.-C. 2010) and control (Boeringer, & Werner, 2004; Khodier, & Christodoulou, 2005; Boeringer, & Werner, 2005; Donelli, Azaro, De Natale, & Massa, 2006), and other interesting practical applications (Adly, & Abd-EL-Haftiz, 2006; Ho, Yang, Ni, & Wong, 2006; Genovesi, Monorchio, Mittra, & Manara, 2007; Cui, & Weile, 2006; Azaro, Donelli, Benedetti, Rocca, & Massa, 2008; Martini, Donelli, Franceschetti, & Massa, 2008). Among them, the development of CAD tools, aimed at providing the designer with an environment where a given circuit or microwave device can be characterized, investigated and also modified to obtain desired requirements, represents a significant example. As far as the modification of a given component is concerned, a key point is the development of microwave CAD environments (Azaro, De Natale, Donelli, & Massa, 2006) where the device or the circuit is modified automatically in an unsupervised way by the microwave tool itself with an improvement of performance. Such kind of CAD tools can be used to reduce the time to market of devices, and microwave circuits, and keep the commercial predominance. The unsupervised CAD tool do not require expert microwave engineers and they

are more efficient with respect to standard trial errors methodologies. Such useful automatic design tools have been the object of research since some years (Azaro, De Natale, Donelli, & Massa, 2006; Caorsi, Donelli, Massa, & Raffetto, 2002). The importance of the subject is widely recognized. In fact, it is well known that the design of new microwave devices and circuit are needed in several important areas, e.g., civil and military telecommunication systems, industrial and medical equipments. Standard microwave synthesis techniques are quite effective for the design of standard devices such as microwave filter, combiners, broad band coupler. Unfortunately, such approaches sometimes led to the development of impracticable, very expensive networks or unrealizable devices especially for very high frequencies. Moreover these techniques are not suitable when new technologies or materials such as meta-materials are considered. For these reasons, the design problem is usually recast as a global optimization problem. Formulated in such a way, the problem can be efficiently handled by the Particle Swarm Optimizer by defining a suitable cost function that represent the distance between the requirements and the obtained solution. In this framework and in recent years, many efforts have been devoted to the developments of unsupervised CAD tools for the control of smart antenna, the design and optimization of non-conventional antennas (Azaro, Donelli, Francheschini, Zeni, & Massa, 2006), and the development of passive microwave devices (Azaro, De Natale, Donelli, & Massa, 2006; Caorsi, Donelli, Massa, & Raffetto, 2002). Despite the successful results reported in scientific literature, the availability of effective CAD tools able to automatically develop microwave circuits and devices is still a challenge. In the following, Section 2, presents the solution methodology based on the use of PSO optimizer for the optimization and synthesis of microwave circuits and devices. A set of selected and representative synthesis results concerned microwave passive devices such as Hybrid ring, filters and frequency doublers are presented in Section 3. Section 4 presents a brief discussion concerning the use of parallel computation. Finally some conclusions are drawn (Section 5).

2 APPLICATION OF THE PSO TO THE DESIGN OF MICROWAVE CIRCUITS AND DEVICES

In this section the application of the PSO algorithm to the synthesis of microwave circuits and devices will be described. In particular this section starts with a detailed description of how to define the cost function to manage the problems at hand then a brief description of the customized particle swarm algorithm used to solve the proposed problems will be reported.

2.1 The Synthesis Method

Let us consider the problem of designing generic passive microwave devices or circuits. Without loss of generality, we assume that the synthesis is aimed at determining the optimal values of different geometrical or circuital parameters to obtain the required values of the scattering parameters by means the minimization of a suitable defined cost

function. In particular starting from the scattering parameters constraints, valid for a given frequency range, and usually expressed in Equation 1 (see Box 1) where i and j are the indexes of the device ports, and f is the working frequency.

The solution space of the optimization problem is defined as an array of unknowns $\underline{x} = \left(x_1, ..., x_N \right)$, being N the unknowns number. The array elements represent the descriptive parameters of the circuit or the geometrical parameters that describe the structure of a microwave device. The solution of the optimization problem is a vector determined by minimizing the cost function in Equation 2 (see Box 2). Δf and f are the frequency step and index respectively, $S_{ij}^{act}(h \cdot \Delta f)$ i = 1, ... I, j = 1, ... J, h = 1, ..., H, are the scattering parameters requirements that have to be specified, by the user, for each port and each frequency step.

The cost function represents the distance between the requirements, given by the user, and a trial solution. Generally, due to the complexity of the microwave devices, characterized by a lots of parameters, the cost function results highly nonlinear and presents a large number of "local minima" where classical gradient based optimization methods can be trapped, moreover the estimation of the cost function gradient is not so simple

Box 1.

$$S_{ij}^{\min}(f) \le S_{ij}(f) \le S_{ij}^{\max}(f) \quad i = 1, ..., I \ j = 1, ..., J \tag{1}$$

Box 2.

$$\Phi(\underline{x}_s^{(k)}) = \sum_{h=1}^{H} \sum_{i=1}^{I} \sum_{j=1}^{J} \frac{\left| S_{ij}^{act}(h \cdot \Delta f) - S_{ij}(\underline{x}_s^{(k)}, h \cdot \Delta f) \right|^2}{\left| S_{ij}^{act}(h \cdot \Delta f) \right|^2} \tag{2}$$

for imaginary cost function. To efficiently explore the solution space and obtaining the minimization of the cost function, Equation 2, a suitable optimization technique is needed. Toward this end, a customized implementation of the PSO optimizer (Kennedy, Eberhart, & Shi, 2001; Robinson, & Rahmat-Samii, 2004; Clerc, & Kennedy, 2002) has been integrated with a circuital generator and then interfaced with a circuital simulator. The particle swarm is a robust stochastic optimization procedure, inspired by social behavior of smart swarms of bees, birds, and fishes. It was developed by Kennedy and Eberarth in 1995. Thanks to the collaborative philosophy of such algorithm, PSO can avoid the problem of stationary solutions, typical of genetic algorithms. Thanks to its features in exploring high dimensional solution spaces, to avoid local minima and to handle complex cost functions characterized with strong nonlinearities, PSO has been applied with success in the framework of applied and computational problems (Robinson, & Rahmat-Samii 2004). For the problem at hand, starting from a set of trial solutions $\underline{x}_s^{(k)}$ (being s the trial solution index, and k the iteration index k=1, ... K_{max}), following the PSO strategy, the circuits/devices generator defines the corresponding trial solutions. Then the circuital simulator provides to estimate the scattering parameters S_{ij} and consequently the cost function (according to Equation 2). The iterative process continues until the stopping criteria are reached.

In particular when $k = K^{max}$ or $\Phi\left(\underline{x}_s^{(k)}\right) < \beta$,

where K^{max} and β are the maximum number of iterations and a convergence threshold respectively. The schema of the optimization/design loop is reported in Figure 1.

2.2 Optimization Approach: The Particle Swarm Algorithm

To develop an integrated system for the unsupervised synthesis of microwave circuits and devices,

an evolutionary algorithm namely the particle swarm optimizer (PSO) has been considered (Kennedy, Eberhart, & Shi, 2001; Robinson & Rahmat-Samii, 2004; Clerc, & Kennedy, 2002). The PSO with respect to genetic algorithms (GAs) and other heuristic techniques, PSO presents the following interesting advantages that make it particularly suitable for the solution of applied electromagnetic problems:

- The PSO is simpler, both in formulation and computer implementation, than the GA, which considers three-genetic operators (the *selection*, the *crossover*, and the *mutation*). PSO considers one simple operator that is the *velocity updating*. In general, heuristic techniques require a careful tuning of the *control parameters*, which strongly influence the behavior of the optimization method. PSO allows an easier setting of the calibration parameters and, in general, a standard configuration turns out to be adequate for a large class of problems and problem sizes. Consequently, unlike regularization parameters, there is no need to perform a PSO calibration for every experiment.

- PSO generally requires a smaller population size, which turns out in a reduced computational cost of the overall minimization, since the most of computational burden due to the cost function estimation of each particle of the swarm.

- PSO has a flexible and well-balanced mechanism to enhance the global (i.e., the exploration capability of the social term) and the local (i.e., the exploitation capability of the cognitive term) exploration of the search space. Such a feature allows one to overcome the premature convergence (or *stagnation typical of the GAs*) moreover it enhances the search capability of the optimizer.

Figure 1. Schema of the design/optimization loop

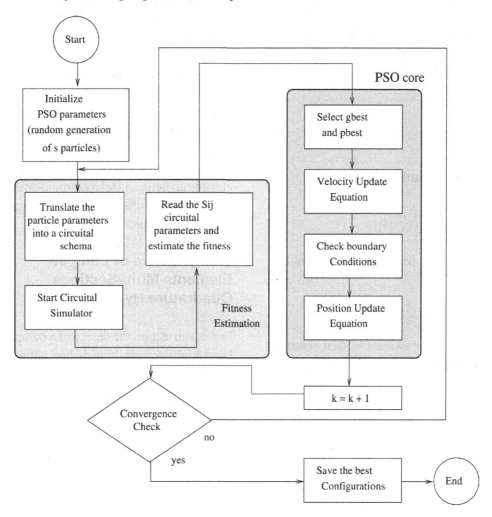

In the following a brief description of the PSO will be reported. In particular a standard PSO version with constant inertial weight has been considered. The PSO is a stochastic multiple-agent algorithm able to find optimal regions in complex search spaces through the interaction of individuals in a population of particles. The algorithm is based on a metaphor of social interactions and simulates the collective behavior of swart swarms where simple individuals interacting with their environment and each other. In the same way of other population based optimization approaches, PSO updates the set of individuals according to the cost function (or *fitness function*, Φ) information so that the population move towards better solution areas. Instead of using complex evolutionary operators to manipulate the individuals, each individual (called *particle*, $\boldsymbol{x} = \{x_n; n = 1, ..., N\}$ being N the unknowns number flies as a "bird" in the search space with a velocity dynamically changed according to its experience and the experience of other particles. The PSO, when applied in the framework of microwave imaging to minimize (2), requires the definition of a population of particles (called *swarm*) $\mathbf{S}_0 = \left\{ \underline{x}_0^s; s = 1, \cdots, S \right\}$ and a corresponding flying velocity array $\mathbf{V}_0 = \left\{ \underline{v}_0^s ; s = 1, \cdots S \right\}$ *S being* the dimension of the trial solution population. Iteratively (being *k* the iteration number), the trial solutions are ranked according their fitness measures.

$$F_k = \{\Phi(\underline{x}_s^{(k)}); s = 1, \cdots S\} \qquad (3)$$

The best swarm particle

$$\underline{x}_k^{(opt)} = \arg\left[\min_s \left(\Phi\left\{\underline{x}_k^{(s)}\right\}\right)\right]$$

and the best previous position of each particle

$$\underline{x}_k^{(opt)} = \arg\left[\min_{h=1,\cdots,k} \left(\Phi\left\{\underline{x}_k^{(s)}\right\}\right)\right]$$

are stored. Then, new swarms of trial solutions are obtained updating each particle according to the following equation:

$$x_{i,k+1}^{(s)} = x_{i,k}^{(s)} + v_{i,k+1}^{(s)} \qquad (4)$$

being

$$v_{i,k+1}^{(s)} = \omega v_{i,k+1}^{(s)} + \beta\left(x_{i,k}^{(\overline{s})} - x_{i,k}^{(s)}\right) + \gamma\left(x_{i,k}^{(opt)} - x_{i,k}^{(s)}\right) \qquad (5)$$

where β and γ are two random values included in the range (0, 2) with a uniform distribution generated by a pseudo-random number generator; ω is the *inertia* weight. The iterative process stops when the termination criterion, based on a maximum number of iterations, K (i.e., $k = K$) or on a threshold on the fitness measure, δ (i.e., $\underline{x}_k^{(opt)} \leq \delta$), is verified and $\underline{x}_k^{(opt)}$ is assumed as the problem solution.

3 NUMERICAL RESULTS: SYNTHESIS OF MICROWAVE CIRCUITS AND DEVICES

In this Section a selected set of examples concerning the synthesis of microwave circuits and devices will be proposed and discussed to assess the capabilities and the potentialities of the proposed PSO-based design technique. In particular, three typical microwave devices widely used in electromagnetic and telecommunication applications will be synthesized and the obtained results analyzed and commented.

3.1 Synthesis of Lumped Elements Multi-Sections Quadrature Hybrid Ring

In this subsection the design of a compact broadband multi-section branch line coupler will be investigated. In particular the branch line coupler will be synthesized considering lumped elements. Directional couplers are widely used for microwave integrated circuits (MIC's), and microwave transmitters/receivers systems (Pozar, 1998). In particular the -3 dB couplers are used for balanced amplifier, mixers and other microwaves devices. Branch line couplers are the most used in practical applications because they are simple and suitable for printed circuit board process. A traditional branch line coupler is composed by four quarter wavelength transmission lines connected together and it is characterized by a narrow bandwidth of 10%-15%. Recently several methods have been presented to improve the bandwidth of such devices; in particular multi-sections devices are widely used (Chiang, & Chen, 2001). In the following, a two arms branch line coupler will be synthesized. In particular, to reduce the size of the device, lumped capacitors and inductors will be considered instead of conventional transmission line. The four quarter wavelength transmission lines that compose the hybrid rings can be implemented using one inductors and two capacitors.

Figure 2. Circuit layout of the lumped elements two arms branch line coupler

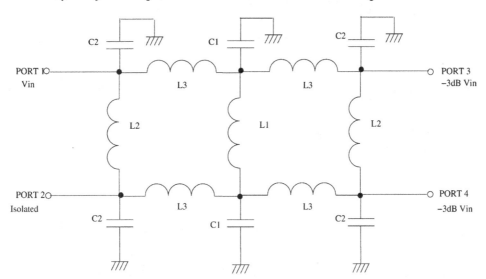

Following this way, a two arms hybrid ring can be implemented as shown in Figure 2, considering seven inductors and six capacitors. The hybrid ring structure is uniquely determined by the following descriptive vector $\underline{x} = \left\{ C_1, C_2, L_1, L_2, L_3 \right\}$ As far as hybrid ring requirements are concerned, a return loss S_{ij} i=j and an isolation $S_{12}=S_{34}$ less than -10dB, a pass through values of $S_{13}=S_{14}=0.707*V_{in}$ (for a -3dB standard hybrid ring), and a frequency band of 4.5-7.5 GHz have been considered. To satisfy the design requirements, the unknowns array $\underline{x} = \left\{ C_1, C_2, L_1, L_2, L_3 \right\}$ is determined by the minimization of the cost function (2) in the range 4.5-7.5 GHz. The circuital simulations for the estimation of the scattering parameters and the related cost function were carried out using the Ansoft Nexim 3.5 circuital simulator, considering a ceramic dielectric substrate characterized by a dielectric constant of $\varepsilon = 3.38$, a loss tangent of $\delta = 0.003$, and a thickness of 0.8 mm. Concerning the lumped elements, a model for surface mount devices (SMD) of dimensions $1 \times 0.5\ mm^2$ has been considered. As far as the PSO parameters are concerned, a swarm of S=5 particles, a maximum

number of iterations equal to K=100, and a threshold $\beta = 10^{-5}$ have been assumed. The remaining parameters of the PSO have been chosen according the guidelines indicated in the reference literature (Donelli, & Massa, 2005), in particular the acceleration terms $C_1=C_2=2.0$, and the constant inertial weight w=0.2. Starting from a completely random swarm of five particles, the PSO explores the solution space searching for a vector $\underline{x}^{opt}$ that satisfies the input requirements for the scattering parameters. In particular after k=68 iterations the optimizer reaches a good configuration. Figure 3 reports the magnitude of the simulated scattering parameters of the optimal configuration. As can be noticed from the data reported in Figure 3, the return loss S_{11} and the isolation $S_{12}=S_{24}$ parameters are always below -10 dB for the whole frequency range of interest. The pass through parameter S_{13} is quite satisfactory and equal to -3 dB in the frequency range from 5 GHz up to 7 GHz, while below 5GHz and above 7 GHz the S_{13} is slightly lower. The other pass through parameter S_{14} is about 1 dB lower respect to S_{13} parameter. However from the data of Figure 3, the values of the scattering parameters in the whole frequency range of interest are quite satis-

Figure 3. Simulated S parameters magnitude of the two arms lumped hybrid ring synthesized with the proposed design methodology

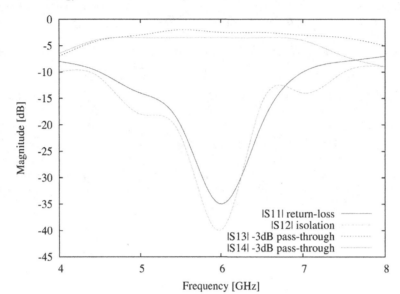

factory and demonstrate the synthesis capabilities of the proposed design methodology.

3.2 Synthesis of a Frequency Doubler

In this subsection, a more complex microwave system with a nonlinear device, a Shottky barrier diode, has been considered. Frequency doublers are usually adopted for transmitters, receivers and telecommunication devices when fixed or tunable oscillators are not able to directly generate the required waveform. The required waveform is obtained by generating harmonics from a very stable low-frequency source such as a crystal oscillator and then multiplying up to reach the desired frequency. Frequency doublers use the nonlinear I/V characteristics of a Shottky-barrier diode to distort an input sinusoidal signal and generate harmonics. The goal of the following experiment is to design the two matching networks of the devices described in Figure 4. The considered frequency doublers, described in (Rajanoronk, Namahoot, & Akkareakthalin, 2006), uses

a single diode and a feed-forward line (the input and output networks detailed in Figure 4) are used to match the characteristic impedance of the ports, (the S_{11} and S_{22} scattering parameters) and to reject unwanted harmonics generated by the nonlinear behavior of the diode. The structure of the considered microwave device is determined by the following descriptive vector

$$\underline{x} = \left\{ W_m, L_n, S \right\} \quad m = 1, \ldots 7; \quad n = 1, \ldots 8 ,$$

being W_m, L_n and S the positions, the heights and the thickness of the network stubs respectively. The unknown space is characterized by sixteen real unknowns that completely describe the two matching networks. Concerning the input requirements, a return loss $S_{11}=S_{22}$ less than -10dB is required for both device ports. The frequency doublers must be able to receive as input a low power sinusoidal signal of $f_i = 1.2$ GHz, and to give as output a sinusoidal signal at $f_o = 2.4$ GHz. Also in this experiment the circuital simulator adopted for the estimation of the scattering parameters and the cost function was the Ansoft Nexim 3.5 circuital simulator, the ceramic dielec-

Figure 4. Schema of the single diode frequency doublers, with the input/output matching networks

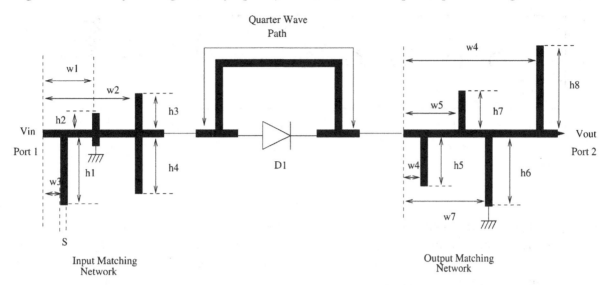

tric substrate of the previous example, $\varepsilon = 3.38$, $\delta = 0.003$, and thickness t = 0.8 mm has been considered. Concerning the nonlinear device, a BAT15 Shottky barrier diode has been considered. To simulate a more realistic scenario the characteristics of the BAT15 have been downloaded from the website of the producer and imported into the circuital simulator. For this experiment the following PSO parameters have

been considered: a swarm of S=5 particles, a maximum number of iterations equal to K=100, a threshold $\beta = 10^{-3}$, $C_1 = C_2 = 2.0$, and a constant inertial weight w=0.4. In Figure 5 are reported the corresponding values of the magnitude of the return loss of the two ports of the device obtained after the synthesis procedure. The synthesized microwave design has been obtained after k=88 iterations. As it can be observed, the synthesized

Figure 5. S parameters magnitude of the frequency doublers obtained at the end of the design methodology

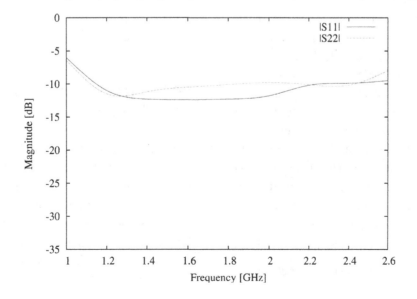

Figure 6. Geometry and descriptive parameters of the band-pass waveguide filter

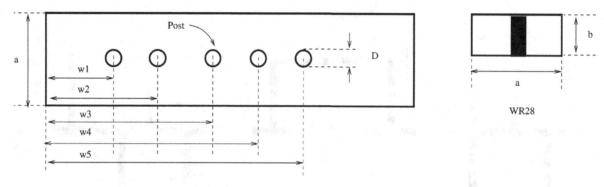

doublers device fits the input requirements; in particular the S11 and S22 are always below -10 dB for the frequencies of interest. The magnitude of the input and output signals at 1.2 GHz are -0.02 dBi and -52 dBi respectively, while for 2.4 GHz the magnitude of input and output signals are -42.12 and -5.43 respectively, confirming the correct behavior of the doublers.

3.3 Synthesis of a WR28 Band-Pass Waveguide Filter

Microwave filters can be found in any kind of microwave devices and telecommunication systems They are two-port networks used to control the frequency response by providing transmission at frequencies within the pass-band of the filter and attenuation in the stop-band of the filter. These filters can be fabricated with lumped, microstrip or waveguide technologies. The last example of this chapter is concerned with a pass-band waveguide filter able to operate in the Ka band (26.5 GHz – 40 GHz). The considered waveguide filter is composed by a segment of rectangular waveguide WR28 (a = 7.11 mm and b = 3.55 mm) and five circular metallic posts. The geometry of the filter and the related descriptive parameters are reported in Figure 6. Also in this case all the descriptive unknown can be written as $\underline{x} = \left\{ W_m, D \right\}$ $m = 1,...5$, being W_m and D the posts position and dimension respectively. In this

experiment the unknown space is characterized by six real unknowns. Concerning the input requirements, a pass band from 31 GHz up to 32 GHz has been required. The circular metallic posts are placed exactly in the middle of the waveguide width to maximize the effect. For this experiment the PSO has been integrated with a finite element method electromagnetic simulator, in particular the HFFS commercial simulator has been considered. Concerning the PSO parameters, the same values of the previous example have been considered. The final filter geometry has been obtained after k=82 iterations. The filter characteristics of the synthesized waveguide filter are reported in Figure 7. In particular Figure 7 reports the return loss (S_{11}) and the S_{12} versus the frequency. As can be noticed from the data of Figure 7 the pass band of the filter is between 31 GHz and 32 GHz, and perfectly matches the initial requirements.

4 COMPUTATIONAL ISSUES OF THE CIRCUITAL DESIGN METHODOLOGY

To give some indications on the computational issues associated to the proposed PSO design method, Figure 8 shows the plots of the optimal value cost function versus the number of iterations for the synthesis of the microwave devices described in the previous Sections. The results reported in Figure 8 have been obtained

Figure 7. S parameters magnitude of the synthesized pass-band waveguide filter

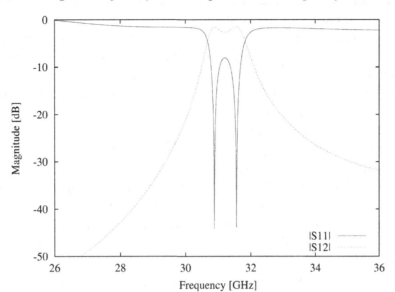

Figure 8. Behavior of the cost function versus the iteration number during the synthesis of the micro-wave devices

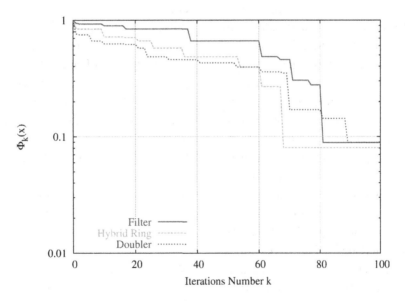

averaging the results of 100 runs for each considered problem.

As it can be noticed, whatever the considered test case, the convergence value of the cost function turns out to be lower than 10^{-1} pointing out an accurate matching with the design requirements, after about one hundred iterations. The CAD tool described in the previous sections integrates an optimization program and a numerical simulator. In particular the circuital and planar electromagnetic simulator (namely Ansoft Designer and Nexim) have been integrated with the optimizer (developed in MATLAB code) using the suitable subroutines written in visual-script language and integrated into the MATLAB code core. This represents one of the most attractive issues related to the computer-aided design (CAD) for the design and optimization of microwave circuits and devices. Following this way, it is possible to obtain unsupervised CAD tools which do not require the interaction with expert engineers, and this can contributes to the reduction of the costs and of the time to market of a single design cycle. Unfortunately to efficiently use such design tools, some problems have to be solved. The main problem is the computational burden required: to carry out the automatic design of microwave circuits and devices, usually standard computers and workstations are not satisfactory since the minimization algorithm requires the evaluation of several parametric configurations of the circuits and devices under development. Moreover, each evaluation is usually carried out by means of a computationally heavy commercial simulators (such as FEM, FDTD or MoM electromagnetic engine). Recently, several software packages that permit the exploitation of heterogeneous network of computers as if it was a single powerful parallel computer, have been developed (MPI, Linda, PVM, etc). Moreover it can be noted that the particle swarm algorithm is a multi-agents procedure that presents an intrinsic parallelism. At each iteration, different trial solutions, the particles, explore the solution space contemporarily, and consequently the evaluation of the cost function can be carried out in parallel. By using this intrinsic characteristic of the algorithm, a single fitness evaluation can still be regarded as single indivisible operation. Thus the numerical electromagnetic simulator does not need changes and the parallel implementation of the CAD tool is quite easy if the two building blocks namely the optimization algorithm and the electromagnetic simulator have already been developed. Such kind of approach called master-slave model have been efficiently adopted for solving complex electromagnetic problems like in (Donelli, Azaro, Massa, & Raffetto, 2006; Massa, Franceschini, France-

schini, Pastorino, Raffetto, & Donelli, 2005), and it seems to be promising for such kind of CAD tools. Following these guidelines the parallelism is exploited in natural way since it is applied to carry out tasks which are conceptually separated entities in any optimization problems.

CONCLUSION

In this work, an unsupervised methodology for the design of microwave devices and circuits has been described. Because of several electromagnetic and geometrical constraints, the synthesis process has been faced with a PSO-based optimization procedure. The design process has been assessed through a selected set of numerical experiments, in particular the design of a multi-section lumped elements hybrid ring, a waveguide band-pass filter and a microwave frequency doublers have been presented. The obtained results in terms of scattering parameters, performances and geometrical compactness confirmed the effectiveness of the proposed unsupervised procedure as effective CAD tool for the design of microwave circuits and devices for modern telecommunication system applications. Future work will be aimed at the improvement of the CAD tool in particular the integration between different kind of electromagnetic simulators able to manage wired as well wireless devices such as antenna.

REFERENCES

Adly, A., & Abd-El-Hafiz, S. (2006). Using the particle swarm evolutionary approach in shape optimization and field analysis of devices involving nonlinear magnetic media. *IEEE Transactions on Magnetics*, *42*(10), 3150–3152. doi:10.1109/TMAG.2006.880103

Azaro, R., Boato, G., Donelli, M., Massa, A., & Zeni, E. (2006). Design of a prefractal monopolar antenna for 3.4–3.6 GHz wi-max band portable devices. *IEEE Antennas and Wireless Propagation Letters, 5*(1), 116–119. doi:10.1109/LAWP.2006.872427

Azaro, R., De Natale, F., Donelli, M., & Massa, A. (2006). PSO-based optimization of matching loads for lossy transmission lines. *Microwave and Optical Technology Letters, 48*(8), 1485–1487. doi:10.1002/mop.21738

Azaro, R., De Natale, F., Donelli, M., & Zeni, E. (2006). Optimized design of a multi-function/multi-band antenna for automotive rescue systems. *IEEE Transactions on Antennas and Propagation, 54*(2), 897–904. doi:10.1109/TAP.2005.863387

Azaro, R., De Natale, F., Zeni, E., Donelli, M., & Massa, A. (2006). Synthesis of a pre-fractal dual-band monopolar antenna for GPS applications. *IEEE Antennas and Wireless Propagation Letters, 5*(1), 361–364. doi:10.1109/LAWP.2006.880695

Azaro, R., Donelli, M., Benedetti, M., Rocca, P., & Massa, A. (2008). A GSM signals based positioning technique for mobile applications. *Microwave and Optical Technology Letters, 50*(4), 2128–2130. doi:10.1002/mop.23568

Azaro, R., Donelli, M., Franceschini, D., Zeni, E., & Massa, A. (2006). Optimized synthesis of a miniaturized SARSAT band pre-fractal antenna. *Microwave and Optical Technology Letters, 48*(11), 2205–2207. doi:10.1002/mop.21922

Bevelacqua, P. J., & Balanis, C. A. (2009). Geometry and weight optimization for minimizing sidelobes in wideband planar arrays. *IEEE Transactions on Antennas and Propagation, 57*(4), 1285–1289. doi:10.1109/TAP.2009.2015853

Boeringer, D., & Werner, D. (2004). Particle swarm optimization versus genetic algorithms for phased array synthesis. *IEEE Transactions on Antennas and Propagation, 52*(3), 771–779. doi:10.1109/TAP.2004.825102

Boeringer, D., & Werner, D. (2005). Efficiency-constrained particle swarm optimization of a modified bernstein polynomial for conformal array excitation amplitude synthesis. *IEEE Transactions on Antennas and Propagation, 53*.

Caorsi, S., Donelli, M., Lommi, A., & Massa, A. (2004). Location and imaging of two-dimensional scatterers by using a Particle Swarm algorithm. *Journal of Electromagnetic Waves and Applications, 18*(4), 481–494. doi:10.1163/156939304774113089

Caorsi, S., Donelli, M., Massa, A., & Raffetto, M. (2002). A parallel implementation of an evolutionary-based automatic tool for microwave circuit synthesis: Preliminary results. *Microwave and Optical Technology Letters, 35*(3). doi:10.1002/mop.10547

Chamaani, S., Mirtaheri, S. A., & Abrishamian, M. S. (2011). Improvement of time and frequency domain performance of antipodal Vivaldi antenna using multi-objective particle swarm optimization. *IEEE Transactions on Antennas and Propagation, 59*(5), 1738–1742. doi:10.1109/TAP.2011.2122290

Chiang, Y. C., & Chen, C. Y. (2001). Design of a wideband lumped-element 3-dB quadrature coupler. *IEEE Transactions on Microwave Theory and Techniques, 9*, 476–479. doi:10.1109/22.910551

Clerc, M., & Kennedy, J. (2002). The particle swarm—Explosion, stability, and convergence in a multidimensional complex space. *IEEE Transactions on Evolutionary Computation, 6*(1), 58–73. doi:10.1109/4235.985692

Cui, S., & Weile, D. (2006). Application of a parallel particle swarm optimization scheme to the design of electromagnetic absorbers. *IEEE Transactions on Antennas and Propagation, 54*(3), 1107–1110.

Donelli, M., Azaro, A., De Natale, F., & Massa, A. (2006). An innovative computational approach based on a particle swarm strategy for adaptive phased-arrays control. *IEEE Transactions on Antennas and Propagation, 54*(3), 888–898. doi:10.1109/TAP.2006.869912

Donelli, M., Azaro, R., Massa, A., & Raffetto, M. (2006). Unsupervised synthesis of microwave components by means of an evolutionary-based tool exploiting distributed computing resources. *Progress in Electromagnetics Research, 56,* 93–108. doi:10.2528/PIER05010901

Donelli, M., Martini, A., & Massa, A. (2009). A hybrid approach based on PSO and Hadamard difference sets for the synthesis of square thinned arrays. *IEEE Transaction on Antennas and Propagation Letters, 57*(8), 2491–2495. doi:10.1109/TAP.2009.2024570

Donelli, M., & Massa, A. (2005). Computational approach based on a particle swarm optimizer for microwave imaging of two-dimensional dielectric scatterers. *IEEE Transactions on Microwave Theory and Techniques, 53*(5), 1761–1776. doi:10.1109/TMTT.2005.847068

Fimognari, L., Donelli, M., Massa, A., & Azaro, R. (2007). A planar electronically reconfigurable wi-fi band antenna based on a parasitic microstrip structure. *IEEE Antennas and Wireless Propagation Letters, 6,* 623–626. doi:10.1109/LAWP.2007.913274

Genovesi, S., Monorchio, A., Mittra, R., & Manara, G. (2007). A subboundary approach for enhanced particle swarm optimization and its application to the design of artificial magnetic conductors. *IEEE Transactions on Antennas and Propagation, 55*(3), 766–770. doi:10.1109/TAP.2007.891559

Genovesi, S., Salerno, E., Monorchio, A., & Manara, G. (2009). Permittivity range profile reconstruction of multilayered structures from microwave backscattering data by using particle swarm optimization. *Microwave and Optical Technology Letters, 51*(10), 2390–2394. doi:10.1002/mop.24642

Grimaccia, F., Mussetta, M., & Zich, R. E. (2007). Genetical swarm optimization: self-adaptive hybrid evolutionary algorithm for electromagnetics. *IEEE Transactions on Antennas and Propagation, 55*(3), 781–785. doi:10.1109/TAP.2007.891561

Ho, S., Yang, S., Ni, G., & Wong, H. (2006). A particle swarm optimization method with enhanced global search ability for design optimizations of electromagnetic devices. *IEEE Transactions on Magnetics, 42*(4), 1107–1110. doi:10.1109/TMAG.2006.871426

Huang, C.-H., Chen, C.-H., Chiu, C.-C., & Li, C. L. (2010). Reconstruction of the buried homogenous dielectric cylinder by FDTD and asynchronous particle swarm optimization. *Applied Computational Electromagnetics Society Journal, 25*(8), 672–681.

Huang, T., & Mohan, A. S. (2007). A microparticle swarm optimizer for the reconstruction of microwave images. *IEEE Transactions on Antennas and Propagation, 55*(3), 568–576. doi:10.1109/TAP.2007.891545

Ismail, T. H., & Hamici, Z. M. (2010). Array pattern synthesis using digital phase control by quantized particle swarm optimization. *IEEE Transactions on Antennas and Propagation*, 58(6), 2142–2145. doi:10.1109/TAP.2010.2046853

Jin, N., & Rahmat-Samii, Y. (2005). Parallel particle swarm optimization and finite-difference time-domain (PSO/FDTD) algorithm for multiband and wide-band patch antenna designs. *IEEE Transactions on Antennas and Propagation*, 53(11), 3459–3468. doi:10.1109/TAP.2005.858842

Jin, N., & Rahmat-Samii, Y. (2007). Advances in particle swarm optimization for antenna designs: real-number, binary, single-objective and multi-objective implementations. *IEEE Transactions on Antennas and Propagation*, 55(3), 556–567. doi:10.1109/TAP.2007.891552

Kennedy, J., Eberhart, R. C., & Shi, Y. (2001). *Swarm intelligence*. San Francisco, CA: Morgan Kaufmann.

Khodier, M., & Christodoulou, C. (2005). Linear array geometry synthesis with minimum sidelobe level and null control using particle swarm optimization. *IEEE Transactions on Antennas and Propagation*, 53(8), 2674–2679. doi:10.1109/TAP.2005.851762

Lin, C., Zhang, F.-S., Zhao, G., Zhang, F., & Jiao, Y.-C. (2010). Broadband low-profile microstrip antenna design using GPSO based on mom. *Microwave and Optical Technology Letters*, 52(4), 975–979. doi:10.1002/mop.25069

Martini, A., Donelli, M., Franceschetti, M., & Massa, A. (2008). Particle density retrieval in random media using a percolation model and a particle swarm optimizer. *IEEE Antennas and Wireless Propagation Letters*, 921140, 213–216. doi:10.1109/LAWP.2008.921140

Massa, A., Franceschini, D., Franceschini, G., Pastorino, M., Raffetto, M., & Donelli, M. (2005). Parallel GA-based approach for microwave imaging applications. *IEEE Transactions on Antennas and Propagation*, 53(10), 3118–3127. doi:10.1109/TAP.2005.856311

Migliore, M., Pinchera, D., & Schettino, F. (2005). A simple and robust adaptive parasitic antenna. *IEEE Transactions on Antennas and Propagation*, 53(10), 3262–3272. doi:10.1109/TAP.2005.856361

Mikki, S., & Kishk, A. (2006). Quantum particle swarm optimization for electromagnetics. *IEEE Transactions on Antennas and Propagation*, 54(10), 2764–2775. doi:10.1109/TAP.2006.882165

Mussetta, M., Selleri, S., Pirinoli, P., Zich, R. E., & Matekovits, L. (2008). Improved particle swarm optimization algorithms for electromagnetic optimization. *Journal of Intelligent and Fuzzy Systems*, 19(1), 75–84.

Nanbo, J., & Rahmat-Samii, Y. (2010). Hybrid real-binary particle swarm optimization (HPSO) in engineering electromagnetics. *IEEE Transactions on Antennas and Propagation*, 58(12), 3786–3794. doi:10.1109/TAP.2010.2078477

Pozar, D. (1998). *Microwave engineering*. New York, NY: John Wiley & Sons.

Rajanaronk, P., Namahoot, A., & Akkareakthalin, P. (2006). A single diode frequency doubler using a feed-forward technique. *Proceeding of Asia-Pacific Microwave Conference*.

Robinson, J., & Rahmat-Samii, Y. (2004). Particle swarm optimization in electromagnetics. *IEEE Transactions on Antennas and Propagation*, 52(2), 397–407. doi:10.1109/TAP.2004.823969

Robinson, J., Sinton, S., & Rahmat-Samii, Y. (2002). Particle swarm, genetic algorithm, and their hybrids: Optimization of a profiled corrugated horn antenna. *IEEE Antennas Propagation Society International Symposium Digest, 1,* 314–317.

Yilmaz, A. E. (2010). Swarm behavior of the electromagnetics community as regards using swarm intelligence in their research studies. *Acta Polytechnica Hungarica, 7*(2), 81–93.

ADDITIONAL READING

Afshinmanesh, F., Marandi, A., & Shahabadi, M. (2006). Design of a single-feed dual-band dual-polarized printed microstrip antenna using a Boolean particle swarm optimization. *IEEE Transactions on Antennas and Propagation, 56*(7), 1845–1852. doi:10.1109/TAP.2008.924684

Akdagli, A., & Guney, K. (2006). New wide-aperture-dimension formula obtained by using a particle swarm optimization for optimum gain pyramidal horns. *Microwave and Optical Technology Letters, 48*(6), 1201–1205. doi:10.1002/mop.21580

Ali, F. A., & Selvan, K. T. (2009). A study of PSO and its variants in respect of microstrip antenna feed point optimization. *APMC 2009 - Asia Pacific Microwave Conference 2009,* (pp. 1817-1820).

Chauhan, N. C., Kartikeyan, M. V., & Mittal, A. (2009). CAD of RF windows using multiobjective particle swarm optimization. *IEEE Transactions on Plasma Science, 37*(6, part 2), 1104–1109. doi:10.1109/TPS.2009.2020589

Chauhan, N. C., Kartikeyan, M. V., & Mittal, A. (2009). A modified particle swarm optimizer and its application to the design of microwave filters. *Journal of Infrared, Millimeter, and Terahertz Waves, 30*(6), 598–610. doi:10.1007/s10762-009-9474-x

Dib, N., & Khodier, M. (2008). Design and optimization of multi-band Wilkinson power divider. *International Journal of RF and Microwave Computer-Aided Engineering, 18*(1), 14–20. doi:10.1002/mmce.20261

Fei, X., Xiao-Hong, T., Ling, W., & Tao, W. (2008). Application of the particle swarm optimization in microwave engineering. *Proceeding IEEE MTT-S International Microwave Workshop Series IMWS on Art of Miniaturizing RF and Microwave Passive Components,* (pp. 187-189).

Gangopadhyaya, M., Mukherjee, P., & Gupta, B. (2010). Resonant frequency optimization of coaxially fed rectangular microstrip antenna using particle swarm optimization algorithm. *Proceedings of the 2010 Annual IEEE India Conference: Green Energy, Computing and Communication,* INDICON 2010, art. no. 5712677.

Goudos, S. K., & Sahalos, J. N. (2006). Microwave absorber optimal design using multi-objective particle swarm optimization. *Microwave and Optical Technology Letters, 48*(8), 1553–1558. doi:10.1002/mop.21727

Goudos, S. K., Zahairs, Z. D., Baltzis, K. B., Hilas, C. S., & Sahalos, J. N. (2009). A comparative study of particle swarm optimization and differential evolution on radar absorbing materials design for EMC applications. *International Symposium on Electromagnetic Compatibility - EMC Europe,* art. no. 5189697.

Goudos, S. K., Zaharis, Z. D., Salazar-Lechuga, M., Lazaridis, P. I., & Gallion, P. B. (2007). Dielecric filter optimal design suitable for microwave communications by using multiobjective evolutionary algorithms. *Microwave and Optical Technology Letters, 49*(10), 2324–2329. doi:10.1002/mop.22755

Hao, W., Junping, G., Ronghong, J., Jizheng, Q., Wei, L., Jing, C., & Suna, L. (2009). An improved comprehensive learning particle swarm optimization and its application to the semiautomatic design of antennas. *IEEE Transactions on Antennas and Propagation, 57*(10), 3018–3028. doi:10.1109/TAP.2009.2028608

Huang, G.-R., Zhong, W.-J., & Liu, H. W. (2010). Microwave imaging based on the AWE and HPSO incorporated with the information obtained from born approximation. *International Conference on Computational Intelligence and Software Engineering, CiSE,* art. no. 5676968.

Mahanfar, A., Bila, S., Aubourg, M., & Verdeyme, S. (2008). Cooperative particle swarm optimization of passive microwave devices. *International Journal of Numerical Modelling: Electronic Networks. Devices and Fields, 21*(1-2), 151–168.

Mhamdi, B., Grayaa, K., & Aguili, T. (2009). An inverse scattering approach using hybrid PSO-RBF network for microwave imaging purposes. *16th IEEE International Conference on Electronics, Circuits and Systems, ICECS 2009,* art. no. 5410978, (pp. 231-234).

Mhamdi, B., Grayaa, K., & Aguili, T. (2011). Hybrid of particle swarm optimization, simulated annealing and tabu search for the reconstruction of two-dimensional targets from laboratory-controlled data. *Progress in Electromagnetics Research B, 28,* 1–18.

Modiri, A., & Kiasaleh, K. (2011). Modification of real-number and binary PSO algorithms for accelerated convergence. *IEEE Transactions on Antennas and Propagation, 59*(1), 214–224. doi:10.1109/TAP.2010.2090460

Ninomiya, H. (2009). A hybrid global/local optimization technique for robust training of microwave neural network models. *IEEE Congress on Evolutionary Computation, CEC 2009,* art. no. 4983315, (pp. 2956-2962).

Ülker, S. (2008). Particle swarm optimization application to microwave circuits. *Microwave and Optical Technology Letters, 50*(5), 1333–1336. doi:10.1002/mop.23369

Wang, D., Zhang, H., Xu, T., Wang, H., & Zhang, G. (2011). Design and optimization of equal split broadband microstrip Wilkinson power divider using enhanced particle swarm optimization algorithm. *Progress in Electromagnetics Research, 118,* 321–334. doi:10.2528/PIER11052303

KEY TERMS AND DEFINITIONS

Evolutionary Algorithm: Is an algorithm incorporating aspects of natural selection.

Microwave Circuits: Any particular grouping of physical elements which are arranged or connected together to produce certain desired effects on the behavior of microwave signals.

Microwave Devices: Any device capable of generating, amplifying, modifying, detecting, or measuring microwave signals.

Optimization Technique: Is a mathematic method refers to choosing the best element from some set of available alternatives.

Particle Swarm Optimizer (PSO): Is a population based stochastic optimization technique inspired by behavior of social animals.

Unsupervised CAD Tools: Is a design tool where the component is modified automatically by the microwave tool itself with an improvement of the component performances without any contribution by the designer.

Chapter 2
Microwave Circuit Design

Gabriel Cormier
Université de Moncton, Canada

Tyler Ross
Carleton University, Canada

ABSTRACT

Circuit models play an important role in the design and optimization of microwave circuits (circuits in the GHz frequency range). These circuit models contain many parameters, including parasitic elements, necessary to correctly model the behavior of transistors at high frequencies. These models are often designed based on a series of measurements. Because of its ability to efficiently locate the global optimum of an objective function, particle swarm optimization (PSO) can be a useful tool when matching a model to its measurements. In this chapter, PSO will be used to calculate a transistor's small-signal model parameters, determine the noise parameters of the transistor, and design a microwave mixer. The mixer is designed at 39.25 GHz, and a comparison between measurements and simulation results shows good agreement.

INTRODUCTION

Particle swarm optimization is very useful technique that can be used when developing models for microwave components. Proper models are important in microwave engineering, as any microwave designer will need to account for the parasitic components of the transistors and passive components used in their design. Good models reduce the number of design/fabrication cycles required to meet a circuit's specifications. This is of considerable benefit to engineers designing microwave circuits, especially monolithic microwave integrated circuits (MMICs), where circuit fabrication is costly and turn-around times are lengthy.

DOI: 10.4018/978-1-4666-2666-9.ch002

The first part of this chapter will introduce and explore the subject of small-signal transistor modeling. For the interested reader, references to important background papers treating small-signal transistor modeling in greater detail will be provided.

In many transistor models, it is possible to obtain the same response for different sets of model parameters. As a result, it is very difficult for conventional optimization to converge to a global optimum. In addition, it is difficult to calculate the model parameters accurately using a single frequency or narrowband measurements. Therefore, a method for finding the global optimal solution using broadband measurements is required. The particle swarm algorithm is a suitable method, since it allows the minimization of an objective function with discontinuities, local optima, while not requiring its derivatives. Since the algorithm is less likely to be trapped by local optima, it is usually be more effective than traditional gradient-based algorithms, which often require estimates of model parameters which should be very close to their actual values. Other approaches and the advantages the particle swarm optimization offers are discussed in greater detail in the coming sections.

As a first example of the use of particle swarm optimization in microwave circuit design, we will use a discrete, commercially-available gallium arsenide (GaAs) FET with a 0.3 μm gate length from Excelics Semiconductor (part number EFA018A), and determine the small-signal model and noise parameters. The manufacturer provides S-parameters as well as large and small-signal models for this device.

One use of a small-signal model is for determining noise parameters, required for the design of a low-noise amplifier. Given a transistor's small-signal model parameters, it is possible to calculate the four noise parameters required for low-noise amplifier design: the source resistance needed to achieve the minimum noise figure (R_{opt}), the source reactance needed to achieve the minimum noise figure (X_{opt}), the noise resistance (R_n) and the noise minimum noise figure (F_{min}). As an example, the minimum noise figure will be determined for this transistor, which would be one of the first steps when designing a low-noise amplifier.

Since the algorithm is less likely to be trapped by local optima, it can be more effective than traditional gradient-based algorithms, which often require estimates of model parameters which should be very close to their actual values. Particle swarm optimization can also be helpful in the design of circuits. While the first half of the chapter deals with transistor modeling, in the second half of this chapter, a differential mixer operating at an RF frequency of 39.25 GHz will be optimized to yield maximum conversion gain. The algorithm will be used to determine the optimal transistor size, transmission line lengths and impedances needed to match and stabilize the mixer.

Conventional approaches to this process involve considerable iterations consisting of transistor stabilization, matching and checking the circuit for conformance to design specifications. Using particle swarm optimization allows for all of these goals to be achieved much more quickly and ensures that the *best* circuit is found.

Fitting a Model to S-Parameters

There have been many studies on how to extract a transistor model from a set of data, often measured S-parameters. Most circuits used to model FETs, including MESFETs and HEMTs at microwave frequencies, are similar to the model shown in Figure 1, with somewhere between 12 and 18 circuit parameters (Curtice, 1980; Maas, 1993; Kayali et al., 1996).

Having a set of measurements of a transistor (typically S-parameters), one might be tempted to directly apply particle swarm optimization to the problem to determine all circuit parameters simultaneously. However, the particle swarm

Figure 1. A typical high-frequency FET model

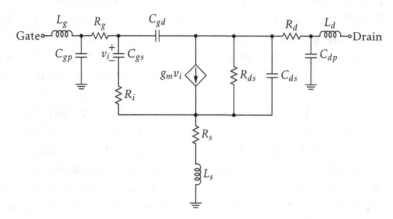

algorithm yields many different solutions, all appearing to yield a good fit between the measured and modeled S-parameters. The non-uniqueness of solutions is not a problem specific to the particle swarm optimization algorithm, but is a well-known occurrence when applying optimization algorithms (of any kind) to small-signal model extraction (Vaitkus, 1983).

To overcome this problem, a "hot"/"cold" approach is usually applied, where a transistor is measured with the gate biased at pinch-off, with zero drain-source voltage (cold), and also at the bias point of interest (hot) (Lin and Kompa, 1994; Rorsman et al., 1996; Dambrine et al., 1988). The bias-independent, extrinsic transistor parameters can then be obtained, usually analytically, but sometimes through optimization, followed by the intrinsic transistor parameters. Other approaches also exist, including multi-bias point extraction algorithms (van Niekerk et al., 2000). Similar approaches using genetic algorithms and particle swarm optimization have also been discussed in the scientific literature (Menozzi et al., 1996; Sabat et al., 2008).

For our example, we are interested in determining small-signal model parameters at different bias points. We will therefore use the extrinsic component values supplied by the manufacturer. The manufacturer's large-signal model will be used to generate reference S-parameters, and the remaining small-signal model parameters will be optimized using particle swarm optimization to match the reference S-parameters. If measurements were available, these could also be used as reference parameters.

Objective Function

At microwave frequencies, S-parameters are typically used to characterize circuits, including transistors. Intuitively, the best fit is obtained when the two sets of S-parameters are closest. For a two-port network, the objective function of the particle swarm algorithm can be defined as in Equation 1 (see Box 1) where N is the number of points to be compared, and ε is the uncertainty on each measured data point, and $S_{i,j}$ are the S-parameters being optimized.

If the uncertainty is known, then the objective function is said to be biased, and $\chi^2 \approx 1$ if the uncertainty has been estimated correctly. If the uncertainty is unknown, a value of $\varepsilon = 1$ can be used, and the final value of the objective function will only provide an indication of the closeness of the fit.

In this case, the real and imaginary parts of the S-parameters are used to determine fit. It is possible to measure fit using different quantities, such as magnitude and phase, magnitude in dB and phase, as long as the numerical magnitudes of

Box 1.

$$\chi^2 = \frac{1}{4N} \sum_{k=1}^{N} \sum_{S_{i,j}} \left[\frac{\left(\mathrm{Re}\left\{S_{i,jk_{ref}} - S_{i,jk_{calc}}\right\}\right)^2 + \left(\mathrm{Im}\left\{S_{i,jk_{ref}} - S_{i,jk_{calc}}\right\}\right)^2}{\varepsilon} \right] \qquad (1)$$

these two quantities (especially if the angle is in degrees) are considered. If the uncertainty ε of the S-parameters is not available, then the parameters can be normalized, so each parameter has equal weight. Otherwise, S_{12}, which tends to be very small, could end up having a very poor fit, since its contribution to the objective function will be dominated by the other three S-parameters.

Implementation of the Particle Swarm Optimization Algorithm

A standard implementation of the particle swarm algorithm was used to fit the model. While the algorithm can be customized in many different ways in an effort to enhance its convergence properties, a very simple implementation of the algorithm was used in this case.

At each time step of the algorithm (or, at each iteration), the velocity of a particle k is updated for the next time step using the following:

$$\vec{v}_k[t+1] = w\vec{v}_k[t] + c_1\vec{r}_1(\hat{x}_g - \vec{x}_k[t]) + c_2\vec{r}_2(\hat{x}_k - \vec{x}_k[t]), \qquad (2)$$

where $\vec{v}_k$ is the velocity of the particle, $\vec{x}_k$ is the position of a particle, $\hat{x}_k$ is the best local solution found so far for particle k, $\hat{x}_g$ is the best solution found for the entire swarm, $\vec{r}_1$ and $\vec{r}_2$ are vectors with random numbers between 0 and 1. w, c_1 and c_2 are constants controlling the behaviour of the swarm with regards to exploring new space instead of remaining close to the optimal solutions found so far (w), the tendency of a particle to stay explore

near the global optimum (c_1) and the tendency of a particle to explore near its local optimum (c_2).

This implementation allows the algorithm to be controlled using only three parameters. In our work, $c_1 = c_2 = 1.2$ and $w = 0.3$ worked well.

Convergence is not highly dependent on the value of w, and values up to 0.9 also work well.

After the new velocities for the particles are calculated, their positions are updated:

$$\vec{x}_k[t+1] = \vec{x}_k[t] + \vec{v}_k[t+1] \qquad (3)$$

Calculation of S-Parameters from the Small-Signal Model

To calculate the objective function given by Equation 1, the S-parameters of the FET model shown in Figure 1 must be known. There are several ways of performing this calculation. One relatively easy way is by determining the Z-parameters of the circuit, and then transforming them to S-parameters. The Z-parameters of a given two-port circuit are:

$$Z_{11} = \frac{V_a}{I_a}\bigg|_{I_b=0}$$

$$Z_{21} = \frac{V_b}{I_a}\bigg|_{I_b=0} \qquad (4)$$

$$Z_{12} = \frac{V_a}{I_b}\bigg|_{I_a=0}$$

$$Z_{22} = \frac{V_b}{I_b}\bigg|_{I_a=0}$$

Figure 2. Small-signal FET model with $I_b = 0$

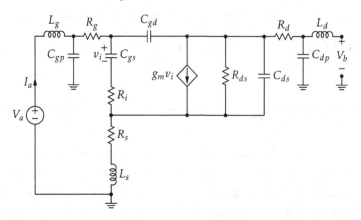

Box 2.

$$
\begin{pmatrix}
Z_g + Z_{GP} & -Z_{GP} & 0 & 0 & 0 & 0 & 0 \\
-Z_{GP} & Z_{X1} & -Z_{GS} - R_i & 0 & 0 & -R_S - Z_S & 0 \\
0 & -Z_{GS} - R_i & Z_{GD} + R_i + Z_{GS} & 0 & 0 & 0 & -1 \\
0 & 0 & 0 & R_{DS} & -R_{DS} & 0 & 1 \\
0 & 0 & 0 & -R_{DS} & R_{DS} + Z_{DS} & -Z_{DS} & 0 \\
0 & -R_S - Z_S & 0 & 0 & -Z_{DS} & Z_{X2} & 0 \\
0 & g_m Z_{GS} & -1 - g_m Z_{GS} & 1 & 0 & 0 & 0
\end{pmatrix}
$$

$$
\cdot \begin{pmatrix} I_1 \\ I_2 \\ I_3 \\ I_4 \\ I_5 \\ I_6 \\ V_{VCCS} \end{pmatrix} = \begin{pmatrix} V_a \\ 0 \\ 0 \\ 0 \\ 0 \\ 0 \\ 0 \end{pmatrix}
$$

(5)

where V_a is the test voltage applied at the terminal, I_a is the current at that terminal, and V_b and I_b are the voltage and current at the other terminal. Since I_a is sometimes zero, and at other times I_b is zero, both cases must be analyzed to determine the four Z-parameters. With $I_b = 0$, we obtain the circuit shown in Figure 2.

To determine the required currents, we can use basic mesh-current analysis, in matrix form. Equation 5 is the matrix representation of the mesh-current problem (See Box 2) where $Z_{X1} = R_G + Z_{GS} + R_i + R_S + Z_S + Z_{GP}$ and $Z_{X2} = R_D + Z_{DP} + R_S + Z_S + Z_{DS}$. Likewise, when $I_a = 0$, we obtain the circuit in Figure 3.

Figure 3. Small-signal FET model with $I_a = 0$

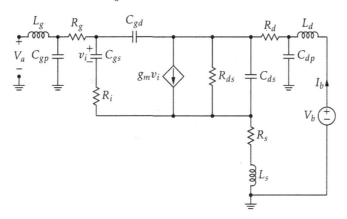

Box 3.

$$
\begin{pmatrix}
Z_{Y1} & -Z_{GS} - R_i & 0 & 0 & -R_S - Z_S & 0 & 0 \\
-Z_{GS} - R_i & Z_{GD} + R_i + Z_{GS} & 0 & 0 & 0 & 0 & 0 \\
0 & 0 & R_{DS} & -R_{DS} & 0 & 0 & -1 \\
0 & 0 & -R_{DS} & R_{DS} + Z_{DS} & -Z_{DS} & 0 & 1 \\
-R_S - Z_S & 0 & 0 & -Z_{DS} & Z_{Y2} & -Z_{DP} & 0 \\
0 & 0 & 0 & 0 & -Z_{DP} & Z_D + Z_{DP} & 0 \\
g_m Z_{GS} & -1 - g_m Z_{GS} & 1 & 0 & 0 & 0 & 0
\end{pmatrix}
$$

$$
\cdot
\begin{pmatrix}
I_1 \\ I_2 \\ I_3 \\ I_4 \\ I_5 \\ I_6 \\ V_{VCCS}
\end{pmatrix}
=
\begin{pmatrix}
0 \\ 0 \\ 0 \\ 0 \\ 0 \\ -V_b \\ 0
\end{pmatrix}
\tag{6}
$$

The matrix equation representing this circuit is given by Equation 6 in Box 3 where $Z_{Y1} = R_G + Z_{GS} + R_i + R_S + Z_S + Z_{GP}$ and $Z_{Y2} = R_D + Z_{DP} + R_S + Z_S + Z_{DS}$. Once the vector containing the solutions is found, then the Z-parameters may be calculated, the first two from the results of the first equation.

$$
Z_{11} = \frac{V_a}{I_a} = \frac{V_a}{I_1}
\tag{7}
$$

$$Z_{21} = \frac{V_b}{I_a}$$

$$= \frac{Z_{DP}I_6}{I_1} = \frac{\frac{1}{j\omega C_{DP}}I_6}{I_1} \tag{8}$$

The remaining two parameters are obtained from the results of the second matrix equation:

$$Z_{12} = \frac{V_a}{I_b}$$

$$= \frac{-Z_{GP}I_1}{-I_6} = \frac{\frac{1}{j\omega C_{GP}}I_1}{I_6} \tag{9}$$

$$Z_{22} = \frac{V_b}{I_b}$$

$$= \frac{V_b}{-I_6} \tag{10}$$

Once the Z-parameters have been calculated, they are readily transformed into S-parameters using well-known relationships (Pozar, 2005):

$$S_{11} = \frac{\left(Z_{11} - Z_0\right)\left(Z_{22} + Z_0\right) - Z_{12}Z_{21}}{\left(Z_{11} + Z_0\right)\left(Z_{22} + Z_0\right) - Z_{12}Z_{21}} \tag{11}$$

$$S_{12} = \frac{2Z_0 Z_{12}}{\left(Z_{11} + Z_0\right)\left(Z_{22} + Z_0\right) - Z_{12}Z_{21}} \tag{12}$$

$$S_{21} = \frac{2Z_0 Z_{21}}{\left(Z_{11} + Z_0\right)\left(Z_{22} + Z_0\right) - Z_{12}Z_{21}} \tag{13}$$

$$S_{22} = \frac{\left(Z_{11} + Z_0\right)\left(Z_{22} - Z_0\right) - Z_{12}Z_{21}}{\left(Z_{11} + Z_0\right)\left(Z_{22} + Z_0\right) - Z_{12}Z_{21}} \tag{14}$$

Table 1. Extrinsic parameters for the small-signal model

Parameter	Value
R_g	2.5Ω
R_d	1.3Ω
R_s	4.3Ω
C_{gp}	0.05pF
C_{dp}	0.05pF
L_g	0.2nH
L_d	0.3nH
L_s	0.04nH

where Z_0 is the reference impedance of the system (50Ω for this problem). The small-signal model's parameters having been calculated, the value of the objective function can be calculated to determine the quality of the fit.

Deriving a Small-Signal Model

As discussed earlier, it is generally not possible to fit all a model's parameters to a single data set. Either several bias points must be used in the fitting process simultaneously, or some of the parameters (e.g., the extrinsic parameters) should be analytically determined beforehand. Since the analytical calculation is not pertinent to the usage of the particle swarm algorithm and various techniques are discussed at length in the scientific literature (e.g, (Lin and Kompa, 1994; Rorsman et al., 1996; Dambrine et al., 1988)), this will not be discussed in detail here and the values provided by the manufacturer of the transistor will be used. These values are given in Table 1.

With a class A bias point ($V_{DS} = 6$ V and $V_{GS} = -0.63$ V), the model parameters were restricted to the range specified in Table 2.

The model is fitted to S-parameters from 1 GHz to 40 GHz at 1 GHz steps. Three different bias points are used as a demonstration:

Table 2. Allowed ranges for the intrinsic model parameters, used during the particle swarm optimization process

Parameter	Minimum	Maximum
R_i (Ω)	0	5
R_{ds} (Ω)	100	1000
C_{gs} (pF)	0.1	1
C_{gd} (fF)	1	100
C_{ds} (fF)	1	100
g_m (S)	0	0.5

1. $V_{DS} = 6V$, $V_{GS} = -0.63V$ (bias point for class A operation)
2. $V_{DS} = 5V$, $V_{GS} = -0.8V$
3. $V_{DS} = 2V$, $V_{GS} = -0.9V$

After the model-fitting procedure, the model parameters given in Table 3 were obtained. The value of the objective function (Equation 1) is also given. $\varepsilon = 1$ was used in the evaluation of the objective function.

In general, computation time is low: about 30 seconds is required on an Intel Core i5 desktop computer using 500 particles, far more than is required for the particle swarm algorithm to identify the solution. In several tests, 100 particles were sufficient to obtain the same solution after 1000 iterations. The optimal number of particles required will depend in part on the size of the search space and the number of model parameters to fit (if a different model than that in Figure 1 is used). Given the fast computation time, however, there is little reason not to choose a large number of particles to increase the convergence rate. This is especially true on modern computers, as it is possible to calculate the objective function for many particles in parallel. Figure 4 illustrates this by showing the value of the objective function after each iteration for three tests. As can be seen, the algorithm converges quickly, remaining relatively constant after 20 iterations.

Table 3. Small-signal model parameters for the three bias points

Parameter	$V_{DS} = 6V$ $V_{GS} = -0.63V$	$V_{DS} = 5V$ $V_{GS} = -0.8V$	$V_{DS} = 2V$ $V_{GS} = -0.9V$
R_g (Ω)	2.5	2.5	2.5
R_d (Ω)	1.3	1.3	1.3
R_s (Ω)	4.3	4.3	4.3
R_i (Ω)	2.1	2.1	2.1
R_{ds} (Ω)	281	281	263
C_{gp} (fF)	50	50	50
C_{dp} (fF)	50	50	50
C_{gs} (fF)	146	144	140
C_{gd} (fF)	10.3	10.8	12.2
C_{ds} (fF)	30.4	29.0	27.0
L_g (nH)	0.2	0.2	0.2
L_d (nH)	0.3	0.3	0.3
L_s (nH)	0.04	0.04	0.04
g_m (mS)	42	37.3	33.5
χ^2	**0.00833**	**0.00820**	**0.00775**

Figure 4. Value of the objective function after each iteration in three separate tests. Convergence is achieved quickly with a large number of particles.

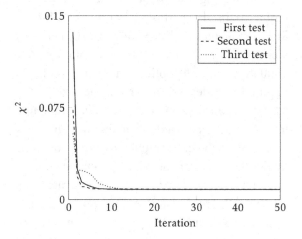

Using the Small-Signal Model to Predict Noise Parameters

One use of a small-signal model (See Figure 5) is to determine noise parameters to be used in the design of a low-noise amplifier. The subject of noise modeling is related to small-signal transistor modeling, but can also be considered its own large research area. Just as there many small-signal models and extraction techniques, there are a multitude of noise models and noise model extraction techniques.

Some noise models also require noise figure measurements, sometimes at different source impedances. However, some noise models are able to predict noise parameters using only a transistor's bias conditions and its small-signal

Figure 5. Shows a comparison between the newly-fitted model and the reference S-parameters from the manufacturer, at the class A bias point (V_{DS} = 6V, V_{GS} = -0.63V). As can be seen, the fit between the two sets of S-parameters is good. S-parameters of the manufacturer's large-signal model (—) and the small-signal model fit to the manufacturer's model using particle swarm optimization (- - -), from 1 GHz to 40 GHz.

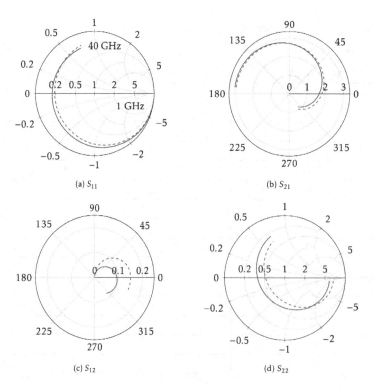

model. One such model was proposed by Sanabria (Sanabria et al., 2006). This model was developed to account for noise associated with gate leakage current present in many gallium nitride transistors and assumes that FET noise sources are uncorrelated, although this simplifying approximation has been validated in some studies (Podell, 1981; Pospiezalski, 1989; Danneville et al., 1994; Sanabria et al., 2006).

We will assume that gate leakage is low in these GaAs FETs and can therefore be ignored. In this case, the noise figure of a FET is given:

$$F = 1 + \frac{R_{in}}{R_{source}} + \frac{a}{R_{source}}$$
$$\left| R_{in} + R_{source} + j \left(X_{source} - \frac{1}{\omega C'_{gs}} \right) \right|^2 \qquad (15)$$

Where

$$a = g'_m \Gamma \left(\frac{\omega}{g'_m / C'_{gs}} \right)^2 \qquad (16)$$

$$g'_m = \frac{g_m}{1 + g_m R_s} \qquad (17)$$

$$C'_{gs} = \frac{C_{gs}}{1 + g_m R_s} \qquad (18)$$

$$R_{in} = R_g + R_i + R_s \qquad (19)$$

where the circuit parameters (R_s, R_i, R_g, C_{gs}) are those given in Figure 3, ω is the operating frequency (in rad/s), and R_{in} is the input impedance of the FET. The value of Γ is approximately 2/3 for a transistor in saturation (Sanabria et al., 2006). The optimal source impedance is given by:

$$R_{opt} = \sqrt{R_{in} \frac{1 + a R_{in}}{a}} \qquad (20)$$

$$X_{opt} = \frac{1}{\omega C'_{gs}} \qquad (21)$$

The noise resistance is calculated as:

$$R_n = \frac{Z_0}{4} \left(F_{Z_0} - F_{min} \right) \left| 1 + \frac{1}{\Gamma_{opt}} \right|^2 \qquad (22)$$

The minimum noise figure F_{min} can be determined by substituting R_{opt} and X_{opt} into the expression for F. For more details on the derivation of these equations, the interested reader is encouraged to refer to (Sanabria et al., 2006), which presents this model in greater detail, and where gate leakage current is not neglected.

Using this model and the small-signal circuit parameters previously calculated, the noise figure and minimum noise figure were calculated. Since gate leakage was neglected for these large devices, the calculated noise figure will be somewhat optimistic. Figure 6 shows the noise parameters for the first bias point (class A operation). While the minimum noise figure is unrealistically low at low frequencies (due to noise associated with leakage current being neglected), the noise figure at 12 GHz (approximately 1 dB) is in agreement with the value provided on the transistor's data sheet (1.1 dB).

This information can be readily used to design a low-noise amplifier, since all noise parameters have been determined. If S-parameters are available at many frequencies, it is also possible to study the behavior of the minimum noise figure as a function of bias current, for example. The same approach discussed here was used to accomplish precisely this, for gallium nitride HEMTs of varying dimensions (Ross et al., 2011). For gallium nitride transistors, unlike here, the gate leakage

Figure 6. Noise parameters calculated from the small-signal model found using particle swarm optimization

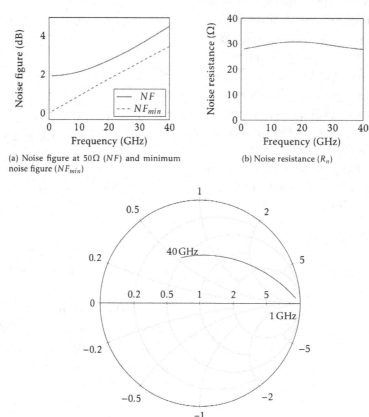

(a) Noise figure at 50Ω (*NF*) and minimum noise figure (*NF$_{min}$*)

(b) Noise resistance (*R$_n$*)

(c) Optimal source impedance to obtain minimum noise figure (*Z$_{opt}$*)

current was not neglected, and the full noise model was applied.

Differential Mixer

As another example showing the use of particle swarm optimization in microwave circuit design, a differential mixer (sometimes known as a half Gilbert cell) was optimized. A differential mixer is a 3-transistor circuit, in a configuration similar to a differential amplifier. In this case, one switching transistor was grounded (Dearn and Devlin, 1999). The operation of the differential mixer with one switching transistor grounded is roughly the same as one with two LO inputs (LO+ and LO-), while giving a simpler layout (since only one LO input is needed) (Lee, 2004; Maas, 2003). The circuit will

be designed to achieve maximum gain. The mixer will be down-converting, with an input frequency of 39.25GHz, an LO frequency of 38.25GHz, for an output frequency of 1GHz. A schematic of the circuit is given in Figure 7.

The three transistors form the central core of the mixer. Transistor H_1 serves to amplify the RF input. Transistors H_2 and H_3 are switched by the LO. In this design, a commercially available process with 0.2 μm gate length pHEMTs was used.

When designing a differential mixer, many parameters are present. These parameters must be optimized according to specific goals: conversion gain, linearity, or noise. The objective of this design is to optimize and automate the calculation of these parameters using particle swarm optimization. Us-

Figure 7. Differential mixer with one LO input grounded

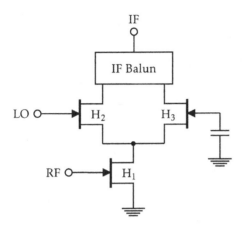

ing the FET model, particle swarm optimization will be used to match the input impedance of the top two switching transistors of the differential mixer to the output impedance (the drain) of the bottom amplifying transistor (as in Figure 7).

PHEMT Model

A first step in optimizing the centre cell is to correctly model the FETs used in the cell (pHEMTs in the current process). Although this modeling will result in a small-signal model, this is sufficiently precise to serve as a starting point in the design. The objective of this section is to compare the pHEMT model and its extracted S-parameters to the simulation results. The pHEMT will be

modeled using a standard high-frequency FET model, and foundry data will be used to complete the model.

Impedance Seen Looking into the Drain

To get an equation of the output impedance of the pHEMT, the first step is to determine which parameters from the model are present. The model used for the small-signal equivalent circuit is given in Figure 8. This model is similar to the standard FET model presented earlier. Although many models ignore some of the parasitics presented in Figure 8, they are included here for completeness.

In order to use this model to optimize the impedance between the bottom FET of the differential mixer and the top two FETs, an equation is needed giving the impedance seen looking into the drain (for the bottom FET) and the source (for the top two FETs). From the circuit given in Figure 8, certain parasitics can be grouped together to simplify the analysis. The following impedances are defined:

$$
\begin{aligned}
Z_s &= R_s + L_s \parallel C_s \\
Z_g &= \left(Z_{in} + L_g \right) \parallel C_g + R_g \\
Z_s &= C_{gs} + R_{gs} \\
Z_{ds} &= R_{ds} \parallel C_{ds}
\end{aligned}
\qquad (23)
$$

Figure 8. Small-signal model of the pHEMT

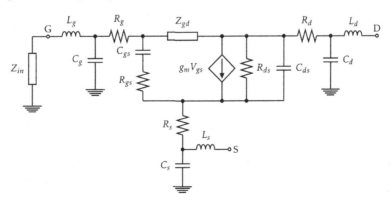

Box 4.

$$
\mathbf{A} = \begin{pmatrix}
Z_{cd} + Z_{ld} & -Z_{cd} & 0 & 0 \\
-Z_{cd} & Z_s + Z_{ds} + R_d + Z_{cd} & -Z_{ds} - g_m Z_{gs} Z_{ds} & -Z_s + g_m Z_{gs} Z_{ds} \\
0 & -Z_{ds} & Z_{gs} + Z_{gd} + Z_{ds} + g_m Z_{gs} Z_{ds} & -g_m Z_{gs} Z_{ds} - Z_{gs} \\
0 & -Z_s & -Z_{gs} & Z_g + Z_{gs} + Z_s
\end{pmatrix} \qquad (24)
$$

Where it is of course understood that the capacitances and inductances in the above equations represent their actual impedance. All the parasitics given in Figure 8 depend on the size of the pHEMT (number of fingers and finger width). With these simplifications done, standard circuit analysis (using a mesh current technique) is used to get a matrix for the impedances (matrix A) in Equation 24 (see Box 4) where Z_{cd} represents the impedance of C_d and Z_{ld} is the impedance of L_d.

The voltage matrix B is given by:

$$
\mathbf{B} = \begin{pmatrix}
V_x \\
0 \\
0 \\
0
\end{pmatrix} \qquad (25)
$$

where V_x is a test voltage applied to the drain (terminal under study).

The impedance seen looking into the drain is then given by:

$$
Z_x = \frac{V_x}{I_x} \qquad (26)
$$

where I_x is the first current obtained from solving the system:

$$
\mathbf{C} = \mathbf{A}^{-1} \cdot \mathbf{B} \qquad (27)
$$

The output impedance calculated by the above model was compared to the values given by an

Figure 9. Smith chart showing the calculated output impedance and simulated S-parameters

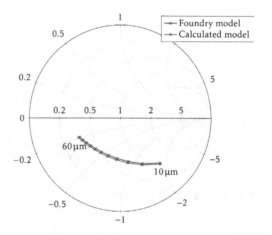

Agilent ADS simulation (small-signal S-parameter linear test bench). A 6-finger FET is simulated, with the finger length varied from 10 μm to 60 μm. The results are given in Figure 9.

There is a small difference between the calculated output impedance and the simulated S-parameters. The calculated values show good agreement with the pHEMT simulation.

The impedance can be obtained from the S-parameters by using the following equation (Pozar, 2005):

$$
\Gamma_o = S_{22} + \frac{S_{12} S_{21} \Gamma_s}{1 - S_{11} \Gamma_s} \qquad (28)
$$

In a standard S-parameter simulation, the input impedance $Z_{in} = 50$, and so $\Gamma_s = 0$. Hence $\Gamma_o = S_{22}$, and the reflection coefficient can be transformed

Box 5.

$$\mathbf{A} = \begin{pmatrix} Z_{cs} + Z_{ls} & -Z_{cs} & 0 & 0 \\ -Z_{cs} & Z_s + Z_{gs} + R_s + Z_{cs} & -Z_{gs} & -Z_g \\ 0 & -Z_{gs} - g_m Z_{gd} Z_{gs} & Z_{gs} + Z_{gd} + Z_{ds} + g_m Z_{gd} Z_{gs} & -Z_{gd} \\ 0 & -Z_g & -Z_{gd} & Z_g + Z_{gd} + Z_d \end{pmatrix} \tag{31}$$

into an impedance, using the well-known relation (Pozar, 2005):

$$Z_x = Z_0 \frac{1 + \Gamma_o}{1 - \Gamma_o} \tag{29}$$

where Z_0 is the reference impedance, 50Ω.

Impedance Seen Looking into the Source

For the top two pHEMTs of the differential mixer, the impedance used for matching is the source impedance. Although these pHEMTs operate in large-signal, the small-signal equivalent circuit (Figure 8) is used for simplification. In this case, the simplified impedances Z_g, Z_{gs}, Z_{dg} and Z_{ds} given by Equation 23 will remain the same. A new impedance Z_d will be added:

$$Z_d = \left(Z_{dd} + L_d\right) \| C_d + R_d \tag{30}$$

where Z_{dd} is the impedance connected to terminal D. The impedance Z_s (Equation 23) will not be used here, and using the same analysis technique in Equation 31 (see Box 5).

The voltage matrix B is the same as for the calculation of the drain impedance (Equation 25). The impedance is given in the same manner by Equation 26.

Again, for clarification purposes, the inductances and capacitances used in Equations 30 and 31 represent the actual impedances. The imped-

Figure 10. Smith chart showing the calculated output impedance at the source of pHEMT H_2

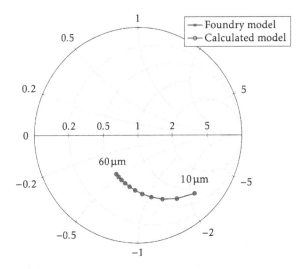

ance Z_{cs} represents the impedance of C_s and Z_{ls} is the impedance of L_s. As was the case for the drain impedance, a 6-finger FET is simulated, with the finger lengths varied from 10 μm to 60 μm. Figure 10 shows the calculated and simulated values of the source impedance. There is very little difference between both results.

Problem Definition

Now that the output impedance of the lower pHEMT and the input impedance of the top pHEMTs are known, the problem becomes one of optimization. The pHEMTs must be biased, and so far the biasing circuitry has been neglected. For H_1, a drain bias must be applied to supply

Figure 11. Diagram of optimization problem

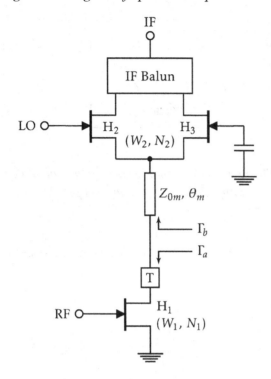

Figure 12. Layout of coplanar bias "T" circuit

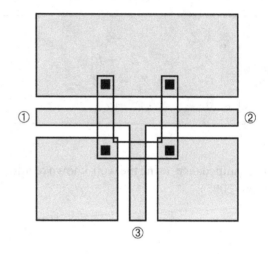

Figure 13. Stability elements for transistor H_1

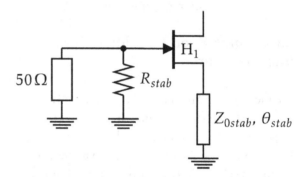

$$Z_a = Z_b^* \tag{32}$$

the required DC voltage. A bias T will therefore be placed between H_1 and the matching line. The circuit will resemble that of Figure 11.

The bias T was simulated as a 2-port circuit, assuming the impedance of the third branch, connected to the biasing circuit, would present a very high impedance at the frequency of interest. This is a reasonable assumption in this case, since a first-order approximation is all that is needed. The bias T is represented in Figure 12. Port 3 is connected to the DC bias line.

The output impedance of the pHEMTs is a function of their width and number of fingers (as well as the bias values, but these are not considered here). So $Z_1 = f(W_1, N_1)$, and $Z_{2,3} = f(W_2, N_2)$ because the top two pHEMTs will be of the same dimensions. The reflection coefficient of the matching line is a function of its characteristic impedance Z_{0m} and its length θ_m.

Maximum power transfer is achieved if $\Gamma_a = \Gamma_b^*$, or

Since transistor H_1 operates at a relatively high bias current and at a high frequency, stability of the transistor must be verified. A grounded transmission line is added to the source of H_1, and a shunt resistor is added to the gate of H_1, as shown in Figure 13. A stability analysis for H_1 gives a range of values for the transmission line impedance and length and the shunt resistor. The input resistance Z_{in} of Figure 8 will be modified as follows:

$$Z_{in} = Z_{ref} \| R_{stab} \tag{33}$$

where Z_{ref} is the reference impedance (50Ω) and R_{stab} is the stabilizing resistor.

At the source, instead of inductance L_s being grounded, it is now in series with a grounded transmission line, whose impedance is given by (Pozar, 2005):

$$Z_{stab} = jZ_{0stab} \tan\left(\theta_{stab}\right) \tag{34}$$

where Z_{0stab} is the impedance of the stabilizing transmission line, and θ_{stab} is the electrical length of the line.

After the bias T circuit was simulated, the S-parameters were extracted, and then converted to an ABCD matrix. The FET impedance was calculated and then converted to an ABCD matrix as well. Since these two components are in cascade, it is easy to multiply the two ABCD matrices to obtain a single matrix.

$$\left[ABCD\right]_1 = \left[ABCD\right]_{H_1} \cdot \left[ABCD\right]_T \tag{35}$$

This ABCD matrix is then converted to a Z impedance matrix, which is used to calculate the optimum matching.

Optimization was done using a particle swarm algorithm. There are nine variables to optimize: W_1, N_1, Z_{0m}, θ_m, W_2, N_2, Z_{0stab} and θ_{stab}. The objective function for this problem is such that $Z_a - Z_b{}^* = 0$, or

$$f_{obj} = \mathrm{Re}\left\{Z_a - Z_b^*\right\}^2 + \mathrm{Im}\left\{Z_a - Z_b^*\right\}^2. \tag{36}$$

Optimization Results

Although the model does not correspond exactly to the calculated values, the results calculated by the particle swarm will be a good starting point for the design of the mixer. Electromagnetic simulations must still be done, especially on the

Table 4. Upper and lower limits for the optimization parameters

Parameter	Minimum	Maximum
W_1 (μm)	10	60
N_1	2	8
Z_{0m} (Ω)	30	90
θ_m (°)	10	70
W_2 (μm)	10	60
N_2	2	8
R_s (Ω)	500	1500
Z_{0stab} (Ω)	30	110
θ_{stab} (°)	10	70

length of line since the effective permittivity (ε_{eff}) used is just an approximation. Also, no loss was considered in the line, and the pHEMT models will not correspond exactly to measured data. Nonetheless, the steps presented here can give a designer a good reference point from which to start his design. The upper and lower limits used to perform the calculations are given in Table 4.

The parameters are chosen according to the limitations of the design process. The individual finger widths are limited to 60 μm, and the maximum number of fingers is 8. A design could be made to have more fingers, but the scaling model is only accurate up to 8 fingers. The impedances and electrical length chosen for the matching and stability lines are values that can be readily created on-chip without taking up too much space.

Calculated Results

Using the particle swarm algorithm, the following parameters were calculated as the optimum (see Table 5). These values are compared here to the values obtained by simulation with ADS. The values calculated by the particle swarm algorithm are very close to the optimum values obtained by ADS simulation.

The particle swarm algorithm arrived at these values consistently, with an error function of 0 (or

Table 5. Optimized parameters for the differential mixer

Parameter	Particle Swarm	ADS Simulated Value
W_1 (μm)	42	38
N_1	2	2
Z_{0m} (Ω)	80	75
θ_m (°)	40	41
W_2 (μm)	50	50
N_2	6	6
R_s (Ω)	1200	1200
Z_{0stab} (Ω)	40	40
θ_{stab} (°)	10	10

$f_{obj} = 0$). The error was of the order of magnitude of the precision of the *double* variable type in Matlab®.

Layout of the Circuit

Once the parameters for the transistors and interstage matching line have been chosen, the next step is designing the matching circuits and doing the layout of the circuit. First, however, the bias circuits for the mixer core must be designed, before any matching is done. Designing the layout and matching circuits follows an 'inside-out' approach: the layout of the circuit between the top and bottom transistors is done first, and then steadily moving 'outwards', finishing with the matching circuits. All matching circuits were done in coplanar technology and simulated to ensure proper function.

A schematic of the completed circuit is shown in Figure 14. The two IF outputs will be matched off chip using a jig circuit.

A microphotograph of the mixer is shown in Figure 15. The mixer has a height of 1.614 mm, and a width of 1.875 mm.

Experimental Setup

The RF and LO power levels were measured at the probe, using an HP 437B Power Meter. The probing station is a Karl Suss PA200 Semi-Automatic Probing Station. The probes are from GBB Industries Inc. The rest of the equipment used is as follows:

Figure 14. Circuit schematic of differential mixer

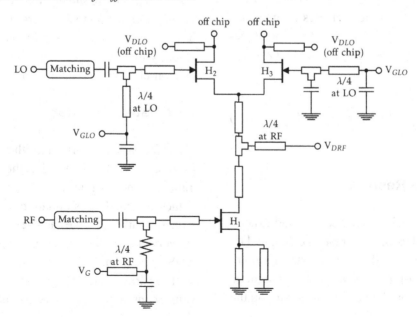

Figure 15. Microphotograph of fabricated differential mixer

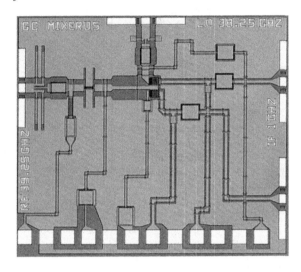

- RF and LO probes: RF is a Picoprobe model 40A-GSG-150-P, with 0.65 dB loss at 39.25 GHz. LO probe is also a Picoprobe model 40A-GSG-150-P, with 0.7 dB loss at 38.25 GHz.
- The IF dual probe is a Picoprobe model 40A-GSG-150/40A-GSG-150-D-800, with 0.13 dB loss at 1 GHz.
- Cables are type K, with approximately 0.3 dB loss at 1 GHz, and 4.5 dB loss at 39 GHz.

Measured Results

The conversion gain as a function of the RF input power is given in Figure 16. The measured gain of the mixer is close to the simulated gain. The gain is relatively stable until around 10 dBm of input power. The input LO power was +10 dBm at the probe, at a frequency of 38.25 GHz. The RF input is at 39.25 GHz. With matching at the output, the conversion gain would be approximately 2.5 dB better.

The 3rd-order intercept point (IP3) is a measure of the linearity of an amplifier. It refers to the power where a 3rd-order product reaches the same amplitude as the fundamental frequency. It can be referenced at the input (IIP3) or the output (OIP3). It is measured by applying two close frequencies (for example, separated by 10MHz), and measuring the output of the fundamental frequency and the 3rd-order component ($2f_1 - f_2$), which will be close to the fundamental. The IIP3 of the top output is given in Figure 17. The measured data are marked with circles, while the linear extrapolation of that data is shown in dashed lines. The input third order intercept point, where the fundamental and 3rd-order products have the same amplitude, is 10.6 dBm.

Figure 16. Conversion gain vs RF input power (no matching at output)

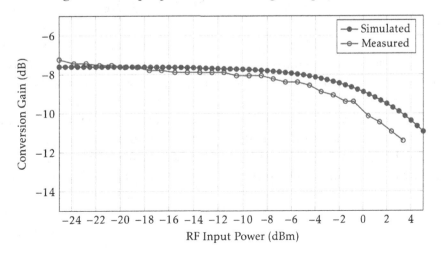

Figure 17. 3rd order intercept point

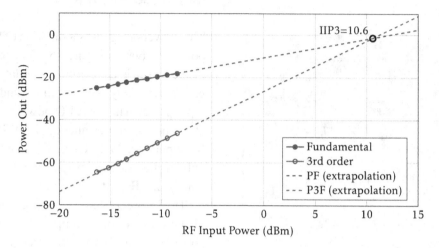

Figure 18. RF and LO return loss, measured

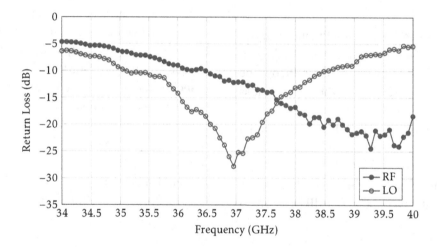

The measured return loss for the RF and LO inputs are given in Figure 18. The return loss shows the ratio of power reflected back to the input, if there is a mismatch between the source impedance and the input impedance. A good value of return loss is 20 dB. The measured RF return loss at 39.25 GHz is slightly less than 20 dB, which is close to the simulated value of 24 dB. The measured LO return loss is approximately 11 dB, close to the simulated value of 12 dB.

CONCLUSION

Particle swarm optimization was used to develop small-signal models for microwave FETs, and to obtain noise parameters. These are important steps in the design of many microwave circuits, such as low noise amplifiers, mixers, and power amplifiers. To further illustrate the application of particle swarm optimization, a differential mixer was designed and fabricated. Nine variables were optimized to achieve the best matching between the bottom transistor and the top two switching transistors of the mixer core. The mixer was built on a GaAs substrate using a commercial 0.2 μm pHEMT process.

REFERENCES

Curtice, W. R. (1980). A MESFET model for use in the design of GaAs integrated circuits. *IEEE Transactions on Microwave Theory and Techniques*, *28*(5), 448–455. doi:10.1109/TMTT.1980.1130099

Dambrine, G., Cappy, A., Heliodore, F., & Playez, E. (1988). A new method for determining the FET small-signal equivalent circuit. *IEEE Transactions on Microwave Theory and Techniques*, *36*(7), 1151–1159. doi:10.1109/22.3650

Danneville, F., Happy, H., Dambrine, G., Belquin, J.-M., & Cappy, A. (1994). Microscopic noise modeling and macroscopic noise models: How good a connection? *IEEE Transactions on Electron Devices*, *41*(5), 779–786. doi:10.1109/16.285031

Dearn, A., & Devlin, L. (1999). A 40-45 GHz monolithic Gilbert cell mixer. In *MM-Wave Circuits and Technology for Commercial Applications* (Ref. No. 1999/007), (pp. 7/1–7/6). IEE Colloquium.

Kayali, S., Ponchak, G., & Shaw, R. (1996). *GaAs MMIC reliability assurance guideline for space applications*. Technical report, NASA, Jet Propulsion Laboratory. Retrieved from http://parts.jpl.nasa.gov/mmic/contents.htm

Lee, T. H. (2004). *Planar microwave engineering* (2nd ed.). Cambridge, UK: Cambridge University Press.

Lin, F., & Kompa, G. (1994). FET model parameter extraction based on optimization with multiplane data-fitting and bidirectional search–A new concept. *IEEE Transactions on Microwave Theory and Techniques*, *42*(7), 1114–1121. doi:10.1109/22.299745

Maas, S. A. (1993). *Microwave mixers*. Norwood, MA: Artech House Publishers.

Maas, S. A. (2003). *Nonlinear microwave and RF circuits* (2nd ed.). Norwood, MA: Artech House Publishers.

Menozzi, R., Piazzi, A., & Contini, F. (1996). Small-signal modeling for a microwave FET linear circuits based on a genetic algorithm. *IEEE Transactions on Circuits and Systems*, *43*(10), 839–847. doi:10.1109/81.538990

Podell, A. F. (1981). A functional GaAs FET noise model. *IEEE Transactions on Electron Devices*, *28*(5), 511–517. doi:10.1109/T-ED.1981.20375

Pospiezalski, M. W. (1989). Modeling of noise parameters of MES-FET's and MODFET's and their frequency and temperature dependence. *IEEE Transactions on Microwave Theory and Techniques*, *37*(9), 1340–1350. doi:10.1109/22.32217

Pozar, D. M. (2005). *Microwave engineering* (3rd ed.). Hoboken, NJ: John Wiley & Sons.

Rorsman, N., Garcia, M., Karlsson, C., & Zirath, H. (1996). Accurate small-signal modeling of HFETs for millimeter-wave applications. *IEEE Transactions on Microwave Theory and Techniques*, *44*(3), 432–437. doi:10.1109/22.486152

Ross, T., Cormier, G., Hettak, K., & Amaya, R. E. (2011). Particle swarm optimization in the determination of the optimal bias current for noise performance of gallium nitride HEMTs. *Microwave and Optical Technology Letters*, *53*(3), 652–656. doi:10.1002/mop.25758

Sabat, S. L., Raju, V., & Ali, L. (2008). MESFET small signal model parameter extraction using particle swarm optimization. In *International Conference on Microelectronics* (pp. 208-211).

Sanabria, C., Chakraborty, A., Xu, H., Rodwell, M. J., Mishra, U. K., & York, R. A. (2006). The effect of gate leakage on the noise figure of AlGaN/GaN HEMTs. *IEEE Electron Device Letters*, *27*(1), 19–21. doi:10.1109/LED.2005.860889

Vaitkus, R. L. (1983). Uncertainty in the values of GaAs MESFET equivalent circuit elements extracted from measured two-port scattering parameters. In *IEEE/Cornell Conference on High-Speed Semiconductor Devices & Circuits*, (pp. 301–308).

van Niekerk, C., Meyer, P., Schreurs, D. M. M.-P., & Winson, P. B. (2000). A robust integrated multibias parameter-extraction method for MESFET and HEMT models. *IEEE Transactions on Microwave Theory and Techniques*, *48*(5), 777–786. doi:10.1109/22.841871

ADDITIONAL READING

Anholt, R. (1995). *Electrical and thermal characterization of MESFETs, HEMTs, and HBTs*. Norwood, MA: Artech House Publishers.

Berroth, M., & Bosch, R. (1990). Broad-band determination of the FET small-signal equivalent circuit. *IEEE Transactions on Microwave Theory and Techniques*, *38*(7), 891–895. doi:10.1109/22.55781

Curtice, W. R. (1980). A MESFET model for use in the design of GaAs integrated circuits. *IEEE Transactions on Microwave Theory and Techniques*, *28*(5), 448–456. doi:10.1109/TMTT.1980.1130099

Curtice, W. R., & Ettenberg, M. (1985). A nonlinear GaAs FET model for use in the design of output circuits for power amplifiers. *IEEE Transactions on Microwave Theory and Techniques*, *33*(12), 1383–1394. doi:10.1109/TMTT.1985.1133229

Dambrine, G., Cappy, A., Heliodore, F., & Playez, E. (1988). A new method for determining the FET small-signal equivalent circuit. *IEEE Transactions on Microwave Theory and Techniques*, *36*(7), 1151–1159. doi:10.1109/22.3650

Dickson, T. O., Yau, K. H. K., Chalvatzis, T., Mangan, A. M., Laskin, E., & Beerkens, R. (2006)... *IEEE Journal of Solid-State Circuits*, *41*(8), 1830–1845. doi:10.1109/JSSC.2006.875301

Follman, R., Borkes, J., Waldow, P., & Wolff, I. (2000). Extraction and modeling methods for FET devices. *IEEE Microwave Magazine*, *1*(3), 49–55. doi:10.1109/6668.871187

Fukui, H. (1979). Design of microwave GaAs MESFET's for broad-band low-noise amplifiers. *IEEE Transactions on Microwave Theory and Techniques*, *27*(7), 643–650. doi:10.1109/TMTT.1979.1129694

Gao, J., & Werthof, A. (2009). Scalable small-signal and noise modeling for deep-submicrometer MOSFETs. *IEEE Transactions on Microwave Theory and Techniques*, *57*(4), 737–744. doi:10.1109/TMTT.2009.2015075

Ladbrooke, P. H. (1989). *MMIC design: GaAs FETs and HEMTs*. Norwood, MA: Artech House Publishers.

Lane, R. Q. (1969). The determination of device noise parameters. *Proceedings of the IEEE*, *57*(8), 1461–1462. doi:10.1109/PROC.1969.7311

Lee, S., Webb, K. J., Tilak, V., & Eastman, L. F. (2003). Intrinsic noise equivalent-circuit parameters for AlGaN/GaN HEMTs. *IEEE Transactions on Microwave Theory and Techniques*, *51*(5), 1567–1577. doi:10.1109/TMTT.2003.810140

Liu, W. (1999). *Fundamentals of III-V devices: HBTs, MESFETs, and HFETs/HEMTs*. New York, NY: Wiley-Interscience.

Maas, S. A. (2003). *Nonlinear microwave and RF circuits* (2nd ed.). Norwood, MA: Artech House Publishers.

Maas, S. A. (2005). *Noise in linear and nonlinear circuits* (1st ed.). Norwood, MA: Artech House Publishers.

Niu, G. (2005). Noise in SiGe HBT RF technology: Physics, modeling, and circuit implications. *Proceedings of the IEEE*, *93*(9), 1583–1597. doi:10.1109/JPROC.2005.852226

Pospieszalski, M. W. (2010). Interpreting transistor noise. *IEEE Microwave Magazine*, *11*(6), 61–69. doi:10.1109/MMM.2010.937733

Statz, H., Haus, H. A., & Pucel, R. A. (1974). Noise characteristics of gallium arsenide field-effect transistors. *IEEE Transactions on Electron Devices*, *21*(9), 549–562. doi:10.1109/T-ED.1974.17966

van der Ziel, A. (1962). Thermal noise in field-effect transistors. *Proceedings of the IRE*, *50*(8), 1808–1812. doi:10.1109/JRPROC.1962.288221

Voinigescu, S. P., Maliepaard, M. C., Showell, J. L., Babcock, G. E., Marchesan, D., & Schroter, M. (1997). A scalable high-frequency noise model for bipolar transistors with application to optimal transistor sizing for low-noise amplifier design. *IEEE Journal of Solid-State Circuits*, *32*(9), 1430–1439. doi:10.1109/4.628757

Wood, J., & Root, D. (2005). *Fundamentals of nonlinear behavioral modeling for RF and microwave design* (1st ed.). Norwood, MA: Artech House Publishers.

KEY TERMS AND DEFINITIONS

Circuit Optimization: Determination of one or more circuit parameters, such as component values and dimensions to yield the best performance, as defined by a designer.

Mixer: Non-linear circuit that shifts an input frequency to another frequency.

Model Extraction: Calculation of the parameters needed to model (represent) the behavior of a circuit or device.

Noise Parameters: Transistor parameters useful for calculating the noise added to a circuit by a transistor.

Particle Swarm Optimization: Optimization method based on the behavior of swarm animals.

pHEMT: Pseudomorphic high electron mobility transistor; a type of transistor used in microwave circuits.

Small Signal Model: Linear transistor model at a specific bias point.

S-Parameters: Circuit parameters commonly used in microwave circuits.

Chapter 3
MO–TRIBES for the Optimal Design of Analog Filters

Mourad Fakhfakh
University of Sfax, Tunisia

Patrick Siarry
University of Paris-Est Créteil, France

ABSTRACT

The authors present the use of swarm intelligence in the analog design field. MO-TRIBES, which is an adaptive user-parameter less version of the multi-objective particle swarm optimization technique, is applied for the optimal design of a versatile building block, namely a CMOS current conveyor trans-conductance amplifier (CCTA). The optimized CCTA is used for the design of a universal filter. Good reached results are highlighted via SPICE simulations, and are compared to the theoretical ones. The use of such adaptive optimization algorithm is of great interest in the analog circuit design, as it is highlighted in this chapter.

INTRODUCTION

Up to date, automated design methodologies and tools for analog, mixed-signal and radiofrequency (AMS/RF) circuits, still lag behind the digital ones (Fakhfakh, M., Tlelo-Cuautle, M., & Fernandez F.V., 2011; Barros, M.F.M., Guilherme, J.M.C., & Horta, N.C.G., 2010). This is due to the fact that the latter is relatively easier to transform into different levels of abstraction so that algorithmic approaches can reduce the complexity of the de-

sign process. Besides, digital design is based on practices that are already established, i.e. the use of IPs (digital intellectual property) that has led to a considerable increase in the design productivity (Fakhfakh, M., Tlelo-Cuautle, M., & Fernandez F.V., 2011; Barros, M.F.M., Guilherme, J.M.C., & Horta, N.C.G., 2010). AMS/RF design has thus become the 'bottleneck' in the design flow of the integrated circuit industry. This has caused the so-called 'productivity gap', i.e. the difference between what technology can offer and what can be manufactured (Barros, M.F.M., Guilherme, J.M.C., & Horta, N.C.G., 2010). As a mean of fact, tremendous efforts are being deployed by

DOI: 10.4018/978-1-4666-2666-9.ch003

researchers and R&D engineers to develop new design methodologies in the AMS/RF domains.

Actually, these domains form a trilogy in the realm of AMS/RF circuit and system design, namely: Synthesis, Design Methodologies and Optimization. Endeavors are being made to develop new synthesis techniques, design methodologies and sizing/optimization techniques.

This Chapter deals with the third aggregate, i.e. Optimization. Indeed, due to the wide complexity of AMS/RF circuits and systems, the sizing and optimization task in the AMS/RF design field has always been considered as a knowledge intensive, iterative and multiphase chore that highly relies on the designer's experience (Barros, M.F.M., Guilherme, J.M.C., & Horta, N.C.G., 2010; Tlelo-Cuautle, E., Guerra-Gómez, I., de la Fraga, L.G., Flores-Becerra, G., Polanco-Martagón, S., Fakhfakh, M., Reyes-García, C.A., Rodríguez-Gómez, G., & Reyes-Salgado, G., 2010, Barros, M., Guilherme, J., & Horta, N., 2010). This is mainly due to the large number of constraints, companion formula, and objective functions that have to be handled.

Thus, optimization techniques have been explored, mainly the statistic-based sizing approaches and related ones, see for instance (Toumazou, C., Moschytz, G., & Gilbert, B., 2010; Medeiro, F., Rodríguez-Macías, R., Fernández, F.V., Domínguez-Astro, R., Huertas, J.L., & Rodríguez-Vázquez, A., 1994; Su, H., Michael, C., & Ismail, M., 1994; O'connor, I., & Kaiser, A., 2000; Graeb, H., Zizala, S., Eckmueller, J., & Antreich, K., 2001; Jespers, P.G., 2009).

Some heuristic-based mathematical approaches were also used, such as simulated annealing, tabu search, genetic algorithms, etc. (Barros, M.F.M., Guilherme, J.M.C., & Horta, N.C.G., 2010; Grimbleby, J.B., 2000 ; Dinger, R. H., 1998 ; Marseguerra, M., &. Zio, E, 2000; Durbin, F., Haussy, J., Berthiau, G., & Siarry, P., 1992; Courat, J.P., Raynaud, G., Mrad, I., & Siarry, P., 1994; Fernandez, F.V., & Fakhfakh, M., 2009; Guerra-Gomez, I., Tlelo-Cuautle, E., McConaghy,

T., & Gielen, G., 2009 ; Han, D., & Chatterjee, A., 2004; Li, Y., 2009; Conca, P., Nicosia, G., Stracquadanio, G., & Timmis, J., 2009). As it is already well known, these techniques do not offer general solution strategies that can be applied to problem formulations where different types of variables, objectives and constraint functions are used (Fernandez, F.V., & Fakhfakh, M., 2009). In addition, their efficiency is also highly dependent on the algorithm parameters, the dimension and the convexity of the solution space, etc.

Few years ago, a new optimization technique was proposed, it is called Swarm intelligence (SI) (Chan, F.T.S., Tiwari, M.K., 2007; http1).

In short, SI is an artificial imitation of the decentralized and self-organized collective intelligence of some homogeneous agents in the environment, such as schools of fish, growth of bacteria, herding of animals, flocks of birds, and colonies of ants. The main basic idea consists of artificially reproducing the social behavior of such animals. These animals are considered as a kind of particles interacting with each other and with their environment. Even though these particles' behaviors obey to simple rules, with no centralized control structure, they present a very intelligent overall behavior that is unknown to the individual particle (Chan, F.T.S., Tiwari, M.K., 2007; http1; Bonabeau, E., Theraulaz, G., & Dorigo, M., 1999).

Among the SI techniques, and due to its simplicity, i.e. little number of user-parameters, facility of integration, and inexpensive requirement of memory, Particle Swarm Optimization (PSO) is relatively the most popular SI technique (Chan, F.T.S., Tiwari, M.K., 2007; http1; Bonabeau, E., Theraulaz, G., & Dorigo, M., 1999; Kennedy, J., Eberhart, R.C., 1995; Clerc, M., 2006; http2; http3; Clerc, M., &. Kennedy, J., 2002; Fakhfakh, M., Cooren, Y., Sallem, A., Loulou, M., & Siarry, P., 2009; Reyes-Sierra, M., & Coello-Coello, C.A., 2006).

In general, metaheuristics (including PSO) are algorithms of which performances can be highly dependent on the chosen user-parameters values.

Therefore, these parameters have to be carefully tuned (Clerc, M., 2006; Clerc, M., &. Kennedy, J., 2002; Fakhfakh, M., Cooren, Y., Sallem, A., Loulou, M., & Siarry, P., 2009). Actually, such tuning is time consuming and generally relies on trial and error basis. Some adaptive algorithms have already been proposed in the literature, see for instance (Shi, Y., Eberhart, R.C., 2001; Murata, Y., Shibata, N., Yasumoto, K. & Ito, M., 2002; Sawai, H., Adachi, S., 1999; Schnecke, V., Vornberger, O. 1996; Ingber, L., 1996; Adra, S.F., Griffin, I.A., Fleming, P.J., 2006; Devireddy, V., & Reed, P., 2004; Knowles, J., Corne, D., 2003; Parmee, I.C., 2001; Suganthan, P.N., 1999; Sawai, H., Adachi, S., 1999). However, these algorithms are not entirely adaptive. Regarding PSO, two main approaches were proposed to alleviate this problem. The first was developed in (Parmee, I.C., 2001) where a dynamic modification of the particle's neighborhood was proposed. However, this approach lacks from rules related to the adaptation of the swarm size. The second approach was introduced by Clerc, the proposed adaptive algorithm is called TRIBES (Clerc, M., 2006). TRIBES is an improvement of the adaptive algorithm given in (Sawai, H., Adachi, S., 1999) and also is an adaptation of the GA-based one to the PSO technique. Details of TRIBES and its multi-objective version MO-TRIBES (Cooren, Y., Clerc, M., & Siarry, P., 2011) will be given in the following.

This Chapter deals with the optimal sizing of CMOS analog circuits using MO-TRIBES. The case of current mode circuits is considered, namely a current conveyor transconductance amplifier, which is a basic building block in analog circuit design. The optimized cell is then used for the realization of a high performance multifunction filter.

The rest of the Chapter is structured as follows. Section II focuses on the AMS/RF optimization problem statement. Section III briefly presents the PSO technique. Section IV gives details about MO-TRIBES and its use to solve such NP-hard

problems. Section V deals with the application of MO-TRIBES to the optimal sizing of a current conveyor based current mode building block and its use for the optimal design of a multifunction filter. Conclusions are reported in section VI.

THE PROBLEM STATEMENT

In general, the AMS/RF design requirements are formulated in terms of bounds of objective functions of which corresponding analytical expressions can be symbolically (or semi-symbolically) expressed in terms of device model parameters, or directly evaluated. Indeed, two main approaches are handled in this field, namely the symbolic technique, and the simulation-based one (Fakhfakh, M., Tlelo-Cuautle, M., & Fernandez F.V., 2011).

In short, the former technique consists of generating and expressing the constraint and performance functions (TF(s)) in the symbolic (or semi-symbolic) form, i.e. components are kept as symbols, as shown by Equation 1.

$$TF(s) = \frac{\sum_{i=1}^{a} \left(s^i \sum_{k=1}^{K} \prod_k x_k \right)}{\sum_{j=0}^{b} \left(s^j \sum_{l=1}^{L} \prod_l x_l \right)} \qquad (1)$$

s is the *Laplace* variable, and x_i represents the ith symbolic variable.

The later, i.e. the simulation based approach, consists of merging a simulator to an optimization routine in such a way that the objectives as well as constraints are directly evaluated by the simulator.

In both cases, and particularly in the second one, the process is very time consuming (Fakhfakh, M., Tlelo-Cuautle, M., & Fernandez F.V., 2011). In addition, the number of the problem's parameters to be handled generally increases exponentially with the complexity of the circuit, thus leading to a real need of rapid, precise, robust, not memory space consuming, and user-parameter-free algorithms.

PSO, more precisely TRIBES, may be one of the most interesting candidates for solving such NP-hard problems. Indeed, and in addition to the aforementioned and highlighted advantages, it has been proven that PSO offers additional interesting advantages in terms of rapidity and convergence stability rates (Ling, S.H., Iu, H.H.C., Chan, K.Y., Lam, H. K., Yeung, B.C.W., & Leung, F.H., 2008; Cooren, Y., Clerc, M., & Siarry, P., 2009), when compared to other classical optimization techniques.

Actually, in AMS/RF optimization problems, the main target it to find 'optimal' values of a set of parameters that optimizes a set of performance functions, while satisfying a set of constraints.

In fact, it is the wide plethora of objectives and constraints in the AMS/RF design field that has made it so complex, and accordingly classified as an NP-hard problem (Fernandez, F.V., & Fakhfakh, M., 2009). Performances, such as signal to noise ratio (SNR), common mode rejection ratio (CMRR), DC-offsets, distortion, slew-rate, parasitic elements, phase margin, frequency operating range etc. are directly dependent on the used technology and the circuit's applications. Constraints, such as the working mode of the transistors, the technology limits, etc. also present a hard and delicate issue.

Without loss of generality, Expression 2 presents this problem which is considered as a minimization problem. It is to be highlighted that the AMS/RF optimization problem is in most cases a multi-objective one, where at least two conflicting and incommensurable objectives are handled. The weighting technique, that mathematicians call 'the naïve technique', and that consists of merging the objectives using (normalized) weightings, was/is widely adopted by the analogue designers. However, it has been proven that transforming a multi-objective problem into a mono-objective one is to be avoided (see for instance (Cooren, Y., Clerc, M., & Siarry, P., 2009)), even though

some published papers have proposed tentative solutions for such weighting approach (Ryu, J., Kim, S., & Wan, H., 2009; Audet, C., Savard, G., Zghal, W., 2008; Das, I., & Dennis, J., 1998).

$$
\begin{vmatrix}
Minimize\ \vec{f}(\vec{x});\quad \vec{f}(\vec{x}) \in R^{k} \\
subject\ to: \\
\vec{g}(\vec{x}) \leq 0; \qquad\qquad \vec{g}(\vec{x}) \in R^{m} \\
and\ \vec{h}(\vec{x}) = 0; \qquad\qquad \vec{h}(\vec{x}) \in R^{n} \\
where\ x_{Li} \leq x_{i} \leq x_{Ui}, \quad i \in \begin{bmatrix} 1, p \end{bmatrix}
\end{vmatrix}
\qquad (2)
$$

$\vec{f}$ is the set of performance functions. $\vec{g}$ and $\vec{h}$ represent inequality and equality constraint functions, respectively. x_{Li} and x_{Ui} are the lower and upper limits of the variable x_i.

k, m, n and p represent the numbers of objective functions, inequality constraints, equality constraints, and variables, respectively.

In the case of multi-objective problems, a set of feasible non-dominated solutions have to be found (For instance, see (Fakhfakh, M., Cooren, Y., Sallem, A., Loulou, M., & Siarry, P., 2009) for details regarding the dominance computing). The non-dominated set of the entire feasible search space is known as the trade-off surface, or the Pareto front. In brief, the Pareto front encompasses solutions in the feasible solution space that can not be compared to each other, i.e. performance on one objective cannot be improved without sacrificing performance on at least one other objective.

THE MULTIOBJECTIVE PARTICLE SWARM OPTIMIZATION TECHNIQUE: CONCEPTS AND FORMULATION

Since its inception by Kennedy and Eberhart in 1995 (Kennedy, J., & Eberhart, R. C., 1995) the particle swarm optimization (PSO) metaheuristic has rapidly become very popular. The literature offers a large plethora of published works dealing with the application of the PSO technique to the engineering problems.

As introduced above, PSO systems are composed of particles 'flying' around a multidimensional search space. During its flight, each particle updates its position according to its own experience, and to the one of its neighbors.

Actually, the aforementioned phenomenon presents a significant advantage of PSO, i.e. its manner of the space exploration. Actually, it combines two search methods:

- A local search method: the *pbest* value that presents the current coordinates associated with the best solution that the particle has achieved so far (Actually, it is the variant local best performance, *lbest*, of PSO),
- A global search method: the *gbest* value (the overall best value and its coordinates), which is the best value that is tracked by the global version of the particle swarm optimizer.

At each algorithm step, *pbest* and *gbest* are computed, updated and used by the particles to adjust their velocities and positions in order to improve their current fitness.

Expressions of a particle velocity at time step *t* and its corresponding position are given by Equations 3 and 4, respectively.

$$\vec{v}_i(t) = \underbrace{\omega\vec{v}_i(t-1)}_{Inertia} + \underbrace{c_1 Rand()(\vec{x}_{pbesti} - \vec{x}_i(t))}_{Personal\ Influence} + \underbrace{c_2 Rand()(\vec{x}_{leaderi} - \vec{x}_i(t))}_{Social\ Influence} \quad (3)$$

$$\vec{x}_i(t) = \vec{x}_i(t-1) + \vec{v}_i(t) \quad (4)$$

where x_{pbesti} is the best personal position of a given particle, $x_{leaderi}$ refers to the best position reached by the particle's neighbourhood, ω is an inertia weight that controls the diversification feature of the algorithm, c_1 and c_2 are related to the intensification feature of the algorithm; they are the acceleration constants that represent the weighting of the stochastic acceleration term that pull each particle towards *pbest* and *gbest*.

Initially, PSO has been proposed for solving mono-objective optimization problems and has been successfully used for both continuous non linear and discrete mono-objective optimization. Then, it was adapted to be able to deal with multi-objective optimization problems. The basic idea consists of using an external archive in which each particle deposits its 'flight' experience at each algorithm step. The archive stores all the non-dominated solutions found during the optimization process. At the end of the algorithm run, the archive contains the non-dominated solutions, i.e. the Pareto front. Figure 1 shows the flowchart of the multi-objective PSO.

It is to be noted that in the original version of PSO, each v_i was bounded in the range $[-V_{MAX}, +V_{MAX}]$. In (Clerc, M., &. Kennedy, J., 2002) was proposed a constriction factor that allows defining V_{MAX} and ensuring a good balance between the intensification and the diversification aspects of the PSO algorithm.

In order to avoid excessive growing of this memory, its size is fixed. The crowding distance technique is used for this purpose. The crowding distance of an element of the archive estimates the size of the largest cuboid enclosing that element without including any other element in the archive. The idea is to maximize the crowding distance of the particles in order to obtain a Pareto front as uniformly spread as possible.

Figure 1. Flowchart of the multi-objective PSO (Difference between the mono-objective flowchart and the multi-objective one is marked in green colour)

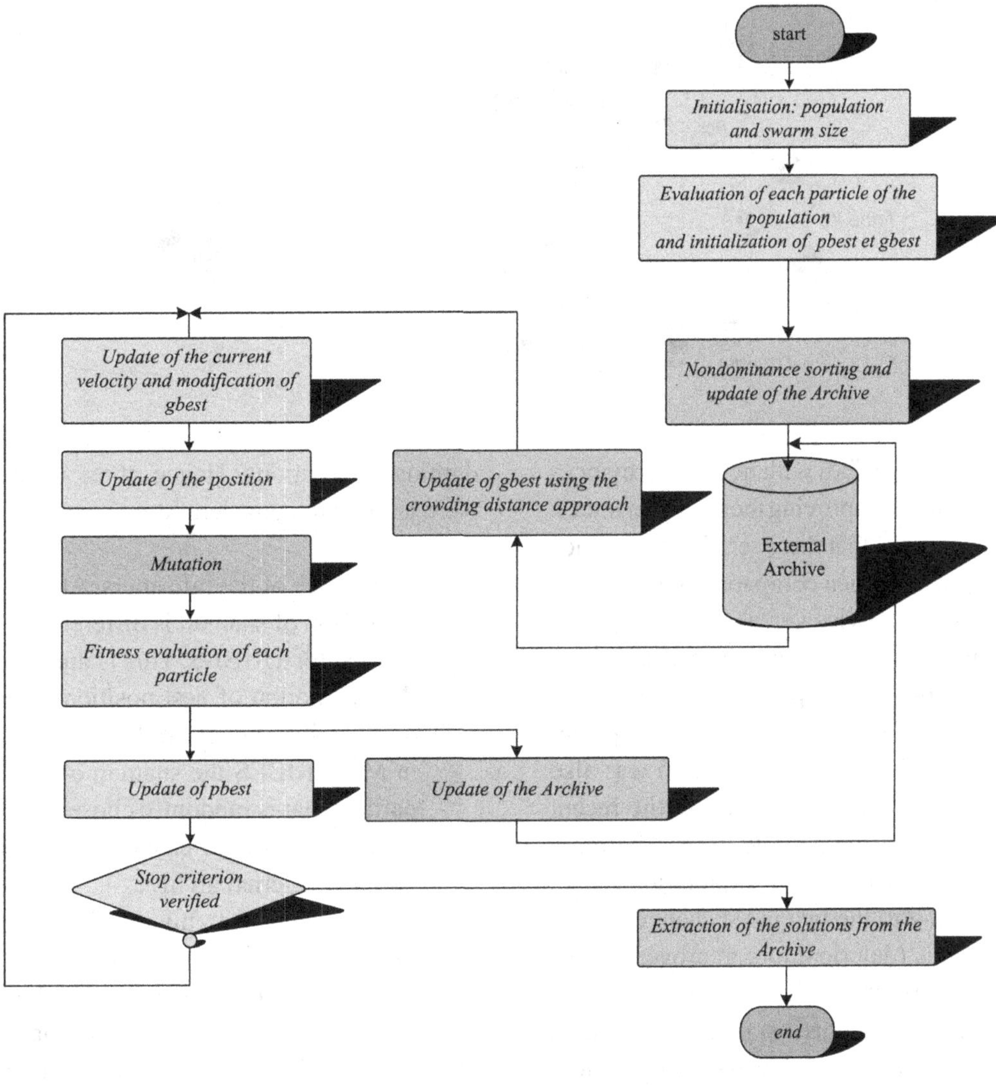

TRIBES AND MO-TRIBES

As it was introduced prior, TRIBES is a PSO user-parameter free algorithm. In TRIBES every particle's movement is conditioned by two or three parameters (See Figure 2) and accordingly is informed by:

- It self, i.e. its own experience,
- Its informers belonging to its tribe,

- The 'shamans' of the other tribes, in case the particle is the 'shaman' of its own tribe.

Figure 2 illustrates the communications intra- and inter-tribes.

The evaluation of the fitness of each particle is the most time consuming task in such algorithms, each iteration of TRIBES removes bad particle(s) from the 'good' swarm. On the other hand, in case a tribe has a 'bad' behavior, TRIBES adds new particle(s) thus forming a new tribe that will

Figure 2. Communications inter-, intra-swarm and with the external archive for a 5-tribe-swarm (: the corresponding tribe's shaman).

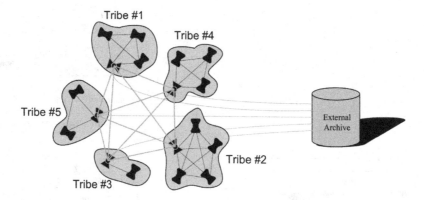

furnish these 'bad' tribes with new performances that may help them improving their performances. (A bad tribe means that the overall performance of that tribe is bad when compared to the ones of the other tribes, and vice versa).

For this purpose, quality quantifiers were defined to evaluate each particle and each tribe performances. Further, a strategy of displacement of the particles inside the swarm was also defined. This displacement relies on the recent past experience of the particle itself. The main objective is to ensure a wider exploration scope. Thus, according to the two latest changes of the particle's status (deterioration, improvement, or even not change of the particle's performance(s)), a movement strategy is established. (Details regarding the quantifiers and the displacement strategies are available in (Cooren, Y., Clerc, M., & Siarry, P., 2008)).

In order to be able to handle more than one objective, a multi-objective version of TRIBES, called MO-TRIBES, was developed by Cooren *et al.* (Cooren, Y., Clerc, M., & Siarry, P., 2011). The main idea consists of generating and finding non-dominated solutions, minimizing the distance between the approximated Pareto front and the true one, and maximizing the spread of the non-

dominated solutions. Hence, three main issues were taken into consideration:

- The choice of the informers. Actually, here the notion of shaman is different to the one defined for TRIBES. This is due to the fact that the notion of best position ever found by the tribe cannot be defined. Therefore, in MO-TRIBES the shaman of a tribe is a particle that is randomly chosen inside that tribe, and that will ensure communication with the external archive.

- The well spread archiving of the non-dominated solutions. Indeed, the main problem here is how to choose the non-dominated solution to keep when the external archive is full. Thus, the size of this archive was adaptively set. First, it is initialized as a function of the number of the considered objectives, then a learning mechanism is integrated and the size of the archive is adapted according to the number of solutions it encompassed at the previous algorithm's run. The crowding distance routine is adopted to maintain diversity inside the archive.

- The insurance of the diversity inside the swarm in order to avoid convergence to-

Figure 3. The MO-CCCCTA proposed in (Singh, S.V., Maheshwari, S., & Chauhan, D.S., 2011)

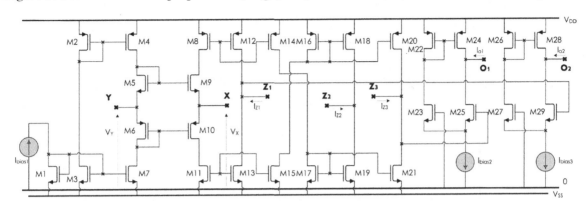

wards a unique solution: actually, MO-TRIBES takes benefit from the drawback of the classical mono-objective PSO that suffers from rapid convergence towards the solution that can be a local one. Thus, in case no new solution is added to the archive after a certain number of iterations, the whole swarm is randomly initialized. Hence, new areas are explored.

APPLICATION OF MO-TRIBES TO THE OPTIMAL DESIGN OF ANALOG FILTERS

In the following, the optimization of a current conveyor based basic building block (CCBBB) is considered. This CCBBB is a versatile current mode building-block that can be used to design various circuits, such as analog active frequency filters and active quadrature oscillators. Among the available CCBBBs, the newly introduced single-input/multi-output (SIMO) CMOS current conveyor transconductance amplifier (CCTA) (Singh, S.V., Maheshwari, S., Mohan, J., & Chauhan, D.S., 2009) is considered. It is obviously composed of a current conveyor (that is formed by transistors (M4-M13,Ibias1) and two transconductance cells ((M22-M25,Ibias2) and (M24-M27,Ibias3)), as it

is shown in Figure 3. the rest of the transistors are mainly current mirror ones.

Basically, a CCTA is a 5-terminal active element. Its properties are given by Equation 5. Current mirrors/followers can be added as needed. In Figure 3, two current mirrors were added to reproduce $-I_{Z1}$ at ports Z_2 and Z_3. These currents will be needed to design the multifunction filter.

$$\begin{pmatrix} I_Y \\ V_X \\ I_Z \\ I_{O+} \\ I_{O-} \end{pmatrix} = \begin{pmatrix} 0 & 0 & 0 & 0 \\ R_X & 0 & 0 & 0 \\ 1 & 0 & 0 & 0 \\ 0 & 0 & g_m & 0 \\ 0 & 0 & -g_m & 0 \end{pmatrix} \begin{pmatrix} I_X \\ V_Y \\ V_Z \\ V_O \end{pmatrix} \qquad (5)$$

gm and R_X represent the transconductance and the X-port parasitic resistance, respectively.

In the following, it is considered the optimization of both main objectives of such current conveyor based cell, i.e. minimizing the X-port parasitic resistance (R_X) and maximizing the X-Z current high cutoff frequency (*fci*). Symbolic expression of R_X is given by Equation 6, whereas expression of *fci* is not provided due to its large number of terms (Both expressions were generated using a symbolic analyzer).

Figure 4. The Pareto front (f_{ci} vs. R_x)

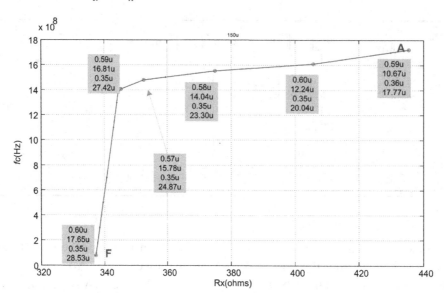

$$R_X = \frac{\left(g_{0P} + g_{0N}\right)}{\left(g_{mN} + g_{mP}\right)\left(g_{0P} + g_{0N} + g_{mP}\right)} \quad (6)$$

with

$$g_{mN,P} = \sqrt{\mu_{N,P} C_{ox} I_{bias1} \frac{W_{N,P}}{L_{N,P}}}$$

and

$$g_{0N,P} = \frac{L_{N,P}}{\alpha_N I_{bias1}}$$

Indexes N and P refer to the NMOS and PMOS transistors, respectively. g_m and g_0 are the grid transconductance and the output conductance of a MOS transistor. μ is the charge mobility parameter, C_{ox} is the oxide capacitance and α is the channel-length modulation parameter.

In addition to the constraints imposed by the technology, the circuit is subject to a set of intrinsic constraints that are due to the saturation working mode of the MOS transistors, mismatch, symmetry, etc. Main such constraints are given by Equation 7.

$$\mu_N \frac{W_N}{L_N} = \mu_P \frac{W_P}{L_P}$$

$$W_{N,P} > L_{N,P}$$

$$W_N L_N = W_P L_P$$

$$\frac{1}{2}V_{DD} - V_{TN} - \sqrt{\frac{I_{bias1}}{\mu_N C_{ox} \frac{W_N}{L_N}}} > \sqrt{\frac{I_{bias1}}{\mu_P C_{ox} \frac{W_P}{L_P}}}$$

$$\frac{1}{2}V_{DD} - V_{TP} - \sqrt{\frac{I_{bias1}}{\mu_P C_{ox} \frac{W_P}{L_P}}} > \sqrt{\frac{I_{bias1}}{\mu_N C_{ox} \frac{W_N}{L_N}}} \quad (7)$$

V_{DD} is the voltage power supply.

MO-TRIBES was used to generate the Pareto tradeoff surface linking these two conflicting performances, i.e. R_X and *fci*. Figure 4 shows this front. NMOS transistors encompassing the circuits were considered to have the same geometric dimensions, i.e. L_N and W_N. Ditto for the PMOS

Figure 5. SPICE simulations of R_X

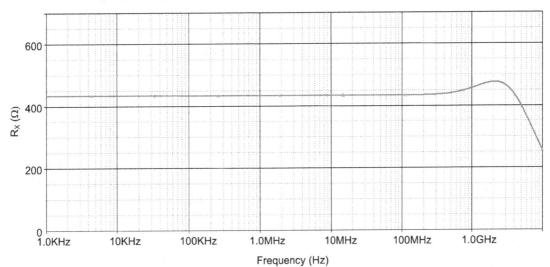

Figure 6. SPICE simulations of the X-Z current transfer

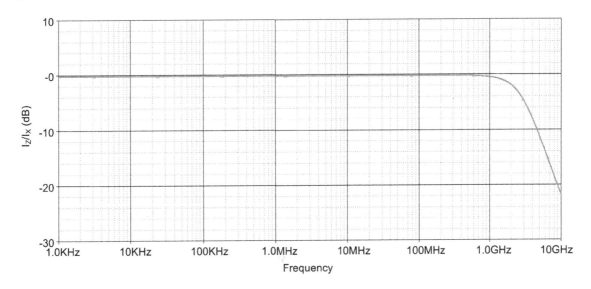

transistors. Main solutions of the compromise surface shown in Figure 4 are marked with blue spots. Corresponding parameters' values are indicated below the spot in the order: L_N, W_N, L_P, W_P.

The used technology is AMS 0.35μm. Voltage (V_{DD}) and current biases (I_{bias1}, I_{bias2}, I_{bias3}) are 2.5V, 150μA, 100μA and 100μA, respectively. Obtained performances were checked via SPICE simulations. Figures 5 and 6 show for instance the cur-

rent X-Z transfer and the parasitic resistance curve ($R_X=V_X/I_X$) corresponding to point A in Figure 4. Figure 5 shows that R_X =439Ω. According to Figure 6, *fci* equals 2.004GHz. The MATLAB value of R_X equals 435Ω. It shows a very good agreement between the simulation results and expected ones. However, a deviation between SPICE simulations of *fci* and MATLAB results that equals 1.721GHz. This deviation is due to

Figure 7. The CCTA based universal filter

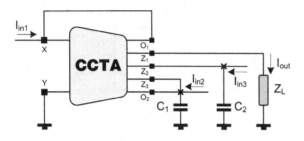

the effect of high frequency parasitic capacitances that are not correctly taken into consideration because of the considered simplified models of the MOS transistors.

In analog signal processing, analog filters are a basic building block. Most used such filters are universal biquadratic filters, since they can provide

all standard functions without altering the circuit's topology. In (Singh, S.V., Maheshwari, S., & Chauhan, D.S., 2011) was proposed the use of the CCTA as a universal filter. This filter is shown in Figure 7. It uses a unique CCTA and only two capacitors, and thus it is suitable for integration.

The output current I_{out} of the filter shown in Figure 7 is expressed in Equation 8 (see Box 1) where s is the *Laplace* parameter.

The CCTA introduced above was used to design the universal filter shown in Figure 7. Optimal parameters' values corresponding to point A shown in the Pareto front of Figure 4 were used. This is argued by the fact that in such design, the parasitic components, mainly R_X, are the most influent non-idealities that affect the overall performances of the circuit.

Box 1.

$$I_{out} = \frac{\left(C_1 C_2 s^2\right) I_{in1} - \left(C_2 \sqrt{\mu_N C_{ox} \frac{W_N}{L_N} I_{bias2}}\right) I_{in2} + \left(C_{ox} \mu_N \frac{W_N}{L_N} \sqrt{I_{bias2} I_{bias3}}\right) I_{in3}}{C_1 C_2 s^2 + C_2 \sqrt{\mu_N C_{ox} \frac{W_N}{L_N} I_{bias2}} s + \mu_N C_{ox} \frac{W_N}{L_N} \sqrt{I_{bias2} I_{bias3}}} \tag{8}$$

Figure 8. I_{out}/I_{in} for the high-pass filter (SPICE simulation vs. MATLAB results); ($I_{in1}=I_{in}$, $I_{in2}=I_{in3}=0$)

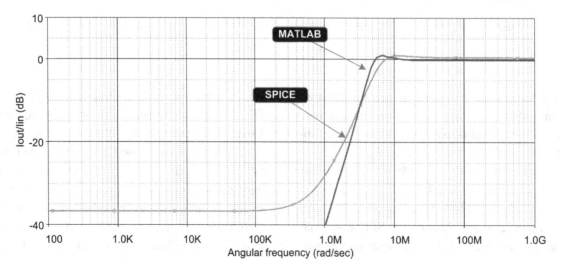

Figure 9. I_{out}/I_{in} for the low-pass filter (SPICE vs. MATLAB simulations); ($I_{in3}=I_{in}$, $I_{in1}=I_{in2}=0$)

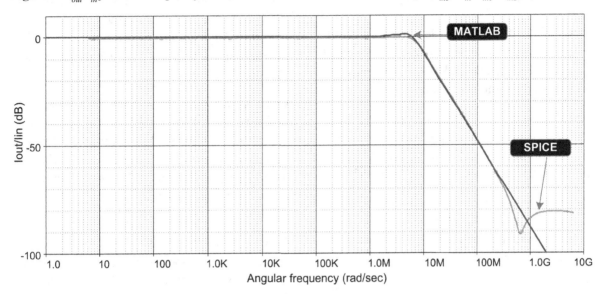

Figures 8 and 9 show a comparison between SPICE and MATLAB simulation results of the current transfer functions I_{out}/I_{in} for the cases of a high-pass filter and a low-pass filter, respectively. $C_1=C_2=100pF$ and $Z_L=50\Omega$.

SUMMARY AND CONCLUSION

It was highlighted the use of an improved version of the muti-objective particle swarm technique, i.e. MO-TRIBES, for the design of analog circuits. A newly introduced current mode basic building cell was first considered: a CMOS current conveyor transconductance amplifier. The Pareto front linking two conflicting objectives was generated thanks to MO-TRIBES. SPICE simulations were given to show the reached good performances of the optimized cell. Then, one solution among the generated non-dominated solutions, that corresponds to the minimum value of a dominant parasitic element, was considered. This 'optimized' building block was used to design a multifunction filter. Performances of the built filters were given and it was shown that they are

with good agreement with the theoretical ones obtained using MATLAB software.

TRIBES and MO-TRIBES are very interesting algorithms that can be used in a simulation-based or an equation-based optimization problem in the analog, mixed-signal and radiofrequency domains. This is due to the simplicity of the implementation of the PSO technique, and to the adaptive tuning of the algorithm parameters. As it was highlighted in the Chapter, it is to be recalled that the analog design realm presents a very large spectrum of complex circuits and systems with a large number of parameters, constraints and performances. Further, and due to the fact that reached performances may highly depend on the algorithm parameters that are generally obtained via a try/errors approach, such techniques are very helpful for the designer. Future works will focus on integrating (MO-)TRIBES into an automated AMS/RF design platform.

REFERENCES

Adra, S. F., Griffin, I. A., & Fleming, P. J. (2006). *An adaptive memetic algorithm for enhanced diversity*. The International Adaptive Computing in Design and Manufacture Conference.

Audet, C., Savard, G., & Zghal, W. (2008). Multi-objective optimization through a series of single-objective formulations. *SIAM Journal on Optimization, 19*(1), 188–210. doi:10.1137/060677513

Barros, M., Guilherme, J., & Horta, N. (2010). Analog circuits optimization based on evolutionary computation techniques. *Integration- The VLSI Journal, 43*, 136–155.

Barros, M. F. M., Guilherme, J. M. C., & Horta, N. C. G. (2010). *Analog circuits and systems optimization based on evolutionary techniques*. Springer-Verlag. doi:10.1007/978-3-642-12346-7

Bonabeau, E., Theraulaz, G., & Dorigo, M. (1999). *Swarm intelligence: From natural to artificial systems*. Oxford University Press.

Chan, F. T. S., & Tiwari, M. K. (2007). *Swarm Intelligence: Focus on ant and particle swarm optimization*. I-Tech Education and Publishing.

Clerc, M. (Ed.). (2006). *Particle swarm optimization*. International Scientific and Technical Encyclopaedia. doi:10.1002/9780470612163

Clerc, M., & Kennedy, J. (2002). The particle swarm: Explosion, stability, and convergence in a multi-dimensional complex space. *IEEE Transactions on Evolutionary Computation, 6*, 58–73. doi:10.1109/4235.985692

Conca, P., Nicosia, G., Stracquadanio, G., & Timmis, J. (2009). *Nominal-yield-area tradeoff in automatic synthesis of analog circuits: A genetic programming approach using immune-inspired operators*. NASA/ESA Conference on Adaptive Hardware and Systems.

Conn, A. R., Coulman, P. K., Haring, R. A., Morrill, G. L., & Visweswariah, C. (1996). *Optimization of custom MOS circuits by transistor sizing*. The International Conference on Computer Aided Design.

Cooren, Y., Clerc, M., & Siarry, P. (2008). Initialization and displacement of the particles in TRIBES, a parameter-free particle swarm optimization algorithm. *Adaptive and Multilevel Metaheuristics Studies in Computational Intelligence, 136*, 199–219. doi:10.1007/978-3-540-79438-7_10

Cooren, Y., Clerc, M., & Siarry, P. (2009). Performance evaluation of TRIBES, an adaptive particle swarm optimization algorithm. *Swarm Intelligence, 3*(2), 149–178. doi:10.1007/s11721-009-0026-8

Cooren, Y., Clerc, M., & Siarry, P. (2011). MO-TRIBES, an adaptive multiobjective particle swarm optimization algorithm. *Computational Optimization and Applications, 49*(2), 379–400. doi:10.1007/s10589-009-9284-z

Courat, J. P., Raynaud, G., Mrad, I., & Siarry, P. (1994). Electronic component model minimization based on Log simulated annealing. *IEEE Transactions on Circuits and Systems, 41*, 790–795. doi:10.1109/81.340841

Das, I., & Dennis, J. (1998). Normal-boundary intersection: A new method for generating the Pareto surface in nonlinear multicriteria optimization problems. *SIAM Journal on Optimization, 8*, 631–657. doi:10.1137/S1052623496307510

Devireddy, V., & Reed, P. (2004). *Efficient and reliable evolutionary multiobjective optimization using epsilondominance archiving and adaptive population sizing*. The Genetic and Evolutionary Computation Conference.

Dinger, R. H. (1998). *Engineering design optimization with genetic algorithm*. The IEEE Northcon Conference.

Durbin, F., Haussy, J., Berthiau, G., & Siarry, P. (1992). Circuit performance optimization and model fitting based on simulated annealing. *International Journal of Electronics, 73*, 1267–1271. doi:10.1080/00207219208925797

Fakhfakh, M., Cooren, Y., Sallem, A., Loulou, M., & Siarry, P. (2009). Analog circuit design optimization through the particle swarm optimization technique. *Analog Integrated Circuits and Signal Processing, 63*(1), 71–82. doi:10.1007/s10470-009-9361-3

Fakhfakh, M., Loulou, M., & Masmoudi, N. (2009). A novel heuristic for multi-objective optimization of analog circuit performances. *Journal of Analog Integrated Circuits and Signal Processing, 61*(1), 47–64. doi:10.1007/s10470-008-9275-5

Fakhfakh, M., Tlelo-Cuautle, M., & Fernandez, F. V. (2011). *Design of analog circuits through symbolic analysis*. Bentham Scientific Publisher.

Fernandez, F. V., & Fakhfakh, M. (2009). *Applications of evolutionary computation techniques to analog, mixed-signal and RF circuit design*. The IEEE International Conference on Electronics, Circuits, and Systems.

Graeb, H., Zizala, S., Eckmueller, J., & Antreich, K. (2001). *The sizing rules method for analog integrated circuit design*. The IEEE/ACM International Conference on Computer-Aided Design.

Grimbleby, J. B. (2000). Automatic analogue circuit synthesis using genetic algorithms. *IEE Proceedings. Circuits, Devices and Systems, 147*(6), 319–323. doi:10.1049/ip-cds:20000770

Guerra-Gomez, I., Tlelo-Cuautle, E., McConaghy, T., & Gielen, G. (2009). Optimizing current conveyors by evolutionary algorithms including differential evolution. *The IEEE International Conference on Electronics, Circuits, and Systems.*

Han, D., & Chatterjee, A. (2004). Simulation-in-the-loop analog circuit sizing method using adaptive model-based simulated annealing. *The IEEE International Workshop on System-on-Chip for Real-Time Applications.*

Ingber, L. (1996). Adaptive simulated annealing (ASA): Lessons learned. *Control and Cybernetics, 25*(1), 33–54.

Intelligence, S. (n.d.). *Codes*. Retrieved from http://www.swarmintelligence.org/codes.php

Jespers, P. G. (Ed.). (2009). *The gm/ID Methodology, a sizing tool for low-voltage analog CMOS Circuits: The semi-empirical and compact model approaches*. Springer.

Kennedy, J., & Eberhart, R. C. (1995). *Particle swarm optimization*. The IEEE International Conference on Neural Networks.

Knowles, J., & Corne, D. (2003). Properties of an adaptive archiving algorithm for storing nondominated vectors. *IEEE Transactions on Evolutionary Computation, 7*(2), 100–116. doi:10.1109/TEVC.2003.810755

Li, Y. (2009). A simulation-based evolutionary approach to LNA circuit design optimization. *Applied Mathematics and Computation, 209*(1), 57–67. doi:10.1016/j.amc.2008.06.015

Ling, S. H., Iu, H. H. C., Chan, K. Y., Lam, H. K., Yeung, B. C. W., & Leung, F. H. (2008). Hybrid particle swarm optimization with wavelet mutation and its industrial applications. *IEEE Transactions on Systems, Man, and Cybernetics. Part B, Cybernetics, 38*(3), 743–763. doi:10.1109/TSMCB.2008.921005

Marseguerra, M., & Zio, E. (2000). *System design optimization by genetic algorithms*. The IEEE Annual Reliability and Maintainability Symposium.

Medeiro, F., Rodríguez-Macías, R., Fernández, F. V., Domínguez-Astro, R., Huertas, J. L., & Rodríguez-Vázquez, A. (1994). Global design of analog cells using statistical optimization techniques. *Analog Integrated Circuits and Signal Processing, 6*(3), 179–195. doi:10.1007/BF01238887

Murata, Y., Shibata, N., Yasumoto, K., & Ito, M. (2002). *Agent oriented self adaptive genetic algorithm* (pp. 348–353). The IASTED Communications and Computer Networks.

O'Connor, I., & Kaiser, A. (2000). Automated synthesis of current memory cells. *IEEE Transactions on Computer-Aided Design of Integrated Circuits and Systems, 19*(4), 413–424. doi:10.1109/43.838991

Parmee, I. C. (Ed.). (2001). *Evolutionary and adaptive computing in engineering design.* Springer. doi:10.1007/978-1-4471-0273-1

Particle Swarm. (n.d.). Retrieved from http://www.particleswarm.info/

Reyes-Sierra, M., & Coello-Coello, C. A. (2006). Multi-objective particle swarm optimizers: A survey of the state-of-the-art. *International Journal of Computational Intelligence Research, 2*(3), 287–308.

Ryu, J., Kim, S., & Wan, H. (2009). *Pareto front approximation with adaptive weighted sum method in multiobjective simulation optimization.* The Winter Simulation Conference.

Sawai, H., & Adachi, S. (1999). *Genetic algorithm inspired by gene duplication.* The Congress on Evolutionary Computing.

Schnecke, V., & Vornberger, O. (1996). *An adaptive parallel genetic algorithm for VLSI-layout optimization.* The International Conference on Parallel Problem Solving from Nature.

Shi, Y., & Eberhart, R. C. (2001). *Fuzzy adaptive particle swarm optimization.* Congress on Evolutionary Computation.

Singh, S. V., Maheshwari, S., & Chauhan, D. S. (2011). Single MO-CCCCTA-based electronically tunable current/trans-impedance-mode biquad universal filter. *Circuits and Systems, 2,* 1–6. doi:10.4236/cs.2011.21001

Singh, S. V., Maheshwari, S., Mohan, J., & Chauhan, D. S. (2009). An electronically tunable SIMO biquad filter using CCCCTA. *Contemporary Computing Communications in Computer and Information Science, 40*(11), 544–555. doi:10.1007/978-3-642-03547-0_52

Su, H., Michael, C., & Ismail, M. (1994). *Statistical constrained optimization of analog MOS circuits using empirical performance models.* The IEEE International Symposium on Circuits and Systems.

Suganthan, P. N. (1999). *Particle swarm optimisation with a neighbourhood operator.* Congress on Evolutionary Computation. doi:10.1109/CEC.1999.785514

Tlelo-Cuautle, E., Guerra-Gómez, I., de la Fraga, L. G., Flores-Becerra, G., Polanco-Martagón, S., & Fakhfakh, M. … Reyes-Salgado, G. (2010). Evolutionary algorithms in the optimal sizing of analog circuits. In M. Köppen, G. Schaefer, & A. Abraham (Eds.), *Intelligent computational optimization in engineering: Techniques & applications.* Springer.

Toumazou, C., Moschytz, G., & Gilbert, B. (2010). *Trade-offs in analog circuit design: The designer's companion.* Kluwer Academic Publishers.

Wikipedia. (n.d.). *Swarm intelligence.* Retrieved from http://en.wikipedia.org/wiki/Swarm_intelligence

ADDITIONAL READING

Beirami, A., Takhti, M., & Shamsi, H. (2009). *Extracting trade-off boundaries of CMOS two-stage op-amp using particle swarm optimization.* International Symposium on Signals, Circuits and Systems.

Clerc, M. (Ed.). (2010). Beyond standard particle swarm optimisation. *International Journal of Swarm Intelligence Research, 1,* 46–61. doi:10.4018/jsir.2010100103

Collette, Y., & Siarry, P. (Eds.). (2003). *Multiobjective optimization: Principles and case studies.* Springer.

Cooren, Y., Nakib, A., & Siarry, P. (2007). *Image thresholding using TRIBES, a parameter-free particle swarm optimization algorithm.* LION. doi:10.1007/978-3-540-92695-5_7

Dorigo, M., Birattari, M., Blum, C., Clerc, M., Stützle, T., & Winfield, A. F. T. (2008). *Ant colony optimization and swarm intelligence.* The International Conference, ANTS.

Kamisetty, S., Garg, J., Tripathi, J. N., & Mukherjee, J. (2011). *Optimization of analog RF circuit parameters using randomness in particle swarm optimization.* World Congress on Information and Communication Technologies.

Kawamura, K., & Saito, T. (2010). *Design of switching circuits based on particle swarm optimizer and hybrid fitness function.* The Annual Conference on IEEE Industrial Electronics Society.

Kotti, M., Benhala, B., Fakhfakh, M., Ahaitouf, A., Benlahbib, B., Loulou, M., & Mecheqrane, A. (2011). *Comparison between PSO and ACO techniques for analog circuit performance optimization.* The International Conference on Microelectronics.

Lazinica, A. (Ed.). (2009). *Particle swarm optimization.* Intech. doi:10.5772/109

Olsson, A. E. (Ed.). (2011). *Particle swarm optimization: Theory, techniques and applications.* Nova Publishers.

Parsopoulos, K. E., & Vrahatis, M. N. (2010). *Particle swarm optimization and intelligence: Advances and applications.* Hershey, PA: IGI Global. doi:10.4018/978-1-61520-666-7

Sabat, S. L., Kumar, K. S., & Udgata, S. K. (2009). *Differential evolution and swarm intelligence techniques for analog circuit synthesis.* World Congress on Nature & Biologically Inspired Computing.

Thakker, R. A., Baghini, M. S., & Patil, M. B. (2009). *Low-power low-voltage analog circuit design using hierarchical particle swarm optimization.* The International Conference on VLSI Design.

Tsai, H.-H., Chang, B.-M., & Lin, X.-P. (2012). Using decision tree, particle swarm optimization, and support vector regression to design a median-type filter with a 2-level impulse detector for image enhancement. *Information Sciences, 195,* 103–123. doi:10.1016/j.ins.2012.01.020

Vural, R. A., & Yildirim, T. (2010). *Component value selection for analog active filter using particle swarm optimization.* The International Conference on Computer and Automation Engineering.

Vural, R. A., Yildirim, T., Kadioglu, T., & Basargan, A. (2012). Performance evaluation of evolutionary algorithms for Optimal filter design. *IEEE Transactions on Evolutionary Computation, 16,* 135–147. doi:10.1109/TEVC.2011.2112664

Zhou, L., & Shi, Y. (2009). *Soft fault diagnosis of analog circuit based on particle swarm optimization, testing and diagnosis.* The IEEE international Conference on Circuits and Systems.

KEY TERMS AND DEFINITIONS

Current Conveyor Transconductance Amplifier: Is a combination of the third–generation current conveyor and balanced-output operational transconductance amplifier.

MO-TRIBES: An adaptive multiobjective Particle Swarm Optimization algorithm.

Multi-Function Filters: Are devices which have the capability to realize simultaneously more than one basic filter function with the same topology.

Particle Swarm Optimization: Is a global optimization, population-based evolutionary algorithm for dealing with problems in which a best solution can be represented as a point of surface in n-dimensional space.

Swarm Intelligence: An artificial-intelligence approach to problem solving using algorithms based on the self-organized collective behaviour of a group of animals, especially social insects such as ants, bees, and termites, that are each following very basic rules.

Chapter 4

Design Automation, Modeling, Optimization, and Testing of Analog/RF Circuits and Systems by Particle Swarm Optimization

Jai Narayan Tripathi
IIT Bombay, India

Jayanta Mukherjee
IIT Bombay, India

Prakash R. Apte
IIT Bombay, India

ABSTRACT

This chapter is an overview of the applications of particle swarm optimization for circuits and systems. The chapter is targeted for the Analog/RF circuits and systems designers. Design automation, modeling, optimization and testing of analog/RF circuits using particle swarm optimization is presented. Various applications of particle swarm optimization for circuits and systems are explained by examples.

1. INTRODUCTION

Analog/RF Circuits are complicated to design as a lot of design constraints are there such as power, area, linearity, noise etc. Due to these complexities, Designing of Analog/RF circuits require experienced designers and design intuitions. To make the design process simple, Design Automation is used. Electronic Design

Automation (EDA) tools are used extensively from many years, for the automated design of such circuits. EDA tools simplifies this design process by facilitating the automatic designs for given design specifications. A goal is set for the simulator and an optimization algorithm is associated with that. The algorithm (defined within the EDA tool) searches for the best possible solution in the given design space with the given design constraints. These algorithms are either deterministic or stochastic. Since

DOI: 10.4018/978-1-4666-2666-9.ch004

deterministic algorithms are not efficient for the complex designs, stochastic algorithms are used majorly.

Particle Swarm Optimization (PSO) is a popular meta-heuristic algorithm which lies in the group of nature inspired algorithms. PSO is based on the social behavior of the birds flocking and fishes schooling. Particle Swarm Optimization (PSO) has also been widely used for design automation, optimization and testing of Analog/RF circuits and systems. This chapter discusses the various applications of PSO for Analog/RF circuits and systems, with supporting examples. The objective of this chapter is to provide the readers an insight of a wide range of applications of PSO for Analog/ RF circuits and systems. Various applications of PSO for circuits and systems, are discussed and supported by examples such as optimization, modeling, design automation and testing etc. As mentioned above, the Analog/RF circuits are complicated with many design constraints, multi-objective optimization is used in most of the examples.

The organization of the chapter is in the following manner. Section 2 describes about RF/ Analog circuit design and automation. There are four subsections which show examples of circuit design and optimization such as in-loop optimization by EDA tools, modeled NP problem solution by PSO etc. Section 3 is regarding the modeling of circuits and elements, with example of an on-chip inductor modeling. In section 4, a system level power integrity problem is discussed for high speed systems. This is an optimum capacitor selection and placement problem. Section 5 is about the application of PSO in testing of circuits by N-terminal based testing method. Section 6 shows the designing of passive circuits by PSO. Section 7 describes variants of PSO and their use for circuits. The last section of the chapter, Section 8 concludes the chapter.

2. RF/ANALOG CIRCUIT DESIGN AND AUTOMATION: VARIOUS APPLICATIONS OF PSO

In this section, different types of applications of PSO for circuit design and automation is described. This is the area where PSO is mostly used, in context of circuits and systems.

2.1 Automatic Circuit Design: In Loop Optimization with CAD

PSO, when clubbed with CAD tools, is used for automatic electronic design of Analog/RF circuits. In this process, an objective function is defined in the design tool and based on this objective function PSO is applied. PSO can be implemented in the design tool itself or in any other computational tool which can communicate with the design tool. Based on the objective function, the design is iterated by PSO algorithm. This can be called as in-loop optimization because for each particle CAD tool needs to be run for each iteration. The number of computations are more in this process. In literature such method is used many times for automatic circuit design and optimization (Fakh-

Figure 1. Design process: In Loop Design and Optimization (Fakhfakh et al., 2009) (permission taken from the author for reuse of figure)

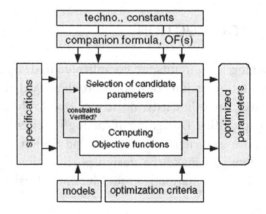

fakh et al., 2009; Thakker et al., 2009; Xuesong et al., 2011; Li, Yu & Li, 2008; Cooren et al., 2007). Design process is shown in Figure 1.

One simple case study from (Fakhfakh et al., 2009) will be helpful to understand in-loop optimization, in which amplifier circuit is designed for a given transconductance gain of an LNA. The design was having 11 design parameters so the particles declared were eleven dimensional. For a 20 particles swarm, 10000 iterations were used and solution was found in 1.57 seconds. Similar examples can be found in (Thakker et al., 2009; Xuesong et al., 2011; Li, Yu & Li, 2008; Cooren et al., 2007).

2.2 Circuit Simulation and Optimization for Emerging Devices

Circuit simulators need extracted device parameters which characterize a device. Based on these parameters only, the design is simulated by the simulator and the circuit simulation output is generated. For new and emerging devices such as FinFETs, the work is going on and the device parameters and analytical models are still under development, and are not available for circuit design. A look-up table (LUT) based approach for automatic design and optimization for FinFET circuits is introduced by R. Thakker (Thakker et al., 2009). A look-up table technique is implemented in circuit simulator and is integrated with PSO. This integration of Look-up tables and PSO, is proved to be very efficient for circuit simulation and automatic design for emerging devices, because the details of device is not required for forming look-up tables (Thakker et al., 2009; Thakker et al., 2010).

For circuit design, design variables, need to be determined such that the circuit meets the desired specifications. Figure 2 shows the steps for the design flow for automatic FinFET circuit design. The FinFET tables are generated before the optimization process begins. The circuit simulator, based on these tables and the design variables,

Figure 2. LUT-PSO based design flow for automatic FinFET circuit design (Thakker et al. 2009) (Permission taken from IEEE, for reuse of figure)

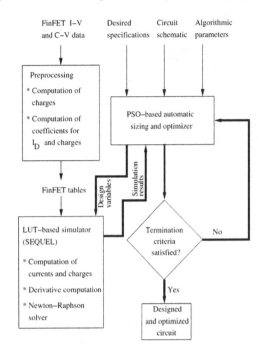

simulates the design and generates output. The optimizer compares the objective function of optimization with the design output and based on that, it defines new variables for next iteration. The iterations are repeated till the stopping criterion is met. In (Navan et al., 2009), transient simulation of a comparatively new device called organic transistor is taken into account using LUT approach of (Thakker et al., 2009) integrated with PSO.

2.3 Parasitic-Aware RF Circuit Design and Optimization

In high data rate communication systems, CMOS RF circuits play the key role. But the parasitics associated with the active and passive components limit the ease of synthesis, as de-tuning effects (because of nonlinear parasitics) come into picture. Thus parasitic-aware RF circuit designing and optimization CAD tools are must, in case of complex circuits. In (Park, Choi & Allsot,

Figure 3. The general parasitic-aware design and optimization flow introduced in (Park, Choi & Allsot, 2004) (Permission taken from IEEE, for reuse of figure)

Figure 4. Flowchart of PSO as used in the core optimization block of a parasitic aware RF circuit synthesis tool shown in Figure 3 (Park, Choi & Allsot, 2004) (Permission taken from IEEE, for reuse of Figure)

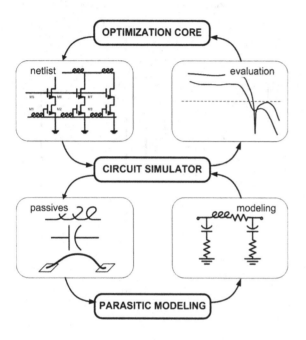

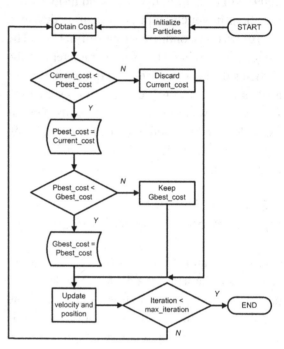

2004), a design methodology is developed, for a fully-integrated high-gain broadband amplifier in CMOS technology, to overcome the degradation due to parasitics. A new parasitic-aware optimization technique based on PSO is introduced as shown in Figure 3.

For the flow given in Figure 3, the performance of PSO is compared with classical Simulated Annealing (SA) and Adaptive Simulated Annealing (ASA) algorithms. Simulated Annealing works on the concept of cooling process in molten metal. High temperature metal is poured into a mold and then is cooled so that the atoms are organized in minimum energy states. In complex problems, SA is trapped in local minima because the hill climbing mechanism is disabled, due to too quick temperature change. It was observed that PSO is more efficient in case of complex problems compared to SA algorithm. In the flow shown in Figure 3, the optimization core used

was made by PSO as shown in Figure 4. This flow was applied to two circuits, one three stage power amplifier and one four stage distributed amplifier. It was found that PSO was better than both SA and ASA. The proposed design and optimization methodology was computationally efficient and robust in searching complex multidimensional design spaces.

2.4 NP Problem for RF Oscillator Circuit: Mathematical Problem

In the in-loop optimization with CAD tools, the number of computations are very large, as in each iteration a number of simulations are run according to the optimization algorithm. This increases the time required for searching the optimal design. For complex circuits and systems, this becomes impractical.

So, equivalent behavioral models can be used for the circuits or systems. One such example is given in (Tripathi, 2011; Tripathi, Mukherjee & Apte, in press), in which nonlinear behavioral models for the objective functions are found by Design of experiments (DOE). Based on these models Nonlinear Programming (NP) problem is defined which is solved by PSO. Such NP problems may be very difficult to be solved by conventional mathematical optimization techniques. In (Tripathi, Mukherjee & Apte, in press), in the oscillator circuit, the desired phase noise (P_d), and desired frequency (F_d) are defined to formulate the problem. Based on the nonlinear models obtained from Design of Experiments (DOE), a multi-objective optimization problem to minimize the area of circuit with the desired phase noise and frequency, can be defined as following:

Minimize the function $\sum_{i=1}^{7} \omega_i^2 + d^2 + l^2$ subjected to the following nonlinear equality constraints:

$$\sum_{i=1}^{9} \chi_{i_0} + \sum_{i=1}^{7} \sum_{j=1}^{4} \chi_{i_j} \omega_i^j + \chi_{8_1} d + \chi_{8_2} d^2 + \chi_{8_3} d^3 \\ + \chi_{8_4} d^4 + \chi_{9_1} l + \chi_{9_2} l^2 + \chi_{9_3} l^3 \\ + \chi_{9_4} l^4 + \mu_F = F_d$$

$$\sum_{i=1}^{9} \nu_{i_0} + \sum_{i=1}^{7} \sum_{j=1}^{4} \nu_{i_j} \omega_i^j + \nu_{8_1} d + \nu_{8_2} d^2 + \nu_{8_3} d^3 \\ + \nu_{8_4} d^4 + \nu_{9_1} l + \nu_{9_2} l^2 + \nu_{9_3} l^3 \\ + \nu_{9_4} l^4 + \mu_p = P_d$$

and following inequality constraints (due to foundry limitations and the levels of DOE)

$$13 \le w_1, w_2, w_3, w_4, w_5 \le 21$$
$$11 \mu m \le w_6 \le 19 \mu m$$
$$40 \mu m \le w_7 \le 60 \mu m$$
$$151 \mu m \le d \le 228 \mu m$$
$$32 \mu m \le l \le 48 \mu m$$

Figure 5. Designing and optimization by behavioral models and PSO (Tripathi, 2011) (Permission taken from the author for reuse, of figure)

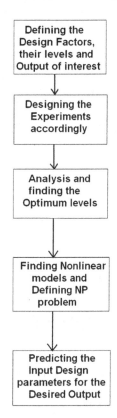

where w_1 to w_7, d and l are design factors. The above NP problem is defined to minimize the objective function, which is sum of square of all the design variables, subjected to the fourth order nonlinear equality constraints. The above nonlinear constraints are defined by the nonlinear models obtained from DOE. This problem was solved by PSO and the final solution (optimized circuit) had the phase noise of 129.1 dBc/Hz at 1.974 GHz oscillation frequency. Figure 5 shows the steps followed for this process.

3. MODELING OF CIRCUITS

PSO is hybridized with other algorithms to improve the efficiency for some applications. However, it

Figure 6. Modeling the input impedance of on-chip inductor by PSO (Bhattacharya, Joshi & Bhattacharya, 2006) - (a) Magnitude (b) Phase (Permission taken from IEEE, for reuse of figure)

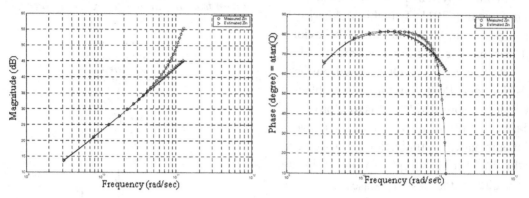

is not always necessary that the hybrid algorithms will be more efficient in all the applications. Two examples of such hybrid algorithms are presented for modeling of RF circuits.

3.1 Modeling of On-Chip Inductor

A PSO-based black box modeling of on-chip inductor is introduced in (Bhattacharya, Joshi & Bhattacharya, 2006). Unlike the conventional modeling methods which require the information of physical structure, this modeling method determines the physical structure and the identification of parameters. A PSO based off-line system identification

Algorithm is proposed, which can be used for arbitrary shaped structures also. In this method, the input impedance of the inductor is estimated based on the transfer characteristics of input current I for driving voltage V. Transfer function $Z_{in}(s)' = \dfrac{V_{in}(s)}{I_{in}(s)}$ is formulated by gain K, poles P_i and zeros Q_i. A cost function in defined in such a way that the modeled transfer function $Z_{in}(s)'$ should be matched with the original impedance of the inductor. By PSO, the positions of poles and zeros are moved in such a way that $Z_{in}(s)'$ should coincide with $Z_{in}(s)$.

Figure 6 shows the comparison between measured and modeled impedance of the on-chip inductor. Both the magnitude and phase are plotted. There is excellent agreement between the measurement and modeled impedance. For this experiment, 100 particles were used with 6000 iterations.

3.2 Analog Behavioral Modeling Based on Both LMNLS and PSO

The analog circuit behavioral models can be extracted directly from their transistor level netlists (Pam, Bhattacharya & Mukhopadhyay, 2010). A Verilog-A in-loop simulation based modeling approach is proposed. This technique uses various test-benches and based on them, their corresponding circuits can be modeled. It eliminates the need of implement structure based estimation tools for each circuit. This technique balances the use of gradient based optimization and search based optimization. It uses the speed of gradient based method and the same time robustness of evolutionary methods as well. There becomes co-operation and switching between these two optimization methods, for achieving faster convergence for global minima.

Levenberg Marquardt nonlinear least square (LM-NLS) is a non-linear least square optimiza-

Figure 7. Impedance profile of a power plane (Tripathi et al., 2012) (permission taken from the first author, for reuse of figure)

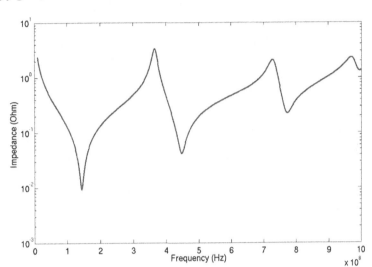

tion algorithm, based on the use of damped Gauss-Newton method. The LM-NLS implementation starts with an initial condition and converges to final solution. The optimization process switches from LM-NLS to

PSO and vice versa, depending upon certain conditions. The efficient method for automatic extraction of behavioral model parameters of analog integrated circuits is presented. This method is compared against standard gradient and search based techniques and is shown to be fast and robust. The method is validated by two distinctly different analog circuits signifying the usefulness of the proposed approach for bottom up behavioral model building.

4. HIGH SPEED SYSTEM DESIGN PROBLEM: POWER INTEGRITY ISSUES

In the above sections, only the circuits were designed by PSO. At higher frequencies, the system is also designed with the circuits, to avoid Signal Integrity (SI) and Power Integrity (PI) problems. One such example of a power plane and decoupling

network design is presented here (Tripathi et al., 2012). In that paper, Power Integrity is maintained by damping the cavity-mode anti-resonances' peaks on a power plane by PSO.

Power Integrity (PI) refers to the uninterrupted, sufficient and efficient distribution of power within a system. Power Integrity is becoming the major issue as the operational frequencies of the integrated circuits are increasing up to GHzs. If not maintained properly, Power Integrity may affect the functionality of a high speed system. Power delivery Networks (PDN) should be designed in such a way that PI should be taken care at a range of frequencies which may affect the system. Power planes are used for off-chip power delivery as a power supplier from Voltage Regulator Module (VRM) and bulk capacitor to the I/O of integrated circuits.

Power planes behave as cavity resonators and thus affect the power delivery of the system. Figure 7 shows the impedance profile of a power plane used in (Tripathi et al., 2012). It can be seen there are resonance and anti-resonance effects due to cavity. Decoupling capacitors are used to nullify the effect of these cavity effects. This example takes in to account the placement of decoupling

Figure 8. Impedance profile of a power plane after placing optimal number of decoupling capacitors (Tripathi et al., 2012) (permission taken from the first author, for reuse of figure)

Figure 8. Impedance profile of a power plane after placing optimal number of decoupling capacitors (Tripathi et al., 2012) (permission taken from the first author, for reuse of figure)

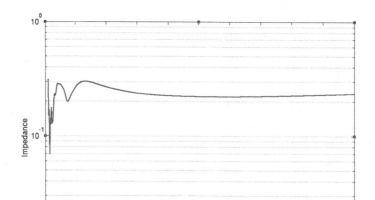

network, to maintain the uninterrupted power supply using PSO for finding the optimum value and position for decoupling capacitor on a power plane of a PDN to meet the target impedance. The power pins of the chip must see target impedance at the power plane. The impedance of the power delivery system is expected to be lesser than this, in order to maintain the PI of the system.

The parameters Equivalent Series Resistance (ESR), Equivalent Series Inductance (ESL), Capacitance, and the position co-ordinates were taken as the design variables. Thus, it was a five dimensional problem. The number of decoupling capacitors needed to meet the target impedance were 5. Their values and positions of capacitors were also found by PSO. The desired target impedance was below 528 m, which was achieved by placing 5 capacitors. The impedance profile of the board loaded with the optimal number of capacitors is shown in Figure 8.

5. TESTING OF ANALOG CIRCUITS

PSO can also be used for effective testing of Analog/RF circuits. Traditional test methods are one dimensional search methods. In addition to that, these methods need partitioning of Circuit Under Test (CUT) in case of large circuits. Oscillating Test Method (OTM) is also very common test method which uses partitioning of the CUT into small sub-circuits which can be converted into oscillators, for the purpose of testing. PSO based testing approach can be used to avoid these problems. PSO based approach is introduced in (Kyziol & Rutkowski, 2010; Kyziol, Rutkowski & Grzechca, 2010), which provides multi-dimensional search without partitioning large circuits. This N-terminal Based Test (N-tBT) approach can be summarized in four stages:

1. In first step, optimization problem is defined. Circuit Under Test (CUT), faults dictionary and Pspice simulation files are specified. After that, N-terminal network is also defined with its structure, value of elements and 10 parameters for PSO.

Figure 9. Multidimensional search testing by PSO (Kyziol & Rutkowski, 2010) (permission taken from IEEE, for reuse of figure)

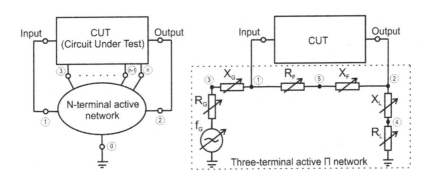

2. Second stage of N-tBT (N-terminal based test) method is searching groups and layouts. During this stage the Pspice simulation for CUT is performed. Data obtained from simulations are used in order to find all groups (states and ambiguity sets of CUT) and layouts (voltage distribution).

3. In third stage, combinatorial optimization problem is formulated using data obtained from stage 2. Then searching for minimal set of testing vectors that assure the best possible identification of CUT states is searching. This problem is also NP-hard and belong to the family of combinatorial optimization problems.

4. The last stage is the test time reduction. The aim is to minimize number of test vectors using 2-terminal networks with specified impedance and multi-tone testing techniques. Figure 9 shows testing schemes using N terminal network and 3 terminal networks.

In pseudoexhaustive test approach, circuit is divided into a number of subcircuits which effectively reduces the number of test vectors for testing. In (Kumar et al., 2009), PSO is used for optimal partitioning of the circuits for efficient testing. Though, this method is for digital circuits testing. It is mentioned here because the parti-tioning is needed in some analog circuits also, as mentioned above in the N-tBT method.

6. DESIGNING PASSIVE CIRCUITS

PSO is applied by researchers, for passive circuit designing also. In (Datta et al., 2010), output matching load network for a dual-band Low Noise Amplifier (LNA) is designed using PSO. A design methodology is developed using PSO, for the selection of the components of output matching load network, for the LNA, designed to work in GSM 1.8 GHz and WLAN 2.4 GHz range. The output matching network is shown in Figure 10. This is connected to a cascode amplifier as shown in Figure 10.

The output seen from the amplifier is

$$Z \approx \frac{sL_1\left(1 + s^2 L_2 C_2\right)}{\left(1 + s^2\left(L_1 C_1 + L_2 C_2 + L_1 C_2\right) + s^4 L_1 C_1 L_2 C_2\right)}$$

(1)

From the above equation, the characteristic equation for frequency can be found from denominator of the right side expression of the equation. This characteristic equation will provide two roots . which will be the two resonant frequencies, and can be given as seen in Equation 2 (see Box 1).

Box 1.

$$\omega_{1,2} \approx \sqrt{\frac{2L_2C_2 \pm \sqrt{\left(2L_2C_2\right)^2 - 4\left(1 + \frac{1}{Q_L^2}\right) + \left(\left(L_2C_2\right)^2 + L_1L_2C_2^2\right)}}{2\left(1 + \frac{1}{Q_L^2}\right)\left(\left(L_2C_2\right)^2 + L_1L_2C_2^2\right)}}$$

(2)

Figure 10. Output matching network design by PSO (Datta et al., 2010) (permission taken from IEEE, for reuse of figure)

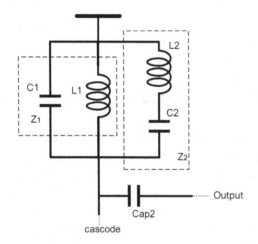

Figure 11. Real part of output impedance (Datta et al., 2010). (permission taken from IEEE, for reuse of figure)

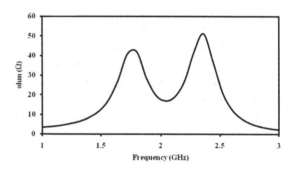

From the previous equations, the four dimensional particles were defined having the random values of . The solution was to meet the target of 50 at desired frequency bands. Figure 11 shows the real part of output impedance, which is near the desired value, at both the frequency bands.

7. MODIFIED PSO ALGORITHMS

In this section, the applications of modified PSO algorithms are presented.

7.1 Elitist Distributed PSO

Elitist Distributed PSO (ED-PSO) is introduced in (Chu & Allstot, 2005), with a case study of optimization of a 5.2 GHz direct-conversion receiver front-end by EDPSO. Figure 12 shows the flow chart of the elitist distributed PSO. Like the basic PSO, a population of N particles with random positions is initialized. The fitness of all the particles are calculated and then sorted based on the concept of non-domination. Two particles will be considered as nondominated particles if each of them is better than the other one, in at least one objective. After generating particles, the dominant particles are copied to an external archive. A particle remains in $G_{best-ext}$ throughout the optimization process and is only replaced if a more dominant particle is found. To prevent particles to converge in a local minima, a distance based approach is used. Crowding distance for

Figure 12. Flow chart of the EDPSO (Chu & Allstot, 2005) (permission taken from IEEE, for reuse of figure)

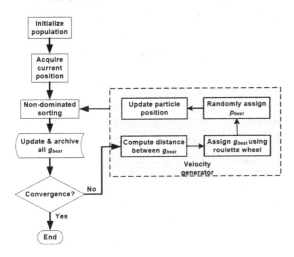

Figure 13. Arrangement of particles in HPSO (Thakker, Baghini, Patil, 2009) (permission taken from IEEE, for reuse of figure)

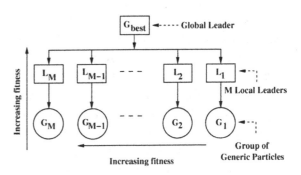

7.2 Hierarchical PSO

each particle in $G_{best-ext}$ is defined, which is the average of the distances of a particle from its nearest two particles. Then, a roulette wheel method selects particles in $G_{best-ext}$ as global optima by normalizing the crowding-distance values into probabilistic values. The particles having the smaller value of crowding distance will have the low probability of selection. This method allows particles to swarm in different directions toward the Pareto optimal front and effectively prevents premature convergence. An additional parameter, which is the individual optimum *'pbest'*, is also needed to update the velocity of the particles. *pbest* is chosen randomly from the non-dominated positions of each individual. Once *pbest* and best are selected, the velocity and position of all particles are updated. This process continues until no new dominant particles are generated or a maximum number of iterations have been executed.

EDPSO is compared with the traditional PSO. It is found that for this optimization problem, EDPSO is more efficient than PSO. A term iteration efficiency (which is defined as the ratio of number of number of solutions provided by the number of iterations) is calculated, which is 0.111 in the case of traditional PSO while the same is 1.31 in the case of EDPSO.

Hierarchical PSO (HPSO) was introduced by S. Janson (Janson & Middendorf, 2005), which works on the basis of arrangement of particles in a hierarchy. After defining the N number of particles, they are arranged in ascending order according to their fitness. The globally best Gbest particle is arranged at position N. The next M particles are categorized as local leaders. The remaining N −M −1 particles ('generic particles') are divided into M groups based on the local leaders. Each remaining particle is associated with one of the M local leaders, which has the closest fitness. The generic particles follow their respective local leaders and the local leaders follow the global leader. This phenomena explores the search space and shows the better consistency in finding the optimum solution (Thakker, Baghini, Patil, 2009).

Figure 13 shows the pictorial representation of HPSO. In (Thakker, Baghini, Patil, 2009), HPSO was used for automatic design of low voltage, low power analog circuits. This was compared with Genetic Algorithm (GA) and PSO for four different circuits and it was observed that the HPSO was better than both and the solution was achieved faster.

8. CONCLUSION

Various applications of PSO for Analog/RF circuits and systems are discussed in this chapter. PSO is proved to useful for Analog/RF circuits and system design in various ways. It can be used to in-loop optimization with CAD tools, can be used to solve NP problems for behavioral models of circuits and for designing the systems also. PSO can be used for modeling, passive circuit design and for testing also. Thus, PSO is very useful for achieving design goals in complex design, modeling and testing problems of circuits and systems.

REFERENCES

Bhattacharya, R., Joshi, A., & Bhattacharya, T. K. (2006). PSO-based evolutionary optimization for black-box modeling of arbitrary shaped on-chip RF inductors. In *Proceedings of 2006 Topical Meeting on Silicon Monolithic Integrated Circuits in RF Systems*, 18-20 Jan. 2006.

Chu, M., & Allstot, D. J. (2005). An elitist distributed particle swarm algorithm for RF IC optimization. In *Proceedings of IEEE Asia and South Pacific Design Automation Conference 2005*, (pp. 671-674).

Cooren, Y., Fakhfakh, M., Loulou, M., & Siarry, P. (2007). Optimizing second generation current conveyors using particle swarm optimization. In *Proceedings of IEEE International Conference on Microelectronics*, December 2007.

Datta, K., Datta, R., Dutta, A., & Bhattacharyya, T. K. (2010). PSO-based output matching network for concurrent dual-band LNA. In *Proceedings of IEEE Microwave and Millimeter Wave Technology 2010*.

Datta, K., Datta, R., Dutta, A., & Bhattacharyya, T. K. (2010). PSO optimized concurrent dual-band LNA with RF Switch for better inter-band isolation. In *Proceedings of 5th European Conference on Circuits and Systems for Communications*, November 23-25, 2010, Belgrade, Serbia.

Fakhfakh, M., Cooren, Y., Sallem, A., Loulou, M., & Siarry, P. (2009). Analog circuit design optimization through the particle swarm optimization technique. *Analog Integrated Circuits and Signal Processing, 63*, 71–82. doi:10.1007/s10470-009-9361-3

Janson, S., & Middendorf, M. (2005). A hierarchical particle swarm optimizer and its adaptive variant. *IEEE Transactions in Systems, Man, and Cybernetics - Part B, 35*, 1272–1282. doi:10.1109/TSMCB.2005.850530

Kumar, K. S., Bhaskar, U. P., Chattopadhyay, S., & Mandal, P. (2009). Circuit partitioning using particle swarm optimization for pseudo-exhaustive testing. In *Proceedings of International Conference on Advances in Recent Technologies in Communication and Computing 2009*, ARTCom '09, (pp. 346-350).

Kyziol, P., & Rutkowski, J. (2010). Searching groups and layouts in n-terminal based test method using heuristic PSO algorithm. In *Proceedings of 2010 International Conference on Signals and Electronic Systems (ICSES)*, (pp. 217-220).

Kyziol, P., & Rutkowski, J. (2010). Searching groups and layouts in n-terminal based test method using heuristic PSO algorithm. In *Proceedings of The International Conference on Signals and Electronic Systems (ICSES 2010)*, Gliwice, Poland, Sep. 7-10, 2010.

Kyziol, P., Rutkowski, J., & Grzechca, D. (2010). Testing analog electronic circuits using N-terminal network. In *Proceedings of 2010 IEEE 13th International Symposium on Design and Diagnostics of Electronic Circuits and Systems (DDECS)*.

Li, Y., Yu, S., & Li, Y. (2008). Electronic design automation using a unified optimization framework. *Mathematics and Computers in Simulation, 79*, 1137–1152. doi:10.1016/j.matcom.2007.11.001

Navan, R. R., Thakker, R. A., Tiwari, S. P., Baghini, M. S., Patil, M. B., Mhaisalkar, S. G., & Rao, V. R. (2009). DC transient circuit simulation methodologies for organic electronics. In *Proceedings of 2nd International Workshop on Electron Devices and Semiconductor Technology 2009*, India.

Pam, S., Bhattacharya, A. K., & Mukhopadhyay, S. (2010). An efficient method for bottom-up extraction of analog behavioral model parameters. In *Proceedings of 23rd International Conference on VLSI Design*, (pp. 363-368).

Park, J., Choi, K., & Allsot, D. J. (2004). Parasitic-aware RF circuit design and optimization. *IEEE Transactions on Circuits and Systems, 51*(10).

Thakker, R. A., Baghini, M. S., & Patil, M. B. (2009). Low-power low-voltage analog circuit design using HPSO. In *Proceedings of IEEE International Conference on VLSI Design 2009*, India.

Thakker, R. A., Baghini, M. S., & Patil, M. B. (2009). Low-power low-voltage analog circuit design using HPSO. In *Proceedings of IEEE International Conference on VLSI Design 2009*, India.

Thakker, R. A., Sathe, C., Baghini, M. S., & Patil, M. B. (2010). A table-based approach to study the impact of process variations on FinFET circuit performance. *IEEE Transactions on Computer-Aided Design of Integrated Circuits and Systems, 29*(4), 627–631. doi:10.1109/TCAD.2010.2042899

Thakker, R. A., Sathe, C., Sachid, A. B., Baghini, M. S., Rao, V. R., & Patil, M. B. (2009). Automated design and optimization of circuits in emerging technologies. In *Proceedings of IEEE Asia and South Pacific Design Automation Conference, 2009*, Japan.

Thakker, R. A., Sathe, C., Sachid, A. B., Baghini, M. S., Rao, V. R., & Patil, M. B. (2009). A novel table based approach for design of FinFET circuits. *IEEE Transactions on Computer-Aided Design of Integrated Circuits and Systems, 28*(7), 1061–1070. doi:10.1109/TCAD.2009.2017431

Tripathi, J. N. (2011). Designing, optimization and modeling of analog/RF circuits by design of experiments. In *Proceedings of Ph.D. Forum, IEEE/IFIP 19th International Conference on VLSI System- On-Chip*, (pp. 457-460). Oct. 2-5, 2011, Hong Kong, China.

Tripathi, J. N., Mukherjee, J., & Apte, P. R. (in press). Nonlinear modeling and optimization by design of experiments: A 2 GHz RF oscillator case study. *International Journal of Design* [in press]. *Analysis and Tools for Integrated Circuits and Systems*.

Tripathi, J. N., Nagpal, R. K., Chhabra, N. K., Malik, R., & Mukherjee, J. (2012). Maintaining power integrity by damping the cavity-mode anti-resonances peaks on a power plane by particle swarm optimization. In *Proceedings of 13th International Symposium on Quality Electronic Design*, March 2012, Santa Clara, USA.

ADDITIONAL READING

Angeline, P. J. (1998). Evolutionary optimization versus particle swarm optimization: Philosophy and performance differences. *Lecture Notes in Computer Science, 1447*, 601–610. doi:10.1007/BFb0040811

Angeline, P. J. (1998). Using selection to improve particle swarm optimization. In *Proceedings of IEEE International Conference on Evolutionary Computation* (pp. 84 – 89). May 1998.

Ciuprina, G., Ioan, D., & Munteanu, I. (2002). Use of intelligent-particle swarm optimization in electromagnetics. *IEEE Transactions on Magnetics*, *38*(2), 1037–1040. doi:10.1109/20.996266

Clerc, M. (2004). Discrete particle swarm optimization illustrated by the travelling salesman problem . In Onwubolu, G. C., & Babu, B. V. (Eds.), *New optimization techniques in engineering* (pp. 219–239). Springer-Verlag.

Eberhart, R. C., & Shi, Y. (2001). Particle swarm optimization: Developments, applications and resources. *Proceedings of the 2001 Congress on Evolutionary Computation, (CEC 2001)* (pp. 81-86).

Higashi, N., & Iba, H. (2003). Particle swarm optimization with Gaussian mutation. In *Proceedings of the IEEE* Swarm Intelligence Symposium 2003, (SIS '03), (pp. 72 – 79).

Kennedy, J., & Eberhart, R. (1995). Particle swarm optimization. In *Proceeding of IEEE International Conference on Neural Networks*, (pp. 1942 – 1948).

Parsopoulos, K. E., & Vrahatis, M. N. (2002). Particle swarm optimization method in multi-objective problems. In *Proceedings of the 2002 ACM Symposium on Applied Computing*, SAC'02.

Poli, R., Kennedy, J., & Blackwell, T. (2007). Particle swarm optimization: An overview. *Swarm Intelligence*, *1*(1), 33–57. doi:10.1007/s11721-007-0002-0

Robinson, J., & Rahmat-Samii, Y. (2004). Particle swarm optimization in electromagnetic. *IEEE Transactions on Antennas and Propagation*, *52*(2), 397–407. doi:10.1109/TAP.2004.823969

Shi, Y., & Eberhart, R. C. (1999). Empirical study of particle swarm optimization. *Proceedings of the 1999 Congress on Evolutionary Computation, (CEC 1999)*.

Sun, J., Xu, W., & Feng, B. (2004). A global search strategy of quantum-behaved particle swarm optimization. In *Proceedings of IEEE Conference on Cybernetics and Intelligent Systems 2004*, (pp. 111-116).

Trelea, I. C. (2003). The particle swarm optimization algorithm: Convergence analysis and parameter selection. *Information Processing Letters*, *85*(6), 317–325. doi:10.1016/S0020-0190(02)00447-7

Tripathi, J. N., Chhabra, N. K., Nagpal, R. K., Malik, R., & Mukherjee, J. (2012). Damping the cavity-mode anti-resonances' peaks on a power plane by swarm intelligence algorithms. In *Proceedings of 2012 IEEE International Symposium on Circuits and Systems,* Seoul, Korea.

KEY TERMS AND DEFINITIONS

Analog/RF Circuits: Is a circuit with a continuous, variable radiofrequency signals.

Design Automation: Automatic software tools for the development and design of circuits and systems.

High Speed Systems: Systems with clock speeds in low megahertz range.

Modeling: The representation, often mathematical, of a process, concept, or operation of a system, often implemented by a computer program.

Testing: Is used at key checkpoints in the overall process to determine whether objectives are being met.

Section 2
Antenna and Optical Devices

In last years, PSO techniques received increasing attention in the electromagnetic community since they are promising algorithms for handling optimization problems in the area of the design of antennas and optical devices. To this purpose, this section illustrates recent results regarding PSO developments aimed at the efficient design of reconfigurable antennas, microwave absorbers, erbium doped fibre amplifiers and Raman amplifiers. Moreover, the proposed contributions show applications of PSO to the synthesis of linear antennas array, the estimation of the permittivity of multilayer structures, the recovering of parameters describing the energy transfer phenomena among the erbium ions.

Chapter 5
Particle Swarm Optimization Algorithm in Electromagnetics Case Studies:
Reconfigurable Radiators and Cancer Detection

Arezoo Modiri
University of Texas at Dallas, USA

Kamran Kiasaleh
University of Texas at Dallas, USA

ABSTRACT

This chapter is intended to describe the vast intrinsic potential of the swarm-intelligence-based algorithms in solving complicated electromagnetic problems. This task is accomplished through addressing the design and analysis challenges of some key real-world problems, ranging from the design of wearable radiators to tumor detection tools. Some of these problems have already been tackled by solution techniques other than particle swarm optimization (PSO) algorithm, the results of which can be found in the literature. However, due to the relatively high level of complexity and randomness inherent to these problems, one has to resort to oversimplification in order to arrive at reasonable solutions utilizing analytical techniques. In this chapter, the authors discuss some recent studies that utilize PSO algorithm particularly in two emerging areas; namely, efficient design of reconfigurable radiators and permittivity estimation of multilayer structures. These problems, although unique, represent a broader range of problems in practice which employ microwave techniques for antenna design and microwave imaging.

DOI: 10.4018/978-1-4666-2666-9.ch005

INTRODUCTION

The main objective pursued in this chapter is to underscore the flexibility and the power of the PSO algorithm in handling variety of analytically intractable problems. To that end, specific scenarios, which are representatives of a larger class of problems, are explored using PSO. We begin by providing a detailed definition of PSO and its modified versions and then dive into the discussions regarding the challenges of solving realistic problems by means of this algorithm. In each case study, along with the problem definition, we bring to the forefront the key characteristics of the problem at hand. This, in fact, tells us what other problems can be solved in the same or similar way. The conventional algorithm, then, is modified, exclusively to fit the requirements of that specific problem, and, finally, the performance of the resulting algorithm is presented and evaluated. The two classes of problems tackled in this chapter are wearable/ switchable antennas and in-vivo cancer detection techniques. Although these topics seem to be quite distinct, they possess major common properties, which make a swarm-based algorithm a viable search mechanism for them. Those properties can be summarized as follows:

- The inherent random nature of the key variables of the problem.
- The large number of unknowns as compared to the available information.
- The necessity of expeditious solution discovery due to the ultimate real-time nature of the corresponding application.

The outline of this chapter is as follows. In the first section, PSO algorithm is introduced, and several modifications to the conventional format of the algorithm are presented and compared in terms of the final result accuracy and optimization cost. In addition to the results presented in (Modiri, et al., 2011), this section brings together

the other discoveries regarding PSO, which are presented in the literature, such as those introduced by (Nakano, et al., 2007; Clerk, *et al,.* 2002). A good list of the articles in this area can be found in the reference list of (Modiri, *et al*, 2011).

When studying optimization algorithms, it is important to bear in mind that, in general, there are two approaches one may take. The first approach is dedicated to a suitable solution found expeditiously, whereas the second approach targets the best possible solution regardless of the optimization time or cost. Given the time-sensitiveness of the emerging electromagnetic problems studied here, we focus on the first approach. Both real-number (RPSO) and binary PSO (BPSO) algorithms are taken into account. In order to rationalize the introduced modifications, well-known benchmark problems are evaluated. This allows us to set the stage for the more challenging problems of later sections.

In the second section, based on our studies presented in (Modiri[2], et al., 2011; Modiri[3], et al., 2011; Modiri, et al., 2010), we illustrate how PSO is utilized in a more challenging (and of course appealing) scenario. In particular, we consider the design of reconfigurable and wearable antennas (Liu, et al., 2005; Jin, et al., 2008). In order to illustrate how PSO as a viable technique enters the picture, let us have a brief description of the problem.

Reconfigurable antenna problems which are studied herein are useful in a variety of applications where software defined control over the antenna performance is required. Particularly, the ability to change the frequency performance of the antenna (in terms of resonance frequency) is investigated. This feature is appropriate in multi-application devices where each application may have its own frequency assignment. Instead of incorporating multiple antennas for supporting individual applications, a single reconfigurable antenna can be designed. The advantage of using PSO algorithm in this problem becomes even more obvious when a level of unpredictability in the start time of an application is present in the overall design; see (Jin, et al., 2008) as an example. In addition to frequency selectivity, the antenna radiation pattern can also be controlled in reconfigurable antennas, resulting in yet another demanding application (Gies, et al., 2003). System designers are often faced with the problem of designing systems in areas which are jam-packed with wireless networks, such as hospitals. In such environments, there is a high chance of interference between adjacent communication links. Therefore, it is more convenient to be able to block the radiation in a certain direction. Here, PSO algorithm is used to control the radiation pattern of the array antennas with the goal of beam blocking, see (Jin, et al., 2007).

Another topic studied here is wearable antennas (WA). Mounting antennas on the fiber of daily clothing in the form of wearable antennas has recently attracted a great deal of interest. This is mainly due to the growing demand for body sensor networks. Sensors are dependent on the unobtrusive radiators in and around human body in order to constantly trace bio signatures over the entire monitoring period. This monitoring state, in some cases, may continue for several months. However, it has been shown in the literature that various seemingly unnoticeable factors, including habitual activities of the user or common daily events can degrade the WA performance substantially, see chapter 6 in (Hall, et al., 2006). More crucially, telecare devices are at risk when used for controlling the vital bio-signs of young, active children remotely. Hence, circumventing the performance degradation caused by the environmental variations can play an important role in preserving a high quality active monitoring. This task is done via applying a real-time, self-tuning algorithm. As a possible variation, in section II, we investigate the problem of bending the antenna and its effect on the antenna performance. A solution technique is introduced for real-time

compensation of this phenomenon using PSO algorithm. The solution to this problem will also pave the way for solving similar problems with antenna modification.

In the third section, the usefulness of PSO in an even more challenging and attractive scenario is discussed. In that scenario, the algorithm is used to estimate the electrical properties of a lossy, multilayer structure. This study becomes more appealing when the medium under investigation is the human body tissue. The main application of this study in real world is tumor detection. Our research on breast cancer detection via microwave imaging, partly published in (Modiri[4], et al. 2011), is the main focus of this section. Dealing with live tissue has its inherent challenges; many articles in the literature concerning live tissue, e.g., (Lazebnik, et al., 2007), have demonstrated the semi-randomness of the electric properties of human tissue. Taking advantage of these studies, in section III, a PSO-based tumor detection method is reviewed, particularly for breast cancer detection through microwave imaging at frequencies below 3GHz. In fact, the proposed detection technique is not dependent on the type of the tissue under investigation, and hence, can be used in other similar applications as well. However, the limitation is imposed by the penetration depth of electromagnetic energy at the desired frequency (herein, microwave) in different tissues. In this scenario, both complex number PSO (CPSO) and discrete PSO (DPSO) algorithms are studied and redesigned in order to address the requirements of the problem at hand. The main goal is to demonstrate that PSO can successfully distinguish between the permittivity values of various layers of the human tissue, enabling one to verify the existence and the possible location of a tumor.

It should be noted that, in addition to real number, binary, complex number and discrete PSO algorithms studied in this chapter, other versions of this algorithm have also been introduced and utilized in distinct electromagnetic problems successfully in the literature. Boolean PSO is an interesting example. However, those PSO versions are out of the scope of this chapter and the interested reader is referred to the available articles in the literature, such as (Afshinmanesh, et al., 2008). It is also noteworthy to mention that, different references do not necessarily use common abbreviations for distinct PSO variations. For example, some articles use BPSO as the abbreviation for Binary PSO, and some for Boolean PSO.

Finally, concluding remarks along with a summary of the potential of PSO and the conditions under which such potentials can be achieved are provided in the future direction and conclusion sections of this chapter.

PARTICLE SWARM OPTIMIZATION: DEVELOPMENT, MODIFICATIONS

Conventional PSO

Similar to genetic algorithm, PSO is a member of the evolutionary algorithm (EA) family. The idea of EAs is originated from the concept of stochastic variation of a group of 'individuals', also named 'particles', in a solution space. Various algorithms have been developed based on this very concept since 1990s. The main difference between the existing algorithms, however, resides in the manner by which their operators are defined. A thorough comparison of EAs in electromagnetics was presented by Hoorfar (2007), where the following classifications of EAs were also introduced: evolutionary programming, evolution strategies, and genetic algorithms. Each of these classes was introduced to address a specific group of problems for which such algorithm yields an effective solution. Therein, the author concluded that the evolutionary programming (EP) class

was a very valuable tool for the efficient design of microwave devices due to its simplicity, ease of implementation, and flexibility. PSO algorithm belongs to EP class, and hence, the aforementioned statement applies to PSO as well.

At this point, let us have a brief review of PSO history. Kennedy and Eberhart (1995) formulated the PSO algorithm, for the first time, by emulating the behavior of group-living creatures during their search activities. The algorithm was further developed by (Shi, et al., 1999; Eberhart, et al., 2001; Clerk, et al., 2002), among others. As shown by Jin (2007), PSO, with the aid of a variety of analytical and numerical tools, becomes a viable technique for various electromagnetic applications. Here are some examples of using PSO; Matekovits (2005), Mikki(2005) and Robinson (2004) have studied how PSO generally can be used in electromagnetics, while Cui (2005), Jin (2005), Hwang (2009), Chamaani (2010) and Boeringer (2004) have focused on some specific applications, such as the design of absorbers, waveguides and antennas. Wang (2005), He (2009), Ko (2009) and Sun (2010) have used PSO in filter design. Benedetti (2006), Bayraktar (2006) and Ismail (2010) have shown how PSO is utilized in array design, and of course, many more interesting articles can be found in the literature that use PSO variants in various electromagnetic applications.

In spite of a relatively large number of articles on EAs, the design of EA algorithms, in general, and PSO, in particular, remains problem dependent. Initial conditions also have a substantial impact on the performance, especially when optimization length is a significant concern. Eberhart (2001), Nakano (2007), and Camci (2008) effectively depicted the abovementioned properties of PSO in their investigations. In other words, it is always recommended to choose the optimization technique and set the parameters of the algorithm accordingly to meet the unique attributes of the problem at hand.

The fundamentals of PSO algorithm have been extensively presented in the literature; for example, see (Eberhart, et al., 2001; Jin, et al. 2007). For the benefit of the reader, a concise, yet complete definition of the algorithm is provided here.

Let us assume a problem with n unknowns or parameters. In fact, n defines the dimensionality of the problem at hand, which means that the solution space is n-dimensional. The optimum solution is a single point in this n-dimensional space. The ranges over which problem parameters vary define how wide the solution space is. A search group, consisting of m individuals, tries to find the optimum solution. Similar to other optimization techniques, a fitness (or objective) function determines how 'good' each location in the solution space is. Although multiple objective functions can exist for one problem, it is often possible to merge all of them into one. Hence, single-objective optimization is studied herein.

In PSO, each particle remembers the best location, in terms of fitness function, that it has discovered during its own exploration. This is named the 'personal best'. At each iteration cycle, the best location found by all particles (the best personal best) is named the 'global best'. Following steps describe the basic building blocks of the PSO algorithm:

- The 'm' particles are spread randomly inside the solution space.
- Fitness values are calculated and personal bests and global best are identified.
- With respect to the personal and global bests, a 'velocity' vector is assigned to each particle.
- The locations of the particles are updated based on the velocity vectors.
- The algorithm loops back to step 2 until the desired fitness value is achieved.

It is important to note that the velocity vector is the only important operator in the PSO algorithm and this is the main reason for the ease of implementation of PSO. A definition for the velocity vector and the impact of its variation on the overall performance will be presented in the ensuing discussion.

Having m particles and n parameters, the positions and velocities are stored in matrices. In the conventional PSO, the velocity is calculated using following equation.

$$\underline{\underline{v}}_t = \omega \underline{\underline{v}}_{t-1} + c_1 \vartheta_1 \left(\underline{\underline{p}}_{best_{t-1}} - \underline{\underline{x}}_{t-1} \right) + c_2 \vartheta_2 \left(\underline{\underline{g}}_{best_{t-1}} - \underline{\underline{x}}_{t-1} \right)$$

(1)

In the above equation, $\underline{\underline{x}}_t$, $\underline{\underline{v}}_t$, $\underline{\underline{p}}_{best}$, and $\underline{\underline{g}}_{best}$ denote the position, velocity, personal best, and global best matrices, respectively. is the iteration number ϑ_1 and ϑ_2 are two random variables uniformly distributed on $[0,1]$, and lastly, ω, c_1 and c_2 are constants used to move the algorithm towards achieving the desired goals. As Eberhart (2001) depicts, ω, c_1 and c_2 have been, in large part, obtained through trial and error. This also means that these parameters can be redesigned according to the problem at hand.

In order to initiate Equation 1, it is necessary to know the initial velocity. In the conventional PSO, the initial velocity matrix, , is composed of the elements shown in Equation 2.

$$v_0(i, j) = \textit{random number uniformly distributed on}$$
$$\left[v^j_{minimum,} \quad v^j_{maximum} \right]$$
$$i\epsilon \left\{ 1, 2, ..., m \right\}, j\epsilon \left\{ 1, 2, ..., n \right\}$$

(2)

In the previous equation, and are the maximum and minimum permissible velocities over dimension. Often, is assumed to be equal to in intensity, but in the opposite direction. is generally assumed to be equal to the parameter range or a fraction of it, for RPSO. As shown by Jin (2007), a mapping function is required in the case of binary and discrete PSO's for the sake of quantization.

Another part of the algorithm is the boundary condition, which is necessary in order to set a bound on the limits of the solution space. Robinson (2004) has explained different boundary conditions and has claimed that 'invisible wall' showed a superior performance. Therefore, invisible wall is mainly used in this chapter. Using this boundary condition, the fitness function calculation will not be performed for the particles that 'fly out' of the solution space. A fairly undesired fitness value is assigned to these particles instead. This way, although all the particles remain intact, the computation complexity is decreased slightly. Upon the return of the escaped particles, the objective function calculation will be restarted for them. Yet as a second option, 'absorbing wall' boundary condition is also taken into account herein. An absorbing wall simply stops particles right at the boundary if they try to fly out of the solution space.

PSO Variations

When deciding to employ PSO in time-sensitive electromagnetic problems, there are two main issues to consider. First, similar to the other global optimization techniques, it is possible that the particles are trapped by a local best solution which, ultimately, prevents them from reaching the actual best solution. Second, in many demanding electromagnetic problems, the optimization cost can get unacceptably high due to the large number of required iterations and/or particles.

Due to the abovementioned two facts, basic PSO has been modified in variety of ways to compensate for its drawbacks. A review of the different modifications existing in the literature can be found in (Nakano, et al., 2007) along with a comparison of their performances. There, the author has also introduced a new method, namely, tabu-searching PSO or TS-PSO. To briefly illustrate, in TS-PSO, a second swarm and a number of 'tabu' conditions are added to the algorithm. Equations 5-7 in (Nakano, et al., 2007) show that the particles of the second swarm are led away from the personal best and/or global best positions which satisfy the tabu condition. In this way, the particles are prevented from getting locked in their local bests. However, optimization length is shown to increase significantly in TS-PSO due to the increased computation complexity.

More modified PSO algorithms are briefly introduced in the ensuing paragraphs in which personal best coefficient (c_1) and initial velocity (v_0) definition are particularly varied. The advantage of the following modifications resides in the non-increasing computation cost.

Looking back at the steps followed in the PSO algorithm, one can observe that, at each iteration cycle, one of the personal bests is qualified to become the global best of that iteration. The other personal bests are actually undesired local bests, which can only be saved in the memory of the particles in order to guide them in their future trajectories. Therefore, the idea of ignoring undesired personal best positions, as an alternate to the move-away procedure in TS-PSO, seems to be reasonable. Hence, the following velocity function was proposed in (Modiri, et al., 2011). There, the authors also changed the manner by which the initial velocity was selected to a bi-state condition, as follows in Equation 4.

$$\underset{=t}{v} = \omega \underset{=t-1}{v} + c_2 \vartheta_2 \left(\underset{=}{g}best_{t-1} - \underset{=t-1}{x} \right) \tag{3}$$

$$v_0(i,j) = randomly\ selected$$
$$to\ be\ equal\ to\ v^j_{minimum}\ \ or\ \ v^j_{maximum}$$
$$i\epsilon\left\{1,2,...,m\right\}, j\epsilon\left\{1,2,...,n\right\} \tag{4}$$

The reason behind the second modification can be found in nature. Bees, as an example, start their initial flower-searching activities with a speed that is higher than that of the subsequent search steps. In the conventional RPSO, ω is decreased from 0.9 to 0.4 during optimization. This way, the algorithm emulates the mentioned reducing search speed concept. However, such an approach has shown no impact on BPSO (Jin, et al., 2007), therefore, ω is generally set to a constant value of 1 in the conventional BPSO. In (Modiri, 2011), however, this large initial velocity was introduced through Equation 4, whereas in the case of BPSO, $v^j_{minimum}$ was set to 0 and $v^j_{maximum}$ was assumed to be 6. The answer to the question "why 6?" can be found in (Jin, 2007).

Many articles in the literature, such as (Eberhart, et al., 2001; Carlisle, et al., 2002; Hoorfar, 2007; Jin, et al., 2007; Camci, 2008), have investigated the best values for c_1 and c_2. According to the most recommended results, both c_1, and c_2 are either set to 1.49 or 2. Jin (2007), however, has shown better convergence performance when c_1 and c_2 are both set to 2. Therefore, $c_1 = c_2 = 2$ is used in this chapter, as well.

As it is shown in details in (Modiri, et al., 2011) and summarized hereafter in Equations 5-8, the modified velocity function reinforces the correct progress of the swarm toward the best solution.

Assuming and as position and velocity of the particle in the dimension during the iteration, Equation 5 shows how the position is updated in PSO.

$$x_{(i,j)_t} = x_{(i,j)_{t-1}} + v_{(i,j)_t} \tag{5}$$

Considering the conventional velocity function in Equation 1, one can find the average particle velocity as follows:

$$\underline{v}_t = \omega \underline{v}_{t-1} + c_1 \left(\frac{\overline{\underline{p}best_{t-1}}}{2} - \frac{\overline{\underline{x}_{t-1}}}{2} \right) + c_2 \left(\frac{\overline{\underline{g}best_{t-1}}}{2} - \frac{\overline{\underline{x}_{t-1}}}{2} \right) \tag{6}$$

where underline sign is averaging operator. To elaborate, for instance, and are the position and velocity vectors averaged over i. Assuming c_1 and c_2 are both equal to 2, Equation 7 can be deduced from Equation 5 and Equation 6.

$$\overline{\underline{x}}_t = \omega \overline{\underline{v}}_{t-1} + \overline{\underline{p}best_{t-1}} + \overline{\underline{g}best_{t-1}} - \overline{\underline{x}}_{t-1} \tag{7}$$

Now, let's consider the scenario in which the particles are leaded correctly. Hence, $\overline{\underline{x}}_{t-1}$ approaches $\underline{p}best_{t-1}$. This means, Equation 7 can be rewritten in the following way:

$$\overline{\underline{x}}_t \cong \omega \underline{v}_{t-1} + \underline{g}best_{t-1} \tag{8}$$

One can simply observe that, Equation 8 could also be deduced by using the velocity function of Equation 3 in Equation 5 and repeating the above averaging steps. In other words, the modified velocity function reinforces a situation in which the particles have communally made a correct progress toward the global best solution. However, this doesn't guarantee that this PSO variant outperforms its conventional counterpart in all scenarios.

In the next step, a brief comparison of the conventional and modified algorithms when used in optimizing a set of well-known benchmark fitness functions is reviewed. The following benchmark functions are studied in many articles in the literature, such as in (Hoorfar, 2007; Nakano, et al., 2007):

Rastigrin Function:

$$f(x) = \sum_{i=1}^{n} \left(x_i^2 - 10 \cos(2\pi x_i) + 10 \right) \tag{9}$$

Griewank Function:

$$f(x) = \frac{1}{4000} \sum_{i=1}^{n} x_i^2 - \prod_{i=1}^{n} \cos\left(\frac{x_i}{\sqrt{i}} \right) + 1 \tag{10}$$

Rosenbrock Function:

$$f(x) = \sum_{i=1}^{n} \left[100\left(x_{i+1} - x_i^2\right)^2 + \left(x_{i-1} - 1\right)^2 \right]$$

$$(11)$$

All these three functions possess a global minimum that occurs at the origin of the solution space. In other words, the goal of optimization is to find through social cooperation of m particles in an n-dimensional space.

The assessment results of the following four methods are reviewed here. It is noteworthy that the initial random solutions should be kept identical for all methods during each run in order to have reliable comparison results.

Method 1: $c_1 = c_2 = 2$, and as shown in Equation 2 (conventional method)

Method 2: $c_1 = c_2 = 2$, and as shown in Equation 4

Method 3: $c_1 = 0, c_2 = 2$, and as shown in Equation 2

Method 4: $c_1 = 0, c_2 = 2$, and as shown in Equation 4

In (Modiri, et al., 2011), the convergence performances of the aforementioned methods were studied in 500 iterations averaged over 1000 runs, and the results were presented in terms of the average final value (AFV), and the average iteration number (AIN) required to attain a given AFV. The average convergence trend over iterations, and the distribution of the final values to which the algorithm has converged were also shown. A 10D scenario ($n=10$) was analyzed with $m= 5, 10$, and 20 particles. Each parameter was allowed to vary either over [-5,5], [-500,500], or [-50000,50000]. Optimization cost which was defined as the time required for the algorithm to converge to an acceptable solution (predefined by a threshold) was claimed to be the highest priority in the study. Due to the fact that decreasing the number of particles can shorten the optimization time considerably, no more than $2n$ particles were employed. A short summary of the conclusions is

depicted in this chapter. For more details, readers are referred to (Modiri, et al., 2011).

It was shown that, Methods 2, 3 and 4 outperformed the conventional one (Method 1) in terms of the convergence speed for Rastigrin function better than the two other functions. In terms of the best final value, Rosenbrock function ended up with more improved results. The very important point is that, the resulting error was only acceptable when the parameters were allowed to vary in a wide variation span. Figure 1 depicts the fitness function trend averaged over 1000 runs when $n=m=10$. Evidently, the modified methods get far better than the conventional one as the variation range becomes wider. The termination criterion was set to the maximum iteration number of 500. Interestingly, for Griewank function, all methods had similar performances. Let us, again, highlight an important fact at this point. That is, the performances of the EAs are problem-dependent.

In terms of the computation complexity, it is obvious that Methods 3 and 4 are lower in complexity due to elimination of a term from fitness function without any increase in the required number of iteration cycles; however, more features are needed to be compared. In Figure 2, the probability density functions of the final fitness values are plotted for Rastigrin. The results are achieved over 1000 runs. Each run is a 500-iteration procedure. In the figure, MFFV denotes the maximum final fitness value. To better illustrate, the four bars at 0.1MFFV show the probability associated to the final fitness value being less than or equal to 0.1MFFV. Each bar represents one of the variants. Similarly, the bars at 0.2MFFV show the probability of the final fitness value being less than or equal to 0.2MFFV and more than 0.1MFFV. By studying the distribution of the final values to which the algorithms converge, Modiri, et al. (2011) concluded that the probability of ending with lower final fitness values increased in the modified methods as the variation span assigned to the parameters increased. This means that the

Figure 1. Fitness values of the four introduced methods averaged over one thousand runs in a ten dimensional problem with ten particles for (a) Rastigrin, (b) Griewank and (c) Rosenbrock functions. The parameter range is considered to be (-50000,50000)

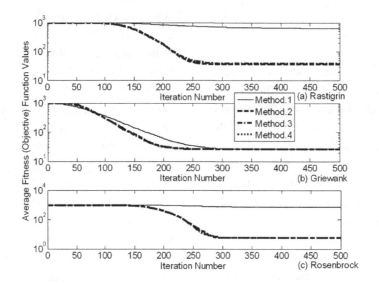

Figure 2. Probability density functions of the final fitness values for the four introduced variants considering one thousand runs in a ten dimensional problem for Rastigrin function with parameter ranges of (a) (-50000,50000), (b) (-500,500), and (c) (-5,5)

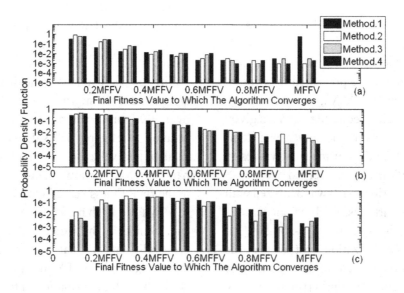

superiority of the convergence rate for the PSO variants is more pronounced in the scenarios where the solution space is wider. The interested reader is referred to (Modiri, et al., 2011), for more details. Similar trend is claimed to be observed by studying the other two functions.

Moreover, in order to have a set of results comparable to that presented by Nakano (2007), particle movements inside the solution space were traced in (Modiri, et al., 2011), and it was shown that the modifications offer the particles a greater degree of freedom in the initial stages of the exploration and faster convergence afterwards. Similar results were achieved when the absorbing wall boundary condition was used.

The aforementioned versions of PSO (Method 1 to Method 4) are further addressed in pursuing case studies. It is worth reminding the reader that the main goal of the chapter is to give a picture of the flexibility of the PSO algorithm and show how it can be redesigned according to the application. We are not trying to prove the viability of a single variant of PSO for all applications. Interested readers are also referred to other variations of PSO introduced in the literature, such as in (Ho, et al., 2008; Gao, et al., 2011; Luan, et al., 2012; Montes de Oca, et al., 2009).

PARTICLE SWARM OPTIMIZATION-RECONFIGURABLE RADIATORS

Case Study 1: PSO in Antenna Beam Forming

Previous section summarized the results achieved by some of the variations of PSO for mathematical benchmark problems. Here, modified BPSO is tested in a more realistic and demanding electromagnetic problem, namely, 'thinned array beam forming', using the velocity function modifications described in the previous section and presented in (Modiri, et al., 2011). The goal was to be able to block the radiation of an array antenna in a given direction in an expeditious manner.

Thinned arrays are periodic antennas with a number of non-illuminating elements. Such antennas are in demand for scenarios where software-defined pattern shaping is of interest; see (Jin, et al., 2007). The applications of this type of beam

forming were described in the Introduction of this chapter.

40-element and 200-element linear arrays, as well as and planar arrays, were studied in the aforementioned paper and the results of this study are summarized in the ensuing paragraphs.

BPSO, as it is obvious from its name, is quite attractive for problems with two-state parameters; see (Camci, 2008). However, in order to extend its utility to real-time smart applications, again, modifications are required. The algorithm is set to minimize the side lobe level (SLL) in the array broadside (as compared to the main-lobe strength). In other words, the following fitness function is to be minimized:

$$f = \max\left\{20 \log \left|array\,factor\right|\right\} \atop at\,desired\,direction\,or\,directions \tag{12}$$

The array factor formula can be found in many references, such as Equation 26 in (Jin, et al., 2007). AFV, AIN, the average convergence trend, and the distribution of the final values averaged over more than 700 runs were studied in (Modiri, et al., 2011) for this problem. Leaving the pure mathematical benchmark problems of previous section behind and tackling the realistic electromagnetic problems, it is shown that the complexity of the problem as well as the necessity of speeding up the design process increases.

Generally speaking, an increase in the number of particles increases the computation time while reducing the probability of failure. Therefore, it is observed that a problem-dependent trade-off is normally taken into account.

Interesting results are achieved when the previously mentioned four modifications are applied to BPSO. The performances of Method 2 and Method 4 in terms of AIN and AFV were shown to be superior to that of Method 1. However, Method 3 ended up with an inconsistent performance advantage over the conventional method. This modification (Method 3) was shown to even degrade the convergence performance. However, by reshaping Method 3 to arrive at Method 4, improved results were observed.

The reader should note that, since the distinction between RPSO and BPSO is mainly originated from the underlying quantization strategy (Jin, et al., 2007), by changing the quantization strategy, the results may also vary to some degree.

Furthermore, it is important to mention that, since in BPSO each parameter is either 1 or 0, the parameter range, which was an underscored factor in the previous section for RPSO, is quite limited.

Figure 3 demonstrates the convergence trend in the case of planar array when $m=2$. The dominance of Method 2 and Method 4 in terms of the convergence speed and final results, over the two other techniques are apparent. The radiation patterns are calculated in various φ-planes. It is also shown in (Modiri, et al., 2011) that a similar set of results emerge for the case of the absorbing wall boundary condition.

An interesting comparison of the optimization speed is also demonstrated in the aforementioned article. There, it is shown that, by adding two

more particles in the 6-φ-plane scenario, the optimization time jumps from 2.8 hours, on a standard 2.83 GHz and 3.23GB personal computer, to 5.25 hours. Therefore, the significance of a method, which is capable of achieving reliable results expeditiously, becomes blaringly clear.

Case Study 2: PSO in Frequency Selectivity

Yet as another demanding application, frequency selectivity of switchable microstrip antennas is studied in this subsection.

Microstrip antennas (MAs) are inexpensive and relatively easy to manufacture. They are also conformal and flexible in shape. Therefore, the demand for MA applications has been constantly growing. However, the design of MA structure using the available analytical techniques becomes too complicated when one goes beyond the regularly shaped patches. This restriction is imposed by the high computational complexity of irregular configurations. In recent years, irregular structures are becoming more attractive due to their poten-

Figure 3. Fitness value trends of the four introduced methods averaged over seven hundred runs in a sixteen dimensional problem considering two particles. Optimization study is performed in (a) six φ-planes, (b) eight φ-planes, and (c) ten φ-planes.

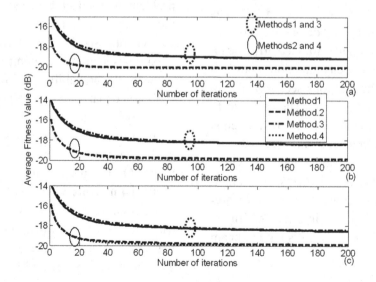

tial of achieving large bandwidths or operating in a multi-band environment. However, existing articles in the literature point to a weakness of the existing approaches; i.e., the lengthy design and optimization processes, see (Balanis, 2008). It is interesting though that, EAs, such as PSO, have also been used for the MA design, see (Jin, et al., 2008) as an example.

As it was shown in the previous section, PSO is easy to implement and control, making it exceptionally appealing for multi-dimensional electromagnetic problems, such as those introduced in (Jin, et al., 2005; Jin, et al., 2007; Camci, 2008). The modifications applied to the velocity vector by Modiri, et al. (2011) and explained in the previous sections are used in this section, as well. Here, the results of the cooperation between modified BPSO (implemented in MATLAB) and EM_SON-NET, which is an MOM[1] electromagnetic solver, are reviewed with the purpose of underscoring the performance improvement of a PSO-based optimization technique in problems where the real-time adjustment of the frequency response of the antenna is of paramount significance.

In fact, this problem possesses a complexity higher than that of the previous case study. Instead

of a relatively simpler fitness function calculation of Equation 12 in the previous problem, here the electromagnetic solver calculates the fitness function. To have a more reasonable analysis, the comparison results of the convergence behavior for the conventional and the modified algorithms are shown in terms of the average optimization trend. For a further discussion of the performance metrics of this technique, such as the particle activity and success rate, the reader is referred to (Modiri[2], et al. 2011; Modiri[3], et al. 2011).

Switchable multi-segment antennas are one of the most versatile reconfigurable antennas due to the fact that their frequency responses can be simply varied by connecting some of the segments to the main antenna body and disconnecting the others. These types of antennas are widely used where frequency-selectivity is required. Yet, the reliability of the switching operation remains an issue.

Figure 4 depicts the 15-segment MA, etched over a *3mm* duroid substrate, which was analyzed in (Modiri[2], et al. 2011). The antenna feeding connection was performed through the middle patch, marked by a dot in the figure. The optimization algorithm does not make any changes to

Figure 4. The fifteen and twenty segment microstrip antennas used for simulation purpose are shown in (a). The best design achieved by the conventional and modified methods for the fifteen-segment scenario is shown in (b). (c) and (d) show the best designs of twenty-segment antenna achieved by the conventional and modified methods, respectively.

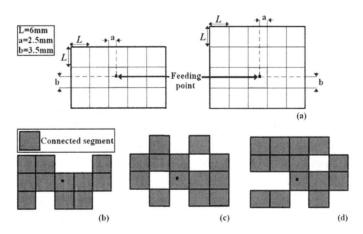

this middle segment. In this article, BPSO was employed to design irregular structures out of both 15-segment and 20-segment MAs (Figure 4). In order to remove a segment, BPSO assigns 0s for both the segment length and width, resulting in the disconnection of the patch from the main antenna body in SONNET simulation. In this example, each segment was assumed to be a *6×6* mm[2] patch. The demonstrated results are obtained by averaging over 50 runs[2]. A magnitude of -10 dB for the reflection coefficient, S_{11}, at the design frequency (or frequencies) of interest is assumed as the success criterion. Also, in order to keep the optimization length reasonable for a real-time application, a maximum iteration number of T=50 is imposed as the termination criterion.

It is noteworthy to mention that Jin (2005) has recommended a parallel strategy in order to expedite the MA design procedure using PSO. There, FDTD[3] solver was used along with PSO. However, the parallel processing, although extremely effective in common applications, may not be valid for most software-defined reconfigurable structures were portable platforms are considered. The reason is that, parallel strategy requires large memory and processing capacity, which are difficult conditions to satisfy for the case of small size portable devices, which actually are the main focus of this chapter.

Both single-frequency and multiple-frequency resonances are studied in (Modiri[2], et al., 2011) using Equation 13 as the objective function.

$$f = \sum_{i=1}^{N} S_{11(at\,f_i)} \tag{13}$$

where N depicts the number of resonant frequencies considered. It should be noted that, although Equation 9 seems to be simple and trivial, S_{11} (reflection coefficient) cannot even be written in the form of closed formulae for irregular shaped patches. This renders this problem mathematically intractable.

Figure 5. (a) Average fitness value trend and (b) the best optimization results achieved by the conventional and the modified BPSO algorithms using seven-particle swarm

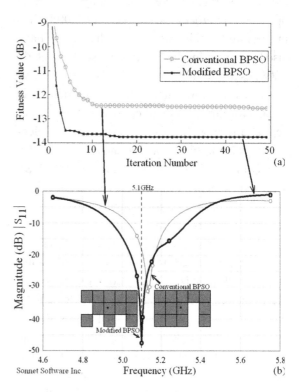

The frequency response and convergence trends of the conventional BPSO and the modified one are shown in Figure 5 when considering single frequency resonance at f_i=5.1 GHz. BPSO employs 7-particle swarm in this example. A close look at Figure 5 reveals the convergence superiority of the modified BPSO over the conventional one. Regardless of the termination point, the modified BPSO offers a gain of at least 1.21 dB over the conventional one. More notably, it meets the design success condition of S_{11}=-10 dB faster than its counterpart with an average speed improvement of 50%. This result is of great interest to those who are engaged in the design of reconfigurable antennas with the computation time as a major constraint.

The final irregular shaped structures associated with the best designs, achieved by the two methods, are also shown in Figure 5. The discrete probability density functions associated with the final fitness values, AFV, and AIN are also studied in (Modiri[2], et al., 2011). In this study, it is shown that the modified methods presented here achieve higher success rate. Furthermore, it was shown that the improvements stay valid for triple resonant design (4.8 GHz, 5 GHz and 5.2 GHz) as well.

Obviously, a triple frequency scenario presents higher levels of computation complexity. For instance, looking back at Equation 13, it is clear that in a multi-resonance scenario more care should be taken in order to avoid the algorithm from getting locked in a cycle which reduces S_{11} at one of the frequencies while leaving the other frequencies with unacceptable S_{11} values. As a solution to this problem, any S_{11} value of less than -15 dB is fixed at -15 dB in (Modiri[2], et al., 2011). As a result, the swarm would stop reducing S_{11} at the frequencies for which -15 dB has already been achieved. It is shown there that the modified algorithm outperforms the conventional one in terms of the convergence speed, assuming a termination criterion of *T=50* iterations. The two algorithms converge to identical configurations for this case. This configuration, along with the two other configurations, achieved in a single-resonance 20-segment MA, is shown in Figure 4b and 4c. For a more comprehensive study of the optimization time, a run-time table was presented in (Modiri[2], et al., 2011), in which the noticeable impact of the initial values on the optimization length was highlighted.

Case Study 3; PSO in Bend Effect Compensation

In this subsection, one more application of PSO, which is an interesting antenna design problem, is investigated. Wearable antennas (WAs), as the name implies, are intended to be a part of everyday clothing. The earliest demand for WAs was estab-lished by military with the purpose of reducing the chance of the identification of a radioman within a squad. The conventional military communications person often has to carry long antennas and such antennas are visible from a distance (Hall, et al., 2006). Later on, WAs received wide interest for sports, emergency, space, and medical applications too. Among these, medical applications are the most demanding so far. As a critical components of body area networks (BANs), WAs need to maintain communications among in-body (implanted), on-body (sensors), and off-body (base stations) devices. Therefore, WAs have been developed with a variety of frequency and radiation specifications. Nevertheless, it has been shown that several factors can deteriorate the performance of a WA, and ultimately, create health hazards in telemedicine applications. Thus, a serious need for real-time reconfigurable WAs has emerged, especially for scenarios where changing posture due to movement imposes serious performance degradation. According to the application requirements, on-body textile antennas need to be flexible and conformal in structure, light in weight, and reasonably small in size. Therefore, MAs seem to be one of the best candidates for such applications.

Considering BANs, two performance criteria are generally expected to be met:

1. The patient comfort should be satisfied, especially in long term monitoring situations.
2. The antenna should preserve an acceptable performance for various performance-degrading scenarios. Wrinkles and bends on the antenna surface or moisture absorbed by substrate textile are some of such scenarios (Hall, et al., 2006).

Textile materials, in general, have low dielectric constants (below 2), and, as a result, a textile MA potentially has a relatively large bandwidth which can compensate for slight variations in the

Figure 6. Measurement arrangements for the wearable antenna in (a) flat, and (b) bent modes. Switchable antenna with compensating parasitic element in bent mode is shown in (c).

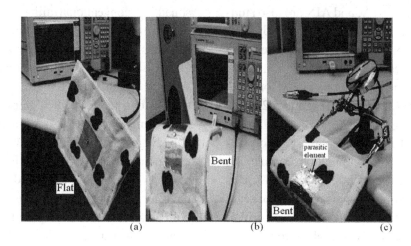

resonant frequency. However, the environmental variations may easily go beyond this limitation.

Here, the effect of one of these variations, i.e., bending, is studied. Other variations can be treated in a similar way. Again, owing to the algorithm's flexibility and high convergence speed, modified BPSO is used here in order to reconfigure a 2.45GHz[4] MA in a real time manner. In (Modiri[3], et al., 2011), a patch antenna was proposed as the 2.45 GHz reconfigurable WA and was simulated in EM Sonnet. BPSO kernel was implemented in MATLAB.

At this point, let us have an in depth discussion of practical details of a few implemented reconfigurable WAs. In (Hall, et al., 2006), a 56mm × 51mm textile patch antenna was fabricated on a 3mm fleece substrate with resonant frequency of 2.45 GHz. The ground plane size was considered to be 76mm × 71mm and the coaxial feeding point was located at the distance of 19 mm from the edge. The same textile antenna was re-fabricated by Modiri[3], et al., and measured using Agilent E5071C VNA[5], see Figure 6. According to the measurements, it was claimed that by creating bends of 45 and 90 degrees on the antenna structure frequency jumps as large as 144 MHz and 250MHz, respectively, would occur in the

frequency behavior of the antenna. It is worth noting that a structure modification (i.e., bending) in only one dimension has resulted in such significant deviation in the frequency response of the MA. In order to compensate for this undesired frequency deviation, parasitic narrow patch elements were appended to the main antenna body. Once bending has occurred, some of these parasitic elements would be connected to the main antenna body via switches in order to enlarge the actual length of the antenna, and thus, compensate for the shrinkage which has occurred in the electrical length of the element.

Experimental verification can be found in (Modiri[3], et al., 2011). Two scenarios were investigated in the aforementioned reference. In one scenario, the reconfigurable antenna consisted of three switches and a single parasitic element. While in the other one, eight switches were considered along with three parasitic elements. Of course, the higher the number of switches and parasitic elements, the better would be the optimization performance of the frequency response. Yet, it is impractical to integrate a large number of switches in a wearable device.

The overall idea of this case study is quite similar to the previous one, although they differ

Figure 7. $|S_{11}|$ measurement results of the flat and bent modes for the eight-switch, three-element scenario

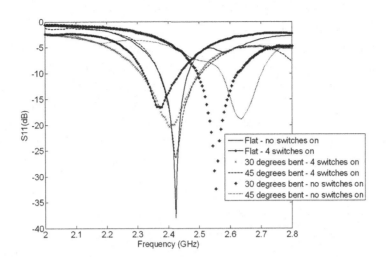

in details and application. In the current scenario, a monitoring system is required to check the signal power at the desired resonant frequency by sampling the power periodically. As soon as the power falls below a predefined threshold (due to some environmental variations), BPSO is invoked in order to find the best switching arrangement to retrieve the original operation frequency. The objective function in these types of problems is typically the reflection coefficient at the desired frequency, which, of course, should be minimized, see Equation 13.

Yet, as another solution to this problem, it seems feasible to generate a bank of predefined switching arrangements to compensate for a set of environmental variations. However, in practice, it is quite possible that an unexpected variation, such as a twisting (multi-dimensional bend), can occur for which no solution is provided. Moreover, such a solution can only be valid if it is customized to a single person due to the exclusive personal body styles. Therefore, the first solution, i.e., real-time random decision making, was chosen by Modiri[3], et al., (2011).

To give the readers an idea of the final results achieved by the algorithm, Figure 7 shows the retrieved frequency response in the case of 30 and 45 degree single-dimensional bends when 8-switch, 3-element scenario is considered. Since, in practice, reflection coefficient magnitude[6] of -10 dB is acceptable in most antenna applications, a success threshold of -10 dB was imposed in (Modiri[3], et al., 2011). It was also shown therein, that, overall, the modified BPSO outperformed the conventional one in maintaining the desired frequency performance. This is a result similar to that of the case studies 1 and 2. More details, such as convergence trends associated with the conventional and modified BPSOs, distribution curves related to the final fitness values, and frequency response curves of the two studied scenarios, can be found in (Modiri[3], *et., al.,* 2011).

PARTICLE SWARM OPTIMIZATION- CANCER DETECTION TECHNIQUES

In the last part of this chapter, PSO algorithm is tested in a more complex and challenging problem where the makeup of an unknown multilayer structure has to be estimated. Gandhi (2010) and Yeung (2009) successfully used PSO to estimate the composition of a multi-layer structure using simulations. The object under the test was exposed to electromagnetic radiation and the scattering field was processed. In these studies, the medium under the test was assumed to either be lossless or low-loss in order to decrease the calculation complexity. However, the application of PSO in these types of problems is not limited to low-loss structures. This concept can, in fact, be utilized in more demanding and, at the same time, interesting fields, such as tumor detection via imaging. As shown by Golnabi (2011), a typical microwave imaging system consists of an array of antennas, which encircle the breast tissue. The antennas radiate the electromagnetic energy, one at a time, and the receivers gather the scattered field. A set of measurements is performed at distinct frequencies and/or different transmitter positions. The inputs to the algorithm are these measurement results. Then, through a challenging inverse problem analysis, the algorithm has to get relatively close to an acceptable guess for the tissue structure. Here, we particularly investigate the breast cancer detection through microwave imaging. The makeup of the tissue layers can be estimated with respect to their dielectric constants. Therefore, the optimization goal can be simplified in words to 'estimating the dielectric constants of the tissue layers'.

Similar to many other cancer cases, it is well proven that the early detection of the breast cancer drastically raises the survival rate of the patient. Moreover, the treatment and recovery becomes more tolerable, both physically and financially, when the tumor is detected soon enough. Although, X-ray mammography is available as the most recommended breast cancer detection technique, it has a dismal record of 20% detection failure, which cannot be easily ignored (Lazebnik, et al., 2007). In addition, X-ray is ionizing, rendering this modality somewhat unsafe for scenarios where frequent checkups are required or highly recommended, see (Hassan, et al., 2011; Nikolova, 2011).

There have been thorough attempts toward developing more convenient, non-invasive breast cancer detection techniques that pose little risk to the patient when routine checkups are performed. Ultrasound and MRI seem to be two promising examples. However, these two techniques are either low in accuracy, in the case of ultrasound, or exceedingly expensive, in the case of MRI, when compared with X-ray mammography. In other words, they cannot compete with X-ray in cost/accuracy, especially when routine checkups come into the picture. That is why they are not used as widely as X-ray in clinical practice for breast cancer.

Moreover, remarkable initial studies have been published, introducing more innovative methods of cancer detection at microwave, near-infra-red, optical, and tera-hertz frequency bands, see (Sabouni, 2010; Gamagami, 1997; Lazebnik, 2008; Khan, 2007). Reviewing these articles sheds light on the promise of microwave imaging as the diagnostic tool of choice for a non-invasive, safe, and accurate detection of breast cancer. The reason behind this fact is, simply, the practical aspects of electromagnetic imaging techniques. For instance, one important factor is the detection of tumors deep within the tissue. It is well known that, higher is the radiation frequency, smaller will be the penetration depth inside the human body. Therefore, higher frequency radiation (optics and terahertz) can hardly reach subcutaneous tumors. Also, the standard sizes of the electrical components at microwave frequencies are small enough to make a non-invasive, portable detector feasible. Yet as another fact, microwave radiation is quite responsive to the liquid content of the tissue. This would be of significant importance since the overall ratio of the blood flow in the breast tumor is between 4.7 and 5.5 times larger than that of the normal tissue (Dellile, et al., 2002). As a result, employing non-ionizing microwave radiation in cancer detection and treatment monitoring has received a great deal of attention. A list of research studies in this area can be found in (Lazebnik, et al., 2007). However, a hybrid technique, may ultimately be shown to be the most effective means of detecting breast cancers, see (Paulsen, et al., 2005).

Reviewing the existing articles on the subject of breast cancer detection using microwave radiation, one can identify the following key observations:

1. A lack of large-scale, in-vivo measurements for both validation and/or revision of the theoretically proposed techniques.
2. Low success rates of malignancy detection in non-superficial areas of the breast, even in the simulation stage, especially, when distinguishing between glandular tissue and tumor.

A closer examination of the above factors reveals that the first factor is partly created by the second one. That is, in order to conduct extensive in-vivo measurement studies, one has to obtain approval from regulatory bodies, which rely heavily on theoretical viability of the propose techniques.

In Modiri[4], et., al. (2011), it is argued that the inadequacy in modeling the tissue is the main culprit behind one arriving at low success rate in detecting cancerous tissues. In particular, non-negligible variations in the tissue characteristics from person to person, and even in a single person at different times of the day or under diverse experimental conditions, must be taken into account for the purpose of theoretical modeling. In one of the largest scale studies, reported by Lazebnik (2007), 807 samples[7] were investigated over the frequency band of 0.5 GHz-20 GHz. There, it was practically demonstrated that the electrical properties of the tissue extend over a wide range. Cho (2006) also illustrated that these electrical properties were significantly influenced by the measurement setup. Therefore, Kiasaleh, in Modiri[4], et al., (2011), proposed modeling permittivities as random variables. In addition, PSO, as a random-based EA algorithm, was chosen to find the breast tissue composition through solving the inverse scattering problem. It should be noted that dealing with inverse problems with a large number of unknowns has its inherent difficulty of arriving at non-unique solutions. Hence, the unknowns have to be managed carefully.

In order to determine the solution technique, we need to model the measurement setup. It should be noted that, typically, array radiators are used for cancer detection, see (Golnabi, et al., 2011). Therefore, the electromagnetic analysis can be simplified to the interaction of a narrow beam with a multi-layer structure, and this can be further simplified to planar wave scattering analysis in a lossy multilayer structure (Orfanidis, 2008; Thakur, et al., 2002).

As it is shown in Equation 14, the accumulated error, considering both amplitudes and phases of the reflection and transmittance coefficients, are utilized as the objective function (f) of the search algorithm. N_m is the total number of measurements. R and T are the reflection and transmittance coefficients, respectively.

$$f_R = \sum_{1}^{N_m} \{ | \, |R_{measured}| - |R_{estimated}| \, | + | \, arg(R_{measured}) - arg(R_{measured}) \, |$$

$$f_T = \sum_{1}^{N_m} \{ | \, |T_{measured}| - |T_{estimated}| \, | + | \, arg(T_{measured}) - arg(T_{measured}) \, |$$

$$f = f_R + f_T \qquad (14)$$

where 'arg' refers to the phase operator.

To illustrate how PSO can be used in such a problem, let us review the methodology and summarize the results demonstrated in (Modiri[4], et al., 2011) in the next subsection.

Modeling and Estimation Methodology

The first step in solving this problem is to generate a model for the object under the test. The breast tissue model was generated with respect to the breast physiology, available on Internet in (Modiri[4], et al.,2011). The breast model structure is composed of glandular tissue, fat, and skin. For the case of cancerous tissue, tumor should be also added to the picture. As shown by Lazebnik (2007), the electrical properties of normal breast tissue are primarily dependent on its adipose[8] content. Therefore, the model shown in Figure 8 takes the adipose contents into account. According to the breast physiology, various models of breast, in terms of the number of layers, can be suggested. This, in fact, defines the accuracy and computation simplicity levels. It is also required to assume a value as the total thickness of the tissue. Of course, this parameter is not a fixed value for all subjects; nevertheless, 4.4 cm was used in (Modiri[4], et al., 2011). Since the breast is typically placed between the transmitter and receiver, which would slightly squeeze the tissue, 4.4 cm was determined to be reasonable. The microwave radiation was analyzed in the frequency band of 1-2.25 GHz.

The following two scenarios were studied:

1. Assuming there was no a priori knowledge available of the type and composition of the tissue under the test.
2. Assuming there was some reasonable a priori knowledge of the tissue under the test.

Of course, the first scenario was the most difficult one in terms of the computation complexity. It should be noted that, the complexity of the problem is not due to the fitness function calculation. It is mainly due to the difficulty of solving the inverse problem in such a large solution space.

Two PSO variants were proposed for the two desired scenarios. In both scenarios c_1 and

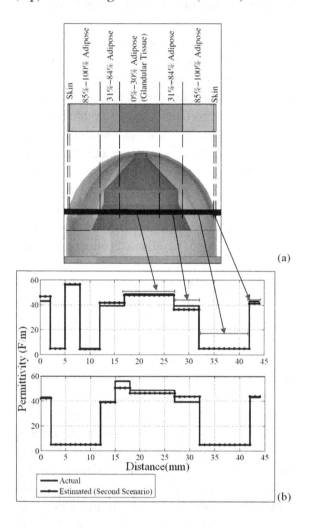

Figure 8. (a) Simplified two dimensional model of the breast tissue; (b) Actual permittivity values and the estimated ones in the second scenario, assuming a 3 mm thick tumor in the fatty area (top) and in the glandular area (bottom)

c_2 are assumed to be equal to 2. ω is set to 1 for discrete PSO and gradually decreasing from 0.9 to 0.4 for the complex one. Here, it is noteworthy to emphasize again the problem dependency of PSO variants. The variant introduced for the previous case studies was not the best choice for this case study.

To better illustrate, in scenario 1, due to the lack of a priori knowledge, the random variables representing the dielectric constants could possess any values between the lowest possible and largest possible constants related to the human body tissues. In this case, Complex number PSO (CPSO) was used for which the variation range was relatively wide; CPSO is similar to RPSO, except for dealing with complex numbers instead of real ones. In fact, by implementing this scenario, it was shown that PSO could be utilized in estimation of the composition of any tissue. This can be achieved if the penetration depth and receiver sensitivity satisfy the detection requirements.

In the second scenario, it was known that the tissue under the test is the breast tissue. There, at each iteration cycle, two optimization procedures were performed. First, for each layer, a tissue type was randomly assigned out of a set of six tissues, including blood, skin, fat-dominated tissue, glandular tissue, 31-84% fatty tissue, and malignant tissue (Lazebnik[2], et al.,2007). These are, in fact, the six tissue types, which can exist in human breast. Each of these six classes was represented by a random variable whose variation range was defined according to the measurements reported in the literature.

In the second optimization step of scenario 2, after defining the arrangement of the tissues, the exact value of the dielectric constant related to each tissue layer was chosen out of its corresponding variation range, so that the optimum matching to the measured scattering fields could be achieved. For this scenario, in order to better manage the problem, discrete PSO (DPSO) was proposed. In principle, DPSO is similar to BPSO. The differences reside in the following two points:

1. The number of data levels is two for BPSO, while it is six for the aforementioned DPSO.
2. The values, which can be assumed at each level, are restricted to 0 and 1 in BPSO, while it is extended to a continuous variation span in the above DPSO.

For more detailed information, the interested reader is referred to (Modiri[4], et., al.,2011). The permittivity estimation results in the case of 7-layer modeling when DPSO is employed are shown in Figure 8. A 3 mm thick tumor was added to the model in two distinct positions. First, the tumor was added inside the fatty tissue, and then it was moved to the glandular tissue. Obviously, due to similarity of the permittivity values of the glandular tissue and the tumor, it is somewhat difficult for the swarm to make an accurate estimation for the second case. Nonetheless, DPSO can successfully identify the abnormal heterogeneity in the tissue characteristic in both cases. Figure 9 shows the comparison results for the two algorithms along with the convergence trend. As expected, it was shown that not only DPSO algorithm finished the estimation procedure much faster than CPSO, but also it offered a more acceptable estimation of the tissue composition.

Figure 9. (a) Optimization cost and (b) average fitness function trends for the two scenarios

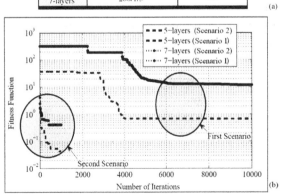

Layers (Scenario)	Number of agents/frequencies/angles	Convergence time (min)
Scenario 1 5-layers	40/2/2	7.81
Scenario 1 7-layers	280/6/3	185.3
Scenario 2 5-layers	40/1/2	0.7
Scenario 2 7-layers	280/1/3	9.2

FUTURE RESEARCH DIRECTIONS

The authors and other research groups have studied a variety of complex electromagnetic problems using swarm-based algorithms and a variety of modifications to the conventional format of the algorithm have been introduced. The complexity of the investigated problems and similar problems, which were not mentioned herein, resides in their wide solution space, random nature of the parameters and lack of closed analytical formulae.

A set of electromagnetic problems, which belong to the emerging applications, were analyzed and reviewed in this chapter. An appropriate PSO variant was also introduced for each case. Although it was shown that PSO, in its various versions, could successfully handle the desired problems, further studies to optimize the algorithm for distinct classes of problems independently is needed. Particularly, in the case of cancer detection scenario described in the last section of this chapter, there is a lot of room for improvement. PSO variants are problem-dependent; therefore, in the cancer detection scenario, for instance, the algorithm may need to be re-optimized each time the tissue model is updated to a more accurate one. Moreover, PSO variants can be used in other data processing applications in a similar way.

A comprehensive comparison study of the performance efficiency of various evolutionary algorithms in a single scenario is another invaluable research topic. Although some studies have been done in the field of electromagnetic component design in this regard, there is a lot more to be done, especially, in the more recent field of cancer detection.

CONCLUSION

In this chapter, particle swarm optimization algorithm was used in several emerging electromagnetic applications. The main idea of this study was to demonstrate the flexibility and strength of the PSO-based algorithms in handling complex electromagnetic problems. What makes these problems 'complicated' is that a large number of unknowns with random nature are present, leading to intractable analytical analyses. In those scenarios, numerical electromagnetic methods have to be used instead of the analytical ones. A group of problems and their PSO-based solutions were briefly summarized. It was shown that, for each problem, the algorithm could be exclusively modified from its original format in order to better satisfy the requirements of the problem at hand. Each of the problems investigated here possessed different levels of complexity and was a representative of a larger class of problems. The imposed modifications were presented in order to offer more desirable results for the case of benchmark mathematical optimizations, as well as the reconfigurable antennas.

In the case of reconfigurable antennas, beam-forming, frequency-selectivity, and self-tuning were studied in thinned antenna arrays, switchable multi-segment antennas, and wearable antennas, respectively. This study, in fact, clearly demonstrated the strength of a swarm-based algorithm in handling analytically intractable problems, as well as the superiority of the modified PSO algorithm over the conventional one. The modifications resulted in better performance, particularly in terms of the convergence speed and, in most applications, the final optimized results. A comparative study of the optimization cost was performed as well.

In the case of wearable antennas, this chapter demonstrated the reliability of the PSO algorithm in compensating for the bend effect, as a sample environmental variation, through real-time self-tuning operation.

Finally, in the last section of this chapter, PSO was used in tissue composition estimation with the goal of cancer detection. Two versions of the algorithm, complex and discrete PSOs, i.e., CPSO and DPSO, were introduced and used to handle two different scenarios; namely 1) when the algorithm is given no a priori knowledge of the tissue under investigation, and 2) when the algorithm is given some reasonable knowledge of the tissue (for instance, when it is known that the tissue under the test is breast and, consequently, nonrelated tissues, such as bone, should not be included in modeling). It was shown that PSO successfully identified the difference between the permittivity values of the tissue layers in both cases. Obviously, the accuracy increases in the second scenario where side information is available.

To summarize, it can be concluded that a swarm-based algorithm can effectively handle problems with the following properties:

- Possessing variables with intrinsic random nature.
- Large number of unknowns as compared to the known data.
- Time-sensitive.

Moreover, through analyzing different types of problems, it was shown that the performance of PSO algorithm, in particular, is problem-dependent and, hence, for an optimum performance, one has to carefully redesigned the algorithm to meet the unique challenges of the problem at hand.

REFERENCES

Afshinmanesh, F., Marandi, A., & Shahabadi, M. (2008, July). Design of a single-feed dual-band dual-polarized printed microstrip antenna using a Boolean particle swarm optimization. *IEEE Transactions on Antennas and Propagation*, *56*(7), 1845–1852. doi:10.1109/TAP.2008.924684

Balanis, C. A. (2008). *Modern antenna handbook*. Hoboken, NJ: Wiley-Interscience. doi:10.1002/9780470294154

Bayraktar, Z., Werner, P. L., & Werner, D. H. (2006, December). The design of miniature three-element stochastic Yagi-Uda arrays using particle swarm optimization. *IEEE Antennas and Wireless Propagation Letters*, *5*(1), 22–26. doi:10.1109/LAWP.2005.863618

Benedetti, M., Azaro, R., Franceschini, D., & Massa, A. (2006, December). PSO-based real-time control of planar uniform circular arrays. *IEEE Antennas and Wireless Propagation Letters*, *5*(1), 545–548. doi:10.1109/LAWP.2006.887553

Boeringer, D. W., & Werner, D. H. (2004). Particle swarm optimization versus genetic algorithms for phased array synthesis. *IEEE Transactions on Antennas and Propagation*, *52*(3), 771–779. doi:10.1109/TAP.2004.825102

Camci, F. (2008, August). *Analysis of velocity calculation methods in binary PSO on maintenance scheduling*. Paper presented at the First International Conference on the Applications of Digital Information and Web Technologies, Ostrava, Czech Republic.

Carlisle, A., & Dozler, G. (2002, December). *Tracking changing extrema with adaptive particle swarm optimizer*. Paper presented at the 5th Biannual World Automation Congress, Orlando, USA.

Chamaani, S., Abrishamian, M. S., & Mirtaheri, S. A. (2010, May). Time-domain design of UWB Vivaldi antenna array using multiobjective particle swarm optimization. *IEEE Antennas and Wireless Propagation Letters*, *9*, 666–669. doi:10.1109/LAWP.2010.2053691

Chen, W. N., Zhang, J., Chung, H. S. H., Zhong, W. L., Wu, W. G., & Shi, Y. H. (2010, April). A novel set-based particle swarm optimization method for discrete optimization problems. *IEEE Transactions on Evolutionary Computation*, *14*(2), 278–300. doi:10.1109/TEVC.2009.2030331

Cho, J., Yoon, J., Cho, S., Kwon, K., Lim, S., & Kim, D. (2006, August). In-vivo measurements of the dielectric properties of breast carcinoma xenografted on nude mice. *International Journal of Cancer, 119*(3), 593–598. doi:10.1002/ijc.21896

Clerc, M., & Kennedy, J. (2002, February). The particle swarm—Explosion, stability and convergence in a multidimensional complex space. *IEEE Transactions on Evolutionary Computation, 6*(1), 58–73. doi:10.1109/4235.985692

Cui, S., & Weile, D. (2005, November). Application of parallel particle swarm optimization scheme to the design of electromagnetic absorbers. *IEEE Transactions on Antennas and Propagation, 53*(11), 3616–3624. doi:10.1109/TAP.2005.858866

Delille, J. P., Slanetz, P. J., Yeh, E. D., Kopans, D. B., & Garrido, E. (2002, May). Breast cancer: Regional blood flow and blood volume measured with magnetic susceptibility-based MR imaging-initial results. *Radiology, 223*(2), 558–565. doi:10.1148/radiol.2232010428

Eberhart, R., & Shi, Y. (2001, May). *Particle swarm optimization: Developments, applications, and resources.* Paper presented at the Congress on Evolutionary Computation, Seoul, Korea.

Gamagami, P., Silverstein, M. J., & Waisman, J. R. (1997, November). *Infra-red imaging in breast cancer.* Paper presented at the proceedings of the 19th Annual International Conference of the IEEE Engineering in Medicine and Biology Society, Chicago, USA.

Gandhi, K. R., Karnan, M., & Kannan, S. (2010, February). *Classification rule construction using particle swarm optimization algorithm for breast cancer data sets.* Paper presented at the International Conference on Signal Acquisition and Processing, Singapore, Singapore.

Gao, H., & Xu, W. (2011, October). A new particle swarm algorithm and its globally convergent modifications. *IEEE Transactions on Systems, Man, and Cybernetics. Part B, Cybernetics, 41*(5), 1334–1351. doi:10.1109/TSMCB.2011.2144582

Gies, D., & Rahmat-Samii, Y. (2003, August). Particle swarm optimization for reconfigurable phase-differentiated array design. *Microwave and Optical Technology Letters, 38*(3), 172–175. doi:10.1002/mop.11005

Golnabi, A. H., Meaney, P. M., Epstein, N. R., & Paulsen, K. D. (2011, August). *Microwave imaging for breast cancer detection: Advances in three-dimensional image reconstruction.* Paper presented at the Annual International Conference of the IEEE on Engineering in Medicine and Biology (EMBC), Boston, MA.

Hall, P., & Hao, Y. (2006). *Antennas and propagation for body-centric wireless communications.* Norwood, MA: Artec House, INC.

Hassan, A., & El-Shenawee, M. (2011, September). Review of electromagnetic techniques for breast cancer detection. *IEEE Reviews in Biomedical Engineering, 4*, 103–114. doi:10.1109/RBME.2011.2169780

He, N., Xu, D., & Huang, L. (2009, August). The application of particle swarm optimization to passive and hybrid active power filter design. *IEEE Transactions on Industrial Electronics, 56*(8), 2841–2851. doi:10.1109/TIE.2009.2020739

Ho, S. Y., Lin, H. S., Liauh, W. H., & Ho, S. J. (2008, March). OPSO: Orthogonal particle swarm optimization and its application to task assignment problems. *IEEE Transactions on Systems, Man, and Cybernetics. Part A, Systems and Humans, 38*(2), 288–298. doi:10.1109/TSMCA.2007.914796

Hoorfar, A. (2007, March). Evolutionary programming in electromagnetic optimization: A review. *IEEE Transactions on Antennas and Propagation, 55*(3), 523–537. doi:10.1109/TAP.2007.891306

Hwang, K. C. (2009, September). Design and optimization of a broadband waveguide magic-T using a stepped conducting cone. *IEEE Microwave and Wireless Components Letters, 19*(9), 539–541. doi:10.1109/LMWC.2009.2027052

Ismail, T. H., & Hamici, Z. M. (2010, June). Array pattern synthesis using digital phase control by quantized particle swarm optimization. *IEEE Transactions on Antennas and Propagation, 58*(6), 2142–2145. doi:10.1109/TAP.2010.2046853

Jin, N., & Rahmat-Samii, Y. (2005, November). Parallel particle swarm optimization and finite difference time-domain (PSO/FDTD) algorithm for multiband and wide-band patch antenna designs. *IEEE Transactions on Antennas and Propagation, 53*(11), 3459–3468. doi:10.1109/TAP.2005.858842

Jin, N., & Rahmat-Samii, Y. (2007, March). Advances in particle swarm optimization for antenna designs: real-number, binary, single-objective and multiobjective implementations. *IEEE Transactions on Antennas and Propagation, 55*(3), 556–567. doi:10.1109/TAP.2007.891552

Jin, N., & Rahmat-Samii, Y. (2008, July). *Particle swarm optimization for multi-band handset antenna designs: A hybrid real-binary implementation.* Paper presented at IEEE International Symposium of Antennas and Propagation, Taipei, Taiwan.

Kennedy, J., & Eberhart, R. (1995, December). *Particle swarm optimization.* Paper presented at IEEE International Conference of Neural Networks, Piscataway, NJ.

Khan, U. A., Al-Moayed, N., Nguyen, N., Korolev, K. A., Afsar, M. N., & Naber, S. P. (2007, December). Broadband dielectric characterization of tumorous and nontumorous breast tissues. *IEEE Transactions on Microwave Theory and Techniques, 55*(12), 2887–2893. doi:10.1109/TMTT.2007.909621

Ko, C. N., Chang, Y. P., & Wu, C. J. (2009, February). A PSO method with nonlinear time-varying evolution for optimal design of harmonic filters. *IEEE Transactions on Power Systems, 24*(1), 437–444. doi:10.1109/TPWRS.2008.2004845

Lazebnik, M., Changfang, Z., Palmer, G. M., Harter, J., Sewall, S., Ramanujam, N., & Hagness, S. C. (2008, October). Electromagnetic spectroscopy of normal breast tissue specimens obtained from reduction surgeries: Comparison of optical and microwave properties. *IEEE Transactions on Bio-Medical Engineering, 55*(10), 2444–2451. doi:10.1109/TBME.2008.925700

Lazebnik, M., Popovic, D., McCartney, L., Watkins, C. B., Lindstorm, M. J., & Harter, J. (2007, April). A large-scale study of the ultrawideband microwave dielectric properties of normal, benign and malignant breast tissues obtained from cancer surgeries. *Physics in Medicine and Biology Journal, 52*(20), 6093–6115. doi:10.1088/0031-9155/52/20/002

Liu, W. (2005, October). Design of multiband CPW-fed monopole antenna using a particle swarm optimization approach. *IEEE Transactions on Antennas and Propagation, 53*(10), 3273–3279. doi:10.1109/TAP.2005.856339

Luan, F., Choi, J. H., & Jung, H. K. (2012, February). A particle swarm optimization algorithm with novel expected fitness evaluation for robust optimization problems. *IEEE Transactions on Magnetics, 48*(2), 331–334. doi:10.1109/TMAG.2011.2173753

Matekovits, L., Mussetta, M., Pirinoli, P., Selleri, S., & Zich, R. (2005, July). *Improved PSO algorithms for electromagnetic optimization*. Paper presented at IEEE Antennas and Propagation Society International Symposium, Washington, DC.

Mikki, S., & Kishk, A. (2005, July). *Investigation of the quantum particle swarm optimization technique for electromagnetic applications*. Paper presented at IEEE Antennas and Propagation Society International Symposium, Washington, DC.

Modiri, A., & Kiasaleh, K. (2010, May). *Efficient design of microstrip antennas using modified PSO algorithm*. Paper presented at the 14th Biennial IEEE Conference on Electromagnetic Field Computation (CEFC), Chicago, IL.

Modiri, A., & Kiasaleh, K. (2011, January). Modification of real-number and binary PSO algorithms for accelerated convergence. *IEEE Transactions on Antennas and Propagation, 59*(1), 214–224. doi:10.1109/TAP.2010.2090460

Modiri, A., & Kiasaleh, K. (2011, May). Efficient design of microstrip antennas for SDR applications using modified PSO algorithm. *IEEE Transactions on Magnetics, 47*(5), 1278–1281. doi:10.1109/TMAG.2010.2087316

Modiri, A., & Kiasaleh, K. (2011, January). *Real time reconfiguration of wearable antennas*. Paper presented at IEEE Topical Conference on Biomedical Wireless Technologies, Networks & Sensing Systems, Phoenix, AZ, USA.

Modiri, A., & Kiasaleh, K. (2011, September). *Permittivity estimation for breast cancer detection using particle swarm optimization algorithm*. Paper presented at the 33rd Annual International Conference of the IEEE Engineering in Medicine and Biology Society (EMBC), Boston, MA.

Montes de Oca, M. A., Stutzle, T., Birattari, M., & Dorigo, M. (2009, October). Frankenstein's PSO: A composite particle swarm optimization algorithm. *IEEE Transactions on Evolutionary Computation, 13*(5), 1120–1132. doi:10.1109/TEVC.2009.2021465

Nakano, S., Ishigame, A., & Yasuda, K. (2007, September). *Particle swarm optimization based on the concept of Tabu search*. Paper presented at IEEE Congress on Evolutionary Computation (CEC), Singapore.

Nikolova, N. K. (2011, December). Microwave imaging for breast cancer. *IEEE Microwave Magazine, 12*(7), 78–94. doi:10.1109/MMM.2011.942702

Orfanidis, S. J. (2008). *Electromagnetic waves and antennas*. eBook, Retrieved January 3, 2012, from http://www.ece.rutgers.edu/~orfanidi/ewa/

Poli, R. (2009, August). Mean and variance of the sampling distribution of particle swarm optimizers during stagnation. *IEEE Transactions on Evolutionary Computation, 13*(4), 712–721. doi:10.1109/TEVC.2008.2011744

Ratnaweera, A., Halgamuge, S. K., & Watson, H. C. (2004, June). Self-organizing hierarchical particle swarm optimizer with time-varying acceleration coefficients. *IEEE Transactions on Evolutionary Computation, 8*(3), 240–255. doi:10.1109/TEVC.2004.826071

Robinson, J., & Rahmat-Samii, Y. (2004, February). Particle swarm optimization in electromagnetic. *IEEE Transactions on Antennas and Propagation, 52*(2), 397–407. doi:10.1109/TAP.2004.823969

Sabouni, A., Noghanian, S., & Pistorius, S. (2010, Winter). A global optimization technique for microwave imaging of the inhomogeneous and dispersive breast. *Canadian Journal of Electrical and Computer Engineering, 35*(1), 15–24. doi:10.1109/CJECE.2010.5783380

Sharaf, A. M., & El-Gammal, A. A. A. (2009, November). *A discrete particle swarm optimization technique (DPSO) for power filter design.* Paper presented at 4th International Design and Test Workshop (IDT).

Shi, Y., & Eberhart, R. (1999, July). *Empirical study of particle swarm optimization.* Paper presented at Congress on Evolutionary Computation, Washington, DC.

Sun, J., Fang, W., & Xu, W. (2010, February). A quantum-behaved particle swarm optimization with diversity-guided mutation for the design of two-dimensional IIR digital filters. *IEEE Transactions on Circuits and Wystems. II, Express Briefs, 57*(2), 141–145. doi:10.1109/TCSII.2009.2038514

Tao, Q., Chang, H. Y., Yi, Y., Gu, C. Q., & Li, W. J. (2010, July). *An analysis for particle trajectories of a discrete particle swarm optimization.* 3rd IEEE International Conference on Computer Science and Information Technology (ICCSIT).

Thakur, K. P., Chan, K., Holmes, W. S., & Carter, G. (2002, June). *An inverse technique to evaluate thickness and permittivity using reflection of plane wave from inhomogeneous dielectrics.* Paper presented at 59th Automatic RF Techniques Group (ARFTG) Conference, Seattle, WA.

Vural, R. A., Yildirim, T., Kadioglu, T., & Basargan, A. (2012, February). Performance evaluation of evolutionary algorithms for optimal filter design. *IEEE Transactions on Evolutionary Computation, 16*(1), 135–147. doi:10.1109/TEVC.2011.2112664

Wang, W., Lu, Y., Fu, J. S., & Xiong, Y. Z. (2005, May). Particle swarm optimization and finite-element based approach for microwave filter design. *IEEE Transactions on Magnetics, 41*(5), 1800–1803. doi:10.1109/TMAG.2005.846467

Yeung, C. W., Leung, F. H., Chan, K. Y., & Ling, S. H. (2009, June). *An integrated approach of particle swarm optimization and support vector machine for gene signature selection and cancer prediction.* Paper presented at the International Joint Conference on Neural Network, Atlanta, Georgia.

ADDITIONAL READING

Afshinmanesh, F., Marandi, A., & Rahimi-Kian, A. (2005, November 21-24). *A novel binary particle swarm optimization method using artificial immune system.* Paper presented at the EUROCON 2005 – The International Conference on "Computer as a Tool", Belgrade, Serbia and Montenegro.

Chew, W. C. (1995). *Waves and fields in inhomogeneous media.* New York, NY: IEEE Press.

Clerc, M. (1999, July 6-9). *The swarm and the queen: Towards a deterministic and adaptive particle swarm optimization.* Paper presented at the 1999 Congress on Evolutionary Computation, Washington.

Coello Coello, C. A., Pulido, G. T., & Lechuga, M. S. (2004). Handling multiple objectives with particle swarm optimization. *IEEE Transactions on Evolutionary Computation, 8*(3), 256–279. doi:10.1109/TEVC.2004.826067

Deligkaris, K. V., Zaharis, Z. D., Kampitaki, D. G., Goudos, S. K., Rekanos, I. T., & Spasos, M. N. (2009). Thinned planar array design using Boolean PSO with velocity mutation. *IEEE Transactions on Magnetics, 45*(3), 1490–1493. doi:10.1109/TMAG.2009.2012687

Demarcke, P., Rogier, H., Goossens, R., & De Jaeger, P. (2009). Beamforming in the presence of mutual coupling based on constrained particle swarm optimization. *IEEE Transactions on Antennas and Propagation, 57*(6), 1655–1666. doi:10.1109/TAP.2009.2019923

Fear, E. C., & Stuchly, M. A. (2000). Microwave detection of breast cancer. *IEEE Transactions on Microwave Theory and Techniques, 48*(11), 1854–1863. doi:10.1109/22.883862

Genovesi, S., Monorchio, A., Mittra, R., & Manara, G. (2007). A sub-boundary approach for enhanced particle swarm optimization and its application to the design of artificial magnetic conductors. *IEEE Transactions on Antennas and Propagation, 55*(3), 766–770. doi:10.1109/TAP.2007.891559

Gibson, W. C. (2008). *The method of moments in electromagnetics.* Boca Raton, FL: CRC Press.

Golnabi, A. H., Meaney, P. M., Gheimer, S., & Paulsen, K. D. (2011, January 16-19). *Microwave imaging for breast cancer detection and therapy monitoring.* Paper presented at the 2011 IEEE Topical Conference on Biomedical Wireless Technologies, Networks, and Sensing Systems (BioWireleSS), Phoenix, AZ.

Goudos, S. K., Moysiadou, V., Samaras, T., Siakavara, K., & Sahalos, J. N. (2010). Application of a comprehensive learning particle swarm optimizer to unequally spaced linear array synthesis with sidelobe level suppression and null control. *IEEE Antennas and Wireless Propagation Letters, 9,* 125–129. doi:10.1109/LAWP.2010.2044552

Goudos, S. K., Rekanos, I. T., & Sahalos, J. N. (2008). EMI reduction and ICs optimal arrangement inside high-speed networking equipment using particle swarm optimization. *IEEE Transactions on Electromagnetic Compatibility, 50*(3), 586–596. doi:10.1109/TEMC.2008.924389

Huang, T., & Mohan, A. S. (2007). A microparticle swarm optimizer for the reconstruction of microwave images. *IEEE Transactions on Antennas and Propagation, 55*(3), 568–576. doi:10.1109/TAP.2007.891545

Kennedy, J., & Eberhart, R. (1997, October 12-15). *A discrete binary version of the particle swarm algorithm.* Paper presented at the 1997 IEEE International Conference on Systems, Man and Cybernetics, Orlando.

Khodier, M. M., & Christodoulou, C. G. (2005). Linear array geometry synthesis with minimum sidelobe level and null control using particle swarm optimization. *IEEE Transactions on Antennas and Propagation, 53*(8), 2674–2679. doi:10.1109/TAP.2005.851762

Lanza, M., Pérez, J. R., & Basterrechea, J. (2009). Synthesis of planar arrays using a modified particle swarm optimization algorithm by introducing a selection operator and elitism. *Progress in Electromagnetics Research, 93,* 145–160. doi:10.2528/PIER09041303

Li, W., Hei, Y., Shi, X., Liu, S., & Lv, Z. (2010). An extended particle swarm optimization algorithm for pattern synthesis of conformal phased arrays. *International Journal of RF and Microwave Computer-Aided Engineering, 20*(2), 190–199.

Liang, J. J., Qin, A. K., Suganthan, P. N., & Baskar, S. (2006). Comprehensive learning particle swarm optimizer for global optimization of multimodal functions. *IEEE Transactions on Evolutionary Computation, 10*(3), 281–295. doi:10.1109/TEVC.2005.857610

Marandi, A., Afshinmanesh, F., Shahabadi, M., & Bahrami, F. (2006, July 16-21). *Boolean particle swarm optimization and its application to the design of a dual-band dual-polarized planar antenna.* Paper presented at the 2006 IEEE Congress on Evolutionary Computation, Vancouver, Canada.

Paulsen, K. D., Meaney, P. M., & Gilman, L. C. (2005). *Alternative breast imaging, four model-based approaches.* Springer Science Business Media Inc.

Pérez, J. R., & Basterrechea, J. (2009). Hybrid particle swarm-based algorithms and their application to linear array synthesis. *Progress in Electromagnetics Research, 90*, 63–74. doi:10.2528/ PIER08122212

Sill, J. M., & Fear, E. C. (2005). Tissue sensing adaptive radar for breast cancer detection: Study of immersion liquids. *Electronics Letters, 41*(3), 113–115. doi:10.1049/el:20056953

Zaharis, Z. D. (2008). Radiation pattern shaping of a mobile base station antenna array using a particle swarm optimization based technique. *Electrical Engineering, 90*(4), 301–311. doi:10.1007/ s00202-007-0078-y

KEY TERMS AND DEFINITIONS

Finite-Difference Time-Domain (FDTD): Is a popular grid-based differential time-domain numerical modeling technique. Since it is a time-domain method, a single simulation run can cover a wide frequency range, provided the time step is small enough to satisfy the Nyquist–Shannon sampling theorem for the desired highest frequency. In FDTD, Maxwell's equations (in partial differential form) are modified, discretized, and implemented in software. The equations are solved in a cyclic manner: the electric field is solved at a given instant in time, then the magnetic field is solved at the next instant in time, and the process is repeated over and over again.

Fitness Function or Objective Function: Is used to summarize the desired optimization features, as a single figure of merit, and shows how close a given design solution is to achieving the set of goals.

Method of Moments (MoM): Is a numerical computational method of solving linear partial differential equations which have been formulated as integral equations. In electromagnetics, this technique has been mostly used for planar structures, such as microstrip antennas.

Microwave Imaging: Uses nonionizing microwave radiation in order to create images of inside-body tissue. In microwave imaging, the sensors are the transmitter and receiver antennas. The transmitter antennas illuminate the tissue with microwave signals and the signals scattered back from body are collected by the receiver antennas.

Reconfigurable Antennas: Are defined as the antennas for which it is possible to alter the shape and electric features of the antenna using either mechanical controllers or electrical ones.

Scattering: Is a general physical process where some forms of radiation, such as electromagnetic fields, are forced to deviate from their normal trajectory by one or more localized non-uniformities in the medium through which they pass.

Wearable Antennas: Are meant to be incorporated inside the daily clothing. These antennas are particularly used where preserving continuous communication link is required.

ENDNOTES

[1] Method of moments
[2] The reason for choosing 50 runs is explained in (Modiri[2], et al., 2011).
[3] Finite Difference Time Domain
[4] ISM band
[5] Vector network analyzer
[6] $|S_{11}|$
[7] 488 samples from reduction surgeries and 319 samples from cancer surgeries
[8] Fat

Chapter 6
Particle Swarm Optimization Algorithms Applied to Antenna and Microwave Design Problems

Sotirios K. Goudos
Aristotle University of Thessaloniki, Greece

Zaharias D. Zaharis
Aristotle University of Thessaloniki, Greece

Konstantinos B. Baltzis
Aristotle University of Thessaloniki, Greece

ABSTRACT

Particle Swarm Optimization (PSO) is an evolutionary optimization algorithm inspired by the social behavior of birds flocking and fish schooling. Numerous PSO variants have been proposed in the literature for addressing different problem types. In this chapter, the authors apply different PSO variants to common antenna and microwave design problems. The Inertia Weight PSO (IWPSO), the Constriction Factor PSO (CFPSO), and the Comprehensive Learning Particle Swarm Optimization (CLPSO) algorithms are applied to real-valued optimization problems. Correspondingly, discrete PSO optimizers such as the binary PSO (binPSO) and the Boolean PSO with velocity mutation (BPSO-vm) are used to solve discrete-valued optimization problems. In case of a multi-objective optimization problem, the authors apply two multi-objective PSO variants. Namely, these are the Multi-Objective PSO (MOPSO) and the Multi-Objective PSO with Fitness Sharing (MOPSO-fs) algorithms. The design examples presented here include microwave absorber design, linear array synthesis, patch antenna design, and dual-band base station antenna optimization. The conclusion and a discussion on future trends complete the chapter.

DOI: 10.4018/978-1-4666-2666-9.ch006

INTRODUCTION

In the past decade, several evolutionary algorithms (EAs) that mimic the behavior and evolution of biological entities emerged. Among others, Particle Swarm Optimization (PSO) is a popular evolutionary algorithm which is based on the intelligence and movement of swarms (birds, fishes, bees, etc.) and resembles their behavior (Kennedy & Eberhart, 1995).

Many similarities exist between PSO and other evolutionary computation techniques such as Genetic Algorithms (GAs). In general, PSO does not have any evolution operators like crossover and mutation. Compared to Genetic Algorithms, PSO has fewer parameters to adjust and is easier to implement in any programming language. PSO is also computationally more efficient than a GA with the same population size. The algorithm has been successfully applied in many engineering disciplines: function optimization, artificial neural network training, fuzzy system control and other areas where GAs are also applied. The fact that particle swarm optimizers can handle efficiently arbitrary optimization problems has also made them popular for solving problems in electromagnetics, especially in electromagnetic design ones (Baskar, Alphones, Suganthan, & Liang, 2005; Deligkaris et al., 2009; Goudos, Moysiadou, Samaras, Siakavara, & Sahalos, 2010; Goudos, Rekanos, & Sahalos, 2008; Goudos & Sahalos, 2006; Goudos, Zaharis, Kampitaki, Rekanos, & Hilas, 2009; Khodier & Christodoulou, 2005; Robinson & Rahmat-Samii, 2004; Zaharis, 2008; Zaharis, Kampitaki, Lazaridis, Papastergiou, & Gallion, 2007). Numerous different PSO variants exist in the literature. The most common algorithms include the classical Inertia Weight PSO (IWPSO) and the Constriction Factor PSO (CFPSO) (Clerc, 1999). However, in order to further improve the performance of PSO on complex multimodal problems, a PSO variant was proposed (Liang, Qin, Suganthan, & Baskar, 2006). This variant is the Comprehensive Learning Particle

Swarm Optimizer (CLPSO) which utilizes a new learning strategy. The CLPSO algorithm accelerates the convergence of the classical PSO. It has been applied successfully to Yagi-Uda antenna design by Baskar et al. (2005) and to linear array synthesis by Goudos et al. (2010).

The PSO algorithm is inherently used only for real-valued problems but can easily expand to discrete-valued problems. A simple modification of the real-valued PSO called binary PSO (binPSO) has been presented by Kennedy & Eberhart (1997). In (Marandi, Afshinmanesh, Shahabadi, & Bahrami, 2006) the Boolean PSO is introduced and applied to dual-band planar antenna design. The Boolean PSO is based on the idea of using exclusively Boolean update expressions in the binary space. An extension to Boolean PSO, that improves the algorithm performance, is the Boolean PSO with velocity mutation (BPSO-vm) which has been applied successfully to patch antenna design (Deligkaris et al., 2009).

Multi-objective extensions of PSO include the Multi-objective Particle Swarm Optimization (MOPSO) (Coello Coello, Pulido, & Lechuga, 2004) and the Multi-objective Particle Swarm Optimization with fitness sharing (MOPSO-fs) (Salazar-Lechuga & Rowe, 2005). The above algorithms have also been used in antenna and microwave design problems (Goudos & Sahalos, 2006; Goudos et al., 2009). The purpose of this chapter is to briefly describe the aforementioned algorithms and present their application to antenna and microwave design problems. Within this context, we present results from design cases using the IWPSO, CFPSO, CLPSO, binPSO, BPSO-vm and the multi-objective variants MOPSO and MOPSO-fs. The examples comprise the design of high-absorption planar multilayer coatings, the synthesis of unequally spaced linear arrays with sidelobe level (*SLL*) suppression under mainlobe beamwidth and null control constraints, the design of thinned planar microstrip arrays under constraints of impedance matching, low *SLL* and null control, and finally the optimization of dual-band base station antennas for wireless networks. Comparisons with optimization methods in the literature validate the efficacy of PSO. Our study completes with a brief discussion of future directions and perspectives in the area and it is supported with an adequate number of references.

The chapter is subdivided into four sections. First, we present the different PSO algorithms. In the next Section, we describe the design cases and present the numerical results. An outline of future research directions is provided in the following Section while in the "Conclusion" Section we conclude the chapter and discuss the advantages of using a PSO-based approach in the design and optimization of microwave systems and antennas. Finally, an "Additional Reading Section" gives a list of readings to provide the interested reader with useful sources in the field.

PARTICLE SWARM OPTIMIZATION: CLASSICAL ALGORITHMS AND VARIANTS

Particle swarm optimization is based on the intelligence and movement of swarms. In the nature, the swarm intelligence helps the swarm to find the place where there is more food than anywhere else. For example, let us consider a swarm of bees. The bees search for the place with the highest density of flowers. In general, the direction and speed of their motion are random quantities. However, each bee remembers the places where it found the highest concentration of flowers and it also knows the places where the rest of the bees have found plenty of flowers. As a result, and despite the random characteristics of their flights, the bees take into account the best positions encountering by them and the rest of the bees and thus make an attempt to balance exploration and exploitation. This behavior finally leads the bees to the place with the highest density of flowers. A similar approach is followed by the PSO method.

In the PSO algorithm, each individual in the swarm is called "particle" or "agent" and moves in an N-dimensional space trying to find an even better position by adjusting its direction. It learns from its own experience and the experiences of the neighboring particles and updates its velocity based on them. The position and velocity of the ith particle is respectively, $\bar{x}_i = \left(x_{i1}, x_{i2}, ..., x_{iN} \right)$ and $\bar{v}_i = \left(v_{i1}, v_{i2}, ..., v_{iN} \right)$, $i = 1, ..., M$ and $n = 1, ..., N$, where M is the "population size", i.e., the number of the particles that compose the swarm. The particle positions may be limited in the respective (nth) dimension between an upper and a lower boundary (U_n and L_n, respectively), i.e., $x_{in} \in \left[L_n, U_n \right]$, restricting the search space within these limits. The difference $R_n = U_n - L_n$ is the "dynamic range" of the nth dimension.

In general, the system is initialized and searches for optima by updating the particles positions in any iteration. Each particle position is updated by finding two optimum values, the best position, $\bar{p}_i = \left(p_{i1}, ..., p_{iN} \right)$, achieved so far by the particle (*pbest* position) and the best position, $\bar{g} = \left(g_1, ..., g_N \right)$, obtained so far by any particle of the swarm (*gbest* position). By taking into account these two values respectively as cognitive and social information, each particle updates its position and velocity. The algorithm is executed repeatedly until a specified number of iterations is reached or the velocity updates are close to zero. The particle quality is measured according to a predefined mathematical function $F\left(\bar{x}_i \right)$ called "fitness function", which is related to the problem to be solved and reflects the optimality of a particular solution. In each iteration, the change of $\bar{x}_i$ is $\Delta \bar{x}_i = \bar{v}_i \Delta t$, where Δt is the time interval. Thus, by setting $\Delta t = 1$ the new position of the ith particle is

$$\bar{x}_i^t = \bar{x}_i^{t-1} + \bar{v}_i^t \tag{1}$$

where $\overline{x}_i^t$ and $\overline{x}_i^{t-1}$ denote the particle position at the current and the previous iteration, respectively, and $\overline{v}_i^t$ is the velocity at the current iteration.

At this point, we have to notice that $\overline{g}$ corresponds to the maximum fitness value found so far by the swarm. The *gbest* neighborhood is equivalent to a fully connected social network in which every individual is able to compare the performance of every other member of the population, imitating the very best. In this case, every particle is attracted to the *gbest* position. In a similar way, we may consider that each (*i*th) particle is affected by the best performance of K_i neighbors. The last are particles near the individual in a topological rather than in the parameter space. In this case, the best position is called the " *ℓbest* position" and it is represented as $\overline{\ell}_i = \left(\ell_{i1}, ..., \ell_{iN} \right)$. Equivalently, in the first approach, the neighborhood of a particle is the whole swarm, while in the second one it consists of a specific number of certain particles. The optimal connectivity pattern between individuals depends on the problem to be solved. In general, the *gbest* approach convergences faster but it is more susceptible to convergence on local optima. This behavior is similar to GAs remaining stagnant to local optima.

Classical PSO Algorithms

As mentioned above, in particle swarm optimization, individuals are influenced by both their own previous behavior and the successes of their neighbors. In the IWPSO algorithm, the particle's velocity depends on its previous velocity value, on the distance between the particle's position and the position *pbest*, and on the distance between the particle's position and *gbest* or *ℓbest* position depending on the neighborhood model applied by the user. In the following analysis, we will refer only to the *gbest* case; the extension to the *ℓbest* case is straightforward.

In the IWPSO algorithm, the *i*th particle's velocity in the *n*th dimension is updated as in Equation 2 (see Box 1) where w is a positive parameter called "inertia weight", c_1 and c_2 are positive parameters known respectively as "cognitive coefficient" and "social coefficient", and rnd_{in1}^t, rnd_{in2}^t are random numbers uniformly distributed in the interval from 0 to 1. The index t denotes the current iteration, while $t-1$ denotes the previous one. The parameter w controls the impact of previous velocity values on the current velocity and usually has fixed values between 0.0 and 1.0. Large values of w facilitate global exploration, while a smaller w tends to facilitate local exploration to fine-tune the current search area. A proper choice of w achieves a balance between global and local exploration and results in faster convergence (Shi & Eberhart, 1998). A linear decrease of w from 0.9 to 0.4 during the simulation is usually employed (Shi & Eberhart, 1999). It has to be noticed, that the same value of w is used for all dimensions of all particles in a given population. The parameter c_1 represents the impact of the particle's distance from its best position, while c_2 determines the influence of the swarm. Usually, the two parameters are both set equal to 2.0 (Kennedy & Eberhart, 1995) or 1.49 (Robinson & Rahmat-Samii, 2004).

The stochastic changes in the velocity of the particles may result in an expansion of a particle's trajectory into wider cycles through the problem space. A solution to this problem is to set a

Box 1.

$$v_{in}^t = w \cdot v_{in}^{t-1} + c_1 \cdot rnd_{in1}^t \cdot \left(p_{in}^t - x_{in}^{t-1} \right) + c_2 \cdot rnd_{in2}^t \cdot \left(g_n^t - x_{in}^{t-1} \right) \tag{2}$$

Box 2.

$$v_{in}^{t} = K \cdot \left[v_{in}^{t-1} + \varphi_1 \cdot rnd_{in1}^{t} \cdot \left(p_{in}^{t} - x_{in}^{t-1} \right) + \varphi_2 \cdot rnd_{in2}^{t} \cdot \left(g_n^{t} - x_{in}^{t-1} \right) \right] \qquad (3)$$

maximum allowed velocity $\overline{v}_{\max} = \left(v_{\max,1}, \ldots, v_{\max,N} \right)$.

Thus, we set $v_{in} \equiv v_{\max,n}$, $\forall i, n : v_{in} > v_{\max,n}$ and $v_{in} \equiv -v_{\max,n}$ $\forall i, n : v_{in} < -v_{\max,n}$. As a result, we prevent explosion and scale the exploration of the particle. The choice of $\overline{v}_{\max}$ differs from problem to problem. In general, a small maximum velocity limits the global exploration. On the contrary, large $v_{\max,n}$ values may reduce the local exploration ability (Shi & Eberhart, 1998). If a step larger than $\overline{v}_{\max}$ is required to escape a local optimum then the particle will be trapped; on the other hand, it is better to take smaller steps as the solution approaches to an optimum. A good choice is to set each $v_{\max,n}$ coordinate around 10-20% of the dynamic range of the respective dimension when $w = 1$ and equal to this range if $w < 1$ (Eberhart & Shi, 2001). Another option as suggested in (Jin & Rahmat-Samii, 2007) is to set its value equal to the dynamic range in each dimension of the particle. For example, if a variable is allowed to be optimized within $\left(L_n, U_n \right)$, the maximum velocity in this dimension is $\left| L_n - U_n \right|$ in both directions.

In another classical PSO algorithm, the CFPSO, a different velocity update rule is proposed. In that case, the velocity components are updated as in Equation 3 (Box 2) where K is the "constriction coefficient" defined as:

$$K \equiv \frac{2}{\varphi - 2 + \sqrt{\varphi^2 - 4\varphi}}, \quad \varphi \geq 4 \qquad (4)$$

The parameters $\varphi_{1,2}$ have a similar meaning to $c_{1,2}$ and the parameter φ, known as "accel-

eration constant", is their sum. A good choice for both φ_1 and φ_2 is 2.05 (Eberhart & Shi, 2001). The introduction of the constriction coefficient eliminates the need of $\overline{v}_{\max}$. However, it has been concluded from (Eberhart & Shi, 2000) that it is still better to use $\overline{v}_{\max}$ and to set it equal to the dynamic range of each variable on each dimension.

Nevertheless, the parameters w, K and $\overline{v}_{\max}$ are not always able to confine the particles within the search space. In order to solve this problem, three boundary conditions have been suggested: (i) The absorbing walls condition: When x_{in} becomes greater than U_n or less than L_n, the respective velocity component becomes zero and the *i*th particle is pulled back toward the search space, i.e., if $x_{in} > U_n$ then $x_{in} = U_n$ and $v_{in} = 0$, and also if $x_{in} < L_n$ then $x_{in} = L_n$ and $v_{in} = 0$. (ii) The reflecting walls condition: When x_{in} becomes greater than U_n or less than L_n, v_{in} is set to $-v_{in}$, and the particle is reflected back towards the search space. (c) The invisible walls condition: The particles are allowed to move outside the search space without any restriction but a large predefined fitness value is assigned to them. This condition saves computational time, because the fitness function is calculated only for the particles inside the search space.

Comprehensive Learning Particle Swarm Optimizer (CLPSO)

In order to improve the convergence speed of the classical PSO algorithms, a different learning strategy has been proposed. This strategy is involved in the CLPSO algorithm and ensures that the diversity of the swarm is preserved in order to discourage premature convergence. To achieve this, the algorithm adopts a velocity update rule that considers all particles' previous experiences, i.e., best solutions, as a potential during the calculation of the particle's new velocity. As a result, a particle can learn from a different exemplar in each dimension.

In the CLPSO algorithm, the velocity update equation is given below:

$$v_{in}^t = w \cdot v_{in}^{t-1} + c \cdot rnd_{in}^t \cdot \left(p_{f_i(n)n}^t - x_{in}^{t-1} \right) \qquad (5)$$

where the $f_i = \left[f_i(1), ..., f_i(N) \right]$ defines the particle's *pbest* that should by followed by the *i*th particle and the $p_{f_i(n)n}^t$ is the corresponding dimension of all particles' *pbest*. In practice, a random number is generated for each dimension; this number is compared with a parameter called "learning probability" that takes different values for different particles. On the basis of the previous comparison, a decision whether the particle will learn in the corresponding dimension from its own *pbest* (if the random number is larger than the learning probability) or from another particle's *pbest* is made. In the second case, we first choose two particles of the population using a uniform random distribution (obviously, we exclude the particle whose velocity is to be updated). Next, we compare the fitness values of these two particles *pbest's* and select the best one. The selected particle's *pbest* is used as the exemplar to learn for that dimension. If all the exemplars of a particle are its own *pbest*, then we randomly choose one dimension, in order to learn from another particle's *pbest*. The updating strategy used in CLPSO provides a larger potential to explore more of the search space (it searches more promising regions to find the global optimum) than classical PSO algorithms. Obviously, the diversity is also increased with the particles' potential search. As it is reported in Liang et al. (2006) the

improved learning strategy has enabled CLPSO to make use of the information in the swarm more effectively and to generate frequently better quality solutions when compared to eight PSO variants on numerical test problems.

The CLPSO algorithm can be easily implemented. However, its drawbacks compared to the classical particle swarm optimizers are the increased complexity and the higher computational load.

Discrete PSO Variants: The binPSO and BPSO-vm Algorithms

PSO algorithm is inherently used only for real-valued problems. However, in many real-world applications the variables to be optimized take integer values. Several PSO variants for discrete-valued problems exist in the literature (Chen et al., 2010; Modiri & Kiasaleh, 2011a; Sharaf & El-Gammal 2009; Tao, Chang, Yi, Gu, & Li, 2010). In the following paragraphs, we present two options to expand PSO for discrete-valued problems. These popular discrete PSO variants are the binary PSO (binPSO) and the Boolean PSO with velocity mutation (BPSO-vm). The binPSO algorithm is a simple modification of IWPSO that allows the algorithm to operate using binary representations. In binPSO, the location of each particle is a binary string with length D while the particle velocities remain real-valued and are updated as in IWPSO. The *i*th coordinate of each particle's position is a bit updated as

$$x_i^{t+1} = \begin{cases} 1, & \text{if } \rho < s(v_i^{t+1}) \\ 0, & \text{otherwise} \end{cases} \qquad (6)$$

where ρ is random uniformly distributed number within the interval [0, 1] and $s(\cdot)$ is a sigmoid function that maps all the real values of velocity to the range $\left[0, 1 \right]$. Such a function is given below:

$$s(x) = \frac{1}{1 + e^{-x}} \tag{7}$$

Another discrete PSO algorithm is the Boolean PSO (BPSO) (Marandi et al., 2006). The BPSO is based on the idea of using exclusively Boolean expressions to update both the velocity and the position of a particle. These expressions are given below:

$$v_{id}^t = w \bullet v_{id}^{t-1} + c_1 \bullet \left(p_{id}^t \oplus x_{id}^{t-1}\right) + c_2 \bullet \left(g_d^t \oplus x_{id}^{t-1}\right) \tag{8}$$

$$x_{id}^t = x_{id}^{t-1} \oplus v_{id}^t \tag{9}$$

where $d = 1, ..., D$, g_d is the dth bit of the position $\bar{g}$, while the parameters w, c_1 and c_2 are random bits chosen with probability of being "1" respectively equal to W, C_1 and C_2, which are user-selected parameters. Also the notations ($\bullet$), (Å) and (+) represent respectively the *and*, *xor* and *or* Boolean operators.

In order to control the convergence speed of the optimization process, the algorithm sets a maximum allowed velocity $v_{\max}$ defined as the maximum allowed number of '1's in the velocity vectors $\bar{v}_i = [v_{i1}, ..., v_{iD}]$. The number of '1's in the binary vector $\bar{v}_i$ is the "velocity length" L_i. This parameter is controlled using "negative selection" (NS), a biological immunity process responsible for eliminating the T-cells that recognize self antigens in the thymus (Afshinmanesh, Marandi, & Rahimi-Kian, 2005). According to NS, if $L_i \leq v_{\max}$, the $\bar{v}_i$ is a non-self antigen and remains the same. If $L_i > v_{\max}$, $\bar{v}_i$ is considered as a self antigen and thus the NS is applied to it by changing randomly chosen '1's into '0's until the condition $L_i = v_{\max}$ is met.

The BPSO-vm extends the BPSO by using a mutation operator applied to the particle velocities. In particular, the bits of $\bar{v}_i$ are allowed to change with probability m_r (mutation probability) from '0' to '1', but are not allowed to change from '1' to '0'. This mutation scheme diversifies the population and thus results in an increase of the exploration ability of the particles. It should be mentioned that the mutation probability must be relatively small to avoid producing a pure random-search algorithm.

The structure of the BPSO-vm algorithm is described by the following steps:

1. The population size, the dimension of the binary solution space, the maximum allowed velocity, the mutation probability and the total number of iterations are set.
2. The initial particle positions $\bar{x}_i$ and their corresponding velocities $\bar{v}_i$ are randomly generated.
3. The NS process is applied to correct the particle velocities $\bar{v}_i$.
4. The cost functions, $F(\bar{x}_i)$, and the minimum one among all, $F_{\min}$, are evaluated. The particle with the minimum cost function is considered as the global best, $\bar{g}$.
5. The initial positions of the particles are set as the individual *pbest* ones, i.e., $\bar{p}_i = \bar{x}_i$.
6. The particles velocities are updated using Equation 8.
7. The NS process is applied to correct the particle velocities.
8. Velocity mutation is applied to every '0' bit of $\bar{v}_i$.
9. The positions $\bar{x}_i$ are updated using Equation 9.
10. The cost functions, $F(\bar{x}_i)$, are evaluated.
11. The *pbest* and *gbest* positions, respectively $\bar{p}_i$ and $\bar{g}$, are updated.
12. The algorithm is repeated from step 6 until the total number of iterations is reached.

Multi-Objective PSO Extensions: The MOPSO and MOPSO-fs Methods

PSO seems to be suitable for multi-objective optimization due to its speedy convergence in single objective problems. Over the last few years, several Multi-Objective PSO algorithms have been proposed.

The MOPSO algorithm proposed in Coello Coello et al. (2004) has been validated against highly competitive evolutionary multi-objective algorithms like the Non-dominated Sorting Genetic Algorithm II (NSGA-II) (Deb, Pratap, Agarwal, & Meyarivan, 2002). The key elements of MOPSO are:

1. The historical record of the best solutions found by a particle is used to store non-dominated solutions generated in the past. Therefore, a repository is introduced and the positions of the particles that represent non-dominated solutions are stored. The parameter that has to be adjusted is the "repository size".

2. A new mutation operator is also introduced. This operator intends to produce a highly explorative behavior of the algorithm. The effect of mutation decreases as the number of iterations increase. The parameter of mutation probability is therefore used.

3. The idea of an adaptive grid is introduced. An external archive is used to store all the solutions that are non-dominated with respect to the contents of the archive. Into the archive the objective function space is divided into regions. The adaptive grid is a space formed by hypercubes. Each hypercube can be interpreted as a geographical region that contains a number of particles. The adaptive grid is used to distribute in a uniform way the largest possible amount of hypercubes. It is necessary therefore to provide the parameter of grid subdivisions.

As reported in Coello Coello et al. (2004), the MOPSO algorithm is relatively easy to implement. The exploratory capabilities of PSO are improved using the mutation operator. Additionally, the authors have found that MOPSO requires low computational cost, which is a key issue in cases of electromagnetic optimization.

An improvement of the previous variant, the MOPSO-fs utilizes both particle swarm optimization and fitness sharing. The last aims to spread the solutions along the Pareto front. Fitness sharing is used in the objective space and enables the algorithm to maintain diversity between solutions. This means that particles within highly populated areas in the objective space are less likely to be followed.

An external repository stores the non-dominated particles found. The best particles found in each iteration are inserted into the repository. The role of the repository is twofold: it helps the search for the next generations and it maintains a set of non-dominated solutions until the end of the process. This set of solutions forms the Pareto front. The structure of the MOPSO-fs algorithm follows.

We consider an M-size population of N-dimensional particles $\bar{x}^i$, $i = 1, ..., M$ (N is the number of design parameters). First, the algorithm initializes with random population from a uniform distribution. The external repository is filled with all the non-dominated particles. A fitness sharing value is calculated for each particle in the repository. A high value of f_{sh}^i suggests that the vicinity of the ith particle is not highly populated. The fitness sharing value for the ith particle is

$$f_{sh}^i = \frac{10}{\displaystyle\sum_{j=1}^{R_s} sharing^{i,j}} \qquad (10)$$

where R_s is the number of the particles in the repository and

$$sharing^{i,j} = \begin{cases} 1 - \left(d^{i,j}/\sigma_{share}\right)^2, & \text{if } d^{i,j} < \sigma_{share} \\ 0, & \text{otherwise} \end{cases}$$

(11)

with $d^{i,j}$ the Euclidean distance between the ith and the jth particle and σ_{share} the radius of the vicinity area of a particle. According to the fitness sharing principle, the particles that have more particles in their vicinity are less fit than those with fewer particles around it.

Provided that fitness sharing is assigned to each particle in the repository, some particles are chosen as leaders according to roulette wheel selection, i.e., the particles with higher levels of fitness are more likely to be selected. In the next iteration, the leaders are going to be followed by the rest of the particles. The velocity update rule for the ith particle shows similarities it is given by

$$v_{in}^t = w \cdot v_{in}^{t-1} + c_1 \cdot rnd_{in1}^t \cdot \left(p_{in}^t - x_{in}^{t-1}\right) + c_2 \cdot rnd_{in2}^t \cdot \left(g_{hn}^t - x_{in}^{t-1}\right)$$

(12)

where p_{in}^t is the best position found by the ith particle so far, g_{hn}^t is the leader particle position along the nth dimension, and

$$x_{in} = \begin{cases} L_n + r_3 \cdot (U_n - L_n), & \text{if } x_{in} < L_n \\ U_n - r_4 \cdot (U_n - L_n), & \text{if } x_{in} > U_n \end{cases},$$

rnd_{in2}^t are uniformly distributed numbers within 0 and 1. The position of each particle is updated using Equation 1.

Next, the repository is updated with the current solutions found by the particles according to the dominance and fitness sharing value criteria. In particular, the particles that dominate those inside the repository are inserted while all dominated solutions are deleted. This operation allows the repository to be maintained as the Pareto front found so far. In case of a repository full of non-

dominated particles, if a particle non-dominated by any in the repository is found then fitness sharing values are compared; if its value is better than the worst fitness sharing in the repository, it replaces the respective particle. The fitness sharing of all particles is updated when a particle is inserted in or deleted from the repository. Finally, the memory of each particle is updated according to the criterion of dominance, i.e., if the current particle position dominates the previous one in the particle's memory it replaces it.

This algorithm can be improved by adding a constraint handling mechanism that assigns to each particle a constraint violation number (CVN) which carries the sum of the constraint violations produced by the solution. Particles with smaller CVN dominate the ones with larger CVN while feasible solutions have zero CVN. Also, if the ith particle position in the nth dimension is found to be out of bounds, then it is updated as

$$x_{in} = \begin{cases} L_n + rnd_{in3} \cdot (U_n - L_n), & \text{if } x_{in} < L_n \\ U_n - rnd_{in4} \cdot (U_n - L_n), & \text{if } x_{in} > U_n \end{cases}$$

(13)

where rnd_{in3}, rnd_{in4} are uniformly distributed numbers within 0 and 1.

Control Parameter Selection

It must be pointed out that several PSO variants exist in the literature. In order to select, the best algorithm for every problem one has to consider the problem characteristics. For example, micro-PSO performs very well for microwave image reconstruction (Huang & Mohan, 2007). Another key issue is the selection of the algorithm control parameters, which is also in most cases problem-dependent. The control parameters selected here for these algorithms are those that commonly perform well regardless of the characteristics of the problem to be solved.

For real-coded GAs typical values are 0.9 for the crossover probability and 1/N for the mutation probability. The same values apply also for NSGA-II (Deb et al., 2002). For the binary-coded GA we set the crossover and mutation probabilities equal to 0.8 and 0.25, respectively.

In the PSO algorithms c_1 and c_2 are set equal to 2.05. For CFPSO, these values result in $K = 0.7298$. For IWPSO, it is common practice to linearly decrease the inertia weight starting from 0.95 to 0.4. The velocity is updated asynchronously, which means that the global best position is updated at the moment it is found. In CLPSO the parameter c in Equation 5 is set equal to one as it is suggested in Liang et al. (2006).

In BPSO, it is $W = 0.1$, $C_1 = C_2 = 0.5$ and $v_{max} = 2$. In BPSO-vm, we additionally set $m_r = 0.02$.

The main characteristics of the MOPSO algorithm are; the repository size, the mutation operator and the grid subdivisions. The repository is the archive where the positions of the particles that represent non-dominated solutions are stored. Therefore the parameter that has to be adjusted is the repository size. This is usually set equal to the swarm size. MOPSO introduces a mutation operator that intends to produce a highly explorative behavior of the algorithm. The effect of mutation decreases as the number of iterations increase. A setting of 0.5 has been found suitable after experiments for mutation probability; similarly, a value of 30 has been found suitable for the parameter of grid divisions (Coello Coello et al., 2006). The above parameters are those selected for our problem given below for the MOPSO algorithm.

MOPSO-fs also uses a repository to store all the all the non-dominated solutions (Salazar-Lechuga & Rowe, 2005). As in MOPSO the repository size parameter is set equal to swarm size. Another parameter that has to be set in MOPSO-fs is the sigma share value. This is set empirically after several trials to 2.0.

It must be pointed out that several modifications were made to MOPSO and MOPSO-fs algorithms. Constraint handling was added to both algorithms. Furthermore in case of discrete valued variables like the material number the velocity update rules given by the Binary PSO were used (Kennedy & Eberhart, 1997). More details about these modifications can be found in Goudos and Sahalos (2006) and in Goudos et al. (2009).

Finally, another issue for all the above algorithms is the definition of the stopping criterion. Usually, this is the iteration number or the number of objective-function evaluations. In all the design cases presented here the algorithms stop when the maximum iteration number is reached.

APPLICATIONS, RESULTS, AND DISCUSSION

In this Section, we provide representative microwave and antenna design problems that are solved with the aforementioned optimizers. First, we present an example case of a multilayer planar microwave absorber. The design method is a hybrid of the IWPSO algorithm that uses both binary and real variables. In the second example, we address the common problem of linear array synthesis. In this case, we apply the Comprehensive Learning PSO variant, the classical IWPSO and CFPSO algorithms and real-coded GA. The design of a thinned planar microstrip array under constraints on the radiation pattern and the impedance matching between the array elements is the subject of the third example. We show that the BPSO-vm variant is the most suitable optimizer for the problem. The obtained solutions from other discrete PSO variants are also presented. In the last example, we apply the MOPSO-fs variant in the design of a dual band base station antenna array for mobile communications under constraints on return loss, sidelobe level and half-power beamwidth.

Multilayer Planar Microwave Absorber Design

The microwave absorbers are used in several applications, which range from electromagnetic interference mitigation to anechoic chambers. Usually, microwave absorbers are constructed by tiling material in a periodic way to simplify construction. Multilayer microwave absorbers are frequently combined with other devices to reduce the radar cross section of a wide range of objects. Such absorbers need not only to suppress reflection over a wide frequency band but also to be thin, practical, and economical.

The problem of a planar microwave absorber design lies in the minimization of the reflection coefficient of an incident plane wave in a multilayer structure for a desired range of frequencies and angles of incidence. The reflection coefficient depends on the thickness and the electric and magnetic properties of each layer. Several studies, which address this problem, exist in the literature. For example in (Cui & Weile, 2005) a parallel PSO algorithm is applied.

The multilayer planar microwave absorber, see Figure 1, comprises M layers of different materials backed by a perfect electric conductor (PEC) such as the ground plane (this layer's number is $M+1$). In this structure, the unknown parameters are the number of layers and their physical characteristics. The last are the thickness d_i and the frequency dependent permittivity and permeability of each layer given by

$$\varepsilon_i(f) = \varepsilon_0 \left(\varepsilon_i'(f) - j\varepsilon_i''(f) \right) \qquad (14)$$

$$\mu_i(f) = \mu_0 \left(\mu_i'(f) - j\mu_i''(f) \right) \qquad (15)$$

where ε_0 and μ_0 are the free space permittivity and permeability and $i = 1,...,M$. All layers are assumed infinite. Obviously, the total thickness of the absorber is

Figure 1. Multilayer planar absorber

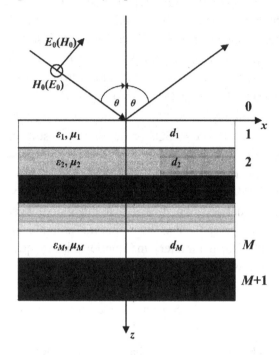

$$d_{tot} = \sum_{i=1}^{M} d_i \qquad (16)$$

The incident plane wave is either TE or TM and it is normal to the absorber. The incident media is the free space. In this multilayer structure, the general expression of the reflection coefficient at the interface between layers i and $i+1$ for an incident plane wave, $R_{i,i+1}{}^{TE}$ and $R_{i,i+1}{}^{TM}$, can be calculated (Chew, 1995) from the recursive formula

$$R_{i,i+1}{}^{TE/TM} = \frac{r_{i,i+1}^{TE/TM} + R_{i+1,i+2}^{TE/TM} e^{-2jk_{i+1}d_{i+1}}}{1 + r_{i,i+1}^{TE/TM} R_{i+1,i+2}^{TE/TM} e^{-2jk_{i+1}d_{i+1}}}$$

$$(17)$$

with

$$r_{i,i+1}^{TE} = \begin{cases} \dfrac{\mu_{i+1}k_i - \mu_i k_{i+1}}{\mu_{i+1}k_i + \mu_i k_{i+1}}, & i \leq M \\ -1, & i = M+1 \end{cases}$$

and

$$r_{i,i+1}^{TM} = \begin{cases} \dfrac{\varepsilon_{i+1}k_i - \varepsilon_i k_{i+1}}{\varepsilon_{i+1}k_i + \varepsilon_i k_{i+1}}, & i \leq M \\ 1, & i = M+1 \end{cases}$$

(18)

where k_i is the wavenumber of the ith layer, that is,

$$k_i = 2\pi f \sqrt{\varepsilon_i \mu_i}$$

(19)

The design of the absorber is defined as the minimization problem of $R_1^{TE/TM}$ (in dB) given below:

$$R_{0,1}^{TE/TM} = 20\log\left(\max\left|R(f,\theta)\right| \ f \in \mathrm{B}, \ \theta \in \mathrm{A}\right)$$

(20)

where $\max\left|R(f,\theta)\right|$ is the maximum reflection coefficient of the first layer over the desired frequency and angle range for a given polarization and A, B are the desired sets of angles of incidence and frequencies, respectively. Free space is assumed to be layer 0.

Several design approaches exist for the above problem. An effective way to do so is to use an exact penalty method and to combine the two objective functions in a single one. The objective function can be expressed as

$$F(f,\theta) = R(f,\theta) + \Xi \cdot \max\left\{0, d_{tot} - d_{des}\right\}$$

(21)

where Ξ is a very large number and d_{des} is the desired maximum total thickness. Parameter Ξ is chosen large enough to ensure that solutions which do not fulfill the constraint result in large fitness values. By minimizing the above formula with a global optimizer a solution can be found. The same algorithm can run for different values of d_{des} and different designs for different total thicknesses can be found.

We solve the above problem using a slightly modified hybrid variant of IWPSO. This uses the IWPSO position and velocity update rules for the real variables and binPSO for the discrete variables. We select a swarm size of 100 particles and set the number of iterations up to 3000.

A five-layer broadband absorber in the frequency range from 200MHz to 2GHz for normal incidence and TE polarization is considered. The design parameters are the material number ID_i and the layer thickness d_i. Here, we use a materials database in the literature (Michielssen, Sajer, Ranjithan, & Mittra, 1993). This database comprises several types of materials such as lossless and lossy dielectrics, lossy magnetics and relaxation type magnetic ones. The maximum layer thickness is 2mm. The calculated results are compared against the solutions obtained from a binary coded GA (Michielssen et al., 1993). Two design examples are considered. Table 1 presents the optimal design parameters found by the hybrid PSO variant and those reported in the literature using the GA. The characteristics of the materials are presented in Table 2; for brevity, we give the characteristics of the finally selected ones only. Finally, Figure 2 illustrates the frequency response of the structures. We notice that the two algorithms obtain absorber structures with similar performance. However, the PSO gives slightly thinner geometries.

Linear Antenna Array Synthesis

Several applications of modern wireless communications systems require radiation characteristics such as higher directivity that cannot be achieved by using a single antenna. The use of antenna arrays is the obvious solution for such a case. Among others, antenna arrays are employed for several radar and wireless communications applications in space and on earth. Their advantages include the possibility of fast scanning and precise control of the radiation pattern.

Table 1. Optimal design parameters (the distances are in mm)

Layer	Design example #1				Design example #2			
	GA		PSO		GA		PSO	
	ID_i	d_i	ID_i	d_i	ID_i	d_i	ID_i	d_i
1	14	0.966	14	0.96180	16	0.516	16	0.02335
2	8	1.002	8	0.94795	4	1.092	16	0.45153
3	4	1.182	6	0.04390	4	1.440	4	1.99725
4	4	0.984	4	1.69495	4	0.306	4	1.01406
5	4	1.380	4	1.85884	4	0.234	4	0.01202
6	Ground plane (PEC)							
d_{tot}		5.514		**5.494**		3.588		**3.498**

Table 2. Materials characteristics

Lossy Magnetic Materials ($\varepsilon' = 15$, $\varepsilon'' = 0$) $\mu = \mu' - j\mu''$ $\mu'(f) = \mu'(1\text{GHz})f^{-a}$ $\mu''(f) = \mu''(1\text{GHz})f^{-b}$				
#	$\mu'(1GHz)$	a	$\mu''(1GHz)$	b
4	3	1.000	15	0.957
Lossy Dielectric Materials ($\mu' = 1$, $\mu'' = 0$) $\varepsilon = \varepsilon' - j\varepsilon''$ $\mu'(f) = \mu'(1\text{GHz})f^{-a}$ $\varepsilon''(f) = \varepsilon''(1\text{GHz})f^{-b}$				
#	$\varepsilon'(1GHz)$	a	$\varepsilon''(1GHz)$	b
6	5	0.861	8	0.569
8	10	0.778	6	0.861
Relaxation-type Magnetic Materials ($\varepsilon' = 15$, $\varepsilon'' = 0$) $\mu = \mu' - j\mu''$ $\mu'(f) = \dfrac{\mu_m f_m^2}{f^2 + f_m^2}$ $\mu''(f) = \dfrac{\mu_m f_m f}{f^2 + f_m^2}$				
#	μ_m		f_m	
14	30		2.5	
16	25		3.5	
f is measured in GHz				

Figure 2. Plots of absorbers' frequency response; solid (dashed) lines refer to the PSO- (GA-) based solutions

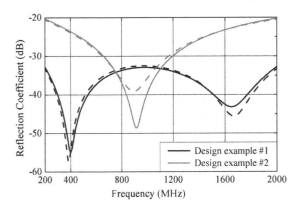

Figure 3. Geometry of the antenna array

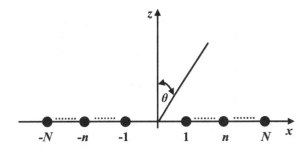

Synthesis of linear antenna arrays has been extensively studied in the last decades using several analytical or stochastic methods (Bayraktar, Werner, & Werner, 2006; Benedetti, Azaro, Franceschini, & Massa, 2006; Boeringer & Werner, 2004; Ismail, & Hamici, 2010; Modiri & Kiasaleh, 2011b). The linear array design problem is multimodal and therefore requires the use of an optimization method that does not easily get trapped in local minima. Common optimization goals in array synthesis are the sidelobe level suppression (while preserving the main lobe gain) and the null control to reduce interference effects. For a uniformly excited linear array the above goals can be achieved by finding the optimum element positions.

We assume a $2N$-element linear array which is symmetrically placed along the x-axis, see Figure 3. The array elements are isotropic sources.

The antenna array factor in the xz-plane can be written as

$$AF(\theta) = 2\sum_{n=1}^{N} I_n \cos\left(\frac{2\pi}{\lambda} x_n \sin\theta + \varphi_n\right) \quad (22)$$

where I_n and φ_n are the excitation amplitude and phase, respectively, of the nth element, x_n is its position and λ is the wavelength. In a uniform excited array, it is $I_n = 1$ and $\varphi_n = 0$, $n = 1,...,N$. As a result, Equation 22 is simplified into

$$AF(\theta) = 2\sum_{n=1}^{N} \cos\left(\frac{2\pi}{\lambda} x_n \sin\theta\right) \quad (23)$$

Our objective is to find the optimum element positions that minimize *SLL* while setting the mainlobe to a desired beamwidth BW_d within $\pm\Delta\theta$ deg. The use of an exact penalty method provides an effective way to combine the above objectives into a single cost function. In this case, the design problem is defined by the minimization of the objective function

$$F(\overline{x}) = \max_{\theta \in S}\left\{AF_{dB}^{\overline{x}}(\theta)\right\} + \\ \Xi \cdot \max\left\{0, \left|BW_c - BW_d\right| - \Delta\theta\right\} \quad (24)$$

with $\overline{x}$ the vector of the element positions, S the space spanned by the angle θ excluding the main lobe and BW_c the calculated beamwidth.

The additional requirement of desired null level at specific directions modifies the objective function as:

$$F'(\overline{x}) = F(\overline{x}) + \Xi \cdot \left(\sum_{k=1}^{K} \max\left\{0, AF_{dB}^{\overline{x}}(\theta_k) - C_{dB}\right\}\right) \quad (25)$$

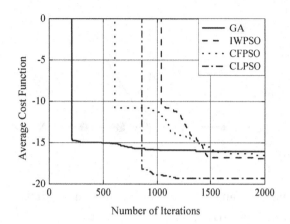

Figure 4. Convergence rate plot for the 28-element array case

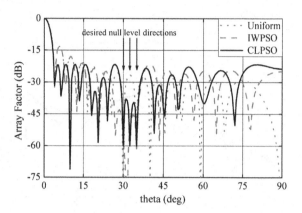

Figure 5. Optimized and uniform arrays patterns

where K denotes the number of the null directions, C_{dB} gives the desired null level (in dB) and θ_k is the direction of the kth null.

In a representative example taken from Khodier and Christodoulou (2005), we design a 28-element array with desired nulls at 30, 32.5, and 35 degrees. The desired null level and main-lobe beamwidth are –60dB and 8.35deg, respectively. The beamwidth tolerance is set to ±12%. We apply three different PSO variants and a real-coded GA to the above-mentioned problem. The PSO algorithms are the IWPSO, the CFPSO and the CLPSO. In all the cases, the population size is 40 and the number of iterations is 2000. Each algorithm is initialized by using random values and it is executed twenty times. The best results are compared. Parameter Ξ is set to 10^6.

The convergence rate plot is shown in Figure 4. Clearly, CLPSO outperforms IWPSO and CFPSO methods in convergence speed and cost function value. The GA algorithm convergences faster than the PSO variants, but at the expense of a higher cost function value. Table 3 holds the optimal element positions in the array resulted from the CLPSO. The peak *SLL* is –21.63dB and the null levels are –60.04, –60.01 and –60dB at 30, 32.5 and 35deg, respectively. Khodier and Christodoulou (2005) solved the same problem with IWPSO. In that case, the optimized antenna's peak *SLL* was –13.27dB and the respective null levels were –52.74dB, –51.66dB and –61.46dB. The obtained radiation pattern is plotted in Figure 5. Notice that the array resulted from CLPSO gives a lower sidelobe level at least for the first two sidelobes closer to the main lobe. For comparison reasons, the radiation pattern of a uniform linear array with 28 elements is also drawn.

Table 3. Elements positions (normalized with respect to $\lambda/2$)

Element #	1	2	3	4	5	6	7
Position	0.470	1.322	2.263	3.178	4.142	5.369	6.212
Element #	8	9	10	11	12	13	14
Position	7.135	8.313	9.794	11.192	12.792	14.360	15.960

Thinned Planar Microstrip Array Design

Aperiodic antenna arrays have received great attention with the advances in both radio astronomy and radar techniques. Thinned arrays are a special type of aperiodic arrays in which a fraction of the elements in a uniformly spaced array are turned off to reduce the grating lobes resulted from the periodic grid. Thinned arrays are used in modern wireless communications ranging from cellular systems like UMTS to satellite communications.

In the third example, we focus on the constrained design of thinned planar microstrip arrays. Let us consider a $M \times M$ planar array with elements rectangular microstrip patches placed on a substrate of relative dielectric constant ε_r and thickness h. The array elements are uniformly excited with equal phase, simplifying the structure of the feeding network and providing a broadside radiation pattern.

In order to describe the array geometry, we consider a binary string of $D = M \times M$ bits in which "1" denotes the existence and "0" the absence of element. This string can be considered as the position of a particle in a binary PSO variant, a chromosome in a binary GA method, etc. The binary string that corresponds to the minimum cost function provides the optimum antenna array geometry. Next, we apply the method-of-moments (MoM) (Gibson, 2008) to calculate the far-field gain, $g(\theta, \varphi)$, and SLL of the array and the input impedance Z_{in} of each element. The last is used for the calculation of the return loss RL at the input of each element that is equal to

$$RL = 20 \log \left| \frac{Z_{in} - Z_0}{Z_{in} + Z_0} \right| \quad (\text{in dB}) \qquad (26)$$

with Z_0 the characteristic impedance of the feeding network (a typical value is 50 Ohm). The RL calculation for all the elements allows the derivation of the average return loss among them, RL^{avg}. Low RL^{avg} values ensure that the impedance matching condition is well-satisfied for all the array elements.

Here, we desire a maximum far-field gain g_m at $\theta = 0°$, a side lobe level threshold SLL_{des} and a maximum average return loss RL_{des}^{avg}. Therefore, the cost function may be formed as a weighted sum of three terms, i.e., it is

$$F = W_g \, g_m + W_s F_s + W_r F_r \qquad (27)$$

with

$$F_s = R\left(SLL - SLL_{des}\right) \qquad (28)$$

$$F_r = R\left(RL^{avg} - RL_{des}^{avg}\right) \qquad (29)$$

where $R(\cdot)$ is the ramp function.

In case we require null control, i.e., a gain below a threshold value g_{des} at a specific direction (θ, φ), Equation 27 is modified into

$$F = W_g \, g_m + W_s F_s + W_r F_r + W_n F_n \qquad (30)$$

where

$$F_n = R\left(g\left(\theta, \varphi\right) - g_{des}\right) \qquad (31)$$

Here, we solve this design problem with the BPSO-vm variant. In order to validate our approach, comparisons with the BPSO, the binPSO, and a binGA variant (Dorica & Giannacopoulos, 2007a, 2007b) are also performed. In all cases, the population size is 30 and the maximum number of iterations in each run is 200. The algorithms are executed 20 times.

Two antenna design examples are studied. In both, it is $\varepsilon_r = 4$ and $h = 0.3$cm while the side length of each element is 3.4cm. Thus, the resonant

Table 4. Cost function characteristics (the best values are in bold)

Optimization Method	Minimum Value		Maximum Value		Mean Value		Standard Deviation	
	Case 1	Case 2	Case 1	Case 2	Case 1	Case 2	Case 1	Case 2
BPSO-vm	**234.1**	**509.9**	**365.4**	**721.7**	**304.9**	**619.0**	**17.3**	**35.8**
binPSO	526.7	625.1	911.9	1017.5	728.2	820.8	53.9	66.4
BPSO	458.5	622.7	731.6	988.4	609.3	805.9	36.4	60.1
binGA	290.2	548.2	468.7	841.4	385.1	695.5	27.1	47.7

frequency of each single array element placed on the substrate is 2.14GHz (UMTS downlink). In the first example, we consider a 12×12 thinned planar array composed of elements that are symmetrical arranged with respect to the *x*- and *y*-axis. This symmetry simplifies the problem because the optimization is applied on a quarter of the antenna surface only. Thus, the array geometry is described by a 36-bit string. The distance between the centers of the elements along the *x*- and *y*-axis is 6.8cm. The cost function is described from Equation 27, where $W_g = -1$ and $W_s = W_r = 100$. The side lobe level and impedance matching requirements are $SLL_{des} = -20$dB and $RL_{des}^{avg} = -20$dB, respectively. In the second example, we consider an 8×8 thinned planar microstrip array. Now, the array geometry is de-

scribed from a 64-bit string. The distance between the centers of the elements along the *x*- and *y*-axis is 5.4cm. In this design case, we set (again) $SLL_{des} = RL_{des}^{avg} = -20$dB but we further desire a gain threshold equal to –25dBi at $\theta = 45 \deg$ and $\varphi = 50 \deg$. Therefore, the cost function is the (30) in which we set $W_n = 100$ (the rest of the parameters are kept the same).

In Table 4, the minimum, maximum, mean and standard deviation of the cost function are given for all the optimization algorithms and both design examples. The convergence rate of the algorithms is shown in Figures 6 and 7. It is obvious that the BPSO-vm outperforms the other methods in terms of cost values and standard deviation but has a slightly slower convergence rate. A comparison

Figure 6. Average cost function versus number of iterations; design example 1

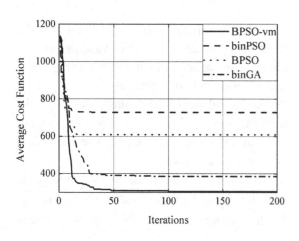

Figure 7. Average cost function versus number of iterations; design example 2

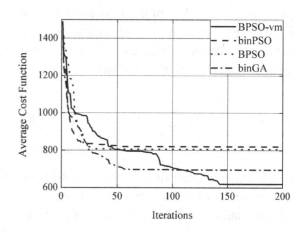

Figure 8. Optimized array geometry: Design example 1 (a) and 2 (b)

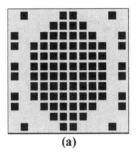 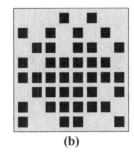

(a)   (b)

Figure 9. Radiation patterns of the optimized arrays: Design example 1 (a) and 2 (b)

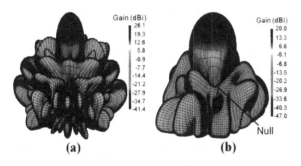

(a)   (b)

Figure 10. Frequency response of the thinned planar microstrip arrays resulted from the BPSO-vm

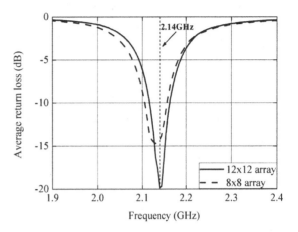

comes that both arrays show an excellent frequency response. Finally, Figure 10 illustrates the frequency response of the two arrays.

Optimal Design of a Dual-Band Base Station Antenna

The simultaneous transmission of radio signals in different frequency bands is a common solution in order to provide a wide variety of services through a wireless network. Among different techniques, dual-band antennas are commonly used. Base stations play a crucial role in any cellular network. The wide spread of 3G networks and wireless LANs imposes new requirements for base station antennas. In the last example of this chapter, we study the design of such a structure. In particular, our objective is the optimal design of a dual-band base station K-element array that operates in the UMTS/WLAN frequency bands. Each array element comprises an active dipole covered by a cylindrical dielectric coating. Adjacent to the coating surface a second (parasitic) dipole is placed, see Figure 11. The system also includes a corner reflector. The array excitation is uniform with equal phases. The geometry parameters to be optimized are the active and the parasitic dipole lengths L_a and L_p, respectively,

between binGA and the other two PSO variants shows the improved performance of the first.

The optimized geometries for the two design examples resulted from the BPSO-vm are shown in Figure 8 while the three-dimensional radiation patterns of these arrays are presented in Figure 9. The maximum far-field gain and sidelobe level values of the first optimized array (design example 1) are $g_m = 26.1$dBi and $SLL = -20$dB, respectively; the return loss of its elements at 2.14GHz varies from −11.3dB to −37dB with $RL_{avg} = -19.7$ dB. The second optimized array (design example 2) has $g_m = 20$dBi and $SLL = -20$dB while the return loss of its elements at 2.14MHz varies from −10.1dB to −27.3dB with $RL_{avg} = -14.7$dB. Its gain at $\theta = 45 \deg$ and $\varphi = 50 \deg$ is −25dBi, i.e., its radiation level is 45dB below the maximum. It obviously

Figure 11. Array element geometry

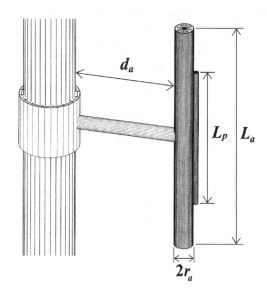

Figure 12. Pareto fronts found by the MOPSO-fs and NSGA-II

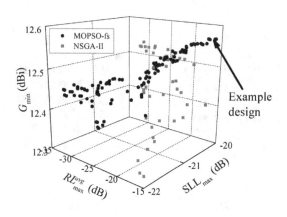

the reflector length L_r, the radius of the dielectric coating r_a, the vertical distance between the closest end points of any two adjacent active dipoles z_a and the distance d_a between an active dipole and the surface of the mast.

In general, this problem requires the optimization of $N = 4K + 1$ design parameters. However, the problem complexity reduces significantly in symmetric structures. For example, if the axis of symmetry is in the middle element normal to the element axis then $N = 2K + 2$ if K is even and $N = 2K + 3$ if it is odd. The problem is defined by three objectives which are subject to six constraints. The objectives are the minimization of the average return loss between all elements RL_f^{avg}, the minimization of the side lobe levels SLL_f, and the third is the gain, G_f, maximization, where the subscript f denotes the operating frequency ($f = 1, 2$).

In order to solve this optimization problem, we apply the MOPSO-fs and the NSGA-II (Deb, Pratap, Agarwal, & Meyarivan, 2002) algorithms. In both cases, the population size is 140 and the number of iterations in an execution run is 2000. Each algorithm is executed 10 times.

The array concerns a six-element symmetrical array design (14 design parameters). The operating frequencies are 2.14GHz and 2.442GHz for UMTS and WiFi, respectively, transmission. The dielectric cover has relative permittivity 2.5 which remains constant in the operating frequency range. The diameter of the mast d_m is 2.54 cm. The imposed constraints are $RL_f^{avg} \leq -15\text{dB}$ and $SLL_f \leq -20\text{dB}$. Also, the horizontal half-power beamwidth for both frequencies is limited to $120 \deg$ with tolerance set to one degree.

Each point of the Pareto front is a feasible design solution with coordinates:

$$RL_{\max}^{avg} = \max_{f=1,2}\{RL_f^{avg}\},$$

$$SLL_{\max} = \max_{f=1,2}\{SLL_f\} \text{ and}$$

$$G_{\min} = \min_{f=1,2}\{G_f\}.$$

The Pareto fronts produced by the two algorithms are shown in Figure 12. In particular, MOPSO-fs has found 140 points of the Pareto front while NSGA-II has found only 88. It must be pointed out the MOPSO-fs required 1345 iterations to find 140 feasible solutions while NSGA-II required 1896 iterations to find 88 feasible solutions. The points on the Pareto front found by the

Table 5. Antenna geometric parameters (units in cm)

*i*th element	$L_a(i)$	$L_p(i)$	$d_a(i)$	$z_a(i, i+1)$
1	4.74	4.49	3.63	3.90
2	4.74	4.24	3.31	1.76
3	6.23	3.76	3.18	2.20
4	5.95	3.76	3.18	1.76
5	4.74	4.24	3.31	3.90
6	4.74	4.49	3.63	
$r_a = 0.4$, $L_r = 3.07$				

Figure 13. Return loss versus frequency

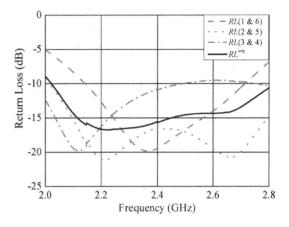

MOPSO-fs have larger dispersion. We select a point (shown with an arrow in Figure 12) of the Pareto front found by MOPSO-fs. This point represents a feasible antenna design with $RL_{\max}^{avg} = -15.1\text{dB}$, $SLL_{\max} = -20\text{dB}$ and $G_{\min} = 12.57\text{dBi}$. The geometric parameters of this antenna are reported in Table 5. We illustrate the return loss of each active dipole and the average return loss versus frequency in Figure 13. The antenna's horizontal and vertical radiation patterns are plotted in Figure 14. At this point, it should be mentioned that MOPSO-fs not only generates a larger number of feasible solutions compared to the NSGA-II but, in average, provides a faster convergence too.

FUTURE RESEARCH DIRECTIONS

The research domain of evolutionary algorithms and especially swarm algorithms is growing rapidly. Specific PSO optimizers that work well for a given problem are introduced. A current and growing research trend in evolutionary algorithms is their hybridization with local optimizers. These algorithms are called Memetic Algorithms (MAs). MAs are inspired by Dawkins' notion of meme. The advantage of such an approach is that the use

Figure 14. Horizontal and vertical antenna radiation patterns

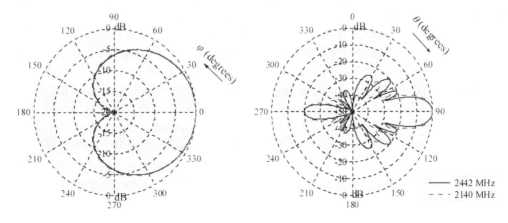

of a local search optimizer ensures that specific regions of the search space can be explored using fewer evaluations and good quality solutions can be generated early during the search. Furthermore, the global search algorithm generates good initial solutions for the local search. MAs can be highly efficient due to this combination of global exploration and local exploitation. So far several PSO-based MAs have been proposed in order to overcome some of the drawbacks and to improve the performance of common PSO optimizers. Recent developments include applications to real world problems like DNA sequence compression (Zhu, Zhou, Ji, & Shi, 2011) and face recognition algorithms (Zhou, Ji, Shen, Zhu, & Chen, 2011). Among others recent Memetic PSO algorithms include multimodal (Wang, Wang, & Wang, 2011) and multi-objective (Wei & Shang, 2011) optimization problems.

Parallelization techniques that use several CPUs or multicore programming can also be used in order to reduce computational cost. The Graphics Processing Unit (GPU) has also been used in electromagnetics combined with a numerical method in order to speed up calculation time (Gao, 2012; Junkin, 2011; Lee, Ahmed, Goh, Khoo, Li, & Hung, 2011; Peng & Nie, 2008; Tao, Lin, & Bao, 2010). PSO-based optimizers can be easily combined with such techniques to reduce computation time.

The application of multi-objective PSO algorithms to antenna and microwave design problems introduces new challenges regarding performance and computational cost. Recently a novel two local bests (lbest) based MOPSO (2LB-MOPSO) version is proposed (Zhao & Suganthan, 2010). This new MOPSO version enables the search around small regions in the parameter space in the vicinity of the best existing fronts. Therefore, new research directions have to be explored.

CONCLUSION

In this chapter, we discussed and evaluated the application of particle swarm optimizers to antenna and microwave design problems. We have presented common and state-of-the-art PSO methods found in the literature. These PSO methods were compared with other evolutionary algorithms on common design problems in electromagnetics. The obtained results exhibit the applicability and efficiency of particle swarm optimizers. The emerging trends and future research directions complete our study.

Within this context, we presented the classical IWPSO and CFPSO algorithms. We discussed about the CLPSO variant, which is a powerful optimizer for solving multimodal optimization problems. Two PSO versions for discrete-valued problems, the binPSO and the BPSO-vm were presented. A brief description of MOPSO and MOPSO-fs algorithms, which are suitable for solving multi-objective optimization problems subject to constraints, was given.

We applied the IWPSO algorithm for the design of a multilayer planar microwave absorber. The problem is inherently bi-objective and the two objective functions were combined into a single-objective one. The results reveal that this algorithm produces at a lower computational cost, slightly thinner geometries compared to GAs. The CLPSO, the common IWPSO and CFPSO algorithms and a real-coded GA were used for solving a constrained linear array synthesis problem. The derived results showed that CLPSO outperforms the two common PSO algorithms and the GA in terms of convergence rate and cost function values. Its major advantage is its updating strategy which results to a larger potential search space. Therefore, CLPSO is best suited for solving complex multimodal problems that are common in electromagnetics. We studied the problem of thinned planar microstrip array design and showed that the BPSO-vm outperforms other discrete PSO variants and a binary-coded GA in terms of best fitness value and standard deviation. We applied the MOPSO-fs to the design of a dual-band base station array. Our study showed that the MOPSO-fs algorithm performs better in finding the Pareto front than the popular NSGA-II algorithm.

Recent PSO algorithms and their application to electromagnetic design will be part of our future work. Furthermore, we plan to explore the applicability of other state-of-the-art algorithms to antenna and microwave design problems such as the recently proposed new MOPSO (Zhao & Suganthan, 2010).

PSO-based optimizers can be used in cases where the computational time plays an important role. Comparisons against popular evolutionary optimization methods in the literature showed that the PSO variants outperform or produce similar results in terms of solution accuracy and convergence speed. The PSO algorithms combined with a numerical method can be valuable tools for constrained single or multi-objective optimization problems in antennas and microwaves.

REFERENCES

Afshinmanesh, F., Marandi, A., & Rahimi-Kian, A. (2005, November 21-24). *A novel binary particle swarm optimization method using artificial immune system*. Paper presented at the EUROCON 2005 – The International Conference on "Computer as a Tool", Belgrade, Serbia and Montenegro.

Baskar, S., Alphones, A., Suganthan, P. N., & Liang, J. J. (2005). Design of Yagi-Uda antennas using comprehensive learning particle swarm optimisation. *IEE Proceedings. Microwaves, Antennas and Propagation*, *152*(5), 340–346. doi:10.1049/ip-map:20045087

Bayraktar, Z., Werner, P. L., & Werner, D. H. (2006). The design of miniature three-element stochastic Yagi-Uda arrays using particle swarm optimization. *IEEE Antennas and Wireless Propagation Letters*, *5*(1), 22–26. doi:10.1109/LAWP.2005.863618

Benedetti, M., Azaro, R., Franceschini, D., & Massa, A. (2006). PSO-based real-time control of planar uniform circular arrays. *IEEE Antennas and Wireless Propagation Letters*, *5*(1), 545–548. doi:10.1109/LAWP.2006.887553

Boeringer, D. W., & Werner, D. H. (2004). Particle swarm optimization versus genetic algorithms for phased array synthesis. *IEEE Transactions on Antennas and Propagation*, *52*(3), 771–779. doi:10.1109/TAP.2004.825102

Chen, W.-N., Zhang, Z., Chung, H. S. H., Zhong, W.-L., Wu, W.-G., & Shi, Y.-H. (2010). A novel set-based particle swarm optimization method for discrete optimization problems. *IEEE Transactions on Evolutionary Computation*, *14*(2), 278–300. doi:10.1109/TEVC.2009.2030331

Chew, W. C. (1995). *Waves and fields in inhomogeneous media*. New York, NY: IEEE Press.

Clerc, M. (1999, July 6-9). *The swarm and the queen: Towards a deterministic and adaptive particle swarm optimization*. Paper presented at the 1999 Congress on Evolutionary Computation, Washington, USA.

Coello Coello, C. A., Pulido, G. T., & Lechuga, M. S. (2004). Handling multiple objectives with particle swarm optimization. *IEEE Transactions on Evolutionary Computation*, *8*(3), 256–279. doi:10.1109/TEVC.2004.826067

Cui, S., & Weile, D. S. (2005). Application of a parallel particle swarm optimization scheme to the design of electromagnetic absorbers. *IEEE Transactions on Antennas and Propagation*, *53*(11), 3616–3624. doi:10.1109/TAP.2005.858866

Deb, K., Pratap, A., Agarwal, S., & Meyarivan, T. (2002). A fast and elitist multiobjective genetic algorithm: NSGA-II. *IEEE Transactions on Evolutionary Computation, 6*(2), 182–197. doi:10.1109/4235.996017

Deligkaris, K. V., Zaharis, Z. D., Kampitaki, D. G., Goudos, S. K., Rekanos, I. T., & Spasos, M. N. (2009). Thinned planar array design using Boolean PSO with velocity mutation. *IEEE Transactions on Magnetics, 45*(3), 1490–1493. doi:10.1109/TMAG.2009.2012687

Dorica, M., & Giannacopoulos, D. D. (2007a). Evolution of wire antennas in three dimensions using a novel growth process. *IEEE Transactions on Magnetics, 43*(4), 1581–1584. doi:10.1109/TMAG.2006.892105

Dorica, M., & Giannacopoulos, D. D. (2007b). Evolution of two-dimensional electromagnetic devices using a novel genome structure. *IEEE Transactions on Magnetics, 43*(4), 1585–1588. doi:10.1109/TMAG.2006.892106

Eberhart, R. C., & Shi, Y. (2000, July 16-19). *Comparing inertia weights and constriction factors in particle swarm optimization.* Paper presented at the 2000 IEEE Conference on Evolutionary Computation, California, CA, USA.

Eberhart, R. C., & Shi, Y. (2001, May 27-30). *Particle swarm optimization: Developments, applications and resources.* Paper presented at the 2001 Congress on Evolutionary Computation, Seoul, Korea.

Gao, P. C., Tao, Y. B., Bai, Z. H., & Lin, H. (2012). Mapping the SBR and TS-ILDCs to heterogeneous CPU-GPU architecture for fast computation of electromagnetic scattering. *Progress in Electromagnetics Research, 122,* 137–154. doi:10.2528/PIER11092303

Gibson, W. C. (2008). *The method of moments in electromagnetics.* Boca Raton, FL: CRC Press.

Goudos, S. K., Moysiadou, V., Samaras, T., Siakavara, K., & Sahalos, J. N. (2010). Application of a comprehensive learning particle swarm optimizer to unequally spaced linear array synthesis with sidelobe level suppression and null control. *IEEE Antennas and Wireless Propagation Letters, 9,* 125–129. doi:10.1109/LAWP.2010.2044552

Goudos, S. K., Rekanos, I. T., & Sahalos, J. N. (2008). EMI reduction and ICs optimal arrangement inside high-speed networking equipment using particle swarm optimization. *IEEE Transactions on Electromagnetic Compatibility, 50*(3), 586–596. doi:10.1109/TEMC.2008.924389

Goudos, S. K., & Sahalos, J. N. (2006). Microwave absorber optimal design using multi-objective particle swarm optimization. *Microwave and Optical Technology Letters, 48*(8), 1553–1558. doi:10.1002/mop.21727

Goudos, S. K., Zaharis, Z. D., Kampitaki, D. G., Rekanos, I. T., & Hilas, C. S. (2009). Pareto optimal design of dual-band base station antenna arrays using multi-objective particle swarm optimization with fitness sharing. *IEEE Transactions on Magnetics, 45*(3), 1522–1525. doi:10.1109/TMAG.2009.2012695

Huang, T., & Mohan, A. S. (2007). A microparticle swarm optimizer for the reconstruction of microwave images. *IEEE Transactions on Antennas and Propagation, 55*(3), 568–576. doi:10.1109/TAP.2007.891545

Ismail, T. H., & Hamici, Z. M. (2010). Array pattern synthesis using digital phase control by quantized particle swarm optimization. *IEEE Transactions on Antennas and Propagation, 58*(6), 2142–2145. doi:10.1109/TAP.2010.2046853

Jin, N., & Rahmat-Samii, Y. (2007). Advances in particle swarm optimization for antenna designs: Real number, binary, single-objective and multi-objective implementations. *IEEE Transactions on Antennas and Propagation, 55*(3), 556–567. doi:10.1109/TAP.2007.891552

Junkin, G. (2011). Conformal FDTD modeling of imperfect conductors at millimeter wave bands. *IEEE Transactions on Antennas and Propagation, 59*(1), 199–205. doi:10.1109/TAP.2010.2090490

Kennedy, J., & Eberhart, R. (1995, November 27 - December 1). *Particle swarm optimization.* Paper presented at the 1995 IEEE International Conference on Neural Networks, Perth, Australia.

Kennedy, J., & Eberhart, R. (1997, October 12-15). *A discrete binary version of the particle swarm algorithm.* Paper presented at the 1997 IEEE International Conference on Systems, Man and Cybernetics, Orlando, USA.

Khodier, M. M., & Christodoulou, C. G. (2005). Linear array geometry synthesis with minimum sidelobe level and null control using particle swarm optimization. *IEEE Transactions on Antennas and Propagation, 53*(8), 2674–2679. doi:10.1109/TAP.2005.851762

Lee, K. H., Ahmed, I., Goh, R. S. M., Khoo, E. H., Li, E. P., & Hung, T. G. G. (2011). Implementation of the FDTD method based on Lorentz-Drude dispersive model on GPU for plasmonics applications. *Progress in Electromagnetics Research, 116*, 441–456.

Liang, J. J., Qin, A. K., Suganthan, P. N., & Baskar, S. (2006). Comprehensive learning particle swarm optimizer for global optimization of multimodal functions. *IEEE Transactions on Evolutionary Computation, 10*(3), 281–295. doi:10.1109/TEVC.2005.857610

Marandi, A., Afshinmanesh, F., Shahabadi, M., & Bahrami, F. (2006, July 16-21). *Boolean particle swarm optimization and its application to the design of a dual-band dual-polarized planar antenna.* Paper presented at the 2006 IEEE Congress on Evolutionary Computation, Vancouver, Canada.

Michielssen, E., Sajer, J.-M., Ranjithan, S., & Mittra, R. (1993). Design of lightweight, broad-band microwave absorbers using genetic algorithms. *IEEE Transactions on Microwave Theory and Techniques, 41*(6), 1024–1031. doi:10.1109/22.238519

Modiri, A., & Kiasaleh, K. (2011a, August 30 - September 3). *Permittivity estimation for breast cancer detection using particle swarm optimization algorithm.* Paper presented at the 2011 Annual International Conference of the IEEE Engineering in Medicine and Biology Society, Boston, USA.

Modiri, A., & Kiasaleh, K. (2011b). Modification of real-number and binary PSO algorithms for accelerated convergence. *IEEE Transactions on Antennas and Propagation, 59*(1), 214–224. doi:10.1109/TAP.2010.2090460

Peng, S., & Nie, Z. (2008). Acceleration of the method of moments calculations by using graphics processing units. *IEEE Transactions on Antennas and Propagation, 56*(7), 2130–2133. doi:10.1109/TAP.2008.924768

Robinson, J., & Rahmat-Samii, Y. (2004). Particle swarm optimization in electromagnetics. *IEEE Transactions on Antennas and Propagation, 52*(2), 397–407. doi:10.1109/TAP.2004.823969

Salazar-Lechuga, M., & Rowe, J. E. (2005, September 2-5). *Particle swarm optimization and fitness sharing to solve multi-objective optimization problems.* Paper presented at the 2005 IEEE Congress on Evolutionary Computation, Edinburgh, UK.

Sharaf, A. M., & El-Gammal, A. A. A. (2009, November 15-17). *A discrete particle swarm optimization technique (DPSO) for power filter design.* Paper presented at the 2009 4th International Design and Test Workshop, Riyadh, Saudi Arabia.

Shi, Y., & Eberhart, R. C. (1998). Parameter selection in particle swarm optimization . In Porto, V. W., Saravanan, N., Waagen, D., & Eiben, A. E. (Eds.), *Evolutionary Programming VII* (*Vol. 1447*, pp. 591–600). Lecture Notes in Computer Science Berlin, Germany: Springer. doi:10.1007/BFb0040810

Shi, Y., & Eberhart, R. C. (1999, July 6-9). *Empirical study of particle swarm optimization*. Paper presented at the 1999 Congress on Evolutionary Computation, Washington DC, USA.

Tao, Q., Chang, H.-Y., Yi, Y., Gu, C.-Q., & Li, W.-J. (2010, July 9-11). *An analysis for particle trajectories of a discrete particle swarm optimization*. Paper presented at the 2010 3rd IEEE International Conference on Computer Science and Information Technology, Chengdu, China.

Tao, Y., Lin, H., & Bao, H. (2010). GPU-based shooting and bouncing ray method for fast RCS prediction. *IEEE Transactions on Antennas and Propagation, 58*(2), 494–502. doi:10.1109/TAP.2009.2037694

Wang, H., Wang, N., & Wang, D. (2011, May 23-25). *A memetic particle swarm optimization algorithm for multimodal optimization problems*. Paper presented at the 2011 Chinese Control and Decision Conference, Mianyang, China.

Wei, J., & Zhang, M. (2011, June 5-8). *A memetic particle swarm optimization for constrained multi-objective optimization problems*. Paper presented at the 2011 IEEE Congress on Evolutionary Computation, New Orleans, USA.

Zaharis, Z. D. (2008). Radiation pattern shaping of a mobile base station antenna array using a particle swarm optimization based technique. *Electrical Engineering, 90*(4), 301–311. doi:10.1007/s00202-007-0078-y

Zaharis, Z. D., Kampitaki, D. G., Lazaridis, P. I., Papastergiou, A. I., & Gallion, P. B. (2007). On the design of multifrequency dividers suitable for GSM/DCS/PCS/UMTS applications by using a particle swarm optimization-based technique. *Microwave and Optical Technology Letters, 49*(9), 2138–2144. doi:10.1002/mop.22658

Zhao, S.-Z., & Suganthan, P. N. (2011). Two-*lbests* based multi-objective particle swarm optimizer. *Engineering Optimization, 43*(1), 1–17. doi:10.1080/03052151003686716

Zhou, J., Ji, Z., Shen, L., Zhu, Z., & Chen, S. (2011, April 11-15). *PSO based memetic algorithm for face recognition Gabor filters selection*. Paper presented at the 2011 IEEE Workshop on Memetic Computing, Paris, France.

Zhu, Z., Zhou, J., Ji, Z., & Shi, Y.-S. (2011). DNA sequence compression using adaptive particle swarm optimization-based memetic algorithm. *IEEE Transactions on Evolutionary Computation, 15*(5), 643–658. doi:10.1109/TEVC.2011.2160399

ADDITIONAL READING

Carro, P. L., De Mingo, J., & Ducar, P. G. (2010). Radiation pattern synthesis for maximum mean effective gain with spherical wave expansions and particle swarm techniques. *Progress in Electromagnetics Research, 103*, 355–370. doi:10.2528/PIER10031808

Chamaani, S., Abrishamian, M. S., & Mirtaheri, S. A. (2010). Time-domain design of UWB Vivaldi antenna array using multiobjective particle swarm optimization. *IEEE Antennas and Wireless Propagation Letters, 9*, 666–669. doi:10.1109/LAWP.2010.2053691

Chamaani, S., & Mirtaheri, S. A. (2010). Planar UWB monopole antenna optimization to enhance time-domain characteristics using PSO. *AEÜ. International Journal of Electronics and Communications, 64*(4), 351–359. doi:10.1016/j. aeue.2008.11.017

Chamaani, S., Mirtaheri, S. A., & Abrishamian, M. S. (2011). Improvement of time and frequency domain performance of antipodal Vivaldi antenna using multi-objective particle swarm optimization. *IEEE Transactions on Antennas and Propagation, 59*(5), 1738–1742. doi:10.1109/ TAP.2011.2122290

Das, S., Abraham, A., & Konar, A. (2008). Particle swarm optimization and differential evolution algorithms: technical analysis, applications and hybridization perspectives. In Y. Liu, A. Sun, E. H. T. Loh, W. F. Lu, & E.-P. Lim (Eds.), *Studies in Computational Intelligence: Vol. 116 - Advances of Computational Intelligence in Industrial Systems* (pp. 1-38). Berlin, Germany: Springer.

Demarcke, P., Rogier, H., Goossens, R., & De Jaeger, P. (2009). Beamforming in the presence of mutual coupling based on constrained particle swarm optimization. *IEEE Transactions on Antennas and Propagation, 57*(6), 1655–1666. doi:10.1109/TAP.2009.2019923

Donelli, M., Martini, A., & Massa, A. (2009). A hybrid approach based on PSO and Hadamard difference sets for the synthesis of square thinned arrays. *IEEE Transactions on Antennas and Propagation, 57*(8), 2491–2495. doi:10.1109/ TAP.2009.2024570

Goudos, S. K., Baltzis, K. B., Bachtsevanidis, C., & Sahalos, J. N. (2010). Cell-to-switch assignment in cellular networks using barebones particle swarm optimization. *IEICE Electronics Express, 7*(4), 254–260. doi:10.1587/elex.7.254

Goudos, S. K., Zaharis, Z., Baltzis, K. B., Hilas, C., & Sahalos, J. N. (2009, June 11-12). *A comparative study of particle swarm optimization and differential evolution on radar absorbing materials for EMC applications*. Paper presented at the EMC Europe Workshop 2009, Athens, Greece.

Khodier, M., & Al-Aqeel, M. (2009). Linear and circular array optimization: A study using particle swarm intelligence. *Progress in Electromagnetics Research B, 15*, 347–373. doi:10.2528/ PIERB09033101

Lanza, M., Pérez, J. R., & Basterrechea, J. (2009). Synthesis of planar arrays using a modified particle swarm optimization algorithm by introducing a selection operator and elitism. *Progress in Electromagnetics Research, 93*, 145–160. doi:10.2528/ PIER09041303

Li, W., Hei, Y., Shi, X., Liu, S., & Lv, Z. (2010). An extended particle swarm optimization algorithm for pattern synthesis of conformal phased arrays. *International Journal of RF and Microwave Computer-Aided Engineering, 20*(2), 190–199.

Marler, R. T., & Arora, J. S. (2004). Survey of multi-objective optimization methods for engineering. *Structural and Multidisciplinary Optimization, 26*(6), 369–395. doi:10.1007/ s00158-003-0368-6

Pan, F., Hu, X., Eberhart, R., & Chen, Y. (2008, September 21-23). *An analysis of bare bones particle swarm*. Paper presented at the 2008 IEEE Swarm Intelligence Symposium, Saint Louis, USA.

Panduro, M. A., Brizuela, C. A., Balderas, L. I., & Acosta, D. A. (2009). A comparison of genetic algorithms, particle swarm optimization and the differential evolution method for the design of scannable circular antenna arrays. *Progress in Electromagnetics Research B, 13*, 171–186. doi:10.2528/PIERB09011308

Pérez, J. R., & Basterrechea, J. (2009). Hybrid particle swarm-based algorithms and their application to linear array synthesis. *Progress in Electromagnetics Research*, *90*, 63–74. doi:10.2528/PIER08122212

Shihab, M., Najjar, Y., Dib, N., & Khodier, M. (2008). Design of non-uniform circular antenna arrays using particle swarm optimization. *Journal of Electrical Engineering*, *59*(4), 216–220.

Wu, H., Geng, J., Jin, R., Qiu, J., Liu, W., Chen, J., & Liu, S. (2009). An improved comprehensive learning particle swarm optimization and its application to the semiautomatic design of antennas. *IEEE Transactions on Antennas and Propagation*, *57*(10), 3018–3028. doi:10.1109/TAP.2009.2028608

Yisu, J., Knowles, J., Hongmei, L., Yizeng, L., & Kell, D. B. (2008). The landscape adaptive particle swarm optimizer. *Applied Soft Computing*, *8*(1), 295–304. doi:10.1016/j.asoc.2007.01.009

Zhang, L., Yang, F., & Elsherbeni, A. Z. (2009). On the use of random variables in particle swarm optimizations: A comparative study of Gaussian and uniform distributions. *Journal of Electromagnetic Waves and Applications*, *23*(5), 711–721. doi:10.1163/156939309788019787

Zhang, S., Gong, S.-X., Guan, Y., Zhang, P.-F., & Gong, Q. (2009). A novel IGA-edsPSO hybrid algorithm for the synthesis of sparse arrays. *Progress in Electromagnetics Research*, *89*, 121–134. doi:10.2528/PIER08120806

Zhang, S., Gong, S.-X., & Zhang, P.-F. (2009). A modified PSO for low sidelobe concentric ring arrays synthesis with multiple constraints. *Journal of Electromagnetic Waves and Applications*, *23*(11-12), 1535–1544. doi:10.1163/156939309789476239

KEY TERMS AND DEFINITIONS

Genetic Algorithms: A stochastic population-based global optimization technique that mimics the process of natural evolution.

Method of Moments (MoM): A method for solving electromagnetic field problems using a full wave solution of Maxwell's integral equations in the frequency domain. The MoM is applicable to problems involving currents on metallic and dielectric structures and radiation in free space.

Non-Dominated Sorting Genetic Algorithm II (NSGA-II): A fast and elitist multi-objective evolutionary genetic algorithm. Its main features include a non-dominated sorting procedure and the implementation of elitism for multiobjective search, using an elitism-preserving approach.

Sidelobe Level (SLL): The ratio, usually expressed in decibels (dB), of the amplitude at the peak of the main lobe to the amplitude at the peak of a side lobe.

Universal Mobile Telecommunications System (UMTS): A third generation mobile cellular technology for networks developed by the 3GPP (3rd Generation Partnership Project).

Wireless Local Area Network (WLAN): A network in which a mobile user can connect to a local area network (LAN) through a wireless (radio) connection. Most modern WLANs are based on IEEE 802.11 standards, marketed under the Wi-Fi brand name.

Chapter 7

Optimum Design and Characterization of Rare Earth–Doped Fibre Amplifiers by Means of Particle Swarm Optimization Approach

Girolamo Fornarelli
Politecnico di Bari, Italy

Antonio Giaquinto
Politecnico di Bari, Italy

Luciano Mescia
Politecnico di Bari, Italy

ABSTRACT

The rapid increasing of internet services requires communication capacity of optical fibre networks. Such a task can be carried out by Er^{3+}-doped fibre amplifiers, which allow to overcome limits of unrelayed communication distances. The development of efficient numerical codes provides an accurate understanding of the optical amplifier behaviour and reliable qualitative and quantitative predictions of the amplifier performance in a large variety of configurations. Therefore, the design and optimization of the optical fibre can benefit of this important tool. This chapter proposes an approach based on the Particle Swarm Optimization (PSO) for the optimal design and the characterization of a photonic crystal fibre amplifier. Such approach is employed to find the optimal parameters maximizing the gain of the amplifier. The comparison with respect to a conventional algorithm shows that the proposed solution provides accurate results. Subsequently, the presented method is used to study the amplifier behaviour by evaluating the curves of optimal fibre length, erbium concentration, gain, and pumping configuration. Finally, the PSO based algorithm is exploited to determine the upconversion parameters corresponding to a desired value of gain. This application is particularly intriguing since it allows recovery of the values of parameters of the optical amplifier, which cannot be directly measured.

DOI: 10.4018/978-1-4666-2666-9.ch007

BACKGROUND

In the 90s the global breakthrough of Internet changed the telecommunications bandwidth requirements permanently. In fact, the rapid increasing of internet services demanded communication capacity which has been more than doubled every year. Such need can be satisfied by optical fibre networks (Azadeh, 2009). Therefore, great efforts have been devoted to the development of dense wavelength division multiplexing technology, which allows the simultaneous transmission of many channels with different wavelengths on a single optical fibre.

Optical amplifier is an essential device which enables to compensate the transmission losses, overcoming the limit of the unrelayed communication distance of the system. Rare-earth doped optical fibre amplifiers represent the most widely used devices to obtain optical amplification. In particular, Er^{3+}-doped fibre amplifiers (EDFAs) are used to realize high-speed, large-capacity and long-haul optical communication systems without the use of optoelectronic and electro-optical conversions of signals. EDFAs allow the exploitation of an uniform dopant concentration and the enhancement of heat dissipation due to the long interaction length. Moreover, several characteristics make EDFAs attractive: their high gain, wide optical bandwidth, high output saturation, near quantum-limited noise, low insertion losses, high reliability and compactness, polarization-independence, immunity to saturation-induced and to crosstalk, possibility of choosing the pumping laser diode at 980 nm or 1480 nm wavelengths. Although this kind of technology is mature and widely employed, further researches are needed to obtain amplifiers with higher efficiency, higher output power and shorter active fibre length. In detail, these devices should be characterized by high power conversion efficiency and high gain coefficient due to two main reasons. Remote-pumping in long transmission links can be carried out without repeaters, even if low pump power levels are col-

lected, as it takes place when the amplification of extremely weak signals is requested in the field of optical fibre sensors/monitoring. Moreover, the advantages can be extended to other application fields where high power optical fibre amplifiers are largely used: booster amplifiers for long haul repeaterless optical links, $1 \times N$ loss-less splitters, CATV distribution architectures, key photonic device for highly-distributed data networks, fibre optic gyroscopes, high-speed intersatellite links and deep-space optical communications (Girard, 2009, Rochat, 2001, Wright, 2005).

The optimization of fibre transversal section is crucial to improve the amplifier performance in terms of gain, noise figure and output power characteristics as well as device compactness and pump power consumption. In fact, in the rare earth doped devices the fibre geometry influences the pump intensity, the overlap of the pump and the signal propagation modes with the doped core. Sophisticated design methods and fabrication techniques have been developed to construct single-mode optical fibre amplifiers. To this aim, a fine control of refractive index profile of both core and cladding as well as more design flexibility of fibre cross section is needed. The conventional optical fibre is not able to satisfy these requirements, but the photonic crystal fibre (PCF) technology seems to be an attractive solution.

A typical PCF is characterized by a transverse crystal lattice (usually periodically arranged) containing air holes running along the fibre axis. Therefore, these fibres are characterized by a different refractive index profile of both core and cladding than the conventional ones. The core and cladding structures of PCF allow more degrees of freedom in tailoring the propagation and dispersion properties, enabling to control the light propagation in a way which is not obtainable by the conventional optical fibres. In particular, the research efforts in rare earth doped PCFs has opened novel opportunities for the enhancement of amplifier performance. In fact, the presence of air holes in the cladding improves the power distribution over the rare earth doped region and makes possible a better control of the overlap between pump and signal modes. This is due to a more accurate route to control the refractive index difference between core and cladding regions. As a consequence, higher gain efficiency, higher output power and shorter active fibre length can be obtained.

MAIN FOCUS OF THE CHAPTER

Accurate simulation of rare-earth doped fibre amplifiers is an important step in the feasibility evaluation of the amplifier configuration. In fact, unwanted degradation mechanisms can take place due to the dopant concentration (cooperative upconversion, cross relaxation and pair-induced quenching), heat dissipation and overall losses (internal scattering, surface scattering, sidewall scattering and Rayleigh scattering). Such phenomena should be controlled and the amplifier performance should be predicted in terms of efficiency and power scaling capability.

Nevertheless, simulations are based on theoretical models requiring the knowledge of the pumping scheme (forward, backward, bidirectional, distributed) and several spectroscopic (radiative and non radiative lifetimes, cross section spectra), operational (pump and signal power levels, pump and signal wavelengths) and fibre parameters. In fact, the effects of numerical aperture (NA), core radius, fibre length, cladding geometry, and dopant concentration have to be considered to obtain efficient coupling of the pump light, high conversion of the pump power, reduction of nonlinear effects, high handling of the thermal load. The fibre geometry affects the modal distribution around both the pump and signal wavelengths, the overlap of the pump and the signal propagation modes with the doped core and the evolution of the optical powers along the fibre length. Large core size reduces the detri-mental effects of nonlinear interaction and further increases the pump absorption. Large NA can make the system significantly less complex and more robust since sophisticated pump coupling optics between the diode laser stacks and fibre can be reduced. In addition, the pump core size is decreased and, consequently, an improvement of the pump absorption can be obtained. A major absorption of the pump power can be promoted by the employment of non-circular inner cladding shapes. This choice ensures shorter fibre lengths, which are attractive to improve the power threshold for nonlinear effects and to obtain higher efficiency of the system. Finally, the good power handling and the power conversion efficiency of the rare-earth doped fibre reduce the thermal heat load of the glass, leading to higher average power systems and requiring a lower cooling.

On the basis of the previous observations, it can be derived that the simulation process is complicated and requires great computational time. Furthermore, tremendous efforts have to be made to find the most efficient design, especially in the optimization stage, in which all the parameters affecting the fibre amplifier performance must be taken into account simultaneously. For this reason, the use of global optimization methods is welcome in order to find the global best solution both in the case of the optimum design and the characterization of rare-earth doped fibre amplifiers.

Generally, local optimization methods are used to find the optimal values of the fibre length, rare earth concentration, pump and signal wavelengths, operational parameters like input pump and signal power, optical fibre parameters such as refractive index profile, core size, and transversal section. This kind of methods cannot optimize all parameters simultaneously, therefore the optimal ones are determined one by one. This operative choice affects the overall time required in design and optimization problem to find the appropriate values of physical and geometrical characteristics of EDFAs.

In recent years, well performing and effective approaches using a genetic algorithm (GA) and a neural network have been proposed to solve optimization and characterization problems of rare-earth doped fibre amplifiers (Cheng, 2004, Fornarelli, 2009, Kim, 2009, Mescia, 2011, Prudenzano, 2005). These approaches give the possibility to simplify the models and reduce the computational time of the simulations conducted by means of conventional approaches. Nevertheless, the advantages offered by Particle Swarm Optimization could be exploited to further simplify the design and optimization of EDFAs. In fact, PSO is a reliable and robust evolutionary algorithm, which performs a global search of parameters over a specified problem space. Moreover, it requires a reduced number of synthesis parameters to be fixed, no complicated evolutionary operators and a simple implementation. Finally, PSO does not require gradient information and its computational burden is reduced, implying a low memory usage and high processing speed.

In this chapter, an approach based on PSO and rate equations analysis is employed to perform a full investigation of PCF amplifiers. In detail, this approach is employed to perform the optimal design and characterization of an Er^{3+}-doped PCF amplifier by considering the striking properties of the PCFs.

The considered model of PCF amplifier takes into account the most relevant active phenomena in Er^{3+}-system such as the radiative and nonradiative rates, at both pump and signal wavelengths, the stimulated emission of the signal, the amplified spontaneous emission (ASE), the lifetimes of the considered energy levels, the ion-ion energy transfers, and the presence of Er^{3+} clusters.

The developed method provides an accurate understanding of the optical amplifier behaviour to obtain reliable qualitative and quantitative predictions of the amplifier performance in a large variety of PCF configurations. Therefore, it constitutes a useful tool for the design and optimization. Moreover, a novel and efficient characterization procedure to recover the most relevant spectroscopic parameters is proposed and illustrated. In detail, the design process is performed by considering the fibre length, the pump power and erbium concentration for which maximum gain is obtained. This choice is due to the fact that the fibre length influences the ASE power and the optimal length. Moreover, it depends on the input pump power and erbium concentration. The fibre amplifier characterization problem is faced by considering two upconversion parameters, whose accurate estimation constitutes a very interesting research activity. In fact, the evaluation of the energy loss mechanisms due to the homogeneous upconversion (HUC) processes between two excited Er^{3+} ions plays a key role for the development of efficient fibre amplifiers.

This chapter is organized as follows. In the first section, basics of the coupling between the fibre amplifier model and the PSO approach are provided. In particular, some difficulties linked to the nonlinear functions describing the optical signal propagation along the fibre amplifier are identified. Subsequently, PSO is applied to solve the optimization and characterization problems in order to obtain intriguing performance in terms of flexibility and accuracy. In the second and third sections numerical results show the optimal design and the recovering of HUC coefficients in PCF, respectively. Conclusions are given in the final section.

THEORETICAL ANALYSIS

Rate and Power Propagation Equations

In this section a brief description of the mathematical model of the fibre amplifier under investigation is given.

Figure 1 illustrates the transitions and energy transfer among the Er^{3+}-energy manifolds, which are numbered by the index $\left| i \right\rangle$, with $i = 1, 2, 3, 4$.

Figure 1. Energy level diagram illustrating the ground state absorption and stimulated emission at pump and signal wavelengths (big arrows), the spontaneous decay (dash arrow), the nonradiative decays (dot arrows), energy transfer processes between Er^{3+} ions (full arrows)

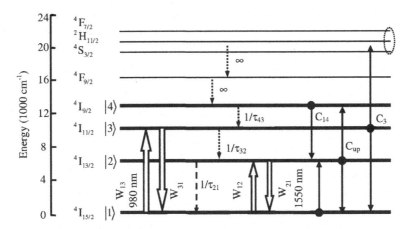

In detail, the following transitions are considered: 1) pump absorption and stimulated emission between the $^4I_{15/2}$ and $^4I_{11/2}$ Er^{3+}-manifolds, 2) signal absorption and stimulated emission between the $^4I_{15/2}$ and $^4I_{13/2}$ Er^{3+}-manifolds, 3) spontaneous decays from the $^4I_{9/2}$, $^4I_{11/2}$, $^4I_{13/2}$ Er^{3+}-manifolds, 4) uniform cooperative up-conversion between a pair of excited erbium ions: $^4I_{13/2}+^4I_{13/2}\rightarrow^4I_{15/2}+^4I_{9/2}$, $^4I_{11/2}+^4I_{11/2}\rightarrow^4I_{15/2}+^4S_{3/2}$, 5) cross-relaxation process taking place between the two neighbouring erbium ions: $^4I_{15/2}+^4I_{9/2}\rightarrow^4I_{13/2}$.

Pump energy at the wavelength $\lambda_p=980$ nm is directly absorbed by the Er^{3+} ions which are in the $^4I_{15/2}$ ground manifold and excited to the $^4I_{11/2}$ manifold. The Er^{3+} ions in the $^4I_{11/2}$ manifold rapidly relax to the $^4I_{13/2}$ metastable manifold by means of a non radiative decay, facilitating the population inversion phenomenon. The transition $^4I_{13/2}\rightarrow^4I_{15/2}$, which is close to the wavelength $\lambda_s=1550$ nm, allows the amplifier gain via the stimulated emission. Unfortunately, two erbium ions of the $^4I_{13/2}$ manifold can exchange energy between them; in fact one ion returns to the erbium ground manifold by a non radiative transition, and the other one raises to the $^4I_{9/2}$ excited manifold. A similar phenomenon occurs between two ions of the $^4I_{11/2}$ manifold, the former belongs to $^4F_{7/2}$ and the latter returns to the ground manifold. In addition, the cross-relaxation of ions belonging to the manifolds $^4I_{15/2}$ and $^4I_{9/2}$ is considered too. Due to an energy transfer such ions come both in the manifold $^4I_{13/2}$, enhancing the population inversion (see Figure 1).

The simulation of the fibre amplifier is performed by implementing a numerical code that is based on the resolution of the nonlinear differential equation system constituted by Er^{3+} multilevel rate equations and the change rate of the propagation of optical power. Both forward and backward amplified spontaneous emission (ASE) in the signal band are taken into account. It is assumed that the $^4F_{7/2}$, $^2H_{11/2}$, $^4S_{3/2}$ and $^4F_{9/2}$ erbium manifolds are empty since their lifetime is almost instantaneous.

By considering the steady state solutions ($\partial / \partial t = 0$), the populations N_1, N_2, N_3, and N_4, of the Er^{3+} ions are calculated according to the formulas

$$-\frac{N_4}{\tau_{43}} + C_{up}N_2^2 + C_3N_3^2 - C_{14}N_1N_4 = 0$$

$$(1)$$

$$W_{13}N_1 - W_{31}N_3 - \frac{N_3}{\tau_{32}} + \frac{N_4}{\tau_{43}} - 2C_3N_3^2 = 0$$

(2)

$$W_{12}N_1 - W_{21}N_2 + \frac{N_3}{\tau_{32}} - \frac{N_2}{\tau_{21}} - 2C_{up}N_2^2 + 2C_{14}N_1N_4 = 0$$

(3)

$$N_1 + N_2 + N_3 + N_4 = N_{Er}$$ (4)

where N_{Er} is the erbium concentration, τ_{21} is the lifetimes of the ${}^4I_{13/2}$ manifold, τ_{32} and τ_{43} are the non-radiative relaxation lifetimes of the ${}^4I_{11/2}$ and ${}^4I_{9/2}$ manifolds, respectively, C_3 and C_{up} are the homogeneous upconversion coefficients, C_{14} is the cross-relaxation coefficient. The absorption, W_{13} and stimulated, W_{31}, transition rates at the pump wavelength are given by

$$W_{ij}(z) = \frac{\Gamma(\lambda_p)\sigma_{ij}(\lambda_p)\left[P_p^+(z) + P_p^-(z)\right]\lambda_p}{hcA_{core}}$$

(5)

where $\Gamma(\lambda_p)$ is the overlap factor between the pump field and the doped core, A_{core} is the doped core area, h is the Planck's constant, P_p^+ and P_p^- are the z-depend forward and backward pump powers, $\sigma_{ij}(\lambda_p)$ is the absorption ($i=1,j=3$) or the emission ($i=3,j=1$) cross section at the pump wavelength, c is the speed of light in vacuum.

Amplifier performance cannot be calculated by means of analytical equations and numerical methods must be applied in the general case, due to the fact that both the forward and backward ASE noise spreads in a continuum wavelength range. In particular, by dividing the wavelength range in M wavelength slots, both the forward and backward ASE noise can be modelled as M optical beams having $\Delta\lambda_M$ bandwidth and centred around λ_M. As a result, the absorption, W_{12}, and stimulated, W_{21}, transition rates at the signal wavelength are given by Equation 6 (see Box 1) where P_s^+ and P_s^- are the z-depend forward and backward signal powers, P_{ASE}^+ and P_{ASE}^- are the z-depend forward and backward ASE powers. The functions $\sigma_{lm}(\lambda_k)$ are the wavelength-depend emission ($l=2$, $m=1$) and absorption ($l=1$, $m=2$) cross sections around the signal wavelength, $\Gamma(\lambda_s)$ is the overlap factor at the signal wavelength, $\Gamma(\lambda_k)$ is the wavelength-depend overlap factor given by

$$\Gamma(\lambda_M) = \int_0^{2\pi}\int_0^R \left|E(r,\phi,\lambda_M)\right|^2 r\,dr\,d\phi$$

(7)

where R is the radius of the doped core and $E(r,\phi)$ is the transverse electric field envelope normalized such that the surface integral of $\left|E(r,\phi)\right|^2$ is equal to one.

The longitudinal z-axis propagation of pump, signal, forward and backward ASE is modelled by first order system shown in Equations 8, 9, and 10 (see Box 2) of nonlinear coupled ordinary differential equations where $\sigma_{12}(\lambda_M)$ and $\sigma_{21}(\lambda_M)$ are the absorption and the emission cross sections at the λ_M wavelength, respectively, and $\alpha(\lambda)$ is the wavelength-depend intrinsic loss of the optical fibre.

The considered EDFA model is accurate but the specific boundary conditions at the two end

Box 1.

$$W_{lm}(z) = \frac{1}{hcA_{core}}\left\{\Gamma(\lambda_s)\sigma_{lm}(\lambda_s)\left[P_s^+(z) + P_s^-(z)\right]\lambda_s + \sum_{k=1}^M \Gamma(\lambda_k)\sigma_{lm}(\lambda_k)\left[P_{ASE}^+(z) + P_{ASE}^-(z)\right]\lambda_k\right\}$$ (6)

points of the integration domain cause a complex numerically treatment especially when $\Delta\lambda_M$ is chosen to be about 1 nm (D'Orazio, 2005, Movassaghi, 2001). In fact, the solution of the general evolution equations requires iterative adjustments of the initial conditions because the backward ASE is not known at the input end. In order to solve these equations, a number of numerical and analytical methods with different complexity have been proposed in literature (Desurvire, 2009, Roudas, 1999). In the conventional approach, shooting and relaxation methods are usually used to solve the evolution equations, even if they may lead to divergence or unstable convergence of the algorithm and are computationally expensive (Desurvire, 2009, Eichhorn, 2008, Roudas, 1999). Therefore, the device modelling can become extremely time consuming, especially when several design parameters have to be simultaneously considered.

PSO Algorithm

In the original version of PSO algorithm (Kennedy, 1995) each particle i is characterized by a position vector $\mathbf{p}_i$, representing a potential solution evaluated by means of a proper fitness function. The swarm is randomly initialized and, subsequently, the position of each particle is updated by applying a proper operator $\mathbf{v}_i$, called velocity. The velocity is computed according to the information of the fitness function, the personal best position of each particle and the global best one among the agents of the whole swarm. Such operator updates the position of the particles with the aim of moving the individuals towards optimal solution areas until an error criterion is satisfied (Kennedy, 2001). In this way, PSO algorithm mimics the behaviour of social animals like bird flocking or school of fishes which is inspired to. These concepts can be formalized as follows.

Let $\mathbf{p}_i = \left[p_{i1}, p_{i2}, ..., p_{iD} \right]^T$ and

$\mathbf{v}_i = \left[v_{i1}, v_{i2}, ..., v_{iD} \right]^T$, i=1,2,...N, be two D-dimensional vector in problem space. Let such two vectors represent the position and velocity of each particle, respectively. The i-th particle updates its velocity and position according to the following equations:

$$\mathbf{v}_i \left(t + 1 \right) = w \times \mathbf{v}_i \left(t \right) + c_1 \times r_1 \times \left[\mathbf{p}_{bi} \left(t \right) - \mathbf{p}_i \left(t \right) \right] + c_2 \times r_2 \times \left[\mathbf{p}_g \left(t \right) - \mathbf{p}_i \left(t \right) \right] \tag{11}$$

$$\mathbf{p}_i \left(t + 1 \right) = \mathbf{p}_i \left(t \right) + \mathbf{v}_i \left(t + 1 \right) \tag{12}$$

Box 2.

$$\frac{dP_p^\pm \left(z \right)}{dz} = \pm\Gamma\left(\lambda_p\right)\left[\sigma_{31}\left(\lambda_p\right) N_3 - \sigma_{13}\left(\lambda_p\right) N_1 - \alpha\left(\lambda_p\right)\right] P_p^\pm \left(z \right) \tag{8}$$

$$\frac{dP_s^\pm \left(z \right)}{dz} = \pm\Gamma\left(\lambda_s\right)\left[\sigma_{21}\left(\lambda_s\right) N_2 - \sigma_{12}\left(\lambda_s\right) N_1 - \alpha\left(\lambda_s\right)\right] P_s^\pm \left(z \right) \tag{9}$$

$$\frac{dP_{ASE}^\pm \left(z, \lambda_M \right)}{dz} = \pm\Gamma\left(\lambda_M\right)\left[\sigma_{21}\left(\lambda_M\right) N_2 - \sigma_{12}\left(\lambda_M\right) N_1 - \alpha\left(\lambda_M\right)\right] P_{ASE}^\pm \left(z, \lambda_M \right) \pm$$
$$\pm \frac{2hc^2}{\lambda_M^3} \Delta\lambda_M \Gamma\left(\lambda_M\right) \sigma_{21}\left(\lambda_M\right) N_2 \tag{10}$$

Box 3.

$$\mathbf{v}_i\left(t+1\right) = \chi \times \left\{\mathbf{v}_i\left(t\right) + \phi_1 \times r_1 \times \left[\mathbf{p}_{bi}\left(t\right) - \mathbf{p}_i\left(t\right)\right] + \phi_2 \times r_2 \times \left[\mathbf{p}_g\left(t\right) - \mathbf{p}_i\left(t\right)\right]\right\} \qquad (13)$$

where t is the iteration counter, w is a constant value, called inertia weight, in the range [0, 1], c_1 and c_2 are two positive constants, called the cognitive and social parameter, respectively; r_1 and r_2 are two random numbers uniformly distributed in the range [0, 1], $\mathbf{p}_{bi}(t)$ is the previous best position of the *i-th* particle and $\mathbf{p}_g(t)$ is the previous best position among all the particles in the population.

The first version of such algorithm has been exploited successfully to solve many practical problems. In fact, when new problems are tackled by making use of PSO-based methods, the original version of this paradigm is the most commonly considered due to its simple implementation (Chen C. Y. (2004), Mendes R. (2002)). Nevertheless, the convergence process of the classical algorithm can be premature, leading to unsatisfactory results especially in processing large data sets and complex problems (Kao I.W. (2007)). This behaviour can be due to two main reasons: the fast information interaction among the agents of the PSO and the stalling of the search. The former drawback reduces the swarm-diversity, whereas the latter one takes place when the boundaries of the solution space are reached (Kao I.W. (2007), Satapathy S.C. (2007)). In order to overcome these problems, modifications of the PSO learning have been proposed (Ratnaweera, 2004, Shi, 1999, Clerc, 2002). In particular, a variant of the learning equation has been introduced by Clerc M. in 2002. In this scheme the velocity is updated by Equation 13 (see Box 3)(Clerc, 2002) where

$$\chi = \frac{2}{\left|2 - \phi - \sqrt{\phi^2 - 4\phi}\right|}, \ \phi = \phi_1 + \phi_2 \text{ and } \phi > 4 \qquad (14)$$

is the *constriction factor*. The introduction of the constriction factor does not assure the best performance necessarily, but it could ensure the convergence of the algorithm (Eberhart, 2000). These improvements proved to be effective on a large collection of problems (van den Bergh, 2004). Furthermore, it presents the advantage to allow a very simple identification of the optimal value of the algorithm parameter; in fact the inertial weight and the cognitive constants are defined when only one value is set. This aspect makes the method very attractive and usable by a not expert user. Therefore, in the present work the reported version of the PSO is investigated.

All the details of the PSO-based algorithm which is used for optimal design and the characterization of an Er^{3+}-doped PCF amplifier are given in the following.

Let the vector T collect the main parameters affecting the characteristics of a PCF amplifier. In this case of study, such parameters correspond to the input pump power P_p, the up-conversion coefficients C_3 and C_{up}, the cross relaxation coefficient C_{14}, the metastable spontaneous emission rate $A_{21}=1/\tau_{21}$, the non-radiative relaxation rate $A_{32}=1/\tau_{32}$, the non-radiative relaxation rate $A_{43}=1/\tau_{43}$, the fibre length L and the erbium concentration N_{Er}. Then the vector is:

$$\mathbf{T} = [P_p, P_s, C_3, C_{up}, C_{14}, A_{21}, A_{32}, A_{43}, L, N_{Er}]^\mathrm{T} \in \Re^D$$

The optimal design consists of determining a subvector $\mathbf{T}_{opt}$ of $\mathbf{T}$ such that the optical gain of the fibre amplifier is maximized. On the contrary, the problem of the characterization corresponds to finding the solution of the inverse problem, i.e. it is required to compute a subvector $\mathbf{T}_{inv}$ of parameters corresponding to a known value of gain.

In detail, in the considered optimal design problem the parameters to be optimized are P_p, L and N_{Er}, whereas the up-conversion parameters C_3 and C_{up} are evaluated in the characterization problem. In both these cases, the computation of $\mathbf{T}_{opt}$ or $\mathbf{T}_{inv}$ is handled as the solution of an optimization problem. To this aim, a swarm composed by N particles, whose position vectors are given by

$$\mathbf{p}_i = \left[p_{i1}, p_{i2}, ..., p_{id}\right]^{\mathrm{T}} \in \Re^d, \text{ with } i = 1, 2, ..., N$$

and $d \leq \mathrm{D}$, is employed.

Let $\mathbf{T}^i \in \Re^d$, with $i = 1, 2, ..., N$ and $d \leq \mathrm{D}$, be a potential solution of the design or characterization problem, then, imposing that $\mathbf{p}_i = \mathbf{T}^i$, the values of the vector $\mathbf{T}_{opt}$ or $\mathbf{T}_{inv}$ are given by the vector $\mathbf{p}_g$ when convergence has been reached by the swarm.

Let $\left[p_k^{\min}, p_k^{\max}\right]$ and $\left[v_k^{\min}, v_k^{\max}\right]$, with $k = 1, 2, ..., d$, be the ranges in which the elements of positions and velocities can vary, respectively. Let N_T be the maximum number of iterates for which the method does not update the $\mathbf{p}_g$ and R_{max} the iteration limit of the algorithm. Let ε be a tolerance which the solution can be affected by, then, the proposed PSO-based method is described by the following steps:

1. Initialize each particle of the swarm with a random position and velocity vector, whose components are in the ranges $\left[p_k^{\min}, p_k^{\max}\right]$ and $\left[v_k^{\min}, v_k^{\max}\right]$, with $k = 1, 2, ..., d$;

2. For each particle calculate the fitness function $F(\mathbf{p}_i)$ by considering the parameters belonging to the position vector $\mathbf{p}_i$ in the rate and propagation Equation Model 1-10;

    a. By assuming that the optimization problem has to be solved, then a global maximum is searched. Therefore if $F(\mathbf{p}_i) > F(\mathbf{p}_{bi})$ then $\mathbf{p}_{bi} = \mathbf{p}_i$, whereas if $F(\mathbf{p}_i) > F(\mathbf{p}_g)$ then $\mathbf{p}_g = \mathbf{p}_i$;

    b. By assuming that the characterization problem has to be solved, then an error has to be minimized. Therefore if $F(\mathbf{p}_i) < F(\mathbf{p}_{bi})$ then $\mathbf{p}_{bi} = \mathbf{p}_i$, whereas if $F(\mathbf{p}_i) < F(\mathbf{p}_g)$ then $\mathbf{p}_g = \mathbf{p}_i$;

3. Update the velocity and position vectors, $\mathbf{v}_i$ and $\mathbf{p}_i$, according to Equations 11 and 12, limiting the elements of the position $\mathbf{p}_i$ and the velocity $\mathbf{v}_i$ in the ranges $\left[p_k^{\min}, p_k^{\max}\right]$ and $\left[v_k^{\min}, v_k^{\max}\right]$, with $k = 1, 2, ..., d$, respectively;

    a. Assuming that a global maximum has to be searched if the vector $\mathbf{p}_g$ has not been changed for the maximum number N_T, then the algorithm ends, else go to step 2.

    b. Assuming that an error has to be minimized if neither $F(\mathbf{p}_g) \leq \varepsilon$ nor R_{max} has been carried out, then go to step 2, else go to step 6;

4. If R_{max} has been reached then go to step 1, else the algorithm ends.

It is straightforward to observe that, due to the potential presence of local maxima/minima of the fitness function, the maximum numbers N_T or R_{max} could be reached, providing solutions which could not yield the maximum/minimum fitness function. For this reason, the algorithm runs more times, increasing the probability to find the required solution. In this way, the PSO procedure converges toward an optimal solution, controlling the satisfaction of requirements for the analyzed device.

Figures 2 and 3 report the flow charts summarizing the algorithm to solve both the optimal design and characterization problem, respectively.

Figure 2. Flow chart of the algorithm for the optimal design problem

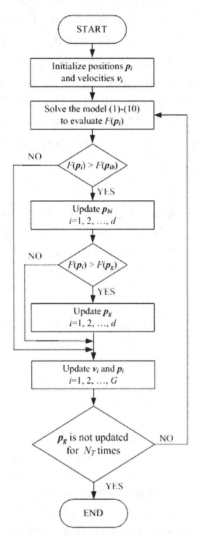

Figure 3. Flow chart of the algorithm for the characterization problem

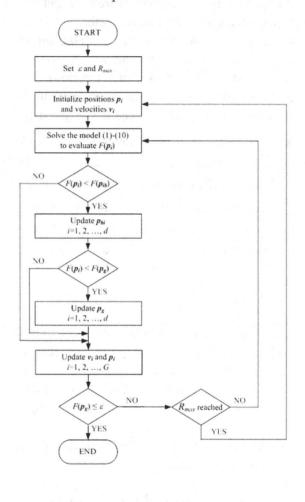

NUMERICAL RESULTS

Optimal Design Problem

This section reports the results showing the capabilities of the proposed method to obtain an optimal design of a PCF or characterize some of its important parameters .

The amplifier properties have been evaluated by means of an ad-hoc numerical code that solves the rate Equations 1-4 and the propagation Equations 8-10 (D'Orazio, 2005, De Sario, 2009,

Prudenzano, 2009). The overlap factor Γ between the dopant and the pump signal, and amplified spontaneous emission field intensities has been calculated by using a full-vector modal solver based on finite element method (FEM), since the PCF presents a complex transversal distribution of the refractive index and high refractive index contrast. This modal solver is flexible and allows to describe the transversal section of PCFs with arbitrary hole sizes and placements. As a consequence, the FEM solver provides a very accurate calculation of both the propagation constant and

Table 1. Parameters used in the numerical calculations

Parameter	Value	Parameter	Value
Signal wavelength λ_s	1538 nm	Cross section $\sigma_{13}(\lambda_p)$	4.78×10^{-25} m²
Pump wavelength λ_p	980 nm	Cross section $\sigma_{31}(\lambda_p)$	3.84×10^{-25} m²
Hole diameter d_h	3.2 μm	Cross section $\sigma_{12}(\lambda_s)$	1.53×10^{-24} m²
Hole-to-hole spacing Λ	8 μm	Cross section $\sigma_{21}(\lambda_s)$	2×10^{-24} m²
Doped region diameter d_d	10 μm	$^4I_{13/2}$ lifetime τ_{21}	4 ms
Loss @1538 nm	2 dB/m	$^4I_{11/2}$ lifetime τ_{32}	1.29 ms
Loss @980 nm	3 dB/m	$^4I_{9/2}$ lifetime τ_{43}	53 μs
Upconversion coefficient C_{up}	3×10^{-23} m³/s	Cross relaxation coefficient C_{14}	5×10^{-24} m³/s
Upconversion coefficient C_3	2×10^{-23} m³/s		

the electromagnetic field distribution of the PCF guided modes. In particular, the electromagnetic field profile has been calculated by solving the wave equation for the magnetic field vector **H**:

$$\nabla \times \left(n^{-2} \nabla \times \mathbf{H} \right) - k_0^2 \mathbf{H} = 0 \qquad (15)$$

where n is the refractive index, $k_0 = 2\pi/\lambda$ is wave-number in the vacuum and λ the wavelength.

The electric field is evaluated by solving the Maxwell's equation:

$$\nabla \times \mathbf{H} = j\omega\varepsilon\mathbf{E} \qquad (16)$$

where ε is the dielectric permittivity. It has been used to evaluate the wavelength-dependent overlap factors as the input data for the rate and propagation equations. These equations have been solved numerically by an iterative procedure based on Runge-Kutta algorithm. Their solution has been optimized in order to reduce the computational time and the wavelength range from 1450 nm to 1599 nm has been divided into M=150 wavelength samples.

A typical PCF having circular air-holes arranged in a triangular lattice is considered. Silica glass has been used as background material and a missing hole at the centre creates a solid core which can be doped with erbium ions. The hole

diameter d_h, hole-to-hole spacing Λ and doped region diameter d_d used in the simulations are reported in Table 1. In the same table, the remaining parameters used for the numerical computations are summarized, whereas the geometric structure of the fibre is illustrated in Figure 4.

The existing six-fold symmetry is exploited in order to reduce the computational time and achieve a more accurate mesh refinement (De Sario M. (2009), McIsaac P.R. (1975)).

In the design problem, the PSO approach has been used to evaluate the optimal values of some fibre and spectroscopic parameters maximizing the optical gain. To this aim, in the first set of

Figure 4. Cross section of the considered Er³⁺-doped photonic crystal fibre

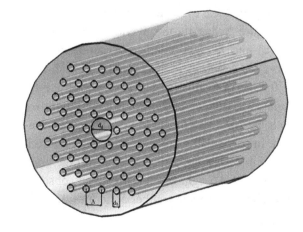

simulations the PSO algorithm is performed by defining a swarm composed by D particles, whose position vector of the *i-th* particle is $\mathbf{p}_i = \left[N_{Er}^i, L^i \right]^T$, where N_{Er}^i and L^i are the erbium concentration and fibre length, respectively. The following expression represents the considered fitness function:

$$f\left(\mathbf{p}_i\right) = 10 \log\left[\frac{P_s^i\left(L\right)}{P_s^i\left(0\right)}\right] \qquad (17)$$

where $P_s^i\left(L\right)$ and $P_s^i\left(0\right)$ are the signal powers at the output and input optical fibre ends, respectively.

The PSO parameters, w, c_1, c_2, have been chosen in order to obtain a balance between global and local exploration by considering the alternative representation of the velocity formulated in Equation 13. A number of simulations has been conducted in order to single out the optimal choice with respect to the simulation time for the parameters ϕ_1 and ϕ_2, being $\phi = 4.1$ as reported in (Carlisle A. (2001)). It should be noted that the choice of ϕ_1 is equivalent to define all the parameters in Equation 13. The other parameters used in the PSO algorithm are the population size $N=20$ and an iteration limit equal to 100. In detail, this choice of the population size is based on previous parametric studies aiming at ensuring a sufficient exploration of the solution space and avoiding excessive fitness evaluations (Carlisle A. (2001)).

Figure 5 depicts the average time (dashed curve) and average number of iterations (full curve) as ϕ_1 varies in the range [1, 3] with step of 0.2. Such evaluation has been performed by means of 20 independent PSO runs to optimize both the fibre length and the erbium concentration. In particular, on the left side of the figure it is possible to estimate the value of average time, whereas on the right side it is possible to observe the values of average number of iterates, accord-

Figure 5. Average time (dash curve) and average number of iterations (full curve) versus the $\phi 1$ parameter

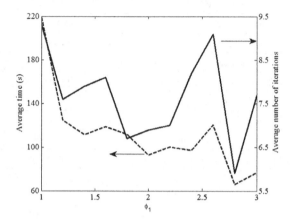

ing to the indication of the arrows in the figure. By an inspection of Figure 5 it can be observed that the choice $\phi_1 = 2.8$ and $\phi_2 = 1.3$ guarantees lower computational time. At the same time, this value corresponds to the minimum number of iterates. This result is in good accordance with those reported in literature (Carlisle, 2001). It is worthwhile to note that such selection of constants corresponds to select inertial weight $w=0.729$, cognitive parameter $c_1 = 2.041$ and social parameter $c_2 = 0.948$.

The quite complex spectroscopic system shows that the amplification process depends on two phenomena: the competition between different emission and absorption mechanisms at different wavelengths, and the nonlinear characteristic of the ion-ion interaction. As a result, the fibre length, the erbium concentration and the pumping scheme influences the amplifier performance strongly. Therefore, a study has been performed to identify the optimal fibre length and the optimal erbium concentration by varying both the total input pump power and the pumping scheme. In this way, it is possible to design amplifiers having high gain efficiency and choose an amplifier configuration minimizing the deleterious effects due to the ion-ion interaction. For this reason, the

Figure 6. 3D plot of the optical gain versus fibre length and erbium concentration, obtained by employing the conventional algorithm. Maximum gain value evaluated by conventional algorithm (cross mark), maximum gain calculated by using PSO based approach (circular mark).

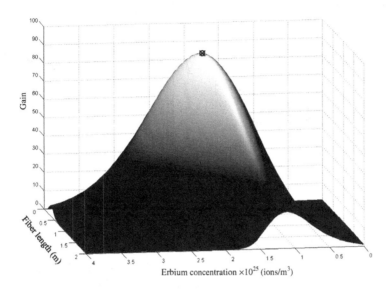

second set of simulations has been devoted to verify the accuracy of the proposed method, whose performance has been evaluated with respect to the conventional algorithm proposed in (De Sario, 2009, D'Orazio, 2005). In detail, the conventional algorithm has been used in order to identify the maximum gain value of a specific PCF amplifier as fibre length and erbium concentration vary. Subsequently, the PSO based-algorithm has been exploited to obtain the same gain value under the same conditions of the conventional algorithm (CA). These simulations have been conducted by assuming the values of the parameters which have been considered in the first set of simulations.

The analyzed PCF has been considered in copropagating configuration with values of pump and signal power equal to 180 mW and 1 μW, respectively; whereas the fibre length and the erbium concentration have been evaluated in the range $0 \div 4$ m and $0 \div 2 \times 10^{25}$ ions/m³, respectively. The performance of the PSO approach is illustrated by Figure 6. Such figure shows the surface of the gain

values versus fibre length and erbium concentration obtained by the conventional algorithm. In particular, the mark "×" at the top of the surface represents the position of the maximum gain value obtained by the conventional algorithm. In the same figure, the mark "○" corresponds to the position of the maximum gain value which has been computed by the PSO based approach. The agreement between the two considered methods is evident in the figure and confirmed in numerical terms since the CA and the PSO method allow to obtain the maximum gain values of 94.70 and 94.78, respectively. The slight difference between the results of the two algorithms is highlighted if considering the values of optimal fibre length and the optimal erbium concentration providing the maximum gain. In fact, the erbium concentration values given by the CA and PSO method are equal to 0.90×10^{25} ions/m³ and 0.91×10^{25} ions/m³, respectively; whereas the optimal fibre length is equal 2.00 m in both cases. It should be noted that such values depend on the variation step

Figure 7. (a) Optimal fibre length, (b) Optimal erbium concentration, (c) optimal gain versus input pump power for forward (full curve), backward (dot curve) and bidirectional (dash curve) pumping scheme. (d) Optimal forward (full curve) and backward (dash curve) power versus input pump power for the bidirectional pumping scheme.

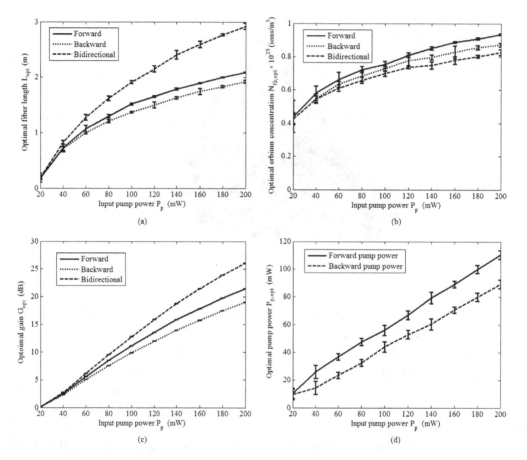

considered in the simulations. It could decrease if a more refined step is employed.

On the basis of the previous calculations, further simulations have been conducted in order to evaluate the behaviour of the PCF amplifier. In detail, the proposed algorithm has been exploited to identify the values of the optimal fibre length, optimal erbium concentration, optimal gain and the pumping configuration versus input pump power.

Figure 7. Reports the computed average (continuous curves) and standard deviations (error bar) for optimal length, optimal erbium concentration,

optimal gain and optimal forward and backward input pump power

This kind of investigation has been carried out by considering 20 independent runs and the PSO parameters which have been evaluated previously. Table 2 summarizes the final parameters used in PSO algorithm.

In detail, Figures 7 (a), 7 (b) and 7 (c) report the values of optimal length, concentration and gain which have been obtained for three different pumping configurations: forward, backward and bidirectional. The abscissas representing the total power of the pump do not change as the pumping scheme varies. In particular, such power value

Table 2. Parameters used in the PSO algorithm for the optimal design problem

Parameter	Value
Cognitive parameter	2.041
Social parameter	0.948
Inertial weight	0.729
Particle number	20
Iteration limit	100
$\left[L_{\min}^{i}, L_{\max}^{i} \right]$ (m)	$0 \div 5$
$\left[N_{Er,\min}^{i}, N_{Er,\max}^{i} \right]$ (ions/m³)	$10^{-27} \div 2 \times 10^{-25}$
$\left[P_{pf,\min}^{i}, P_{pf,\max}^{i} \right]$ (mW)	$0 \div P_p$

corresponds to the power one at input and output ends in the forward and backward pumping scheme, respectively. As a consequence, the optimization has been performed by taking into account the length and erbium concentration as swarm parameters. The total value of the pump power does not vary in the bidirectional configuration, but the values of the power at input and output end have to be determined since the pump signal is injected into both the fibre optic ends. Therefore, in this case such quantities constitute a further degree of freedom. In order to take into account this aspect, the fraction of the pump power at the input end has been added to the set of swarm parameters. For this reason, in the bidirectional pumping scheme the position vector of the *i-th* particle is $\mathbf{p}_i = \left[N_{Er}^{i}, L^{i}, P_{pf}^{i} \right]^T$, where P_{pf}^{i} is the forward input pump power.

Figure 7(d) reports the optimal values of input and output pump power to be used by taking into account that their sum is equal to the total input pump power P_p used in the other pumping schemes. It can be noted that the best performances are obtained when the forward pump power is greater than the backward one rather than when both the values are 50% of the total input pump power. From this analysis it can be derived that, when the total pump power is fixed, the best performance are guaranteed by the bidirectional configuration only if the ratios in Figure 4(d) are satisfied. Figure 7(c) highlights that the use of the bidirectional scheme yields higher optical gains than those obtained by the forward and backward pumping scheme. In fact, this pumping configuration allows a more uniform population inversion along the entire fibre length and, as a consequence, a better signal improvement can be obtained by using lower erbium concentration and higher fibre length.

Characterization Problem

The upconversion phenomena constitute the main mechanisms of energy loss in Er^{3+}-doped glasses. Two different upconversion mechanisms labelled by the phenomenological macroscopic parameters, C_{up} and C_3 have been considered in the following, These upconversion phenomena reduce the Er^{3+} population in the $^4I_{13/2}$ and $^4I_{11/2}$ manifolds and the net effect is a reduction of both the quantum yield of the lasing transition $^4I_{13/2} \rightarrow {}^4I_{15/2}$ and the pump efficiency. The estimation of C_{up} and C_3 is fundamental in any numerical modelling employed for the design and the optimization of Er^{3+}-doped fibre amplifier. Unfortunately, established experimental methods present three drawbacks: i) several steps are required to estimate the value of the C_{up} and C_3 coefficients, ii) the results are strongly sensitive to inaccuracy in the estimation of the population of the excited states, iii) time dependent measurements could be necessary (Lopez 2006, Nikonorov 1999, Jones 2006). Due to the strong difficulties present in direct measurements, these parameters are either taken as adjustable or extracted from the literature regarding different glass composition and/or doping concentration.

The upconversion mechanisms induce depopulation rates $W_{C_{up}}$ and W_{C_3} which depend on the input pump power P_p by the following relations

$$W_{C_{up}} = C_{up} N_2 \left(P_p \right) \qquad (18)$$

$$W_{C_3} = C_3 N_3 \left(P_p \right) \qquad (19)$$

Therefore, it is possible to construct a suitable fitness function which allows to find the right pair (C_{up}, C_3) by taking into account a suitable number of known values of gain evaluated for different values of pump power. In particular, in this computation the following fitness function has been considered:

$$f(\mathbf{p}_i) = \sum_{n=1}^{m} \left(G_n - G_n^i \right)^2 \qquad (20)$$

where G_n^i is the optical gain corresponding to *i-th* particle evaluated by using the rate equation model and G_n, $n = 1, .., m$, is the known gain. The PSO is aimed at evaluating the optimal parameters minimizing the fitness function.

The validity of the proposed method has been tested by considering the Er^{3+}-doped photonic crystal fibre amplifier which has already been illustrated in the optimal design problem section. Table 3 summarizes the used PSO parameters.

Preliminary, simulations have been performed in order to define the suitable number of right values of the optical gain. In particular, it has been proved that, when different independent trials are performed, the PSO algorithm does not converge to the same pair for $m \leq 4$. Moreover, it has been verified that $m \cong 10$ gives a unique solution pair (C_{up}, C_3). This value ensures a good compromise between the results accuracy and the computational cost. Higher values of m does not yield significant

Table 3. Parameters used in the PSO algorithm for the characterization problem

Parameter	Value
Cognitive parameter	2.041
Social parameter	0.948
Inertial weight	0.729
Particle number	20
Iteration limit	100
$\left[C_{up,\min}^i, C_{up,\max}^i \right]$ (m³/s)	$10^{-24} \div 10^{-22}$
$\left[C_{3,\min}^i, C_{3,\max}^i \right]$ (m³/s)	$10^{-24} \div 10^{-22}$

improvements in the accuracy even if the computational time increases. As a consequence, in the evaluation of the fitness function 11 different values of input pump power uniformly distributed in the range from 150 mW to 400 mW have been taken into account, whereas the considered input signal power and erbium concentration are $P_s(0)=1$ µW and $N_{Er}=10^{25}$ ions/m³.

Figure 8 reports the pairs (C_{up}, C_3) which have been obtained by the inversion algorithm applied to different lengths of the optical fibre. The dashed lines represent the ideal value of the pairs. The data have been obtained by 20 different realizations. It can be observed that the yielded C_{up} values are more accurate than the C_3 ones. This behaviour is due to the fact that the population of the energy manifold $|2\rangle$ is greater than the one belonging to the energy manifold $|3\rangle$. As a consequence, it results that $W_{C_{up}} > W_{C3}$, therefore, the fitness is characterized by a higher sensitivity to C_{up}. On the other hand, an analysis of the variance allows to conclude that the PSO provides more reliable results as the length of the fibre increases, especially if considering C_3. In fact, short fibres present a gain

Figure 8. Error bar plot of the pair (C_{up}, C_3) recovered by the PSO algorithm versus the fibre length and by considering for each of them 20 independent computer code runs

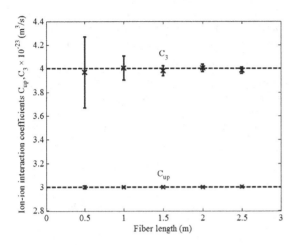

curve which shows a quite constant slope with pump power. In this case, a number of very similar gain curves can be associated to different pairs (C_{up}, C_3). As a consequence, the values of the fitness function result in a reduced range. As far as higher fibre length, the gain curves present more nonlinear trend as the pump power increases. Consequently, the probability to obtain multiple value of (C_{up}, C_3) for fixed gain values decreases. These observations are confirmed in practical cases, in fact it is not convenient to make use of short fibres to perform measures of the gain.

CONCLUSION

In this chapter, a PSO based approach has been proposed to perform the optimal design and the characterization of a PCF amplifier. The presented method has been employed usefully to refine the performance of an amplifier by evaluating the optimal parameters which maximize the gain. The comparison to a conventional

algorithm shows that the solution of the PSO based-technique provides accurate results. Subsequently, the method has been used to map the nonlinear functional relationship between the physical/geometrical characteristics of the considered fibres and the input pump power. In detail, the optimal fibre length, optimal erbium concentration, the optimal gain and the optimal pumping scheme have been calculated. The calculated performance proves that the method can be considered as an useful tool to provide an accurate understanding of the optical amplifier behaviour and predict the optimal parameter values in a large variety of PCF amplifier configurations. In fact, the conducted study allowed to deduce that in these fibres higher values of gains are obtained when a bidirectional pumping scheme is considered.

Finally, the PSO has been applied to solve the inverse problem, consisting of determining unknown fibre parameters which provide known values of gain. In the illustrated case, the algorithm enabled the identification of the upconversion parameters corresponding to the desired values of gain. Numerical results prove that the presented approach finds solutions which are in good accordance with a conventional one. This result has been obtained with the fundamental advantage that the implementation of the algorithm is simple, since it does not require to determine neither complicated evolutionary operators nor a high number of synthesis parameters.

This application of the PSO method seems to be particularly intriguing and could constitute a powerful tool to recover the values of physical parameters of the optical amplifier, which cannot be directly measured. In fact, other design parameters, such as the period and hole radius of the PCFs, could be evaluated, by following the same scheme. Finally, it can be concluded that the proposed technique constitutes an interesting alternative and an efficient tool for design and characterization of this kind of devices.

REFERENCES

Azadeh, M. (2009). *Fibre Optics Engineering*. New York, NY: Springer. doi:10.1007/978-1-4419-0304-4

Carlisle, A., & Doizier, G. (2001). An off- the-shelf PSO. *Proceedings of the Workshop Particle Swarm Optimization*, Indianapolis, IN.

Chen, C. Y., & Ye, F. (2004). Particle swarm optimization algorithm and its application to clustering analysis. *IEEE International Conference on Networking, Sensing and Control* (pp. 789-794).

Cheng, C. (2004). A global design of an erbium-doped fibre and an erbium-doped fibre amplier. *Optics & Laser Technology*, *36*, 607–612. doi:10.1016/j.optlastec.2004.01.006

Clerc, M., & Kennedy, J. (2002). The particle swarm: explosion, stability, and convergence in a multi-dimensional complex space. *IEEE Transactions on Evolutionary Computation*, *6*(1), 58–73. doi:10.1109/4235.985692

D'Orazio, A., De Sario, M., Mescia, L., Petruzzelli, V., & Prudenzano, F. (2005). Refinement of Er^{3+}-doped hole-assisted optical fibre amplifier. *Optics Express*, *13*, 9970–9981. doi:10.1364/OPEX.13.009970

De Sario, M., Mescia, L., Prudenzano, F., Smektala, F., Deseveday, F., & Nazabal, V. (2009). Feasibility of Er^{3+}-doped, $Ga_5Ge_{20}Sb_{10}S_{65}$ chalcogenide microstructured optical fibre amplifiers. *Journal of Optics & Laser Technology*, *41*, 99–106. doi:10.1016/j.optlastec.2008.03.007

Desurvire, E. (2009). *Erbium-doped fibre amplifiers: principles and applications*. Wiley Interscience.

Eberhart, R. C., & Shi, Y. (2000). Comparing inertia weights and constriction factors in particle swarm optimization. *Proceedings of Congress on Evolutionary Computing* (pp. 84–89).

Eichhorn, M. (2008). Quasi-three-level solid-state lasers in the near and mid infrared based on trivalent rare earth ions. *Applied Physics. B, Lasers and Optics*, *93*, 269–316. doi:10.1007/s00340-008-3214-0

Fornarelli, G., Mescia, L., Prudenzano, F., De Sario, M., & Vacca, F. (2009). A neural network model of erbium-doped photonic crystal fibre amplifiers. *Journal of Optics & Laser Technology*, *41*, 580–585. doi:10.1016/j.optlastec.2008.10.010

Girard, S., Ouerdane, Y., Tortech, B., Marcandella, C., Robin, T., & Cadier, B. (2009). Radiation effects on Ytterbium and Ytterbium/Erbium-doped double-clad optical fibres. *IEEE Transactions on Nuclear Science*, *56*, 3293–3299. doi:10.1109/TNS.2009.2033999

Jones, G. C., & Houde-Walter, S. N. (2006). Determination of the macroscopic upconversion parameter in Er^{3+}-doped transparent glass ceramics. *Journal of the Optical Society of America. B, Optical Physics*, *23*, 1600–1608. doi:10.1364/JOSAB.23.001600

Kao, I. W., Tsai, C. Y., & Wang, Y. C. (2007). An effective particle swarm optimization method for data clustering. *Proceedings of the 2007 IEEE Int. Conf. on Industrial Engineering and Engineering Management (IEEM)* (pp. 548–552).

Kennedy, J., & Eberhart, R. (1995). Particle swarm optimization. *Proceedings of IEEE International Conference on Neural Networks* (pp. 1942-1948).

Kennedy, J., & Eberhart, R. C. (2001). *Swarm intelligence*. San Francisco, CA: Morgan Kaufmann Publishers.

Kim, H., Bae, J., & Chun, J. (2009). Synthesis method based on genetic algorithm for designing EDFA gain flattening LPFGs having phase-shifted effect. *Optical Fiber Technology*, *15*, 320–323. doi:10.1016/j.yofte.2009.02.001

Lee, C. H., Shin, S. W., & Chung, J. W. (2006). Network intrusion detection through genetic feature selection. *7th ACIS International Conference on Software Engineering, Artificial Intelligence, Networking and Parallel/Distributed Computing* (pp. 109–114).

Lopez, L., Paez, G., & Strojnik, M. (2006). Characterization of up-conversion coefficient in erbium-doped materials. *Optics Letters, 31,* 1660–1662. doi:10.1364/OL.31.001660

McIsaac, P. R. (1975). Symmetry-induced modal characteristic of uniform waveguide-II: Theory. *IEEE Transactions on Microwave Theory and Techniques, 23,* 429–433. doi:10.1109/TMTT.1975.1128585

Mendes, R., Cortez, O., Rocha, M., & Neves, J. (2002). Particle swarms for feedforward neural network training. *Proceedings of International Joint Conference on Neural Networks* (pp. 1895-1899).

Mescia, L., Fornarelli, G., Magarielli, D., Prudenzano, F., De Sario, M., & Vacca, F. (2011). Refinement and design of rare earth doped photonic crystal fibre amplifier using an ANN approach. *Journal of Optics & Laser Technology, 43,* 1096–1103. doi:10.1016/j.optlastec.2011.02.005

Movassaghi, M., & Jackson, M. K. (2001). Design and compact modeling of saturated Erbium-doped fibre amplifiers with nonconfined doping. *Optical Fiber Technology, 7,* 312–323. doi:10.1006/ofte.2000.0354

Nikonorov, N., Przhevuskii, A., Prassas, M., & Jacob, D. (1999). Experimental determination of the upconversion rate in Erbium-doped silicate glasses. *Applied Optics, 38,* 6284–6291. doi:10.1364/AO.38.006284

Prudenzano, F., Mescia, L., Allegretti, L., De Sario, M., Smektala, F., & Moizan, V. (2009). Simulation of MID-IR amplification in Er^{3+} doped chalcogenide microstructured optical fibre. *Optical Materials, 31,* 1292–1295. doi:10.1016/j.optmat.2008.10.004

Prudenzano, F., Mescia, L., D'Orazio, A., De Sario, M., Petruzzelli, V., Chiasera, A., & Ferrari, M. (2007). Optimization and characterization of rare earth doped photonic crystal fibre amplifier using genetic algorithm. *Journal of Lightwave Technology, 25,* 2135–2142. doi:10.1109/JLT.2007.901331

Ratnaweera, A., Saman, K., & Watson, H. C. (2004). Self–organizing hierarchical particle swarm optimizer with time–varying acceleration coefficients. *IEEE Transactions on Evolutionary Computation, 8,* 240–255. doi:10.1109/TEVC.2004.826071

Rochat, E., Dändliker, R., Haroud, K., Czichy, R. H., Roth, U., Costantini, D., & Holzner, R. (2001). Fibre amplifiers for coherent space communication. *IEEE Journal on Selected Topics in Quantum Electronics, 7*(1), 74–81. doi:10.1109/2944.924012

Roudas, I., Richards, D. H., Antoniades, N., Jackel, J. L., & Wagner, R. E. (1999). An efficient simulation model of the erbium-doped fibre for the study of multiwavelength optical networks. *Optical Fiber Technology, 5,* 363–389. doi:10.1006/ofte.1999.0306

Satapathy, S. C., Naga, B., Murthy, J. V. R., & Reddy, P. (2007). A comparative analysis of unsupervised k-means, PSO and self-organizing PSO for image clustering. *International Conference on Computational Intelligence and Multimedia Applications (ICCIMA)* (pp. 229–237).

Shi, Y., & Eberhart, R. C. (1999). Empirical study of particle swarm optimization. *Proceedings of the IEEE International Conference on Evolutionary Computation,* Vol. 3, (pp. 101–106).

van den Bergh, F., & Engelbrecht, A. P. (2004). A cooperative approach to particle swarm optimization. *IEEE Transactions on Evolutionary Computation, 8*(3), 225–239. doi:10.1109/TEVC.2004.826069

Wright, M. W., & Valley, G. C. (2005). Yb-doped fibre amplifier for deep-space optical communications. *Journal of Lightwave Technology, 23*(3), 1369–1374. doi:10.1109/JLT.2005.843532

ADDITIONAL READING

Banks, A., Vincent, J., & Anyakoha, C. (2007). A review of particle swarm optimization, part I: background and development. *Natural Computing, 6,* 467–484. doi:10.1007/s11047-007-9049-5

Banks, A., Vincent, J., & Anyakoha, C. (2008). A review of particle swarm optimization, part II: Hybridisation, combinatorial, multicriteria and constrained optimization, and indicative applications. *Natural Computing, 7,* 109–124. doi:10.1007/s11047-007-9050-z

Becker, P. C., Olsson, N. A., & Simpson, J. R. (1999). *Erbium-doped fibre amplifiers.* London, UK: Academic Press.

Bjarklev, A. (1993). *Optical fibre amplifiers: Design and system applications.* Artech House Inc.

Bjarklev, A., Broeng, J., & Bjarklev, A. S. (2003). *Photonic crystal fibres.* Kluwer Academic Publishers. doi:10.1007/978-1-4615-0475-7

Bonabeau, E., Dorigo, M., & Theraulaz, G. (1999). *Swarm intelligence: From natural to artificial systems.* Oxford University Press.

Clerc, M. (2006). *Particle swarm optimization.* Wiley-ISTE. doi:10.1002/9780470612163

del Valle, Y., Venayagamoorthy, G. K., Mohagheghi, S., Hernandez, J.-C., & Harley, R. G. (2008). Particle swarm optimization: Basic concepts, variants and applications in power systems. *IEEE Transactions on Evolutionary Computation, 12,* 171–195. doi:10.1109/TEVC.2007.896686

Digonnet, J. F. (2001). *Rare-earth-doped fibre lasers and amplifiers.* Marcel Dekker Inc. doi:10.1201/9780203904657

Engelbrecht, A. P. (2006). *Fundamentals of computational swarm intelligence.* Wiley.

France, P. W. (2000). *Optical fibre lasers and amplifiers.* Boca Raton, FL: CRC Press Inc.

Kennedy, J., & Eberhart, R. C. (2001). *Swarm intelligence.* Morgan Kaufmann Publisher.

Kennedy, J., & Mendes, R. (2006). Neighborhood topologies in fully informed and best-of-neighborhood particle swarms. *IEEE Transactions on Systems, Man and Cybernetics. Part C, Applications and Reviews, 36,* 515–519. doi:10.1109/TSMCC.2006.875410

Lazinica, A. (2009). *Particle swarm optimization.* InTech. doi:10.5772/109

Martens, D., Baesens, B., & Fawcett, T. (2011). Editorial survey: Swarm intelligence for data mining. *Machine Learning, 82,* 1–42. doi:10.1007/s10994-010-5216-5

Mendez, A., & Morse, T. F. (2007). *Specialty optical fibres handbook.* Elsevier.

Pal, B. (2010). *Frontiers in guided wave optics and optoelectronics.* Intech. doi:10.5772/3033

Parsopoulos, K. E., & Vrahatis, M. N. (2010). *Particle swarm optimization and intelligence: Advances and applications.* Hershey, PA: IGI Global. doi:10.4018/978-1-61520-666-7

Poli, F., Cucinotta, A., & Selleri, S. (2007). *Photonic crystal fibres*. Springer.

Premaratne, M., & Agrawal, G. P. (2011). *Light propagation in gain media*. Cambridge University Press.

Quinby, R. S. (2006). *Photonics and lasers: An introduction*. John Wiley & Sons, Inc. doi:10.1002/0471791598

Rana, S., Jasola, S., & Kumar, R. (2011). A review on particle swarm optimization algorithms and their applications to data clustering. *Artificial Intelligence Review*, *35*, 211–222. doi:10.1007/s10462-010-9191-9

Sousa, T., Silva, A., & Neves, A. (2004). Particle swarm based data mining algorithms for classification tasks. *Parallel Computing*, *30*, 767–783. doi:10.1016/j.parco.2003.12.015

Sudo, S. (1997). *Optical fibre amplifiers: Materials, devices, and applications*. Artech House Inc.

Sun, J., Lai, C.-H., & Wu, J. (2011). *Particle swarm optimisation: Classical and quantum perspectives*. CRC Press.

Yang, X.-S. (2010). *Engineering optimization: An introduction with metaheuristic applications*. Wiley. doi:10.1002/9780470640425

KEY TERMS AND DEFINITIONS

Constriction Factor: Factor or parameter which can be necessary to insure convergence of a particle swarm algorithm.

Erbium Doped Fibre Amplifier: Optical device that is used to boost the intensity of optical signals being carried through a fibre optic communications system.

Fibre Amplifier Characterization: Decision action to find input parameters able to implement specific amplifiers.

Fitness Function: A mathematical expression, often used in modelling as part of an optimization process, i.e. the function used to find the parameter values that result in a maximum or minimum of a function.

Global Optimization Method: Finding the global best solution of a problem in the presence of multiple local optima in the search space.

Optimal Design: Improvement of a design to satisfy the original requirements within the available means.

Photonic Crystal Fibre: Specific optical fibre having an arrangement of air holes running along its length.

Chapter 8
Optimum Design of Hybrid EDFA/FRA by Particle Swarm Optimization

Alireza Mowla
K.N. Toosi University of Technology, Iran

Nosrat Granpayeh
K.N. Toosi University of Technology, Iran

Azadeh Rastegari Hormozi
K.N. Toosi University of Technology, Iran

ABSTRACT

In this chapter, the authors introduce the hybrid erbium-doped fiber amplifier (EDFA)/fiber Raman amplifier (FRA) and its optimization procedure by particle swarm optimization (PSO). EDFAs, FRAs, and their combinations, which have the advantages of both, are the most important optical fiber amplifiers that overcome the signal power attenuations in the long-haul communication. After choosing a proper configuration for a hybrid EDFA/FRA, users have to choose its numerous parameters such as the lengths, pump powers, number and wavelengths of pumps, number of signal channels and their wavelengths, the signal input powers, the kind of the fibers and their characteristics such as the radius of the core, numerical apertures, and the density of Er^{3+} ions in the EDFA. As can be seen, there are many parameters that need to be chosen properly. Here, efficient heuristic optimization method of PSO is used to solve this problem.

1. INTRODUCTION

Fiber optic communication is the key element of the information era. The revolutionarily growth of telecommunication industry during the few recent decades are completely indebted to fiber optic communication. One of the fundamental elements of fiber optic communication, beside the optical fiber, transmitter, and receiver, is the optical amplifier. Most important optical amplifiers are optical fiber amplifiers that use optical fibers as the gain media. Erbium-doped fiber amplifier

DOI: 10.4018/978-1-4666-2666-9.ch008

(EDFA) which uses the rare earth ions of erbium as a dopant of the glass fiber to produce the gain and also fiber Raman amplifier (FRA) that uses the nonlinear effect of the stimulated Raman scattering to produce Raman gain are among the most important optical fiber amplifiers. Hybrid EDFA/FRA is also widely used in the telecommunication systems. Hybrid EDFA/FRAs have low noise figure, large bandwidth and it can be used to minimize the nonlinear impairments (Castanon, Nasieva, Turitsyn, Brochier, & Pincemin, 2005). Without optical fiber amplifiers it was impossible to attain such an extremely high capacity fiber optic communication systems on the extra long-haul distances. So, we can consider the invention of the optical fiber amplifiers as the key invention essential for the telecommunication systems.

The EDFA was introduced in 1980s (Poole, Payne, Mears, Fermann, & Laming, 1986; Mears, Reekie, Poole, & Payne, 1986; Mears, Reekie, Jauncey, & Payne, 1987; Desurvire, Simpson, & Becker, 1987) and practically prevalent on fiber optic communications during 1990s (Headley & Agrawal, 2005). Although, Raman amplification as a gain solution for compensating the signal losses was demonstrated in 1980s (Hasegawa, 1983; Mollenauer, Stolen, & Islam, 1985; Smith & Mollenauer, 1989), FRA was not practically used in fiber optic communications till 2000s (Headley & Agrawal, 2005) after realizing some of its advantages over EDFA in the telecommunication systems. During this period of thriving of the EDFAs and FRAs, numerous efforts have been carried on making more efficient amplifiers. Different engineering designs, system configurations, and parameter arrangements have been used to get better results from EDFA ((Yeh, Lee, & Chi, 2004; Lu, Chu, Alphones, & Shum, 2004; Yamada, et al., 1998; Ahn & Kim, 2004; L., et al., 2005; Yi, Zhan, Hu, Tang, & Xia, 2006; Yi, Zhan, Ji, Ye, & Xia, 2004; Singh, Sunanda, & Sharma, 2004; Liang & Hsu, 2007; Lu & Chu, Gain flattening by using dual-core fiber in erbium-doped fiber amplifier, 2000) (Hung, Chen, Lai, & Chi,

2007; Chang, Wang, & Chiang, 2006; Choi, Park, & Chu, 2003; Pal, et al., 2007; Martin, 2001; Cheng & Xiao, Optimization of an erbium-doped fiber amplifier with radial effects, 2005; Cheng & Xiao, Optimization of a dual pumped L-band erbium-doped fiber amplifier by genetic algorithm, 2006), FRA (Zhou, Lu, Liu, Shum, & Cheng, 2001; Perlin & Winful, 2002; Liu, Chen, Lu, & Zhou, 2004) and hybrid EDFA/FRA (Masuda & Kawai, 1999; Carena, Curri, & Poggiolini, 2001; Li, Zhao, Wen, Lu, Wang, & Chen, 2006).

To recall some of the operation criteria of the optical fiber amplifiers we can include low noise figure, large bandwidth, longer telecommunication spans, minimized nonlinearity impairments, and also flattened gain spectrum. Because of the large number of parameters we have to consider the problem as an optimization problem. Therefore, we need an optimization tool to find the best values for the input parameters. Mathematical optimization means to select the best value from a set of available values. In modeling and simulation of any system, we consider the output or different outputs of the system as the objective function and the system parameters or inputs that evaluate the outputs are defined as the system variables. The complexity of the optimization problem has a direct relationship with the number of the variables. There are some computational optimization techniques such as finitely terminating algorithms, convergent iterative methods, and also heuristics methods that can be used in optimization problems. After the invention of computers the extraordinary improvements in the field of the simulation of the systems and also optimization techniques, let the engineers to design much more effective systems. As expected the simulation and modeling of the optical fiber amplifiers and their optimization had been a hot spot of research during the last few decades. The heuristic method of PSO was introduced in 1995 (Kennedy & Eberhart, Particle swarm optimization, 1995) and it has become popular rapidly because of its merits in dealing with a large number of variables, the speed of convergence and straightforwardness. PSO has been used in many engineering problems consisting of the optimization of the optical fiber amplifiers such as EDFA (Mowla & Granpayeh, A novel design approach for erbium-doped fiber amplifiers by particle swarm optimization, 2008), FRA (Mowla & Granpayeh, Design of a flat gain multi-pumped distributed fiber Raman amplifier by particle swarm optimization, 2008) and hybrid EDFA/FRA (Mowla & Granpayeh, Optimum design of a hybrid erbium-doped fiber amplifier/ fiber Raman amplifier using particle swarm optimization, 2009; Afkhami, Mowla, Granpayeh, & Hormozi, 2010).

Swarm intelligence is based on collective behaviour of self-organized or biological systems such as bird flocking, ant colonies, fish schooling, animal herding and also bacterial growth. Swarm intelligence can be considered as a subset of the greater concept of artificial intelligence which is defined as the study and design of intelligent agents which are systems estimating their environments and try to act in a way that approaches them toward success or global optimum. PSO is one of the stochastic global optimization algorithm which is based on swarm intelligence.

In this chapter, after introducing the steps to perform the PSO and the simulation method of the hybrid EDFA/FRA, we will use the PSO to optimize two configurations of hybrid EDFA/ FRA consisting of a single mode fiber (SMF), a dispersion-compensating fiber (DCF), a dispersion-shifted fiber (DSF) as a multi-wavelength pumped FRA with six or ten backward pumps, and a C-band EDFA.

2. IMPLEMENTATION OF PSO ON OPTICAL AMPLIFIERS

Swarm or complex systems are the origin of the definition of many computational methods that can efficiently solve the complicated multi-variable problems. When these swarm systems, resembling the social colonies of ants, bees, birds or fishes are able to handle complicated problems we call them swarm intelligence. In recent year many of these computational methods have proved their efficiencies to solve complicated problems, including ant colony optimization (Colorni, Dorigo, & Maniezzo, 1991; Dorigo, Optimization, Learning and Natural Algorithms, 1992; Dorigo & Stützle, Ant Colony Optimization, 2004), artificial bee colony algorithm (Karaboga & Basturk, 2008), charged system search (Kaveh & Talatahari, 2010), cuckoo search (Yang & Deb, Cuckoo Search via Lévy flights, 2009), artificial immune system (Timmis, Neal, & Hunt, 2000), firefly algorithm (Yang, Nature-Inspired Metaheuristic Algorithms: Second Edition, 2010), gravitational search algorithm (Rashedi, Nezamabadi-pour, & Saryazdi, 2009), intelligent water drop algorithm (Hosseini, 2007), river formation dynamics (Rabanal, Rodríguez, & Rubio, 2007), stochastic diffusion search (Meyer, Nasuto, & Bishop, 2006) and PSO (Kennedy & Eberhart, Particle swarm optimization, 1995; Eberhart & Kennedy, 1995; Kennedy & Eberhart, Swarm Intelligence, 2001) which is one of the best optimization techniques of its kind.

James Kennedy a social psychologist and Russ Eberhart an engineer and computer scientist, developed this method to produce a computational intelligence by using analogues of social interaction between a bird flock searching for food in 1995 (Kennedy & Eberhart, Particle swarm optimization, 1995; Eberhart & Kennedy, 1995). They have done further work on the social nature of the intelligence (Kennedy & Eberhart, Swarm Intelligence, 2001).

In the PSO method a number of individuals, resembling a flock of birds, which are called particles, are spread randomly over the parameter space of a function which is our search space to find the global optimum. The function is our problem and the search space could be multi-dimensional. Each particle is a combination of different variables chosen from a search space or an input variable. We attribute each particle with a fitness value that is the output of the objective function when the input is the particle as an array of input variables. Then the particles will change their positions through the search space at each iteration, which we call it movement and a velocity is defined for each particle. Particles will find their way through the search space according to the historical memory of its best previous position and also its one or more neighbor's best current and previous positions in different neighborhood structures. So, the particles use the information shared by all the swarm to modify their positions along with some random perturbations. Star and ring neighborhood structures are two common different structures in the PSO method (Kennedy & Eberhart, Particle swarm optimization, 1995; Eberhart & Kennedy, 1995; Kennedy & Eberhart, Swarm Intelligence, 2001). We define three features for each particle $P_i(k)$, which are particle's position $x_i(k)$, velocity $v_i(k)$, and fitness values $F(x_i(k))$ or $F(P(x_i(k)))$. Here, k is the number of iteration and i is the indicator of each particle. In the following, steps of PSO are presented with the goal of finding the global minimum;

Step 1: Initialization; to initialize the swarm $P(k)$, we have to distribute the particles P_is randomly within the pre-defined ranges by using $x_i(0)=x_{min}+rand(x_{max}-x_{min})$ where x_{min} and x_{max} are the limits of the variable ranges. Starting velocities $v_i(0)$, personal bests $P_{best\,i}$s and global best G_{best} are initializing to their worst possible values. The number of the particles should not be too small or large. Typically between 20 to 50 based on the expansion of the search space. In this chapter, we pick up 40 random particles P_is.

Step 2: Fitness evaluation; evaluate the fitness function $F(x_i(k))$ for each particle P_i with the position $x_i(k)$.

Step 3: Renew personal bests $P_{best\,i}$s ; if $F(P_i(k))<F(P_{best\,i})$ then $P_{best\,i}=P_i(k)$ and $x_{P\,best\,i}=x_i(k)$.

Step 4: Renew global best G_{best}; if $Min(F(P_{best i}))<G_{best}$ then $G_{best}=P_{best\,i}$ and $x_{G\,best}=x_{P\,best\,i}$.

Step 5: Update the velocities; the particle velocities that show the rate of the particle position variations, will be updated by using $v_i(k)=\varphi v_i(k-1)+\rho_1 C_f(x_{p\,best\,i}-x_i(k))+\rho_2 C_f(x_{G\,best}-x_i(k))$ when $\varphi=1-0.7((1-k)/(1-k_{max}))$, where φ is called the inertia weight and its value will decrease during the following iterations, k is the iteration number, C_f is accelerator constant that will accelerate the optimization searching process, which its value is 2 in our work. Finally, ρ_1 and ρ_2 are random values between 0 and 1 to maintain the random nature of the global optimum searching by PSO.

Step 6: Update the positions; particles' new positions will be found by adding the velocities to the particles' previous positions $x_i(k)=x_i(k-1)+v_i(k)$. If the particles jump out of the valid ranges, those have to be set into the valid domain.

Step 7: Check the ending criterion; the algorithm will repeat the steps until certain termination conditions are met. The termination condition can be the gathering of the particles in a single point.

Flow chart of PSO is shown in Figure 1. In PSO the particles find their ways based on their memories and their learning from other members of the swarm. In the optimization process, the PSO algorithm serves as the main program and the hybrid EDFA/FRA computer model acts as the subprogram. At each iteration, the main program will evaluate the fitness function for all the particles by calling the subprogram. We have to minimize the fitness function to fulfill our expectation to attain a flattened gain spectrum in a broad bandwidth. In this chapter, we present two different fitness functions that each can be useful under some conditions. First fitness function Equation 1 is the maximum of the gain values of the signal channels in decibels. Second fitness function Equation 2 is a combination of $F_{obj\,1}$ and the summation of the signal channel gains in decibels multiplied by an arbitrary constant α. We use $F_{obj\,2}$ when it is important to impose a limitation on the each individual signal channel gain value, not only to control the largest signal gain deviation from zero. In this work $F_{obj\,1}$ is used as the objective function.

$$F_{obj1} = Max \left| 10 \times \log \left(\frac{P_s^k(L)}{P_s^k(0)} \right) \right| \qquad (1)$$

$$F_{obj2} = Max \left| 10 \times \log \left(\frac{P_s^k(L)}{P_s^k(0)} \right) \right| \\ + \alpha \sum_{k=1:n_s} \left| 10 \times \log \left(\frac{P_s^k(L)}{P_s^k(0)} \right) \right| \qquad (2)$$

where n_s is the number of signal channels, k is the kth signal channel, $P_s^k(0)$ and $P_s^k(L)$ are the kth input and output signal powers, respectively. α is an arbitrary constant that determines the weight of each of the components in Equation 2, the value of which is chosen to be 0.05 in this work. The value of α is dependent on the number of signal channels and the average signal gain values. For example, if we have 60 signal channels and we assume that the average gain is almost one third of the maximum gain value. So, by choosing α as 0.05 we have balanced the weights of two components of Equation 2. We can formulate it as $\alpha=1/(n_s \times g_{ave})$, where n_s is the number of signal channels and g_{ave} is the normalized signal average gain value. Furthermore fitness functions can be used based on our expectations.

Signal gain of the hybrid EDFA/FRA is defined over the overall length of the hybrid amplifier, that consists of a lumped EDFA and a distributed FRA. It means that the signal gain of the kth channel in decibels, defined by $g_s^k=10\times log((P_s^k(L))/(P_s^k(0)))$, is desired to be zero. This goal will be

Figure 1. Flow chart of the PSO algorithm

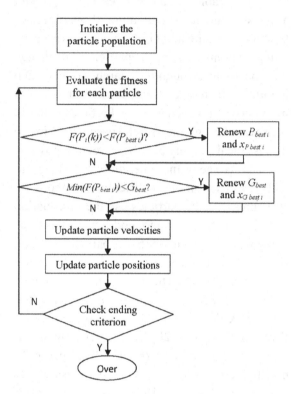

fulfilled when the signal power value at the end of the amplifier with the length of L equals its value at the input. The length of the amplifier is determined based on our design method. We can consider the lengths of different parts as variables or take it as a constant that impose constrains over the other variable parameters. These are the free degree of the engineering problem. In this work, we set the EDFA length as a variable to be optimized and the lengths of FRA parts are constant. Therefore, the optimization program searches for the global or absolute minimum values of $F_{obj\,1}$ and $F_{obj\,2}$ of Equations 1-2 on the valid search space. In the $F_{obj\,1}$ case, the summation of the output power deviations and in the case of $F_{obj\,2}$, both the gain variation and the summation of the output power deviations from the input powers will be minimized. Therefore, the minimum ripple of the signal channel gains around 0 dB is desired. Our convergence ending criterion is based on the gathering of the particles in a single point in the search space with respect to an error limitation. Also, we can set a specified limited number of iterations that has to be done at each round of the optimization process. The number of the iteration should be chosen to cover the estimated necessary time of the convergence.

3. HYBRID EDFA/FRA MODELING

To perform the optimization on the hybrid EDFA/FRA by PSO, we need a fast computer, a computer programming tool and written codes that execute the optimization. The codes consist of the main PSO program. The sub-program is our fitness function and will be included as a function or object in our written codes. We use MATLAB as our programming tool in this work.

First of all, we have to choose a proper configuration for hybrid EDFA/FRA based on the optical communication system engineering standards. In Figure 2 a configuration is proposed for the hybrid EDFA/FRA. After choosing a proper configuration we have to specify the pre-defined parameters and the parameters that have to be optimized. The values of pre-defined parameters are also chosen based on the engineering design expectations. The other parameters that are our variables have important and direct effects on the output or objective function and are the subject of optimization.

The configuration of hybrid EDFA/FRA in Figure 2 consists of a 50 or 60 km standard SMF, a 8.5 or 10.5 km DCF, a 0.1 dB isolator (ISO), a 50 or 60 km DSF, a wavelength division multiplexing (WDM) coupler, another 0.1 dB ISO, and a C-band EDFA. Standard SMF acts as a transmission medium to elongate the transmission length, DCF compensates the dispersion that take places during the signal pulse transmission in SMF, ISO prevents the backward FRA pump powers and amplified spontaneous emission (ASE) powers of the EDFA to enter back into SMF, DCF and

Figure 2. Configuration of the hybrid EDFA/FRA consists of 50 or 60 km standard SMF, 8.5 or 10.5 km DCF, ISO, 50 or 60 km DSF, 6- or 10-backward FRA pumps, ISO, and C-band EDFA

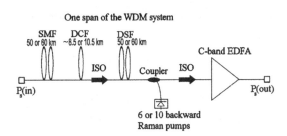

DSF. DSF is the FRA gain medium that has 6- or 10-backward pumps, WDM coupler insets the pump powers into the DSF, and C-band EDFA amplifies the signal channels again. The length of these hybrid EDFA/FRAs are about 108.5 or 130.5 km which is considered as a notable span in the telecommunication systems. To manage dispersion impairments, DCF is used that compensates the dispersion produced by SMF. DSF has almost zero dispersion and dispersion of EDFA is negligible because it is just a few meters. Some of the characteristics of these components are presented in Table 1. C-band EDFA is placed as the last part of hybrid EDFA/FRA because its noise is higher than FRA and the other modules, so placing it at the end will reduce the overall noise figure.

A computer model of hybrid EDFA/FRA is required to complete the optimization task. To obtain a model for hybrid EDFA/FRA, the model of each module is needed. SMF and DCF models are considered by inserting the attenuation and dispersion effects of them on signal powers. Attenuations of SMF, DCF and DSF are variant over the frequency band and can be obtained from Figure 3 (a). Dispersions of SMF and DCF are almost constant over the frequency band and according to Table 1 are 16.5 and -95 ps/(nm.km), respectively. ISO model is considered just by its

Table 1. Characteristics of hybrid EDFA/FRA components

Module	Fiber type	Attenuation (dB/km)	Dispersion (ps/(nm.km))	Length
Transmission	SMF	Fig. 3(a) (Headley & Agrawal, 2005)	16.5 (Islam, 2004)	50 or 60 km
Compensating	DCF	Fig. 3(a) (Headley & Agrawal, 2005)	-95 (Headley & Agrawal, 2005)	8.5 or 10.5 km
FRA	DSF	Fig. 3(a) (Islam, 2004)	0	50 or 60 km
EDFA	SMF	Fig. 3(a) (Headley & Agrawal, 2005)	16.5	Variable (1 to 15 m)

Figure 3. (a) DCF, SMF, and DSF attenuation spectra, (b) the Raman gain efficiency of DSF, and (c) the emission and absorption cross sections of Er^{3+} doped in Al/P-silica SMF (Mowla & Granpayeh, Optimum design of a hybrid erbium-doped fiber amplifier/fiber Raman amplifier using particle swarm optimization, 2009)

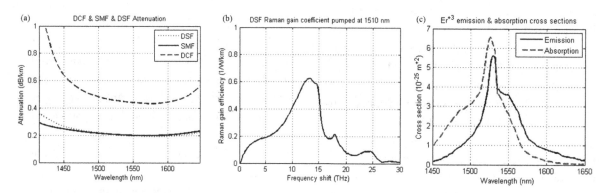

imposed attenuation which is 0.1 dB here. The two remaining modules are the main FRA which is a DSF pumped by 6- or 10-backward pumps and the C-band EDFA which should be modeled based on the equations defining their behaviors. In the following sections we will discuss the methods of modeling EDFA and FRA.

3.1. EDFA Equations and Modeling

EDFA is a device to reinforce the optical signal powers which are passing through it. EDFA use the principles of laser to boost the optical powers. It is like a fiber laser without mirrors. Semiconductor lasers are used to pump the EDFAs with a high power beam at a shorter wavelength. The Er^{3+} ions doped in the EDFA are excited by the pump photons through absorption and these ions will have exited electrons. These exited electrons will release their energy in the form of photons in phase and at the same wavelength with signal photons through stimulated emission. So, the signal optical power will become amplified.

The common model for EDFA is the Giles model, introduced by Giles and Desurvire in 1991 (Giles & Desurvire, 1991). This is a two-level model for Er^{3+} ions with $^4I_{15/2}$ as the ground- and $^4I_{13/2}$ as the excited-levels which are homogeneously broadened. Although, considering the other levels results in more accurate model, but it is enough and acceptable to ignore this complexity and take into account just two levels for Er^{3+} ions. In the two-level model, all the radiative transitions happen between these two

Box 1.

$$
\frac{dP_k(z)}{dz} = u_k \sigma_{ak} \int_0^a i_k(r) n_2(r,z,t) \left[P_k(z) + m h v_k \Delta v_k \right] 2\pi r dr
$$
$$
- u_k \sigma_{ak} \int_0^a i_k(r) n_1(r,z,t) P_k(z) 2\pi r dr - u_k l_k P_k(z)
$$

(3)

levels. Optical light spectrum in the optical fiber is assumed to be continuous. To model numerically this continuous spectrum by computer, we have to make the optical power discrete. Optical light spectrum is divided into discrete frequency bandwidths of Δv_k with the central frequency of f_k and wavelength of λ_k containing the optical power of P_k. Each frequency bandwidth is related to a signal channel or a pumping wavelength. When the total number of signal channels and pumping wavelengths are N, we have to solve an N ordinary differential system of equations that rule the propagation of optical powers in the optical fibers. Also, the rate equations for the population of the upper or excited level of Er^{3+}, $^4I_{13/2}$ will be a summation of the emission and absorption parts of all the optical frequency bandwidths (Cheng & Xiao, Optimization of a dual pumped L-band erbium-doped fiber amplifier by genetic algorithm, 2006). The emission and absorption cross section of Er^{3+} doped in Al/P silica fiber of figure 3 (c) are used (Miniscalco, 1991). There is a limitation on the value of V number when choosing the variable parameters of the EDFA that should be considered to guarantee the signals' single mode propagation in the EDFA. To obtain more simple equations, steady state condition $(\partial n_2(r,z,t)/\partial t = 0)$ and weakly guiding approximation $(\Delta = (n_1 - n_2)/n_2 \ll 1)$ are assumed. After considering the above conditions, the equation that defines the propagation of optical powers along the fibers z-axis is as in Equation 3 (see Box 1) (Cheng & Xiao, Optimization of a dual pumped L-band erbium-doped fiber amplifier by genetic algorithm, 2006); where $P_k(z)$ is the optical power

that propagates in the EDFA along the z axis at the frequency of v_k with the bandwidth of Δv_k.

It stands for pump or signal powers, $u_k = \pm 1$ shows the direction of forward or backward pump powers propagating through the EDFA, σ_{ek} and σ_{ak} are the emission and absorption cross sections at the frequency of v_k, the integration is over the radius of the EDFA core with the radius of a, $i_k(r)$ is the intensity of normalized transverse mode, n_1 and n_2 are the population of the ground $^4I_{15/2}$ and excited $^4I_{13/2}$ levels of the Er^{3+} ions, respectively, $m h v_k \Delta v_k$ is the amount of spontaneous emission from excited level population n_2 with the effective noise bandwidth of Δv_k at the noise frequency of v_k, m is a constant that its value is 2 for the noise and 0 for the pumps and signal powers, h is the plank's constant, and l_k is the excess fiber loss. It is necessary to know the population of the excited level $^4I_{13/2}$ of Er^{3+} ions to solve Equation 3, which can be derived by solving the rate equations (Cheng & Xiao, Optimization of a dual pumped L-band erbium-doped fiber amplifier by genetic algorithm, 2006);

$$
n_2(r,z) = n_t \frac{\sum_k \left(\frac{\tau \sigma_{ak}}{h v_k} P_k(z) i_k(r) \right)}{1 + \sum_k \left(\frac{\tau (\sigma_{ak} + \sigma_{ek})}{h v_k} \right) P_k(z) i_k(r)}
$$

(4)

where $n_2(r,z)$ is the excited level population distribution, $n_t = n_1 + n_2$ is the total erbium ions concentration with a uniform distribution over the fiber core cross section, and τ, the excited level lifetime is *10 ms* here. By assuming the fundamental mode propagating into the EDFA core, the intensity of normalized transverse mode is defined by (Cheng & Xiao, Optimization of a dual pumped L-band erbium-doped fiber amplifier by genetic algorithm, 2006);

$$i_k\left(r\right) = \frac{1}{\pi a^2}\left[\frac{v_k}{V_k}\frac{J_0\left(u_k r/a\right)}{J_1\left(u_k\right)}\right]^2 \qquad (5)$$

where $V_k = 2\pi a NA/\lambda_k$ is the V number at the wavelength of λ_k and $NA = \sqrt{(n_1^2 - n_2^2)}$ is the numerical aperture. When $1 \leq V_k \leq 3$ we can approximately assume that $v_k = 1.1428 V_k$ - *0.9960* and $u_k = \sqrt{(V_k^2 - v_k^2)}$. J_0 and J_1 are the Bessel functions of zero and first order, respectively. In addition, all the optical power propagation is confined into the EDFA core if the normalized condition of Equation 6 is fulfilled;

$$\int_0^a i_k\left(r\right)2\pi r dr = 1 \qquad (6)$$

Assuming that the input values of pump and signal powers are known and the goal is to find their output values, here an initial value differential equation problem must be solved. In the cases of bidirectional pump power propagation, back- and forward-ASE, and pump and signal power propagation in contra directions, the problem is a boundary value problem which is more complicated to be solved. Signal power propagation direction is considered as the forward direction. Pumping method in EDFA can be in forward or backward directions. In this work, to avoid unnecessary complications, forward pumping system is used and backward ASE is ignored, so that the problem will be an initial value problem. Backward ASE has a very slight effect on the optical power flow of signals and pumps in the EDFA and we

can ignore it without any disturbance. A one-step predictor corrector method (PCM) can be used to solve these kinds of problems (Liu & Lee, A fast and stable method for Raman amplifier propagation equations, 2003). To execute the PCM, the length of the EDFA should be divided into small parts or steps. Then the amount of powers will be calculated at each step, starting from the input point of EDFA where the values of signals, pumps, and noises are known. The propagating powers of the signals, pumps, or noises are calculated step by step based on their values at the previous steps that are known and a prediction of their values at the current point. For the all frequency bandwidths of Δv_k, the propagation of powers in the EDFA will be calculated by this method. To use the PCM it is better to write Equation 3 in this format;

$$\frac{dP_k\left(z\right)}{dz} = P_k\left(z\right)F_k\left(z\right) + C_k\left(z\right) \qquad (7)$$

where;

$$F_k\left(z\right) = u_k\left[\sigma_{ek} + \sigma_{ak}\right]$$
$$\int_0^a i_k\left(r\right)n_2\left(r,z,t\right)2\pi r dr - u_k\left[\sigma_{ak} + 1_k\right] \qquad (8)$$

$$C_k\left(z\right) = u_k\sigma_{ek}mhv_k\Delta v_k$$
$$\int_0^a i_k\left(r\right)n_2\left(r,z,t\right)2\pi r dr \qquad (9)$$

It is much easier to use this notation to solve Equation 6 by PCM. Exponential growth is assumed for the amplification and attenuation of optical power in the EDFA (Marhic & Nikonov, 2002), which is used to predict the power at each step. So that, following equation is used to solve Equation 6;

Table 2. Values used in numerical simulation of EDFA

Upper level life time (ms)	10
Numerical aperture NA	0.1
Step distance (m)	L/60
Cladding index	1.445
Fiber core radius a (μm)	4.1
Excess fiber loss (dB/m)	0.03
Start frequency (THz)	184.2
End frequency (THz)	196.2
Number of channels	60
Frequency steps (THz)	0.2

Table 3. Valid ranges of the optimization variables

Pumping wavelength λ_p (nm)	1450-1500
Input pump power (mW)	100 to 500
Er^{3+} concentration n_t	$(1-30) \times 10^{24}$
Fiber length L (m)	1-50

$$P_k\left(z_{j+1}\right) = P_k\left(z_j\right) \exp \left(\left(F_k\left(z_j\right) + C_k\left(z_j\right) / P_k\left(z_j\right)\right)\Delta z\right)$$

(10)

Note that for signal and pump powers $m=0$ so $C_k(z)=0$. Pumping can take place at wavelengths of 980 or 1480 nm when third telecommunication window is adopted. 980 nm is used when three levels of Er^{3+} are used and 1480 nm is speculated for two level Er^{3+} amplification process. To avoid the complexity of radial variations, step index fiber is used instead of graded index fiber and uniform distribution of Er^{3+} is assumed (Cheng & Xiao, Optimization of an erbium-doped fiber amplifier with radial effects, 2005). The data used in the simulation of the EDFA is shown in Table 2 and Figure 3. As it is shown, in the presented model, there are 60 signal channels placed at frequencies 184.4 to 196.2 THz by 200 GHz/channel spaces. ASE noises are also calculated at these frequencies. One forward pump with the power of 50 to 100 mW is used in this model with 1480 nm wavelength. The core of the EDFA with the radius of a=4.1 μm are doped with Er^{3+}(Cheng & Xiao, Optimization of a dual pumped L-band erbium-doped fiber amplifier by genetic algorithm, 2006). The radius of Er^{3-} doped region is assumed to be b. Length of the EDFA is divided into 60 steps that will result in a good accuracy

of power anticipation at each step. By simulation the power propagation in different signal channels, gain spectrum of EDFA is obtained. In fact the input signal powers that enter the EDFA are coming from the previous stages or the FRA. Simulation of the EDFA is a part of the simulation of the optical link. EDFA pump wavelength, Er^{3+} concentration, fiber length, and pump power are four parameters of the EDFA that are variables of the optimization process and ought to be optimized. The acceptable ranges of these variables are given in the Table 3. PSO will select the values of these variables randomly from these valid ranges. These ranges are determined as the spaces that are mostly probable to include the optimum values.

3.2. Raman Equations and Modeling

To model the propagation of the average power of the signals, pumps, ASE noises, and Rayleigh backscattering parts in the steady state, the following ordinary differential system of equations should be solved like in Equation 11 (see Box 2) (Bromage, 2004); where P_i^+ and P_i^- are the average optical powers in the frequency span of Δv centered at v_i propagating in the +z and -z directions, respectively. α_i, γ_i, A_{eff}, h, K and T are the attenuation coefficient, Rayleigh backscattering coefficient, effective area of optical fiber, Planck's constant, Boltzmann constant and temperature, respectively.

Also, Γ is a random value between *1* and *2* that shows the effect of polarization randomization. g_{ij} indicates the Raman gain parameter at v_i resulting from an optical wave at v_j. Summations include both the frequency ranges of $v_j>v_i$ and

Box 2.

$$\pm\frac{dP_i^{\pm}}{dz} = -\alpha_i P_i^{\pm} + \gamma_i P_i^{\mp} + P_i^{\pm}\sum_{v_j>v_i}\frac{g_{ji}}{\Gamma A_{eff}}\Big[P_j^+ + P_j^-\Big] - P_i^{\pm}\sum_{v_j<v_i}\frac{v_i}{v_j}\frac{g_{ij}}{\Gamma A_{eff}}\Big[P_j^+ + P_j^-\Big]$$

$$-2hv_i P_i^{\pm}\sum_{v_j<v_i}\frac{v_i}{v_j}\frac{g_{ij}}{\Gamma A_{eff}}\left[1+\left(e^{\frac{h(v_i-v_j)}{KT}}-1\right)^{-1}\right]\Delta v \qquad (11)$$

$$+2hv_i\sum_{v_j>v_i}\frac{g_{ji}}{\Gamma A_{eff}}\Big[P_j^+ + P_j^-\Big]\left[1+\left(e^{\frac{h(v_j-v_i)}{KT}}-1\right)^{-1}\right]\Delta v$$

$v_j<v_i$ to encounter all the frequency spans from v_0 to v_n which are n frequency spans with the noise bandwidth of Δv. Second term in Equation 11 models Rayleigh backscattering and the last term is related to ASE noises.

Signal powers always propagate in the forward direction, but pump powers can propagate both in the forward and backward directions based on the amplifier configuration. If the pumps are co-propagating with signals, we have an initial value problem to solve. In this case similar to the EDFA modeling, one-step predictor corrector method (PCM) can be used to solve the problem (Liu & Lee, A fast and stable method for Raman amplifier propagation equations, 2003). If we have counter-propagating pumps, our problem is a two-point boundary value problem which is more complicated to be solved. The values of signal and pump powers at *z=0* and *z=L* are the first assumptions to solve the problem, where the RFA length is *L*. In this paper, both co- and counter-propagating pumps are assumed, so that the problem will be a two-point boundary value problem. The predictor-corrector method (PCM) is used with the Adams-Bashforth-Moulton formulation (Liu & Lee, A fast and stable method for Raman amplifier propagation equations, 2003) to solve this problem step by step and the shooting algorithm is used to estimate the signal and pump powers at *z=0* or *z=L*, where their values are not known (Liu & Lee, Effective shooting algorithm

and its application to fiber amplifiers, 2003). The calculation is started at *z=L* where only the initial values of counter-propagating pumps are known. The values for signal and co-propagating pumps are estimated by shooting algorithm at *z=L*. Using the power values of counter-propagating pumps and estimated signal and co-propagating pump powers at *z=L*, we will start the calculation along the fiber in a backward direction step by step by means of a four step PCM. The first round will be over when the step by step calculation arrives at *z=0*. In the steady state, the power at each step is a constant value and our goal is to find these values at *z=L*. After the first round, the calculated values of the signal and co-propagating pump powers are compared with their initial values at *z=0*. When the acceptable accuracy is achieved, the calculation will be over. If not, after modifying the estimated values of signal and co-propagating pump powers at *z=L* by shooting algorithm, the next calculation round will be started again at *z=L*. These process will continue till the desired accuracy is achieved, which is 1% deviation from the initial signal and co-propagating power values at *z=0*. It means an accurate calculation should be in the range of 99 to 101 μW, for an initial value of 100 μW. The typical number of calculation rounds which are required to be done before the accuracy test is satisfied is about 2 to 20 rounds.

Alike the one step PCM that we used in modeling EDFA, four step PCM is used to model FRA

Figure 4. Configuration of fiber Raman amplifier with the direction of signal, co- and counter-propagating powers (Bromage, 2004)

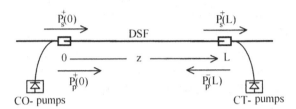

CO- pumps   CT- pumps

with more accurate results. The length of the FRA is divided into some sections or steps. The starting point in power calculation is step 0 at $z=L$ with the initial guess values of signals based on shooting algorithm. The average optical power of each optical beam at different frequency slots will be calculated based on its value at four previous steps, according to the Adams-Bashforth-Moulton formulations. At the first, second, and third steps different formulations are used since at these steps we have less than four previous steps (Liu & Lee, A fast and stable method for Raman amplifier propagation equations, 2003). Figure 4 shows the configuration of FRA with the direction of signals, co- and counter-propagating pump powers. As depicted in this figure, the amplification media of FRA is DSF which its Raman gain efficiency and attenuation are shown in Figure 3 (b) and 3 (a), respectively. The formulation of Raman gain efficiency is g_{ij}/TA_{eff} in Equation 11 where g_{ij} is constant for different pump wavelengths but A_{eff} for the DSF will increase with the frequency (Namiki & Emori, 2001). For the pump wavelengths of about 1420 and 1510 nm the peak values of the Raman gain efficiency will happen which are 0.7612 and 0.6206 l/W/km, respectively (Namiki & Emori, 2001). As pump wavelengths increase, the peak value of the Raman gain efficiency will decrease almost linearly. Additional parameters of the distributed FRA needed for the simulation are given in Table 4.

Table 4. Parameters of the distributed FRA which are needed in the simulation (Mowla & Granpayeh, Design of a flat gain multi-pumped distributed fiber Raman amplifier by particle swarm optimization, 2008)

Rayleigh backscattering γ (m^{-1})	7×10^{-8}
Temperature T (°K)	300
Number of signal channels	60
Signal channels wavelength range (nm)	1529.2-1627.1
Signal channels frequency range (THz)	184.4-196.2
Signal channels spacing (THz)	0.2
Gain bandwidth $\Delta\lambda$ (nm)	97.9
Input signal power per channel (μW)	100
Number of pumps	6 or 10
Pumps wavelength range (nm)	1417.5-1510.3
Pumps frequency range (THz)	198.5-211.5
Fiber length L (km)	50 or 60
Number of steps	23

4. RESULTS OF HYBRID EDFA/FRA OPTIMIZATION

To optimize the hybrid EDFA/FRA we have to run the PSO program. Computer model of the hybrid EDFA/FRA based on the configurations presented in the previous sections will act as a sub-program in the main PSO program. After performing the PSO program, the optimum values of the optimization variables will be attained. There are numerous variables that affect the performance of the hybrid EDFA/FRA. We select just some of these variables to participate in the optimization process which are given in Table 5. PSO can manage to manipulate a larger number of variables in comparison with the other optimization methods like genetic algorithm. Other variables are not involved to avoid more complications and their values are chosen based on the best engineering presumptions. After obtaining the optimum values for the variables listed in Table 5, optimum signal gain spectrum will be obtained.

Table 5. Variables to participate in the optimization process and their ranges (Mowla & Granpayeh, Optimum design of a hybrid erbium-doped fiber amplifier/fiber Raman amplifier using particle swarm optimization, 2009)

EDFA pumping wavelength λ_p (nm)	1450-1500
EDFA Er^{3+} concentration n_t	$(1-30)\times 10^{24}$
EDFA fiber length L (m)	1-50
EDFA pump power (mW)	100-500
FRA pump powers (mW)	5-350

Because of the random nature of the PSO method there is always a possibility that the obtained solutions fall into the local bests instead of the global bests. In writing the PSO program, good contemplations should be made to lessen this possibility. Also, the lack of certainty causes the PSO solutions to be slightly different from the previously obtained solutions or totally irrelevant, each time that it is run. To overcome this problem, the PSO program must be run for a number of times and then the best solutions should be picked up among the other ones. There is no rule to specify this number of times and it depends on the number of variables and the structure of the program. In this work, we run the program for about 20 times and considering the solutions it is almost definite for us that the solutions presented in this chapter are the global bests.

As shown in Figure 2, two configurations of hybrid EDFA/FRA are considered in the optimization process with overall lengths of 108.5 or 130.5 km and each of them have 6- or 10-pumped FRA. Figure 5 and Figure 6 depict the optimized signal gain spectrum for 108.5 and 130.5 km hybrid EDFA/FRA with 6- or 10-Raman-pumped, respectively. As it can be seen from the figures, signal gain spectrum variations of 2.60 and 1.84 dB are attained for 108.5 km hybrid EDFA/FRA with 6- or 10-Raman-pumped, and also signal gain spectrum variations of 2.91 and 2.03 dB are attained for 130.5 km hybrid EDFA/FRA with 6- or 10-Raman-pumped, respectively. Increasing the number of FRA pumps in designing the hybrid EDFA/FRA will decrease the gain spectrum variation. Also, if the FRA pump wavelengths are included in the PSO process, the gain spectrum variation will decrease. The use of higher number of FRA pumps with variable wavelengths is ignored in the suggested configuration to avoid the designing complications and higher costs. In the distributed hybrid EDFA/FRA that is designed, the goal is to reach the network gain of 0 dB. If this goal is obtained, there is no signal gain spectrum variation and the signal power at the output of the link is the same as its input value.

Figure 5. Optimized signal gain spectrum of the 108.5 km hybrid EDFA/FRA with 6- or 10-Raman-pumped

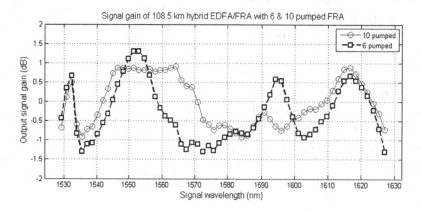

Figure 6. Optimized signal gain spectrum of the 130.5 km hybrid EDFA/FRA with 6- or 10-Raman-pumped

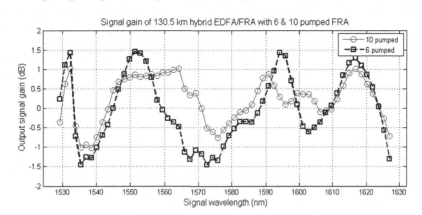

Table 6. Optimized values of the parameters and also fitness values obtained by PSO for 108.5 km hybrid EDFA/FRA with 6- or 10-Raman-pumped

Optimization variables	Optimized values for 10-Raman-pumped hybrid EDFA/FRA	Optimized values for 6-Raman-pumped hybrid EDFA/FRA
EDFA pump wavelength λ_p (nm)	1452.9	1457.1
EDFA Er^{3+} concentration n_t	8.061×10^{24}	6.082×10^{23}
EDFA fiber length L (m)	1.007	14.936
EDFA pump power (mW)	433.7	497.7
FRA pump power P_1 (mW)	72.5 (at 1425.0 nm)	69.7 (at 1425.0 nm)
FRA pump power P_2 (mW)	192.1 (at 1434.5 nm)	250.5 (at 1442.1 nm)
FRA pump power P_3 (mW)	59.2 (at 1444.1 nm)	57.9 (at 1459.7 nm)
FRA pump power P_4 (mW)	86.2 (at 1453.8 nm)	160.4 (at 1477.7 nm)
FRA pump power P_5 (mW)	62.0 (at 1463.6 nm)	37.1 (at 1496.1 nm)
FRA pump power P_6 (mW)	97.0 (at 1473.6 nm)	116.2 (at 1515.0 nm)
FRA pump power P_7 (mW)	45.0 (at 1483.8 nm)	-----
FRA pump power P_8 (mW)	13.7 (at 1494.0 nm)	-----
FRA pump power P_9 (mW)	6.9 (at 1504.4 nm)	-----
FRA pump power P_{10} (mW)	90.5 (at 1515.0 nm)	-----
Fitness (dB)	0.9177	1.3008

The WDM systems that are presented in this work contain 60 channels with the wavelengths of 1529.2 to 1627.1 nm and the bandwidth of 97.9 nm. All of the FRA pumps in the configuration of hybrid EDFA/FRA are considered backward and the lengths of the WDM links are 108.5 or 130.5 km. The optimized values of the parameters and also fitness values obtained by PSO for 108.5 or 130.5 km hybrid EDFA/FRA with 6- or 10-Raman-pumped are listed in Table 6 and Table 7, respectively.

An equidistant scale is considered in defining the FRA pump frequencies. The FRA pump frequencies or wavelengths are not considered as the optimization variables in the optimization process to avoid more complications and high

Table 7. Optimized values of the parameters and also fitness values obtained by PSO for 130.5 km hybrid EDFA/FRA with 6- or 10-Raman-pumped (Mowla & Granpayeh, Optimum design of a hybrid erbium-doped fiber amplifier/fiber Raman amplifier using particle swarm optimization, 2009)

Optimization variables	Optimized values for 10-Raman-pumped hybrid EDFA/FRA	Optimized values for 6-Raman-pumped hybrid EDFA/FRA
EDFA pump wavelength λ_p (nm)	1479.8	1456.4
EDFA Er^{3+} concentration n_t	9.991×10^{24}	9.988×10^{24}
EDFA fiber length L (m)	1.127	1.185
EDFA pump power (mW)	482.1	468.2
FRA pump power P_1 (mW)	147.0 (at 1425.0 nm)	69.6 (at 1425.0 nm)
FRA pump power P_2 (mW)	185.4 (at 1434.5 nm)	254.0 (at 1442.1 nm)
FRA pump power P_3 (mW)	69.8 (at 1444.1 nm)	33.2 (at 1459.7 nm)
FRA pump power P_4 (mW)	86.1 (at 1453.8 nm)	213.5 (at 1477.7 nm)
FRA pump power P_5 (mW)	19.4 (at 1463.6 nm)	48.7 (at 1496.1 nm)
FRA pump power P_6 (mW)	124.1 (at 1473.6 nm)	122.7 (at 1515.0 nm)
FRA pump power P_7 (mW)	82.3 (at 1483.8 nm)	-----
FRA pump power P_8 (mW)	6.7 (at 1494.0 nm)	-----
FRA pump power P_9 (mW)	6.3 (at 1504.4 nm)	-----
FRA pump power P_{10} (mW)	71.9 (at 1515.0 nm)	-----
Fitness (dB)	1.0168	1.4542

capacity of computational works. The numbers of optimization variables are 10 and 14 for 6- and 10-Raman-pumped hybrid EDFA/FRA, respectively. Because one of the optimization variables is FRA pump powers. Although PSO is an efficient optimization method in dealing with higher number of variables, but excessive increase in the number of variables will cause inefficiency in all the optimization methods. PSO can act better in the optimization of hybrid EDFA/FRA in comparison with the other popular optimization methods like genetic algorithm (Mowla & Granpayeh, A novel design approach for erbium-doped fiber amplifiers by particle swarm optimization, 2008; Mowla & Granpayeh, Design of a flat gain multi-pumped distributed fiber Raman amplifier by particle swarm optimization, 2008; Mowla & Granpayeh, Optimum design of a hybrid erbium-doped fiber amplifier/fiber Raman amplifier using particle swarm optimization, 2009; Afkhami, Mowla, Granpayeh, & Hormozi, 2010).

5. CONCLUSION

In this chapter, we present an efficient optimization method, named particle swarm optimization (PSO) to optimize the hybrid EDFA/FRA. Hybrid EDFA/FRA is an optical fiber amplifier that is a combination of EDFA and FRA. EDFA revolutionized the long-haul optical fiber communication systems and can be considered as the most important part of these systems. FRA is also very popular and revives the optical telecommunication systems during the past years. In this chapter, the PSO algorithm is discussed and its implementation in optical amplifiers is formulated. The formulation and modeling of EDFA and FRA are also discussed and their simulations are combined to present the computer simulation of hybrid EDFA/FRA. The computer simulation of hybrid EDFA/FRA will act as a sub-program in the PSO main program. Two configurations for hybrid EDFA/FRA are presented, one with the length of 108.5

km and the other one 130.5 km, each includes 6- or 10-Raman-pumped. Our objective in this optimization is to obtain a flattened signal gain spectrum. Among the numerous variables that have effect on the signal gain spectrum, we chose some of the most effective ones to include in the optimization process. The variables that are optimized are FRA and EDFA pump powers, Er^{3+} concentration, EDFA pump wavelength and length. The numbers of PSO variables are 10 and 14 in the 6- and 10-Raman-pumped hybrid EDFA/FRA, respectively. 108.5 or 130.5 km, 60 signal channel, 97.9 nm bandwidth WDM links are presented with signal gain spectrum variations of 2.60 and 1.84 dB for 108.5 km hybrid EDFA/FRA with 6- or 10-Raman-pumped, and 2.91 and 2.03 dB for 130.5·km hybrid EDFA/FRA with 6- or 10-Raman-pumped, respectively. Using PSO a flattened signal gain spectrum is obtained in this chapter.

REFERENCES

Afkhami, H., Mowla, A., Granpayeh, N., & Hormozi, A. R. (2010). Wideband gain flattened hybrid erbium-doped fiber amplifier/fiber Raman amplifier. *Journal of the Optical Society of Korea*, *14*(4), 342–350. doi:10.3807/JOSK.2010.14.4.342

Ahn, J. T., & Kim, K. H. (2004). All-optical gain-clamped erbium-doped fiber amplifier with improved noise figure and freedom from relaxation oscillation. *IEEE Photonics Technology Letters*, *16*(1), 84–86. doi:10.1109/LPT.2003.818906

Bromage, J. (2004). Raman amplification for fiber communications systems. *Journal of Lightwave Technology*, *22*(1), 79–93. doi:10.1109/JLT.2003.822828

Carena, A., Curri, V., & Poggiolini, P. (2001). On the optimization of hybrid Raman/erbium-doped fiber amplifiers. *IEEE Photonics Technology Letters*, *13*(11), 1170–1172. doi:10.1109/68.959353

Castanon, J. D., Nasieva, I. O., Turitsyn, S. K., Brochier, N., & Pincemin, E. (2005). Optimal span length in high-speed transmission systems with hybrid Raman-erbium-doped fiber amplification. *Optics Letters*, *30*(1), 23–25. doi:10.1364/OL.30.000023

Chang, C. L., Wang, L., & Chiang, Y. J. (2006). A dual pumped double-pass L-band EDFA with high gain and low noise. *Optics Communications*, *267*(1), 108–112. doi:10.1016/j.optcom.2006.06.025

Cheng, C., & Xiao, M. (2005). Optimization of an erbium-doped fiber amplifier with radial effects. *Optics Communications*, *254*(4-6), 215–222. doi:10.1016/j.optcom.2005.05.049

Cheng, C., & Xiao, M. (2006). Optimization of a dual pumped L-band erbium-doped fiber amplifier by genetic algorithm. *Journal of Lightwave Technology*, *24*(10), 3824–3829. doi:10.1109/JLT.2006.881476

Choi, B. H., Park, H. H., & Chu, M. J. (2003). New pumped wavelength of 1540-nm band for long-wavelength-band erbium-doped fiber amplifier (L-band EDFA). *IEEE Journal of Quantum Electronics*, *39*(10), 1272–1280. doi:10.1109/JQE.2003.817582

Colorni, A., Dorigo, M., & Maniezzo, V. (1991). Distributed optimization by ant colonies. *Proceedings of ECAL91, European Conference on Artificial life* (pp. 134-142). Paris, France: Elsevier Publishing.

Desurvire, E., Simpson, J. R., & Becker, P. C. (1987). High-gain erbium-doped traveling-wave fiber amplifier. *Optics Letters*, *12*(11), 888–890. doi:10.1364/OL.12.000888

Dorigo, M. (1992). *Optimization, learning and natural algorithms.* PhD thesis, Politecnico di Milano.

Dorigo, M., & Stützle, T. (2004). *Ant colony optimization.* The MIT Press. doi:10.1007/b99492

Eberhart, R. C., & Kennedy, J. (1995). A new optimizer using particle swarm theory. *Proceedings 6th International Symposium on Micro Machine and Human Science,* (pp. 39-43). Nagoya, Japan.

Giles, C. R., & Desurvire, E. (1991). Modeling erbium-doped fiber amplifiers. *Journal of Lightwave Technology, 9*(2), 271–283. doi:10.1109/50.65886

Hasegawa, A. (1983). Amplification and reshaping of optical solitons in a glass fiber-IV: Use of the stimulated Raman process. *Optics Letters, 8*(12), 650–652. doi:10.1364/OL.8.000650

Headley, C., & Agrawal, G. P. (2005). *Raman amplification in fiber optical communication systems* (pp. 33–163). Burlington, MA: Elsevier.

Hosseini, H. S. (2007). *Problem solving by intelligent water drops.* IEEE Congress on Evolutionary Computation, CEC. Singapore.

Hung, C. M., Chen, N. K., Lai, Y., & Chi, S. (2007). Double-pass high-gain low-noise EDFA over S- and C+L-bands by tunable fundamental-mode leakage loss. *Optics Express, 15*(4), 1454–1460. doi:10.1364/OE.15.001454

Islam, M. N. (2004). *Raman amplifiers for telecommunications I and II.* New York, NY: Springer.

Karaboga, D., & Basturk, B. (2008). On the performance of artificial bee colony (ABC) algorithm. *Applied Soft Computing, 8*(1), 687–697. doi:10.1016/j.asoc.2007.05.007

Kaveh, A., & Talatahari, S. (2010). A novel heuristic optimization method: charged system search. *Acta Mechanica, 213*(3-4), 267–289. doi:10.1007/s00707-009-0270-4

Kennedy, J., & Eberhart, R. C. (1995). Particle swarm optimization. *Proceedings IEEE International Conference on Neural Networks,* (pp. 1942-1948).

Kennedy, J., & Eberhart, R. C. (2001). *Swarm intelligence.* Morgan Kaufmann Publishers, Academic Press.

Li, Z., Zhao, C. L., Wen, Y. J., Lu, C., Wang, Y., & Chen, J. (2006). Optimization of Raman/EDFA hybrid amplifier based on dual-order stimulated Raman scattering using a single-pump. *Optics Communications, 265*(2), 655–658. doi:10.1016/j.optcom.2006.03.049

Liang, T. C., & Hsu, S. (2007). The L-band EDFA of high clamped gain and low noise figure implemented using fiber Bragg grating and double-pass method. *Optics Communications, 281*(5), 1134–1139. doi:10.1016/j.optcom.2007.11.020

Liu, X., Chen, J., Lu, C., & Zhou, X. (2004). Optimizing gain profile and noise performance for distributed fiber Raman amplifiers. *Optics Express, 12*(24), 6053–6066. doi:10.1364/OPEX.12.006053

Liu, X., & Lee, B. (2003). A fast and stable method for Raman amplifier propagation equations. *Optics Express, 11*(18), 2163–2176. doi:10.1364/OE.11.002163

Liu, X., & Lee, B. (2003). Effective shooting algorithm and its application to fiber amplifiers. *Optics Express, 11*(12), 1452–1461. doi:10.1364/OE.11.001452

Lu, Y. B., & Chu, P. L. (2000). Gain flattening by using dual-core fiber in erbium-doped fiber amplifier. *IEEE Photonics Technology Letters, 12*(12), 1616–1617. doi:10.1109/68.896325

Lu, Y. B., Chu, P. L., Alphones, A., & Shum, P. (2004). A 105-nm ultrawideband gain-flattened amplifier combining C- and L-band dual-core EDFAs in a parellel configuration. *IEEE Photonics Technology Letters, 16*(7), 1640–1642. doi:10.1109/LPT.2004.827964

Marhic, M. E., & Nikonov, D. E. (2002). Low third-order glass-host nonlinearities in erbium-doped waveguide amplifiers. *Proceedings of SPIE, 4645*, (p. 193).

Martin, J. C. (2001). Erbium transversal distribution influence on the effectiveness of a doped fiber: Optimization of its performance. *Optics Communications, 194*(4-6), 331–339. doi:10.1016/S0030-4018(01)01312-8

Masuda, H., & Kawai, S. (1999). Wide-band and gain-flattened hybrid fiber amplifier consisting of an EDFA and a multi-wavelength pumped Raman amplifier. *IEEE Photonics Technology Letters, 11*(6), 647–649. doi:10.1109/68.766772

Mears, R., Reekie, L., Jauncey, I., & Payne, D. (1987). Low-noise erbium-doped fibre amplifier operating at 1.54μm. *Electronics Letters, 23*(19), 1026–1028. doi:10.1049/el:19870719

Mears, R., Reekie, L., Poole, S., & Payne, D. (1986). Low-threshold tunable CW and Q-switched fibre laser operating at 1.55 μm. *Electronics Letters, 22*(3), 159–160. doi:10.1049/el:19860111

Meyer, K. D., Nasuto, S. J., & Bishop, M. (2006). Stochastic diffusion search: Partial function evaluation in swarm intelligence dynamic optimisation. *Studies in Computational Intelligence. Stigmergic Optimization, 31*, 185–207. doi:10.1007/978-3-540-34690-6_8

Miniscalco, W. J. (1991). Erbium-doped glasses for fiber amplifiers at 1500 nm. *Journal of Lightwave Technology, 9*(2), 234–250. doi:10.1109/50.65882

Mollenauer, L. F., Stolen, R. H., & Islam, M. N. (1985). Experimental demonstration of soliton propagation in long fibers: Loss compensated by Raman gain. *Optics Letters, 10*(5), 229–231. doi:10.1364/OL.10.000229

Mowla, A., & Granpayeh, N. (2008). A novel design approach for erbium-doped fiber amplifiers by particle swarm optimization. *Progress in Electromagnetic Research M, 3*, 103–118. doi:10.2528/PIERM08061003

Mowla, A., & Granpayeh, N. (2008). Design of a flat gain multi-pumped distributed fiber Raman amplifier by particle swarm optimization. *Journal of the Optical Society of America. A, Optics, Image Science, and Vision, 25*(5), 3059–3066. doi:10.1364/JOSAA.25.003059

Mowla, A., & Granpayeh, N. (2009). Optimum design of a hybrid erbium-doped fiber amplifier/fiber Raman amplifier using particle swarm optimization. *Applied Optics, 48*(5), 979–984. doi:10.1364/AO.48.000979

Namiki, S., & Emori, Y. (2001). Ultrabroad-band Raman amplifiers pumped and gain-equalized by wavelength-division-multiplexed high power laser diodes. *IEEE Journal on Selected Topics in Quantum Electronics, 7*(1), 3–16. doi:10.1109/2944.924003

Pal, M., Paul, M. C., Dhar, A., Pal, A., Sen, R., & Dasgupta, K. (2007). Investigation of the optical gain and noise figure for multichannel amplification in EDFA under optimized pump condition. *Optics Communications, 273*(2), 407–412. doi:10.1016/j.optcom.2007.01.039

Perlin, V. E., & Winful, H. G. (2002). Optimal design of flat-gain wide-band fiber Raman amplifiers. *Journal of Lightwave Technology, 20*(2), 250–254. doi:10.1109/50.983239

Poole, S., Payne, D., Mears, R., Fermann, M., & Laming, R. (1986). Fabrication and characterization of low-loss optical fibers containing rare-earth ions. *Journal of Lightwave Technology, 4*(7), 870–876. doi:10.1109/JLT.1986.1074811

Rabanal, P., Rodríguez, I., & Rubio, F. (2007). Using river formation dynamics to design heuristic algorithms. *Unconventional Computation. Lecture Notes in Computer Science, 4618*, 163–177. doi:10.1007/978-3-540-73554-0_16

Rashedi, E., Nezamabadi-Pour, H., & Saryazdi, S. (2009). GSA: A gravitational search algorithm. *Information Sciences, 179*(13), 2232–2248. doi:10.1016/j.ins.2009.03.004

Singh, R., Sunanda, & Sharma, E. K. (2004). Gain flattening by long period gratings in erbium-doped fibers. *Optics Communications, 240*(1-3), 123–132. doi:10.1016/j.optcom.2004.06.023

Smith, K., & Mollenauer, L. F. (1989). Experimental observation of soliton interaction over long fiber paths: Discovery of a long-range interaction. *Optics Letters, 14*(22), 1284–1286. doi:10.1364/OL.14.001284

Timmis, J., Neal, M., & Hunt, J. (2000). An artificial immune system for data analysis. *Bio Systems, 55*(1-3), 143–150. doi:10.1016/S0303-2647(99)00092-1

Yamada, M., Mori, A., Kobayashi, K., Ono, H., Kanamori, T., & Oikawa, K. (1998). Gain-flattened tellurite-based EDFA with a flat amplification bandwidth of 76 nm. *IEEE Photonics Technology Letters, 10*(9), 1244–1246. doi:10.1109/68.705604

Yang, X. S. (2010). *Nature-inspired metaheuristic algorithms* (2nd ed.). Luniver Press.

Yang, X. S., & Deb, S. (2009). Cuckoo search via Lévy flights. *World Congress on Nature & Biologically Inspired Computing*, (pp. 210-214). Coimbatore.

Yeh, C. H., Lee, C. C., & Chi, S. (2004). S- plus C-band erbium-doped fiber amplifier in parallel structure. *Optics Communications, 241*(4-6), 443–447. doi:10.1016/j.optcom.2004.07.018

Yi, L., Zhan, L., Taung, C. S., Luo, S. Y., Hu, W. S., & Su, Y. K. (2005). Low noise figure all-optical gain-clamped parallel C+L band erbium-doped fiber amplifier using an interleaver. *Optics Express, 13*(12), 4519–4524. doi:10.1364/OPEX.13.004519

Yi, L. L., Zhan, L., Hu, W. S., Tang, Q., & Xia, Y. X. (2006). Tunable gainclamped double-pass erbium-doped fiber amplifier. *Optics Express, 14*(2), 570–574. doi:10.1364/OPEX.14.000570

Yi, L. L., Zhan, L., Ji, J. H., Ye, Q. H., & Xia, Y. X. (2004). Improvement of gain and noise figure in double-pass L-band EDFA by incorporating a fiber Bragg grating. *IEEE Photonics Technology Letters, 16*(4), 1005–1007. doi:10.1109/LPT.2004.823697

Zhou, X., & Lu, C., Liu, Shum, P., & Cheng, T. H. (2001). A simplified model and optimal design of a multiwavelength backward-pumped fiber Raman amplifier. *IEEE Photonics Technology Letters, 13*(9), 945–947. doi:10.1109/68.942655

ADDITIONAL READING

Al-Asadi, H. A., Al-Mansoori, M. H., Hitam, S., Saripan, M. I., & Mahdi, M. A. (2011). Particle swarm optimization on threshold exponential gain of stimulated Brillouin scattering in single mode fibers. *Optics Express, 19*(3), 1842–1853. doi:10.1364/OE.19.001842

Angeline, P. J. (1998). Evolutionary optimization versus particle swarm optimization: Philosophy and performance differences. *Springer Lecture Notes in Computer Science, Evolutionary Programming VII, 1447*, (pp. 601-610).

Deb, K., Pratap, A. K., Agarwal, S., & Meyarivan, T. (2002). A fast and elitist multiobjective genetic algorithm: NSGA-II. *IEEE Transactions on Evolutionary Computation, 6*(2), 182–197. doi:10.1109/4235.996017

Eberhart, R. C., & Shi, Y. (1998). Comparison between genetic algorithms and particle swarm optimization. *Springer Lecture Notes in Computer Science, Evolutionary Programming VII, 1447*, (pp. 611-616).

Eberhart, R. C., & Shi, Y. (2000). Comparing inertia weights and constriction factors in particle swarm optimization. *IEEE Proceedings of the 2000 Congress on Evolutionary Computation,* Vol. 1 (pp. 84-88). La Jolla, CA, USA.

Eberhart, R. C., & Shi, Y. (2001). Particle swarm optimization: developments, applications and resources. *IEEE Proceedings of the 2001 Congress on Evolutionary Computation,* (pp. 81). Seoul, Korea.

Gao, M., Jiang, C., Hu, W., & Wang, J. (2004). Optimized design of two-pump fiber optical parametric amplifier with two-section nonlinear fibers using genetic algorithm. *Optics Express, 12*(23), 5603–5613. doi:10.1364/OPEX.12.005603

Giaquinto, A., Mescia, L., Fornarelli, G., & Prudenzano, F. (2011). Particle swarm optimization-based approach for accurate evaluation of upconversion parameters in Er^{3+}-doped fibers. *Optics Letters, 36*(2), 142–144. doi:10.1364/OL.36.000142

Goldberg, D. E. (1989). *Genetic algorithms in search, optimization, and machine learning.* Addison-Wesley.

Jiang, H., Xie, K., & Wang, Y. (2008). Photonic crystal fibre for use in fibre Raman amplifiers. *IEEE Electronics Letters, 44*(13), 796–798. doi:10.1049/el:20080757

Jiang, H., Xie, K., & Wang, Y. (2010). C band single pump photonic crystal fiber Raman amplifier. *Springer Chinese Science Bulletin, 55*(6), 555–559. doi:10.1007/s11434-009-0296-y

Jiang, H., Xie, K., & Wang, Y. (2010). Pump scheme for gain-flattened Raman fiber amplifiers using improved particle swarm optimization and modified shooting algorithm. *Optics Express, 18*(11), 11033–11045. doi:10.1364/OE.18.011033

Jiang, H., Xie, K., & Wang, Y. (2010). Shooting algorithm and particle swarm optimization based Raman fiber amplifiers gain spectra design. *Optics Communications, 283*(17), 3348–3352. doi:10.1016/j.optcom.2010.04.024

Kurkov, A. S., Paramonov, V. M., Egorova, O. N., Medvedkov, O. I., Dianov, E. M., Zalevskii, I. D., & Goncharov, S. E. (2002). A 1.65-μm fibre Raman amplifier. *Quantum Electronics, 32*(8), 747. doi:10.1070/QE2002v032n08ABEH002283

Mescia, L., Giaquinto, A., Fornarelli, G., Acciani, G., De Sario, M., & Prudenzano, F. (2011). Particle swarm optimization for the design and characterization of silica-based photonic crystal fiber amplifiers. *Journal of Non-Crystalline Solids, 357*(8), 1851–1855. doi:10.1016/j.jnoncrysol.2010.12.049

Poli, R., Kennedy, J., & Blackwell, T. (2007). Particle swarm optimization An overview. *Swarm Intelligence, 1*(1), 33–57. doi:10.1007/s11721-007-0002-0

Robinson, J., & Rahmat-Samii, Y. (2004). Particle swarm optimization in electromagnetics. *IEEE Transactions on Antennas and Propagation, 52*(2), 397–407. doi:10.1109/TAP.2004.823969

Shi, Y., & Eberhart, R. C. (1998). Parameter selection in particle swarm optimization. *Lecture Notes in Computer Science, Evolutionary Programming VII, 1447*, (pp. 591-600).

Shi, Y., & Eberhart, R. C. (1999). Empirical study of particle swarm optimization. *IEEE Proceedings of the 1999 Congress on Evolutionary Computation,* Vol. 3, Washington, DC, USA.

Shi, Y., & Eberhart, R. C. (2001). Fuzzy adaptive particle swarm optimization. *IEEE Proceedings of the 2001 Congress on Evolutionary Computation,* Vol. 1, (pp. 101-106). Seoul, Korea.

Tong, Z., Wei, H., & Jian, S. (2004). Investigation and optimization of bidirectionally dual-order pumped distributed Raman amplifiers. *Optics Express, 12*(9), 1794–1802. doi:10.1364/OPEX.12.001794

Trelea, I. C. (2003). The particle swarm optimization algorithm: Convergence analysis and parameter selection. *Elsevier Information Processing Letters, 85*(6), 317–325. doi:10.1016/S0020-0190(02)00447-7

Wei, H., Tong, Z., & Jian, S. (2004). Use of a genetic algorithm to optimize multistage erbium-doped fiber-amplifier systems with complex structures. *Optics Express, 12*(4), 531–544. doi:10.1364/OPEX.12.000531

Wen, S., & Chi, S. (1992). Distributed erbium-doped fiber amplifiers with stimulated Raman scattering. *IEEE Photonics Technology Letters, 4*(2), 189–192. doi:10.1109/68.122357

Zhang, W., Wang, C., Shu, J., Jiang, C., & Hu, W. (2004). Design of fiber-optical parametric amplifiers by genetic algorithm. *IEEE Photonics Technology Letters, 16*(7), 1652–1654. doi:10.1109/LPT.2004.828541

KEY TERMS AND DEFINITIONS

Erbium-Doped Fiber Amplifier: An optical-fiber amplifier whose fiber core is doped with trivalent erbium ions which absorbs light at pump wavelength and emit at the signal wavelength.

Fiber Raman Amplifier: It is a device that boosts the signal in an optical fiber by transferring energy from a powerful pump beam to a weaker signal beam.

Hybrid Erbium-Doped Fiber Amplifiers/Fiber Raman Amplifiers: Consists of an erbium-doped fiber amplifier and a fiber Raman amplifier connected in cascade to increase the amplifier gain bandwidth.

Swarm Intelligence: Is a kind of artificial intelligence that aims to simulate the behavior of swarms or social insects.

Section 3
Control Optimization

PSO algorithms showed their full potentialities in terms of flexibility, robustness and reliability in modern control theory. In particular, they have been proposed to employ new optimal control techniques in complex systems consisting of a large number of autonomous agents. In this section, the first chapter illustrates the application of swarm intelligence in the emerging field of the swarm robotics. In particular, the distributed bees algorithm is proposed as a solution to distributed multi-robot task allocation. In the second chapter, the use of PSO is proposed to perform the optimal design of fractional order controllers. In the third chapter, a multi-objective particle swarm optimization algorithm is employed to find the optimal structure of a parallel kinematic manipulator-based machining robotic cell, minimizing the power consumed by the manipulator during the machining process in a robotic cell. Finally, the fourth chapter reports a generalized PSO, which is inspired by linear control theory, and its practical engineering applications in two fields, like the fault detection and classification of electrical machines, and the optimal control of water distribution systems.

Chapter 9
Distributed Task Allocation in Swarms of Robots

Aleksandar Jevtić
Robosoft, France

Diego Andina
E.T.S.I.T.-Universidad Politécnica de Madrid, Ciudad Universitaria, Spain

Mo Jamshidi
University of Texas, USA

ABSTRACT

This chapter introduces a swarm intelligence-inspired approach for target allocation in large teams of autonomous robots. For this purpose, the Distributed Bees Algorithm (DBA) was proposed and developed by the authors. The algorithm allows decentralized decision-making by the robots based on the locally available information, which is an inherent feature of animal swarms in nature. The algorithm's performance was validated on physical robots. Moreover, a swarm simulator was developed to test the scalability of larger swarms in terms of number of robots and number of targets in the robot arena. Finally, improved target allocation in terms of deployment cost efficiency, measured as the average distance traveled by the robots, was achieved through optimization of the DBA's control parameters by means of a genetic algorithm.

INTRODUCTION

The initial purpose of swarm intelligence algorithms was to solve optimization problems. However, in recent years, these algorithms have shown their full potential in terms of flexibility and autonomy when it comes to design and control of complex systems that consist of a large number of autonomous agents. In more general terms, these can be referred to as systems of autonomous systems (Jamshidi, 2009). What distinguishes swarm intelligence algorithms in the broad field of soft-computing is that they exploit the decentralizing property of natural swarms in order to create autonomous, scalable, and adaptive multi-agent systems.

DOI: 10.4018/978-1-4666-2666-9.ch009

Swarm robotics emerged as a straight-forward application domain for swarm intelligence due to resemblance of large robot teams to animal swarms (Dorigo & Sahin, 2004; Trianni, Nolfi & Dorigo, 2008; Correll, Cui, Gao & Gross, 2010). In nature, swarming behavior has been studied in ant and bee colonies, bird flocks and fish schools, among others (Engelbrecht, 2005). However, the biological plausibility of swarm intelligence algorithms, and swarm-based systems in general, is not a must; in computer science and engineering researchers are guided by efficiency, flexibility, robustness and cost as main criteria.

In applications that require area coverage, swarms of mobile robots can use their ability to quickly deploy within a large area. Some of the possible applications include planetary exploration, urban search and rescue, communication networks, monitoring, surveillance, cleaning, maintenance, and so forth. In order to efficiently perform their tasks, robots require high level of autonomy and cooperation. They use their sensing abilities to explore an unknown environment and deploy on the sites of interest, i.e. targets. However, the coordination of a robot swarm is not an easy problem, especially when the resources for the deployment task are limited. Such a large group of robots, if organized in a centralized manner, could experience information overflow that can lead to the overall system failure (Gazi, Jevtić, Andina & Jamshidi, 2010). For this reason, the communication between the robots can be realized through local interactions, either directly with one another or indirectly via environment (Beni & Wang, 1989).

As a result of the growing interest in the coordination of swarms of robots, multi-robot task allocation (MRTA) has become an important research topic (Dudek, Jenkin & Milios, 2002; Gerkey & Matarić, 2004). The goal is to assign tasks to robots in a way that, through cooperation, the global objective is achieved more efficiently. In the scenario proposed in this work, tasks are represented by targets defined with their qualities

and their location in the robot arena. For distributed task allocation the Distributed Bees Algorithm (DBA) was proposed (Jevtić, Gazi, Andina & Jamshidi, 2010), which was inspired by the foraging behavior of colonies of bees in nature. In the context of mobile multi-robot systems, scalability refers to the overall system's performance if the number of robots increases in relation to the number of tasks at hand (Rana & Stout, 2000). The resulting effect on the system's performance can be determined in terms of metrics associated with a particular platform or an operating environment, which in this work refers to dispatching a robot to a remote site marked as a target.

The objectives of this chapter are manifold. The following section describes the problem of MRTA and provides a summary of the related work. The Distributed Bees Algorithm (DBA) is then proposed as a solution to distributed MRTA. The DBA's performance is subsequently validated through experiments with physical robots. Moreover, simulator was developed to test the DBA's scalability in terms of number of robots and number of targets. The last experiments present analysis of DBA's performance through optimization of its control parameters. Finally, this chapter provides perspectives on future research directions and gives concluding remarks.

BACKGROUND

Multirobot systems offer the possibility of enhanced task performance, increased task reliability and decreased cost over more traditional single-robot systems. Various architectures for multirobot systems that differ in size and complexity have been proposed. Dudek et al. (2002) provided a taxonomy that categorizes the existing multirobot systems along various axes, including size (number of robots), team organization (e.g., centralized vs. distributed), communication topology (e.g., broadcast vs. unicast), and team composition (e.g., homogeneous vs. heterogeneous).

Box 1. Utility

$$U_{RT} = \begin{cases} Q_{RT} - C_{RT}, \text{if } R \text{ is capable of executing } T, \text{and } Q_{RT} > C_{RT} \\ 0, \text{otherwise} \end{cases} \quad (1)$$

Rather than characterizing architectures, Gerkey and Matarić (2004) categorized instead the underlying coordination problems with a focus on MRTA. The authors distinguish: single-task (ST) and multi-task (MT) robots, single-robot (SR) and multirobot (MR) tasks, and instantaneous (IA) and time-extended (TA) assignment. The authors showed that many MRTA problems can be viewed as instances of well-studied optimization problems in order to analyze the existing approaches, but also to use the same theory in the synthesis of new approaches. In order to estimate a robot's performance, they defined utility that depends on two factors, namely expected quality of task execution and expected resource cost. Given a robot R and a task T one can define Q_{RT} and C_{RT} as the quality and cost, respectively, expected to result from the execution of T by R. The resulting nonnegative utility measure is shown in Equation 1 (see Box 1).

This however is not a strict definition of utility which is a flexible measure of performance and can entail arbitrary computation. The only constraint on utility estimators is that they must each produce a single scalar value that can be compared for the purpose of assigning robots for tasks. The problem addressed in this chapter is categorized as a "single-task robots, multirobot tasks, instantaneous assignment (ST-MR-IA)", which Gerkey and Matarić proposed to be solved as a set partitioning problem. However, this requires the combined utilities of all the robots to be known in advance, which is not the case.

What follows is a survey of various multirobot system architectures that have been proposed for solving different problems. We tend to use the above mentioned taxonomies to categorize them.

One of the common approaches for solving the ST-SR and ST-MR problems is a market-based approach which uses auctioning mechanism for task allocation. Gerkey and Matarić proposed four different strategies for dynamical task allocation in two different emergency-handling scenarios. The robots bid for tasks and decisions are made by auctioning. Authors concluded that there is no overall best strategy and that the success of a strategy is task-related. Michael et al. (2008) proposed a market-based approach for robots formation control. They associate multiple tasks with predefined spatial locations that define a formation.

A thorough overview of market-based approaches for MRTA is given by Dias et al. (2006). A common drawback of these approaches is the underlying auctioning mechanism which requires all the bids from the robots to be gathered at one auctioning point. Sometimes, when resources permit, markets can even behave in a centralized fashion over larger portions of the robot team to improve solution quality. The authors gave a summary of communication costs for various market-based approaches. The main advantage of the method we propose is that, although it imposes certain communication cost for sending the information of the found targets, the robots make decisions autonomously and in a distributed manner. This is not the case with market-based approaches that feature a partial distribution, where robots are divided into sub-teams that take decisions in a centralized manner. For this reason, scalability in market-based approaches is often limited by the computation and communication needs that arise from increasing auction frequency, bid complexity, and planning demands.

Environment exploration and mapping are common applications for multirobot systems. Franchi et al. (2009) proposed a Sensor-based Random Graph (SRG) method for cooperative robot exploration. They addressed the issue of system's performance with respect to exploration time and traveled distance. The authors showed that by adding more robots the system could scale-up, but its performance was highly dependent on the initial team deployment, giving better results when the robots started grouped in a cluster than if scattered in the environment. The robots used broadcast communication with a limited range but the concept of decentralized cooperation was in question since the robots were programmed to gather in sub-teams that had to synchronize for local path-planning and collision-avoidance. This required intensive interchange of information including robot's ID and displacement plan.

Another approach proposed by Burgard et al. (2005) treats the unknown environment exploration as a ST-SR problem, where individual robots select a new target location based on its distance and utility. The map was divided in cells whose size was determined by the robot's visual range. The utility of each target location, i.e. cell, would decrease if more of its neighboring cells were assigned to other robots. To determine appropriate target locations for all the robots, the authors proposed an iterative algorithm. The drawback of this algorithm is its complexity and high computational cost. Although the experimental results show the advantages of collaboration, the proposed centralized approach cannot be applied if not all robots can communicate with each other.

Decentralized coordination of robots has various advantages over more traditional centralized approaches. It can be applied to reduce the communication burden on multirobot system (Ray, Benavidez, Behera & Jamshidi, 2009), especially for large teams of robots. In some applications communication can be difficult to implement or no communication exists at all. Joordens and Jamshidi (2010) proposed a decentralized coordination for a swarm of underwater robots which is based on consensus control. Another decentralized strategy for dynamical allocation of tasks that requires no communication among robots was proposed by Berman et al. (2009). But often, as in case of multirobot area coverage (Schwager, McLurkin, Slotine & Rus, 2009), the decentralized coordination and distributed decision-making is applied having one goal in mind, that the global objective is achieved more efficiently.

Bio-Inspired Coordination of Multirobot Systems

Robot swarms are multirobot systems that typically consist of a large population of simple robots interacting locally with one another and with their environment (Bonabeau, Dorigo & Theraulaz, 1999). These systems draw inspiration from animal swarms in nature but their design is not constrained by biological plausibility. Their main feature is decentralized coordination which results in a desired behavior that emerges from the rules of local interactions.

The self-organizing properties of animal swarms such as insects have been studied for better understanding of the underlying concept of decentralized decision-making in nature (Camazine et al, 2001), but it also gave new approach in applications to multi-agent system engineering and robotics. Bio-inspired approaches have been proposed for multirobot division of labor in applications such as exploration and path formation (Groß et al., 2008), multi-site deployment (Berman et al., 2007), or cooperative transport and prey retrieval (Labella, Dorigo & Deneubourg, 2006; Campo & Dorigo, 2007).

The bottom-up design topology inherent to bio-inspired multirobot systems provides them with one or more of the following features, such as being autonomous, scalable, robust and adaptive to changes in their environment. On the other hand, the collective behavior has emergent properties that give them the ability to produce unpredictable patterns. One way of dealing with the unpredictability issue is statistical analysis through experiments, as proposed in this chapter.

Scalability

Task allocation scenarios include a set of tasks that may have different priorities and require one or more robots to be assigned to their execution. A very important property of multirobot systems is the ability to scale-up with respect to the number of robots or the number of tasks at hand. However, scalability of multirobot systems and multi-agent systems in general has been analyzed from various perspectives including the total number of agents involved, the size of the communication data, the number of rules the agents operate with, or the agents' diversity (Rana & Stout, 2000).

In order to evaluate the scalability of a given multirobot system one needs to identify a performance metrics. Various MRTA methods exist but, to the best of our knowledge, a comprehensive analysis tool for the scalability of such methods has not been given. Some mathematical models that have been proposed could serve as guidelines in multirobot system design, but different scenarios to which these systems are applied usually do not allow staying within the proposed framework.

Lerman et al. (2006) proposed a mathematical model for MRTA in dynamical environments. The authors assumed that robots were able to observe tasks in order to discriminate their types, but also to discriminate the tasks that other robots were assigned to. Robots had limited sensing capabilities and could not directly communicate. The lack of communication made the system more robust to failures, but also more susceptible to noise from the sensors, and requires more time for exploration of available tasks.

Top-down design methodologies apply the classical control theory for performance estimation of distributed agent-based systems. While establishing bounds on the system behavior and provide performance guarantees, they heav-

ily rely on the available bandwidth for robot communication and they are more sensitive to noise. The need for resources becomes even a bigger issue as the number of robots increases. There is therefore a natural tendency to apply bottom-up methodologies that result in autonomous, scalable and adaptable systems requiring minimal communication (Crespi, Galstyan & Lerman, 2008).

Broadcast communication provides quick propagation of tasks' information within the multirobot system but extensive use of communication channel can affect the system's scalability. Previously described market-based approaches suffer from a large requirement in terms of communication bandwidth as they use broadcast messages to auction for the tasks. Farinelli et al. (2006) proposed a mechanism based on token passing for cooperative object retrieval, which scales up for reliable sending of broadcast messages. The authors made a comparison of their method with market-based approaches and the ones based on iterative broadcast communication. Their results show that the ability of the system to adjust to the available communication bandwidth provides guarantees for better performance.

DISTRIBUTED TASK ALLOCATION

Problem Definition

Based on Dudek's taxonomy, the proposed multirobot system can be categorized as homogeneous and distributed, using broadcast communication. The problem addressed in this paper is for singletask robots, multi-robot tasks and instantaneous assignment (ST-MR-IA). The task (i.e., target) allocation scenario is placed in a 2-dimensional robot arena with a preset number of targets that could be of same or different importance. A finite number of robots are allowed to be allocated to any target; still, each robot can only be allocated to one target at any given time. Targets have associated quality values and have their own location coordinates. The quality of a target is an application-specific scalar value that may represent target's priority or complexity, where a higher value requires a higher number of allocated robots. (E.g.: In the cleaning of public spaces, this value may represent the amount of detected garbage on site.) The medium by which these values are obtained is not considered in this paper.

The proposed scenario is presented under the following assumptions:

- All the targets are made available to all the robots. This is done by setting a broadcast communication range of the robots to cover the entire arena.
- Robots take decision once a predefined number of targets in the arena are found. The robots that found a target are automatically allocated to that target.
- Reallocation to another target is not allowed.

These assumptions are taken for simplicity; otherwise, it would be difficult to analyze the performance of the system due to the unpredictability of the robots' distribution prior to target allocation. It is important to mention that the entire swarm is involved in the search for targets. The search phase was used in experiments with physical robots in order to estimate the odometry error rate with respect to the experiment execution time. It was later used in simulation in order to have a realistic scenario. The experimental setup has a limitation that the robots wait for a preset number of targets to be found in order to allocate. This value can be altered or set as a variable, but that is not considered in this study and remains to be a part of future work. Even though the broadcast communication represents a centralized solution, the decision making is executed by the robots in a distributed manner, which is an inherent characteristic of swarms in nature.

The MRTA problem can be described as follows. Consider a population of N robots to be allocated among M targets ($N \geq M$). Let $Q \in \{q_1, ..., q_M\}$ denote the set of normalized qualities of all available targets. We denote the number of robots on the target $i \in \{1, ..., M\}$ by n_i, a nonnegative integer. The population fraction allocated to target i is $f_i = n_i/N$, which represents the target's relative frequency, and the vector of population fraction is $\mathbf{f} = (f_1, ..., f_M)^T$. The expected distribution is the set of desired population fractions for each task, $\mathbf{f^d} = (f_1^d, ..., f_M^d)^T$, where $f_i^d = q_i$. The usage of fractions rather than integers is practical for scaling, but it also introduces a distribution error as the fractions can take only certain values that are defined by the swarm size.

A relevant concept from set theory could be used to observe this as a set partitioning problem. A family X is a partition of a set E if and only if the elements of X are mutually disjoint and their union is E:

Partition set

$$\bigcap_{x \in X} = \varnothing$$

$$\bigcup_{x \in X} = E \tag{2}$$

However, for the proposed scenario the system optimization based on the maximum utility cannot be applied because the combined utilities of the robots are unknown as robots have no knowledge of the decisions taken by other robots. Therefore, the DBA is proposed.

Distributed Bees Algorithm

The DBA (Jevtić et al., 2011) was applied to multi-robot target allocation in the proposed scenario. The robots start a search for the targets from their randomly chosen initial positions in the arena. When a robot finds a target, it broadcasts the message containing the target's quality. When another robot receives information on the predefined number of targets, it calculates the utilities with respect to those targets. The utility depends on the target's quality value and the related deployment cost measured as the robot's distance from the target. The distance to the target is obtained thanks to a local, distributed and situated communication system (Gutiérrez et al., 2008; Gutiérrez et al., 2009). When a robot broadcasts information about the target, a receiver robot obtains the information transmitted together with the range (distance) and bearing (orientation) to the emitter robot. Therefore, the robot is able to calculate the distance and orientate to the emitting robot. The main concepts behind the implementation of the DBA are presented hereafter.

The cost of a target i for robot k is calculated as the Euclidean distance, d_i^k, between the robot and the target in a two-dimensional arena. However, the target's visibility is defined as the reciprocal value of the distance:

Visibility

$$\eta_i^k = \frac{1}{d_i^k} \tag{3}$$

The target's quality is a scalar value that represents its priority or complexity. Normalized qualities are calculated as fractions of the sum of qualities:

Normalized target qualities

$$q_i = \frac{Q_i}{\sum_{j=1}^{M} Q_j} \tag{4}$$

where Q_i is a quality of the target i. In real-world scenarios, the quality of a region of interest is an estimated value that is as a result of sensor-readings or a previously acquired knowledge.

The utility of a robot depends on both visibility and quality of the chosen target. The utility is de-

fined as a probability that the robot k is allocated to the target i, and it is calculated as follows:

Allocation probabilities

$$p_i^k = \frac{q_i^\alpha \eta_i^\beta}{\sum_{j=1}^{M} q_j^\alpha \eta_j^\beta} \tag{5}$$

where α and β are control parameters that allow biasing of the decision-making mechanism towards the quality of the solution or its cost, respectively. ($\alpha, \beta > 0$; $\alpha, \beta \in \mathbf{R}$.) From Equation 5 it is easy to show that:

Sum of the probabilities

$$\sum_{j=1}^{M} p_i^k = 1 \tag{6}$$

The underlying decision-making mechanism of the DBA algorithm adopts the roulette rule, also known as the wheel-selection rule. That is, every target has an associated probability with which it is chosen from a set of available targets. Once all the probabilities are calculated as in Equation 5, the robot will choose a target by "spinning the wheel".

It should be noticed that the resulting robots' distribution depends on their initial distribution in the arena, i.e. their distances from each target prior to target allocation. Therefore, robots' utilities will differ with respect to the same target if their distances from that target are not equal. Since a combined robots utility cannot be computed due to a distributed nature of the proposed algorithm, the quality of the targets is used as the only measure for the expected robots' distribution. Although the overall cost efficiency of the swarm is not analyzed here, target's visibility as used in Equation 5 makes closer targets more attractive to robots.

EXPERIMENTAL EVALUATION

In this section, the results from the experiments with physical robots are presented to validate the performance of the proposed DBA. The overview of robots' hardware and the communication protocol used in experiments is given, followed by the description of the experimental setup. The experiments results are presented and discussed.

Robot Hardware

The robots were assembled with the same selection of hardware components (See Figure 1). The Lynxmotion Terminator Sumo Robot Kit with four Spur Gear Head Motors was used as a base to build the robots, although any platform that could support the listed hardware components would be suitable. The DC motors were powered with 12 VDC, with 200 RPM, torque of 63.89 oz.in (4.6 Kg-cm), 30:1 reduction and 6 mm shaft diameter, and they were paired as left-hand and right-hand, in order to be able to perform rotation on-the-spot. Devantech MD22 Motor Driver was used to control the motors' rotation speed and direction. It averaged the PWM signal received from the Arduino microcontroller board to provide a proportional value of the 12 VDC from the battery. The four switches on the motor

driver's board were used to define the working mode. In the experiments, the analog mode was used which provided satisfactory speed control.

Arduino Duemilanove microcontroller board with ATMEGA328 microcontroller was powered with the 6 VDC battery. ZigBee module used for robot communication was connected using the Arduino Xbee shield. Sensors were programmed for I²C communication protocol with the microcontroller. One ultrasonic sensor was mounted on each robot for obstacle detection. Thermal sensor was used to detect the targets. The odometry error inherent to all mobile robots affected their precise localization. It cannot be eliminated but various methods for its reduction have been proposed, such as the averaging method proposed by Gutierrez et al. (2010).

Coordinator and Communication

A computer with a Pentium IV processor at 3 GHz with 2 GB of RAM was used to program the robots and to connect a ZigBee communication module that created a mesh communication network (the coordinator). The ZigBee modules mounted on robots were able to detect a reserved communication channel and connect with the coordinator. This allowed the communication between the robots and the robots with the computer. The broadcast mode that allowed each module to communicate with any other module in the network was used.

Experimental Results

The main objective of the experimental setup was to test the performance of the proposed algorithm and not the sensing and pattern recognition capabilities of the robots. The robots estimated their position based on the speed, time and direction of displacement. Although the robots performed well in detecting the heat source and sending the estimated location and measured temperature, which was tested in initial experiments, in order

Figure 1. Sumo robot used in the experiments

Figure 2. Results of the first experimental setup: Average odometry error vs. initial random search time

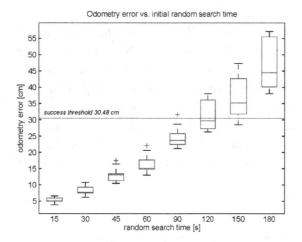

to test the performance of the DBA algorithm a simplified scenario was arranged. A small swarm of three real and two simulated robots was used in search of two targets.

The experiments were performed in 12x12 sq ft (3.65x3.65 m^2) arena, with randomly distributed obstacles. Robots were placed at the preprogrammed initial locations. When the command was sent from the coordinator the robots started the random search. After a certain period of time t_0, the information of two targets was sequentially sent from the coordinator. The information included the targets' estimated locations and their temperature (quality) values. The robots calculated the probabilities to move to each target as in Equation 5. The physical robots are unaware of the message sender identity, so the coordinator was able to simulate two robots that found two different targets.

Two types of experiments were performed. In the first one, the random search time t_0 was changed to test the increase rate of the odometry error over time. Single robot was used in in each run in order to avoid collision with other robots, and in t_0 the information about the found target was sent from the coordinator. Obstacles

were introduced the arena to force the robot to change its motion. Simple obstacle detection algorithm was implemented where the robot randomly rotates left or right by 90 degrees and then continues moving forward. With each of three robots 30 experiments were conducted in order to obtain the average odometry error value. The results of the first experimental setup are shown in Figure 2. The search was considered successful if the robot was able to get as close as 30.48 cm (1 ft) from the target. It can be noted that while increasing the initial random search time of the robot the odometry error increased as well. This happened due to the imperfect calibration of the DC motors, non-constant battery voltage, friction of the ground, etc. It can also be noticed that for the $t_0 < 90$s the experiment success rate was 100%.

The results from the first experimental setup were used to set the parameters for the second experimental setup. The random search initial time value was set to $t_0 = 30$s, which guaranteed that the odometry error would be below the success threshold. The scenario involved all three robots in search for two targets whose information was sent form the coordinator. The choice of having two targets in the scenario was based on the number of real robots that were at our disposal. By having three robots and two targets we could test how the robots distributed in the arena based on the targets' quality values. Four possible events could occur: 1) all robots go for the first target (T1); 2) all robots go for the second target (T2); 3) two robots go for the target T1 and one for the target T2; and 4) one robot goes for the target T1 and two robots go for the target T2.

Because of the relatively small size of the robot arena, the visibility of the targets was set to be constant, $\eta_1 = \eta_2 = 1$, and the target allocation was performed based only on the targets quality values. The values of the control parameters were also set to $\alpha = \beta = 1$. In order to test the self-organized behavior of the swarm of robots, the quality values

Table 1. Robots distribution vs. targets quality values

Quality values	T1:T2	Occurence	Occurrence [%]
$Q_1=Q_2=50$	2:1	11	36.67
$Q_1=Q_2=50$	3:0	4	13.33
$Q_1=Q_2=50$	1:2	13	43.33
$Q_1=Q_2=50$	0:3	2	6.67
$Q_1=70;Q_2=30$	2:1	18	60.00
$Q_1=70;Q_2=30$	3:0	9	30.00
$Q_1=70;Q_2=30$	1:2	2	6.67
$Q_1=70;Q_2=30$	0:3	1	3.33
$Q_1=90;Q_2=10$	2:1	8	26.67
$Q_1=90;Q_2=10$	3:0	21	70.00
$Q_1=90;Q_2=10$	1:2	1	3.33
$Q_1=90;Q_2=10$	0:3	0	0.00

of the two targets sent from the coordinator to the robots were changed. The experimental results are shown in Table 1. It can be noticed that when the quality values were equal, $Q_1 = Q_2 = 50$, in most cases two robots would go for one target and one would go for the other. This was expected since the probabilities of choosing any of the targets were equal. By increasing the difference between the quality values, the distribution would change in favor of the target with the higher quality value because the probability that a robot chooses that target also increased.

The experimental results show that the task allocation was performed according to the targets' quality values in an autonomous and decentralized manner. The targets with higher quality values attracted more robots, which was the objective for the multi-foraging scenario. The odometry error inherent to mobile robots was used as an advantage in order to gather the robots in the vicinity of the found targets and not at their exact locations. Still, there is a necessity to maintain the odometry error within the acceptable limits, and this is planned as a part of the future work.

Figure 3. Simulator screenshot

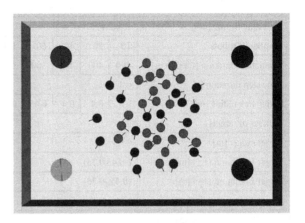

Evaluation Through Simulations

The experiments with real robots could not provide the insight on the multirobot system's scalability because of the small number of available robots. Therefore, the experiments were performed in a simulated environment which provided the results for a thorough analysis of the algorithm's performance. In this section, the simulator and the simulation setup are described, and the simulation results are presented in order to analyze the scalability of the DBA.

Simulator

Simulation platform is a fast, specialized multi-robot simulator for the e-puck robots described in (Gutiérrez et al., 2010). It is a simple and effective simulator implementing 2D kinematics. A screenshot of the simulator is shown in Figure 3. Simulator screenshot shows 40 robots engaged in search for 4 targets of different qualities represented by different grey-level intensity. Robots are programmed for obstacle avoidance; when robot detects an obstacle its color changes from black to blue to mark his new state. Once the robot has taken a new direction, its color goes back to black. In simulations, the e-puck is modeled as a cylindrical body of 3.5 cm in radius that holds 8

Table 2. Parameters describing three arenas used in simulations

	Arena 1					Arena 2					Arena 3				
Area dimensions [m²]	1.5x2.125					1.5x2.125					1.5x2.125				
Number of robots	10	20	40	60	100	10	20	40	60	100	10	20	40	60	100
Simulation duration [time steps]	400	400	400	300	200	400	400	400	300	200	400	400	400	300	200
Time step duration [s]	0.1					0.1					0.1				
Initial area radius [m]	0.4	0.4	0.4	0.4	0.5	0.4	0.4	0.4	0.4	0.5	0.4	0.4	0.4	0.4	0.5
Number of targets	2					4					4				
Target radius [m]	0.09					0.09					0.09				
Target 1 location (x,y) [m]	(-0.45,0.75)					(-0.45,0.75)					(-0.45,0.75)				
Target 2 location (x,y) [m]	(0.45,-0.75)					(0.45,-0.75)					(0.45,-0.75)				
Target 3 location (x,y) [m]	N/A					(-0.45,-0.75)					(-0.45,-0.75)				
Target 4 location (x,y) [m]	N/A					(0.45,0.75)					(0.45,0.75)				
Target 1 quality (q_1)	0.5					0.25					0.1				
Target 2 quality (q_2)	0.5					0.25					0.2				
Target 3 quality (q_3)	N/A					0.25					0.3				
Target 4 quality (q_4)	N/A					0.25					0.4				

infrared (IR) proximity sensors distributed around the body, 3 ground sensors on the lower-front part of the body and a range and bearing communication sensor. IR proximity sensors have a range of 5 cm, while the communication range of the E-puck Range&Bearing module was set to cover the whole arena. For the three types of sensors, real robot measurements were sampled and the data was mapped into the simulator. Furthermore, uniformly distributed noise was added to the samples in order to effectively simulate different sensors; +/- 20% noise is added to the IR sensors and +/- 30% to the ground sensors. In the range and bearing sensor, noise is added to the range (+/- 2.5 cm) and bearing (+/- 20°) values. A differential drive system made up of two wheels is fixed to the body of the simulated robot. At each time step of 100 ms, the robot senses the environment and actuates. The robot's speed was limited to 6 cm/s when moving straight and 3 cm/s when turning.

Simulation Setup

Three different simulation setups have been chosen to compare and study performance and scalability of the proposed DBA algorithm. The setups were carried out in the same arena where the number of robots, number of targets and targets' quality values were changed as shown in Table 2. Additional simulation setup was created in order to analyze the effect of the control parameters α and β on the resulting distribution. Each simulation was repeated 50 times for different initial robot distribution in order to perform an analysis of the results.

Simulation Results and Discussion

In order to test the scalability of the proposed DBA with respect to the size of the swarm, the simulations were performed with 10, 20, 40, 60 and 100 robots for the simulation setup 1, and 20, 40, 60 and 100 robots for the simulation setup 2 and the simulation setup 3. The number of targets was also changed, from two in the simulation setup 1 to four in the simulation setup 2, in order to test the performance of the algorithm with respect to the number of targets. In the simulation setup 3, four targets with different quality values were used in order to show the adaptability of the swarm to a non-uniform distribution of the "food" in the environment. This is also the most realistic scenario. Finally, the simulation setup 4 was created to test how the change in the ratio of the control parameters α and β can affect the resulting robots' distribution.

As the algorithm performance metrics the mean absolute error (*MAE*) of the robots' distribution was used, given by:

MAE of the robots' distribution

$$MAE = \frac{1}{M} \sum_{i=1}^{M} \left| f_i - f_i^d \right| \qquad (7)$$

where $f_i^d = q_i$.

As the name suggests, the *MAE* is the average value of the absolute distribution error (per target) that is the result of discrepancy between the expected and the resulting robots' distribution. For each simulation setup and each swarm size described in Table 2. fifty simulations were performed. The values of *MAE* obtained from the simulations are graphically shown in Figure 4. It can be noticed that the average *MAE* and maximum *MAE* values decrease as the size of the robot swarm increases regardless of the number of targets or their quality values. This was expected because of the probabilistic target allocation mechanism applied in Equation 5. The results from the simulation setup 1, 2, and 3, are shown in Figure 5, Figure 6, and Figure 7, respectively.

The effectiveness of the algorithm in terms of increased number of targets can be seen from the results shown in Figure 5 and Figure 6. The results show that the average and the maximum *MAE* values decreased for larger swarms in case of 4 targets of the same quality. It should be noticed that the allocation of 10 robots to 4 targets pro-

Figure 4. Distribution MAE with respect to the swarm size: a) simulation setup 1; b) simulation setup 2; and c) simulation setup 3. Each box-plot comprises observations ranging from the first to the third quartile. The median is indicated by a horizontal bar, dividing the box into the upper and lower part. The whiskers extend to the farthest data points that are within 1.5 times the interquartile range. Outliers are shown with a plus symbol. The values were obtained from 50 simulations performed for each swarm size within each simulation setup.

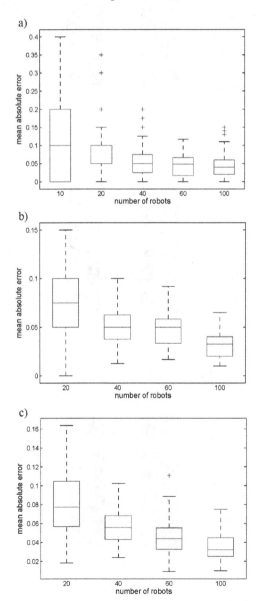

Figure 5. Expected vs. obtained robots distribution for two equal targets. Targets quality values are $q_1 = q_2 = 0.5$. Fifty simulations were performed for each of the following swarm sizes: a) 10 robots; b) 40 robots; and c) 100 robots.

Figure 6. Expected vs. obtained robots distribution for four equal targets. Target quality values are $q_1 = q_2 = q_3 = q_4 = 0.25$. Fifty simulations were performed for each of the following swarm sizes: a) 20 robots; b) 60 robots; and c) 100 robots.

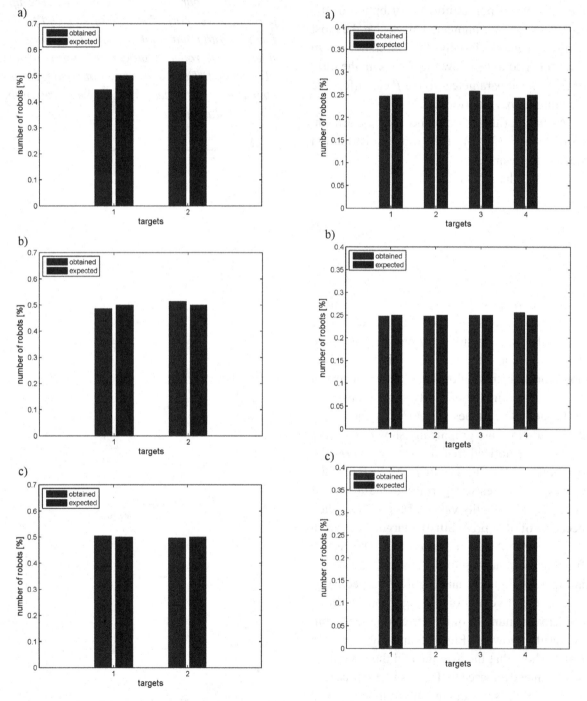

Figure 7. Expected vs. obtained robots distribution for four different targets. Target quality values are $q_1 = 0.1$, $q_2 = 0.2$, $q_3 = 0.3$, and $q_4 = 0.4$. Fifty simulations were performed for each of the following swarm sizes: a) 20 robots; b) 60 robots; and c) 100 robots.

Figure 8. Effects of the DBA's control parameters on the final robots' distribution. Target allocation was performed with 60 robots as described in the simulation setup 3 consisting of 4 targets with different quality values: $q_1 = 0.1$, $q_2 = 0.2$, $q_3 = 0.3$, and $q_4 = 0.4$. The results of the robots' distribution per target are shown for the following values of α/β ratio: a) $\alpha/\beta=1$; b) $\alpha/\beta=2$; c) $\alpha/\beta=5$. The values were obtained from 50 simulations for each scenario.

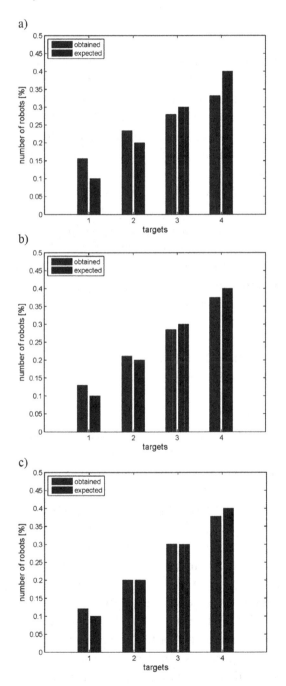

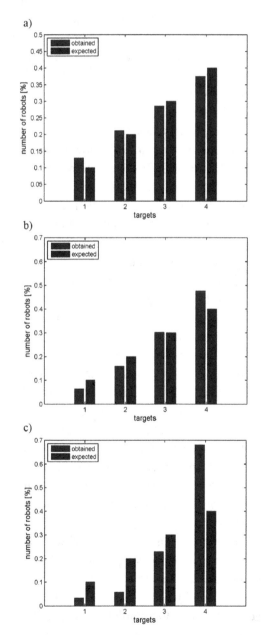

duces an error that is the result of the cardinality of the robot swarm. It is not physically possible to partition the swarm in order to obtain the expected target allocation (2.5 robots per target).

Another inherent source of error results from the assumption that the robots that had found a target are not allowed to reallocate to another target, therefore they are not involved in the decision-making process. Also, it is assumed that the robots wait for a predetermined number of targets to be found before they make a decision, which can result in the same target being found by more than one robot. This fraction of the robot swarm also produces an error in the final distribution because they cannot reallocate to another target. The algorithm's performance is analyzed having these issues mind.

In order to test the ability of the robot swarm to adapt to a non-uniform distribution of "food" in the environment, the simulations were performed for four different targets (simulation setup 3). The robots' distribution changed according to a new set of targets' quality values, as shown in Figure 7. It can be noticed that the resulting distribution is slightly in favor of the less valuable targets. This is another consequence of the robots that had found a target not being able to reallocate, and it is especially evident for smaller robot swarms. For example, let's consider a swarm of 10 robots in search of 4 different targets, as shown in Figure 7. If in the random target search process two robots find the target with the associated quality value of 0.1, then the final relative frequency for this target cannot be less than 0.2 (2 out of 10 robots) which is already above the expected value of 0.1. Although for the larger swarms the effect of the initial robot distribution becomes less relevant, it is always present.

The control parameters α and β were introduced in Equation 5 to compensate for the biased distribution, but also to give more relevance to either the quality of the targets or to the cost of reaching them. In the simulation setup 4, the α/β ratio was increased to give more relevance to the

Table 3. Effects of control parameters on robots' distribution

α/β ratio	Average *MAE*	Maximal *MAE*
1	0.0478	0.1083
2	0.0525	0.1000
5	0.1415	0.2083

quality value of the targets on the expense of their distances from the robots. The resulting robots' distributions per target for different values of the α/β ratio are shown in Figure 8. Results show that, by tuning the control parameters, the final robot distribution can change in favor of the more valuable targets but with an increase in the average *MAE* (see Table 3). It is reasonable to expect that by decreasing the α/β ratio the cost efficiency of the robot swarm would improve in terms of the distance traveled, however, the *MAE* is also expected to increase. More detailed analysis of the effect of the control parameters is given in the following section.

TUNING OF CONTROL PARAMETERS

The probability function in Equation 5 introduced a set of control parameters α and β that can be used to adapt swarm's behavior for different operational objectives. In this work, the parameters are optimized to improve target allocation in terms of deployment cost. Deployment cost is measured as the average distance traveled by the robots in the deployment stage. By changing these parameters' values, robots distribution patterns can be modified. The parameters were optimized by means of a genetic algorithm (GA) (Jevtić & Gutiérrez, 2011). GAs have proven to be powerful optimization tools. They are population-based algorithms, which means that they create a population of solutions (genes) in the data space in order to avoid getting stuck in a local optimum (Golberg, 1989).

Genetic Algorithm

In the DBA optimization both control parameters have been taken into account. The parameters α and β define how distances (i.e. visibilities) and quality values affect distribution of the robots in the arena. The effect of the parameters is exponential; hence, a small change in their values can result in very different robots' distribution patterns, and a larger distribution error. Moreover, since a large number of agents can be found in a swarm, increase in the deployment cost can be significant. Therefore, even though a simple sampling of the solution space would be less computationally demanding, in order to obtain a high accuracy and considering that the parameters optimization is performed offline, a GA was used.

In order to limit the complexity of the exploration process, the following range of possible values was defined for both parameters: $\alpha, \beta \in (0, 5)$. Initially, a population of 30 random genotypes was created, in which values are drawn from uniform distributions in the respective ranges of the parameters. The genetic algorithm was run for 1000 generations, during which new generations of genotypes were bred. The genetic algorithm loop consists of the evaluation, the selection and the reproduction of the genotypes. In order to evaluate the fitness of a given genotype, the controller of 40 simulated robots was parameterized with the values of α and β encoded in the genotype. The total number of 50 simulated experiments was run with different initial conditions. The experiments duration was set to 100 s.

The fitness function $F(g)$, of the evaluated genotype g, is computed as an indicator of the swarm's ability to allocate the robots according to the targets' qualities (q_i) and visibilities (η_i). The fitness F is defined as follows:

Fitness function

Table 4. Parameters describing simulation setups 1, 2, and 3

Parameter	Values range
α	1, 2.65
β	1, 2.55
Area dimension [m²]	2.25x3.1875, 3.0x4.25, 4.5x6.375, 6.0x8.5
Number of robots	40, 100
Simulation duration [time steps]	100, 200, 300
Time step duration [s]	0.1
Initial area radius [m]	0.4, 0.5
Number of targets	2, 4
Target radius [m]	0.09
Target location (x,y) [m]	fixed, random
Target qualities (q)	fixed, random

$$F = \frac{1}{MAE \cdot d_a} \qquad (8)$$

where MAE is the mean absolute distribution error and d_a is the average distance traversed by all the robots.

Generations following the first one are produced by a combination of selection with elitism, recombination and mutation. For each new generation, the two highest scoring individuals ("the elite") from the previous generation are retained unchanged. The remainder of the new population is generated by fitness-proportional selection (also known as roulette selection) from the individuals of the old population. Mutation entails that a random Gaussian offset is applied to each real-valued vector component encoded in the genotype (except the elite), with a probability of 0.5. The mean of the Gaussian is $\mu = 0$, and its standard deviation is $\sigma = 0.1$. During evolution, all vector component values are constrained to remain within the range (0, 1). Once the new population has been created, the genotype parameters are linearly mapped to produce network parameters within the aforementioned ranges.

Figure 9. Box-plot comparison of average distance (a) and MAE (b) for 40 robots. Non-optimal: α=β=1; optimal: α=2.65, β=2.55. Each box-plot comprises observations ranging from the first to the third quartile. The median is indicated by a horizontal bar, dividing the box into the upper and lower part. The whiskers extend to the farthest data points that are within 1.5 times the interquartile range. Outliers are shown with a plus symbol. The values were obtained from 50 experiments.

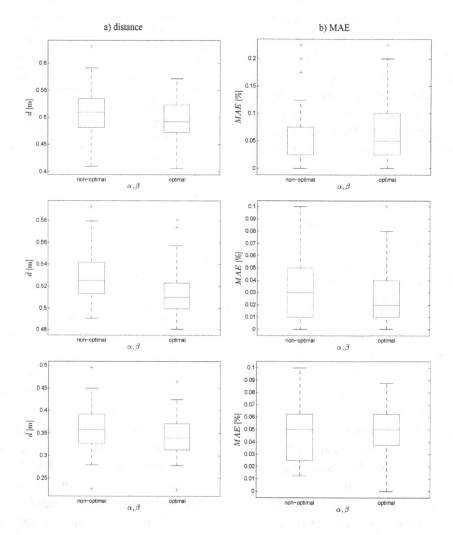

Simulation Setup

The simulation setup 1, 2, and 3 compare the swarm's performance for the new and the initially-used set of α and β values (30), $\alpha = \beta = 1$. These sets are referred to as optimal and non-optimal, in terms of deployment cost. The range of parameters' values is shown in the Table 4.

In the simulation setup 1, the system's robustness was tested with respect to the change of the swarm's size. The number of robots was varied, and the targets' position and quality values were preset. In the simulation setup 2, the size of the robot arena was varied to avoid specialization of the system for a specific environment. Finally, in the simulation setup 3, the performance of the system was tested with respect to different dis-

Figure 10. Box-plot comparison of average distance (a) and MAE (b) for 100 robots. Non-optimal: α=β=1; optimal: α=2.65, β=2.55. Each box-plot comprises observations ranging from the first to the third quartile. The median is indicated by a horizontal bar, dividing the box into the upper and lower part. The whiskers extend to the farthest data points that are within 1.5 times the interquartile range. Outliers are shown with a plus symbol. The values were obtained from 50 experiments.

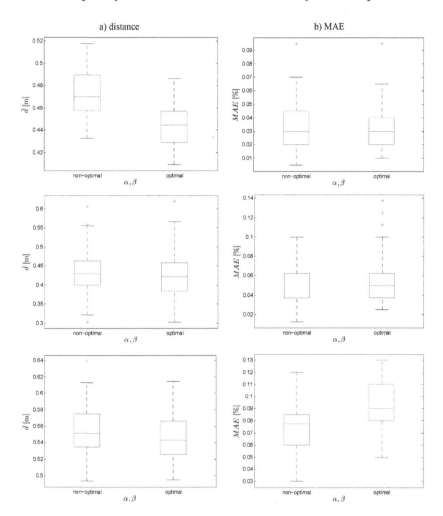

tribution of the targets of random quality values. These simulation setups were proposed in order to perform an indebt system's performance analysis.

Simulation Results and Discussion

In this subsection, the results from three proposed simulation setups are presented and discussed. Each simulation was repeated 50 times.

The simulation setup 1 was proposed to test the swarm's performance when the number of robots (40 and 100) and the number of targets (2 and 4) were changed. The targets' associated quality values were set to $q_1 = q_2 = 0.5$ and $q_1 = q_2 = q_3 = q_4 = 0.25$ for 2 and 4 targets, respectively. Additional experiment was performed with 4 targets that had different, but predefined, associated qualities $q_1 = 0.1$, $q_2 = 0.2$, $q_3 = 0.3$, $q_4 = 0.4$. In order to measure the swarm's performance, median

Figure 11. Box-plot comparison for 4 different robot arenas each involving 100 robots and 4 different randomly distributed targets: a) distance; b) MAE. Non-optimal: $\alpha=\beta=1$; optimal: $\alpha=2.65$, $\beta=2.55$. Each box-plot comprises observations ranging from the first to the third quartile. The median is indicated by a horizontal bar, dividing the box into the upper and lower part. The whiskers extend to the farthest data points that are within 1.5 times the interquartile range. Outliers are shown with a plus symbol. The values were obtained from 50 experiments.

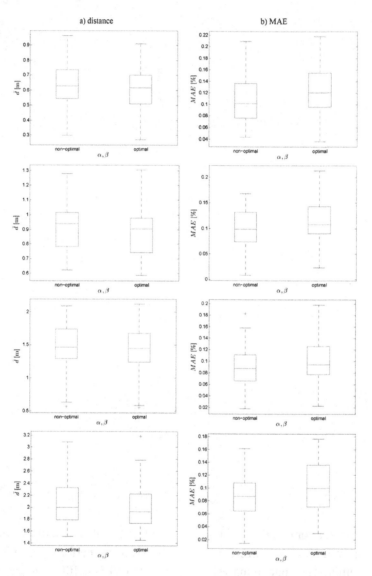

distance value and mean absolute robot distribution error were used. The experimental results for non-optimal ($\alpha = \beta = 1$) and optimal ($\alpha = 2.65$, $\beta = 2.55$) set of values are shown for 40-robot and 100-robot swarms in Figure 9 and Figure 10, respectively. It can be noticed that with the optimal set of control parameters swarm obtains more efficient distribution at a lower deployment cost. Only in case of 100 robots in a search of 4 targets, with equal or different qualities, the deployment cost was decreased at the expense of a higher distribution error.

Figure 12. Box-plot comparison for 100 robots and 4 random-valued, randomly distributed targets: a) distance; b) MAE. Non-optimal: α=β=1; optimal: α=2.65, β=2.55. Each box-plot comprises observations ranging from the first to the third quartile. The median is indicated by a horizontal bar, dividing the box into the upper and lower part. The whiskers extend to the farthest data points that are within 1.5 times the interquartile range. Outliers are shown with a plus symbol. The values were obtained from 50 experiments.

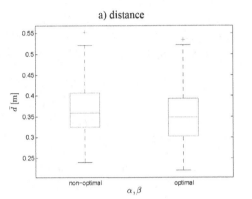
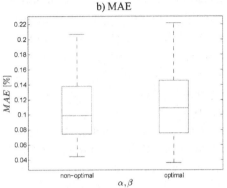

This simulation setup 2 tests swarm's performance in case of a random distribution of the targets in the arena. Four arenas that differ in size were used (see Table 4). The scenario involved 100 robots in the search for 4 different targets with predefined quality values $q_1 = q_2 = q_3 = q_4 = 0.4$. It can be noticed from the Figure 11 that in all the simulations the optimal control parameter values improved the performance of the swarm with respect to the deployment cost. This was achieved at the expense of a higher distribution error.

The simulation setup 3 tests the swarms adaptability when the targets' location and targets' associated qualities are randomly chosen. The scenario considers a case of 100-robot swarm in search for 4 targets. The robot arena used in the simulations is 2.25m x 3.1875m dimension. As it can be noticed from the Figure 12, for the optimal set of control parameters' values the performance of the swarm of robots improved in terms of both deployment cost and distribution error.

CONCLUSION AND FUTURE RESEARCH

Many applications of large multirobot systems require efficient task allocation in terms of individual and combined robots' utilities. The solution quality is analyzed using a defined performance metrics. In the presented study this was the mean absolute error of the robots' distribution in terms of the qualities of the available targets in the robot arena. In case of large, autonomous, multirobot systems, the scalability and the ability to adapt to different environments are the features of great importance. The presented experimental and simulation results showed that the proposed Distributed Bees Algorithm (DBA) provides the robot swarm with scalability in terms of the number of robots and number of targets, but also with the adaptability to a non-uniform distribution of the targets' qualities. The importance of control parameters is that they provide a mechanism to adjust the robot swarm behavior according to the task at hand and the available resources. In this chapter, the values of control parameters were tuned to bias the resulting robots' distribution towards the more favorable targets or to reduce the deployment costs.

Swarm Intelligence is a useful tool for solving a number of real-world problems. The decentralized approach in the design of multi-agent systems offers many advantages such as greater autonomy, scalability, robustness, and adaptability to a dynamically changing environment. This line of research was carried out in the domain of computer science and engineering, but it supports applications that go beyond that as the modeling of multi-agent systems can be carried over to the domains of biology, medicine, sociology, economy, business, etc. The processes in nature and many processes in human society show emergent properties that are the result of multiple interactions between large numbers of individuals. Various scientific disciplines use different approaches to describe this phenomenon. By defining the relation between the stochastic processes on a lower level and the organized complexity on the system's global level, the predictability of such systems could be improved. This would have a high impact in the above-mentioned application domains. Moreover, the research work on Swarm Intelligence models can provide important feedback for the study of the natural swarms from which they were inspired.

REFERENCES

Beni, G., & Wang, J. (1989). Swarm intelligence in cellular robotic systems. *Proceedings of the NATO Advanced Workshop on Robotics and Biological Systems*, Il Ciocco, Tuscany, Italy.

Berman, S., Halasz, A., Hsieh, M. A., & Kumar, V. (2009). Optimized stochastic policies for task allocation in swarms of robots. *IEEE Transactions on Robotics*, 25(4), 927–937. doi:10.1109/TRO.2009.2024997

Berman, S., Halasz, A., Kumar, V., & Pratt, S. (2007). Bio-inspired group behaviors for the deployment of a swarm of robots to multiple destinations. In *Proceedings of 2007 IEEE International Conference on Robotics and Automation*, (pp. 2318–2323).

Bonabeau, E., Dorigo, M., & Theraulaz, G. (1999). *Swarm intelligence: From natural to artificial systems*. New York, NY: Oxford University Press, Inc.

Burgard, W., Moors, M., Stachniss, C., & Schneider, F. E. (2005). Coordinated multi-robot exploration. *IEEE Transactions on Robotics*, 21(3), 376–386. doi:10.1109/TRO.2004.839232

Camazine, S., Franks, N. R., Sneyd, J., Bonabeau, E., Deneubourg, J.-L., & Theraulaz, G. (2001). *Self-organization in biological systems*. Princeton, NJ: Princeton University Press.

Campo, A., & Dorigo, M. (2007). Efficient multiforaging in swarm robotics. In *Proceedings of 9th European Conference on Advances in Artificial Life, Ser. ECAL'07*, (pp. 696–705). Berlin, Germany: Springer-Verlag.

Correll, N., Cui, Z., Gao, X.-Z., & Gross, R. (Eds.). (2010). Swarm intelligence and swarm robotics - SISR 2010: Special issue on swarm robotics. *Neural Computing & Applications, 19*(6).

Crespi, V., Galstyan, A., & Lerman, K. (2008). Top-down vs bottom-up methodologies in multi-agent system design. *Autonomous Robots, 24*(3), 303–313. doi:10.1007/s10514-007-9080-5

Dias, M. B., Zlot, R., Kalra, N., & Stentz, A. (2006). Market-based multirobot coordination: A survey and analysis. *Proceedings of the IEEE, 94*(7), 1257–1270. doi:10.1109/JPROC.2006.876939

Dorigo, M., & Sahin, E. (2004). Swarm robotics - Special Issue. *Autonomous Robots, 17*, 111–113. doi:10.1023/B:AURO.0000034008.48988.2b

Dudek, G., Jenkin, M., & Milios, E. (2002). A taxonomy of multirobot systems. In Balch, T., & Parker, L. (Eds.), *Robot teams: From diversity to polymorphism* (pp. 3–22). Natick, MA: A.K. Peters.

Engelbrecht, A. P. (2005). *Fundamentals of computational swarm intelligence*. New York, NY: John Wiley & Sons.

Farinelli, A., Iocchi, L., Nardi, D., & Ziparo, V. A. (2006). Assignment of dynamically perceived tasks by token passing in multirobot systems. *Proceedings of the IEEE, 94*(7), 1271–1288. doi:10.1109/JPROC.2006.876937

Franchi, A., Freda, L., Oriolo, G., & Vendittelli, M. (2009). The sensor-based random graph method for cooperative robot exploration. *IEEE/ASME Transactions on Mechatronics, 14*(2), 163–175. doi:10.1109/TMECH.2009.2013617

Gazi, P., Jevtić, A., Andina, D., & Jamshidi, M. (2010). A mechatronic system design case study: Control of a robotic swarm using networked control algorithms. In *Proceedings of 4th Annual IEEE Systems Conference (SysCon 2010)*, (pp. 169-173).

Gerkey, B. P., & Matarić, M. J. (2004). A formal analysis and taxonomy of task allocation in multi-robot systems. *The International Journal of Robotics Research, 23*(9), 939–954. doi:10.1177/0278364904045564

Golberg, D. E. (1989). *Genetic algorithms in search, optimization and machine learning*. IN, Lebanon: Addison Wesley USA.

Groß, R., Nouyan, S., Bonani, M., Mondada, F., & Dorigo, M. (2008). Division of labour in self-organised groups. In *Proceedings of 10th international conference on Simulation of Adaptive Behavior: From Animals to Animats, SAB '08*, (pp. 426–436). Berlin, Germany: Springer-Verlag.

Gutiérrez, A., Campo, A., Dorigo, M., Amor, D., Magdalena, L., & Monasterio-Huelin, F. (2008). An Open localization and local communication embodied sensor. *Sensors (Basel, Switzerland), 8*(8), 7545–7563. doi:10.3390/s8117545

Gutiérrez, A., Campo, A., Dorigo, M., Donate, J., Monasterio-Huelin, F., & Magdalena, L. (2009). Open e-puck range and bearing miniaturized board for local communication in swarm robotics. In *Proceedings of the IEEE International Conference on Robotics and Automation*, (pp. 3111–3116).

Gutiérrez, A., Campo, A., Monasterio-Huelin, F., Magdalena, L., & Dorigo, M. (2010). Collective decision-making based on social odometry. *Neural Computing & Applications, 19*(6), 807–823. doi:10.1007/s00521-010-0380-x

Jamshidi, M. (Ed.). (2009). *System of systems engineering - Innovations for the 21st century*. New York, NY: John Wiley & Sons.

Jevtić, A., Gazi, P., Andina, D., & Jamshidi, M. (2010). Building a swarm of robotic bees. *World Automation Congress (WAC 2010)*, (pp. 1–6).

Jevtić, A., & Gutiérrez, Á. (2011). Distributed bees algorithm parameters optimization for a cost efficient target allocation in swarms of robots. *Sensors (Basel, Switzerland), 11*(11), 10880–10893. doi:10.3390/s111110880

Jevtić, A., Gutiérrez, A., Andina, D., & Jamshidi, M. (2011). Distributed bees algorithm for task allocation in swarm of robots. *IEEE Systems Journal, 5*(3), 1–9.

Joordens, M. A., & Jamshidi, M. (2010). Consensus control for a system of underwater swarm robots. *IEEE Systems Journal, 4*(1), 65–73. doi:10.1109/JSYST.2010.2040225

Labella, T. H., Dorigo, M., & Deneubourg, J.-L. (2006). Division of labor in a group of robots inspired by ants' foraging behavior. *ACM Transactions on Autonomous and Adaptive Systems, 1*(1), 4–25. doi:10.1145/1152934.1152936

Lerman, K., Jones, C., Galstyan, A., & Matarić, M. J. (2006). Analysis of dynamic task allocation in multi-robot systems. *The International Journal of Robotics Research, 25*(3), 225–241. doi:10.1177/0278364906063426

Matarić, M. J., Sukhatme, G. S., & Østergaard, E. H. (2003). Multi-robot task allocation in uncertain environments. *Autonomous Robots, 14*(2-3), 255–263. doi:10.1023/A:1022291921717

Michael, N., Zavlanos, M. M., Kumar, V., & Pappas, G. J. (2008). Distributed multi-robot task assignment and formation control. In *Proceedings of IEEE International Conference on Robotics and Automation (ICRA 2008)*, (pp. 128–133).

Rana, O. F., & Stout, K. (2000). What is scalability in multi-agent systems? *Proceedings of the 4th International Conference on Autonomous Agents,* AGENTS '00, (pp. 56–63). New York, NY: ACM.

Ray, A. K., Benavidez, P., Behera, L., & Jamshidi, M. (2009). Decentralized motion coordination for a formation of rovers. *IEEE Systems Journal, 3*(3), 369–381. doi:10.1109/JSYST.2009.2031012

Schwager, M., McLurkin, J., Slotine, J.-J., & Rus, D. (2009). From theory to practice: Distributed coverage control experiments with groups of robots. In, O. Khatib, V. Kumar, & G. Pappas (Eds.), *Experimental robotics, Springer tracts in advanced robotics, 54,* 127–136. Berlin, Germany: Springer.

Trianni, V., Nolfi, S., & Dorigo, M. (2008). Swarm Robotics. *Design, 4433*(31), 163–191.

ADDITIONAL READING

Andina, D., & Pham, D. T. (2007). *Computational intelligence for engineering and manufacturing.* Dordrecht, The Netherlands: Springer. ISBN-10-0-387-37450-7

Bonabeau, E., Dorigo, M., & Theraulaz, G. (1999). *Swarm intelligence: From natural to artificial systems.* New York, NY: Oxford University Press, Inc.

Dahl, T. S., Matarić, M. J., & Sukhatme, G. S. (2006). A machine learning method for improving task allocation in distributed multi-robot transportation. In Braha, D., Minai, A., & Bar-Yam, Y. (Eds.), *Complex engineered systems: Science meets technology.* doi:10.1007/3-540-32834-3_14

Dorigo, M., & Sahin, E. (Eds.). (2004). Special issue on "swarm robotics". *Autonomous Robots, 17,* 111–246. doi:10.1023/B:AURO.0000034008.48988.2b

Dorigo, M., & Stützle, T. (2004). *Ant colony optimization.* Cambridge, MA: MIT Press. doi:10.1007/b99492

Gerkey, B., & Matarić, M. J. (2004). A formal framework for the study of task allocation in multi-robot systems. *The International Journal of Robotics Research, 23*(9), 939–954. doi:10.1177/0278364904045564

Jamshidi, M. (2009). *Systems of systems engineering: Principles and applications.* Boca Raton, FL: CRC Press/Taylor & Francis Group.

Jevtić, A. (2011). Swarm intelligence: Novel tools for optimization, feature extraction, and multi-agent system modeling. *Ph.D. Thesis,* Technical University of Madrid, Spain.

Labella, T. H., Dorigo, M., & Deneubourg, J.-L. (2004). Self-organised task allocation in a group of robots. In R. Alami, R. Chatila, & H. Asama (Eds.), *Proceedings of the 6th International Symposium on Distributed Autonomous Robotic Systems,* (pp. 389-398). Tokyo, Japan: Springer.

Lerman, K., Jones, C., Galstyan, A., & Matarić, M. J. (2006). Analysis of dynamic task allocation in multi-robot systems. *The International Journal of Robotics Research*, *25*(3), 225–242. doi:10.1177/0278364906063426

Matarić, M. J., Sukhatme, G. S., & Ostergaard, E. (2003). Multi-robot task allocation in uncertain environments. *Autonomous Robots*, *14*(2-3), 255–263. doi:10.1023/A:1022291921717

McLurkin, J. (2008). *Analysis and implementation of distributed algorithms for multi-robot systems*. Ph.D. Thesis, M.I.T., USA.

Mclurkin, J., & Yamins, D. (2005). Dynamic task assignment in robot swarms. *Proceedings of the 2005 International Conference on Artificial Intelligence*, (pp. 129-136).

Pini, G., Brutschy, A., Frison, M., Roli, A., Dorigo, M., & Birattari, M. (2011). Task partitioning in swarms of robots: An adaptive method for strategy selection. *Swarm Intelligence*, *5*(3-4), 283–304. doi:10.1007/s11721-011-0060-1

KEY TERMS AND DEFINITIONS

Agent: An entity (hardware or software) that has its own decision and action mechanisms.

Autonomous Robot: A robot that can perform desired tasks without continuous human guidance.

Computational Swarm Intelligence: Mathematical models of Swarm Intelligence.

Distributed Robotic System: A system that consists of autonomous robots that can share resources and coordinate their activities.

Mobile Robot: An automatic machine that is capable of movement in a given environment.

Multi-Robot Task Allocation (MRTA): A problem of assigning a set of given tasks to a team of robots.

Swarm: A population of agents interacting locally with one another and with their environment.

Swarm Intelligence: A problem-solving behavior that emerges from the multiplicity of agents' interactions.

Swarm Robotics: An approach to coordination of multi-robot systems based on swarm intelligence principles.

Chapter 10
Using Swarm Intelligence for Optimization of Parameters in Approximations of Fractional–Order Operators[1]

Guido Maione
Politecnico di Bari, Italy

Antonio Punzi
Politecnico di Bari, Italy

Kang Li
Queen's University of Belfast, UK

ABSTRACT

This chapter applies Particle Swarm Optimization (PSO) to rational approximation of fractional order differential or integral operators. These operators are the building blocks of Fractional Order Controllers, that often can improve performance and robustness of control loops. However, the implementation of fractional order operators requires a rational approximation specified by a transfer function, i.e. by a set of zeros and poles. Since the quality of the approximation in the frequency domain can be measured by the linearity of the Bode magnitude plot and by the "flatness" of the Bode phase plot in a given frequency range, the zeros and poles must be properly set. Namely, they must guarantee stability and minimum-phase properties, while enforcing zero-pole interlacing. Hence, the PSO must satisfy these requirements in optimizing the zero-pole location. Finally, to enlighten the crucial role of the zero-pole distribution, the outputs of the PSO optimization are compared with the results of classical schemes. The comparison shows that the PSO algorithm improves the quality of the approximation, especially in the Bode phase plot.

DOI: 10.4018/978-1-4666-2666-9.ch010

INTRODUCTION: MATHEMATICAL BACKGROUND AND LITERATURE REVIEW

Fractional Calculus (FC) is a topic older than three centuries. Namely, its birth can be dated back to an exchange of letters and ideas between Leibniz and marquis de L'Hôpital in 1695. In this correspondence, de L'Hôpital asked to Leibniz what could be the value and meaning of a non-integer (fractional) order derivative $\dfrac{d^{(\nu)}y(x)}{dx^{(\nu)}}$, more specifically with $v = 0.5$. This could extend the classical integer order derivative $\dfrac{d^{(n)}y(x)}{dx^{(n)}}$, with $n \in \mathbb{N}$, to a more general case, in which the order of differentiation v could be a fractional number. Considering $n \in \mathbb{Z}$ and $v < 0$, an integer order integral could be extended to a fractional order one. Leibniz gave the result and answered on September 30, 1695, that *"It will lead to an apparent paradox, from which one day useful consequences will be drawn"*.

Since then many mathematicians and scientists (Euler, Abel, Lacroix, Fourier, Lagrange, Laplace, Riemann, Liouville, Kellang, Grünwald, Letnikov, Caputo, etc.) have formulated and investigated formal properties of non-integer order differentiation and integration. As an example, Heaviside said that *"there is a universe of mathematics lying in between the complete differentiations and integrations"*. At his time, he faced the problem of a rigorous justification for the square root operation of a partial differentiation operator p. After algebraic manipulations, he obtained p^α, with α non-integer number. Moreover, he first discovered the value of $\sqrt{p} = p^{0.5}$ as an experimental solution of a heat flow problem, by applying classical methods with innovative computations for his time.

The idea of fractional derivative or integral can be described in different ways. The most popular definitions are due to Riemann-Liouville, to Grünwald-Letnikov, and to Caputo. The Riemann-Liouville basic definition of fractional order integral generalizes the repeated integration in the Cauchy formula:

$$_a^{RL}D_t^{-\alpha}f(t) = \frac{1}{\Gamma(\alpha)}\int_a^t (t-\tau)^{\alpha-1}\,f(\tau)\,d\tau$$

$$(1)$$

where $\alpha \in \mathbb{R}$, with $\alpha > 0$, is the non-integer order of integration with respect to t and starting point a, and $\Gamma(x) = \int_0^\infty t^{x-1}\,e^{-t}\,dt$ is the Euler gamma function. Then, the Riemann-Liouville definition of fractional order derivative can be obtained as follows:

$$\begin{aligned}_a^{RL}D_t^{\alpha}f(t) &= \frac{d^{(n)}}{dt^{(n)}}\left\{_a^{RL}D_t^{\alpha-n}f(t)\right\} \\ &= \frac{1}{\Gamma(n-\alpha)}\frac{d^{(n)}}{dt^{(n)}}\left\{\int_a^t (t-\tau)^{n-\alpha-1}f(\tau)d\tau\right\}\end{aligned}$$

$$(2)$$

where $\alpha \in \mathbb{R}$, with $\alpha > 0$ and $n - 1 < \alpha < n$, $n \in \mathbb{N}$, is now the non-integer order of differentiation. The Laplace transform of the fractional order integral operator, provided that $\mathcal{L}\{f(t)\} = F(s)$, gives: $\mathcal{L}\left\{_a^{RL}D_t^{-\alpha}f(t)\right\} = s^{-\alpha}\,F(s)$, that is analogous to the standard transform of integer order integral operator. Whereas, the Laplace transform of the fractional order derivative operator leads to:

$$\mathcal{L}\left\{_a^{RL}D_t^{\alpha}f(t)\right\} = s^{\alpha}\,F(s) - \sum_{k=0}^{n-1} s^k \left[_a^{RL}D_t^{\alpha-k-1}f(t)\right]_{t=0^+}$$

$$(3)$$

However, evaluation of Equation 3 requires initial values of the fractional order derivatives that are of difficult interpretation and measurement.

The Caputo definition avoids physical interpretation of initial conditions in Laplace transform of the fractional order derivative operator by writing:

$$
{}^{C}_{a}D^{\alpha}_{t} f(t) = {}_{a}D^{\alpha-n}_{t}\,{}_{a}D^{n}_{t} f(t)
$$

$$
= \frac{1}{\Gamma(n - \alpha)} \int_{a}^{t} (t - \tau)^{n-\alpha-1}\, f^{(n)}(\tau)\, d\tau.
$$

$$(4)$$

Namely, the Laplace transform requires the knowledge of the initial values of standard integer order derivatives:

$$
\mathcal{L}\left\{ {}^{C}_{a}D^{\alpha}_{t} f(t) \right\} = s^{\alpha} F(s) - \sum_{k=0}^{n-1} s^{\alpha-k-1} \left[{}^{C}_{a}D^{k}_{t} f(t) \right]_{t=0^{+}}.
$$

$$(5)$$

Finally, since it is well-known that an integer order derivative of order n can be obtained by divided differences:

$$
\frac{d^{(n)}}{dt^{(n)}} f(t) = \lim_{h \to 0} \frac{1}{h^{n}} \sum_{i=0}^{n} (-1)^{i} \binom{n}{i} f(t - i\,h)
$$

$$(6)$$

then the Grünwald-Letnikov definition of a fractional order differential or integral operator can be considered as an extension of Equation 6:

$$
{}^{GL}_{a}D^{\alpha}_{t} f(t) = \lim_{h \to 0} \frac{1}{h^{\alpha}} \sum_{i=0}^{\left[\frac{t-a}{h}\right]} (-1)^{i} \binom{\alpha}{i} f(t - i\,h) =
$$

$$
\lim_{h \to 0} \frac{1}{h^{\alpha}} \sum_{i=0}^{\left[\frac{t-a}{h}\right]} (-1)^{i} \frac{\Gamma(\alpha + 1)}{\Gamma(i + 1)\,\Gamma(\alpha - i + 1)} f(t - i\,h)
$$

$$(7)$$

where $[x]$ gives the integer part of x and h is the step time increment. This definition is more complex but is useful for a digital implementation based on a discrete time computational algorithm.

All definitions are equivalent in a wide class of applications in science and engineering and lead to well-known results when the fractional order is $\alpha = 1$. But long-term memory effects can be better described by fractional-order derivatives.

Nowadays, the word "fractional" is commonly accepted in the literature, even if not appropriate, since the order of differentiation or integration can also be an irrational, real, or even a complex number. Many researchers also agree that FC is a promising tool and an opportunity for developing and exploiting useful ideas and new solutions (Oldham, 1974; Podlubny, 1999). Recent advances of FC are stimulated by applications in many fields of science and engineering, e.g. in physics (West, 2003; Hilfer, 2000), electrochemistry (Oldham, 2010), biology (Magin, 2004), probability (Gorenflo, 1988), material science (Torvik, 1984), mechanics (Koeller, 1984), etc.. Namely, non-integer (fractional) order models often give better insight into certain physical dynamic systems and processes, in particular when they are characterized by power-law long-term memory effects. These new mathematical tools are described by fractional-order linear or non-linear differential equations (FODE) and fractional-order transfer functions (FOTF). Some examples of modeling applications are: the non-linear (chaotic, fractal) dynamics of particular structures and materials (Manabe, 1961; Torvik, 1984); visco-elasticity of some mechanical elements (Koeller, 1984; Mainardi, 1988; Bagley, 1991); the rugged surface of a malignant breast cell nucleus (Borredon, 1999); the modeling and prediction of diffusion and transport phenomena (Mainardi, 1996); the use of financial mathematics for analyzing the dynamics of stock prices (Duarte, 2010; Laskin, 2000); the ocean-atmosphere interactions (Saravanan, 1998); the behavior of neurons (Anastasio, 1994), etc..

In electrical and electronic engineering there is an active research in several applications of FC. For example, motion of electronic charges in capacitors was studied by Westerlund (1994); FODE were used to represent Chua-Hartley circuits and non-linear non-integer order circuits (Arena, 2000; Hartley, 1995). In signal processing, fractional order FIR or IIR filters (Chen, 2003; Tseng, 2001), and fractional delay systems and filters (Laakso, 1996; Välimäki, 2000) can be mentioned. A great interest grew for applications in the fields of automatic control (motion control of cutting tables), robotics (trajectory control of robot manipulators, path tracking of mobile robots), and system identification (frequency-domain system identification for flexible structures). For a recent state-of-the-art on modeling and control refer to (Caponetto, 2010; Monje, 2010).

In particular, studies in systems and control engineering are based on the pioneering work of Bode (1945), Tustin (1958), Manabe (1961). Bode had the seminal idea of designing controllers robust to gain and load variations based on an ideal loop transfer function, which is specified by

a fractional-order integrator: $G_B(s) = \left(\dfrac{\omega_{gc}}{s}\right)^{\nu}$

where s is the complex variable of the Laplace transform, ω_{gc} is the gain cross-over angular frequency, and v is the fractional order. Tustin controlled the position of massive objects by approximating the Bode ideal transfer function, with $v = 1.5$, to achieve a constant phase margin of 45° over a wide frequency range between $0.2\ \omega_{gc}$ and $1.4\ \omega_{gc}$. Manabe designed the control of space structures by using a fractional integrator. More recently, the team head by Oustaloup developed design methods of non-integer order robust controllers for automotive suspensions and many other systems (Oustaloup, 1983; Oustaloup, 1991; Oustaloup, 1995, February; Oustaloup, 1995; Oustaloup, 1996).

Consequently FC led to develop and extend standard and classical Proportional-Integral-Derivative (PID) controllers. These extended controllers, in general indicated as Fractional-Order Controllers (FOC) (Vinagre, 2002), could have a deep impact on industrial control loops, in which PI/PID are the most used controllers at the present time, more than 90% as remarked by Åström (1995), Villanova (2012). Chen (2006; 2009) prospected a big change (he called it "a revolution") if FOC could replace existing PI/revolution") if FOC could replace existing PI/

PID controllers and become "ubiquitous", especially if the plant is of noninteger (fractional) order. Namely, the best FOC can outperform the best integer order controller; FOC may allow loop-shaping of phase and gain independently and then provide higher flexibility in adjusting the gain and phase characteristics; and, finally, FOC guarantee similar robustness to the one PID controllers achieve at the expense of a very high integer order (Chen, 2006). To synthesize, FOC can be considered as alternative to PID in many industrial control problems, where it is very important to obtain high robustness to uncertainties of the plant model or to variations of the plant parameters, low sensitivity to load disturbances and high frequency noise, by taking advantage of few tuning knobs and by avoiding high integer order controllers (Lurie, 1994, 2000; Oustaloup, 2006, 2008; Vinagre, 2007; Monje, 2008).

In particular, design and tuning rules were developed to extend differentiation and integration to all non-integer values, therefore leading to $PI^\lambda D^\mu$ controllers, with $0 < \lambda < 1$ and $0 < \mu < 1$. They can be represented by the ideal transfer function $G_c(s) = k_P + \dfrac{k_I}{s^\lambda} + k_D\ s^\mu$, characterized by the proportional, integral, and derivative gains plus the fractional orders (Podlubny, 1999, January). Therefore, FOC are based on irrational fractional derivative or integral operators. The main benefits of FOC are better closed-loop performance, higher robustness to gain variations, and higher disturbance rejection than PID controllers, in much wider frequency ranges. The improvements are more evident when the controlled plants are of non-integer order (Chen, 2006). Moreover, flexibility is enhanced by at least two more design parameters, namely λ and μ.

Then, the main problems are: to develop and standardize design and tuning techniques and to achieve good approximations of the ideal but irrational operator s^v, with $v \in \mathbb{R}$. The last specific problem is the focus of this chapter. Namely, the basic fractional-order differential or integral operator s^v needs a rational approximation for realization purpose. To this aim, many methods were proposed in the literature (Barbosa, 2006; Barbosa, 2005, 4-8 July; Barbosa, 2006, October; Carlson, 1964; Charef, 1992; Chen, 2004; Dutta Roy, 1982; Maione, 2006; Maione, 2008; Maione, 2011, March; Matsuda, 1993; Oustaloup, 2000; Podlubny, 2002; Tenreiro Machado, 1997; Tenreiro Machado, 2001; Tenreiro Machado, 2009; Tenreiro Machado, 2010, March; Vinagre, 2000). Most of the times, the approximations are represented by an N-order transfer function, say $G(s)$, characterized by N pairs of minimum-phase zeros and stable poles, that are simple and interlaced along the negative real half-axis. Therefore, the problem is to minimize the error between s^v and $G(s)$ in a frequency range of interest, i.e. to optimize the location of zeros and poles of $G(s)$. Some results are available but improvement can be achieved by using innovative optimization methods, in particular for high-speed digital implementations (Maione, 2011, March).

PSO can serve this purpose because it is a simple and computationally efficient optimization method. Namely, it is an heuristic that finds sub-optimal solutions in polynomial time (Dorigo, 2004). Generally speaking, to cite a few significant contributions of PSO applied to systems engineering and robotics, consider the problems of designing and coordinating teams of self-organizing and self-assembling robots called swarm-bots (Bonabeau, 1999; Dorigo, 1998; Groß, 2006; Trianni, 2006). The coordination can be achieved even when robots are not sophisticated, thanks to a decentralized control system and to coordination of complementary abilities. Moreover, see (van Ast, 2008) for a good synthesis and review of applications to optimization and control. More in details, few recent contributions used genetic algorithms or PSO to tune or design FOC (Cao, 2006; Meng, 2009; Tenreiro Machado, 2010; Tenreiro Machado, 2010, March; Tricaud, 2009; Zamani, 2009). Here, the aim is to describe how

PSO improves approximations with respect to the standards guaranteed by well-known reference methods. For this purpose, the frequency responses are compared of different rational transfer functions, approximating the fractional-order operators by the different compared methods. Obviously, the better results obtained by PSO are the key to improve performance and robustness of FOC, whose realization implies the approximation of fractional-order operators. Therefore, in the following sections, firstly some of the best and most used methods to approximate fractional-order operators are introduced and revisited. Secondly, the chapter describes how to apply the PSO technique to the specific optimization of stable and minimum-phase approximations of fractional order operators. Then, some simulation results are reported to indicate the improvements obtained by PSO. Finally, some remarks conclude the chapter.

APPROXIMATION OF FRACTIONAL ORDER OPERATORS

Approximation of the infinite dimensional irrational operator s^v by a rational transfer function can be obtained by different methods, as previously recalled. Here, reference is made to some of the most popular and performing ones, namely the Oustaloup's recursive approximation method (Oustaloup, 1983, 1991; Oustaloup, 1995, February; 1996, 2000), the method first introduced by Matsuda and Fujii (Matsuda, 1993; Vinagre, 2000), the Continued Fraction-based method developed by one of the authors (Maione, 2008). In particular, Continued Fraction expansions offer several benefits with respect to power series expansions and other methods, specifically a faster convergence and a larger domain of convergence.

All the four reference approximations are characterized by the interlacing property of stable poles and minimum-phase zeros of the N-order filter $G(s)$ that defines the approximation:

$$G(s) = k \prod_{i=1}^{N} \frac{1 - \dfrac{s}{z_i}}{1 - \dfrac{s}{p_i}} \qquad (8)$$

with $p_N < z_N < p_{N-1} < z_{N-1} < \dots < p_2 < z_2 < p_1 < z_1 < 0$, if the approximation of a fractional order differentiator s^v, with $v > 0$, is considered. Interlacing is important because the quality of the approximation depends on the relative position of alternating zeros and poles.

The gain constant k is determined by imposing a unitary gain cross-over frequency, as it occurs with the fractional order operator $(j\omega)^v$:

$$k = \sqrt{\prod_{i=1}^{N} \frac{(1 + 1/\omega_{p_i}^2)}{(1 + 1/\omega_{z_i}^2)}} \qquad (9)$$

where $\omega_{z_i} = |z_i|$ and $\omega_{p_i} = |p_i|$, for $i = 1, \dots, N$, are the break-away frequency points. The break points are determined in the same, more or less, frequency range to compare the different approximations. Obviously, the better is the approximation, the closer is the frequency behavior to that of $(j\omega)^v$, that gives a gain which is a linear function of the logarithm of frequency (a slope of $+20v$ dB/decade), and a constant phase of $(+v\pi/2)$.

Approximation by Maione's Continued Fraction Method

This method was proposed by Maione (2008). It is based on continued fractions expansion (CFE) of the fractional power $(1+x)^v$:

$$(1 + x)^v = b_0 + \cfrac{a_1}{b_1 + \cfrac{a_2}{b_2 + \cfrac{\dots}{\dots + \cfrac{a_j}{b_j + \dots}}}}$$

$$= b_0 + \frac{a_1}{|\, b_1} + \frac{a_2}{|\, b_2} + \dots + \frac{a_j}{|\, b_j} + \dots \qquad (10)$$

where $b_0 = b_1 = 1$, $a_1 = v\,x$, $a_{2k} = k\,(k-v)\,x$, $b_{2k} = 2\,k$, $a_{2k+1} = k\,(k+v)\,x$, $b_{2k+1} = 2\,k + 1$, with $k \in \mathbb{N}$ (Khovanskii, 1965). Note that the third expression in (10) is the Pringsheim's classical notation to indicate the CFE (Khovanskii, 1963). The CFE converges on the complex plane cut along the real axis from $-\infty$ to $x = -1$, then in a much wider domain that the power series expansion of the same function $(1+x)^v$, that is the open circle of radius 1, centered at the origin of the complex plane. If $x = (s - 1)$ is put and the expansion is truncated to a certain j, then the j-th convergent is obtained, and the consequent rational transfer function that provides the approximation to s^v, with $0 < v < 1$, is derived. More specifically, if $j = 2N$, the N-order approximation is obtained as:

$$G_{MA}(s) = \frac{P_{2N}(\nu, s)}{Q_{2N}(\nu, s)} \qquad (11)$$

where the polynomials $P_{2N}(s)$ and $Q_{2N}(s)$ are computed by applying the following recurrent relations:

$$P_m(s) = b_m P_{m-1}(v,s) + a_m P_{m-2}(v,s) \qquad (12.1)$$

$$Q_m(s) = b_m Q_{m-1}(v,s) + a_m Q_{m-2}(v,s) \qquad (12.2)$$

for $m = 1, \dots, 2N$, with $P_{-1}(v,s) = P_0(v,s) = 1$ and $Q_{-1}(v,s) = Q_0(v,s) = 1$. Put in another form, (11) becomes Box 1, where the coefficients can be easily computed by a closed formula in Box 2

Box 1.

$$G_{MA}(s) = \frac{p_{N,0}(\nu)\ s^N + p_{N,1}(\nu)\ s^{N-1} + \cdots + p_{N,N-1}(\nu)\ s + p_{N,N}(\nu)}{q_{N,0}(\nu)\ s^N + q_{N,1}(\nu)\ s^{N-1} + \cdots + q_{N,N-1}(\nu)\ s + q_{N,N}(\nu)} \qquad (13)$$

Box 2.

$$p_{N,j}(\nu) = q_{N,N-j}(\nu) = (-1)^j \begin{pmatrix} N \\ j \end{pmatrix} (\nu + j + 1)_{(N-j)}\ (\nu - N)_{(j)} \qquad (14)$$

Box 3.

$$G_{MA}(s) = 0.1429\ \frac{\left(1 + \dfrac{s}{0.0521}\right)\left(1 + \dfrac{s}{0.6360}\right)\left(1 + \dfrac{s}{4.3119}\right)}{\left(1 + \dfrac{s}{0.2319}\right)\left(1 + \dfrac{s}{1.5724}\right)\left(1 + \dfrac{s}{19.1957}\right)}$$

Box 4.

$$G_{MA}(s) = 0.1111\ \frac{\left(1 + \dfrac{s}{0.0311}\right)\left(1 + \dfrac{s}{0.3333}\right)\left(1 + \dfrac{s}{1.4203}\right)\left(1 + \dfrac{s}{7.5486}\right)}{\left(1 + \dfrac{s}{0.1325}\right)\left(1 + \dfrac{s \cdot}{0.7041}\right)\left(1 + \dfrac{s}{3.0000}\right)\left(1 + \dfrac{s}{32.1634}\right)}$$

with the Pochammer functions specified by $(v+j+1)_{(N-j)} = (v+j+1)\ (v+j+2)\ \ldots\ (v+N)$ and $(v-N)_{(j)} = (v-N)\ (v-N+1)\ \ldots\ (v-N+j-1)$, with $(v-N)_{(0)} = (v+N+1)_{(0)} = 1$ (Spanier, 1987). The coefficients in (13) and $G_{MA}(s)$ depend on the fractional order v, then the same holds true for zeros and poles of $G_{MA}(s)$, that is finally expressed as

$$G_{MA}(s) = k_{MA} \prod_{i=1}^{N} \frac{1 + \dfrac{s}{\tilde{\omega}_{z_i}}}{1 + \dfrac{s}{\tilde{\omega}_{p_i}}}. \qquad (15)$$

For example, for $N = 3$, the following function is obtained in Box 3.

For $N = 4$, the approximation becomes Box 4.

Approximation by Oustaloup's Recursive Method

The first approximation method is the Oustaloup's recursive approximation, on which the CRONE controller is based (Oustaloup, 1991; Oustaloup, 1996). It is a frequency-domain method that specifies N in advance and applies the interlacing property from the outset. The procedure specifying the zeros and poles uses the following well-known settings:

$$\alpha = \left(\frac{\omega_H}{\omega_L}\right)^{\frac{\nu}{N}} \qquad \eta = \left(\frac{\omega_H}{\omega_L}\right)^{\frac{1-\nu}{N}} \qquad (16)$$

$$\omega_{z_1} = \omega_L \sqrt{\eta} \qquad (17)$$

$$\omega_{p_i} = \omega_{z_i} \, \alpha \qquad \text{for } i = 1, \ldots, N \qquad (18)$$

$$\omega_{z_{i+1}} = \omega_{p_i} \, \eta \qquad \text{for } i = 1, \ldots, N-1 \qquad (19)$$

where $\omega_L = \lambda_1 \, \tilde{\omega}_{z_1} < \tilde{\omega}_{z_1}$ and $\omega_H = \lambda_2 \, \tilde{\omega}_{p_N} > \tilde{\omega}_{p_N}$ are chosen so that $\omega_{z_1} \approx \tilde{\omega}_{z_1}$ and $\omega_{p_N} \approx \tilde{\omega}_{p_N}$. To this aim, the parameters λ_1 and λ_2 are fixed by a rule of thumb. The range $[\omega_L, \omega_H]$ is then wider than the range $[\tilde{\omega}_{z_1}, \tilde{\omega}_{p_N}]$, determined by the lowest and highest break-away frequencies given by Maione's method, but the different methods are compared in the same frequency range specified by the first and last singularities. The Oustaloup's approximation is then expressed as:

$$G_{OU}(s) = k_{OU} \prod_{i=1}^{N} \frac{1 + \dfrac{s}{\omega_{z_i}}}{1 + \dfrac{s}{\omega_{p_i}}}. \qquad (20)$$

For example, for $N = 3$ (with $\lambda_1 = 0.55$ and $\lambda_2 = 1.8$) and $N = 4$ (with $\lambda_1 = 0.61$ and $\lambda_2 = 1.64$), the respective obtained approximations are in Box 5 and Box 6.

Box 5.

$$G_{OU}(s) = 0.1692 \, \frac{\left(1 + \dfrac{s}{0.0518}\right)\left(1 + \dfrac{s}{0.5509}\right)\left(1 + \dfrac{s}{5.8634}\right)}{\left(1 + \dfrac{s}{0.1688}\right)\left(1 + \dfrac{s}{1.7972}\right)\left(1 + \dfrac{s}{19.1293}\right)}$$

Box 6.

$$G_{OU}(s) = 0.1377 \, \frac{\left(1 + \dfrac{s}{0.0311}\right)\left(1 + \dfrac{s}{0.2261}\right)\left(1 + \dfrac{s}{1.6419}\right)\left(1 + \dfrac{s}{11.9237}\right)}{\left(1 + \dfrac{s}{0.0839}\right)\left(1 + \dfrac{s}{0.6093}\right)\left(1 + \dfrac{s}{4.4247}\right)\left(1 + \dfrac{s}{32.1323}\right)}$$

Approximation by Matsuda's Method

Matsuda and Fujii (1993) proposed a method based on the approximation of the gain ω^v of the fractional order operator. The gain is determined at $2N+1$ sampling frequencies $\omega_0, \omega_1, \ldots, \omega_{2N}$, which are taken logarithmically spaced in the approximation interval. Again, the range $[\omega_0, \omega_{2N}]$ is chosen to compare the method in the same approximation interval as the two preceding methods. More in details, a parameter λ is chosen so that $\omega_0 = \tilde{\omega}_{p_N} / \lambda$ and $\omega_{2N} = \lambda \tilde{\omega}_{z_1}$. In this way, the lowest break-away frequency $\hat{\omega}_{z_1}$ and the highest break-away frequency $\hat{\omega}_{p_N}$ in the obtained Matsuda's approximating function satisfy $\hat{\omega}_{z_1} \approx \tilde{\omega}_{z_1}$ and $\hat{\omega}_{p_N} \approx \tilde{\omega}_{p_N}$, respectively.

The approximation is based on the functions $m_k(\omega) = \dfrac{\omega - \omega_{k-1}}{m_{k-1}(\omega) - m_{k-1}(\omega_{k-1})}$ that are recurrently defined for $k = 0, \ldots, 2N$, starting from $m_0(\omega) = \omega^v$. Namely, the parameters $\alpha_k = m_k(\omega_k)$, for $k = 0, \ldots, 2N$, may be computed to develop the following CFE:

$$s^v = \alpha_0 + \cfrac{s - \omega_0}{\alpha_1 + \cfrac{s - \omega_1}{\alpha_2 + \cfrac{s - \omega_2}{\alpha_3 + \cdots}}}$$

$$= \alpha_0 + \cfrac{s - \omega_0 \,|}{|\, \alpha_1} + \cfrac{s - \omega_1 \,|}{|\, \alpha_2} + \cfrac{s - \omega_2 \,|}{|\, \alpha_3} + \cdots. \tag{21}$$

Truncating the expansion to α_{2N} gives the convergent that defines the approximation. The convergent is then expressed as a rational transfer function:

$$G_{MF}(s) = k_{MF} \prod_{i=1}^{N} \cfrac{1 + \cfrac{s}{\hat{\omega}_{z_i}}}{1 + \cfrac{s}{\hat{\omega}_{p_i}}}. \tag{22}$$

For example, for $N = 3$ (with $\lambda = 45$) and $N = 4$ (with $\lambda = 39$), the respective obtained approximations are in Box 7 and Box 8.

Box 7.

$$G_{MF}(s) = 0.1373 \, \frac{\left(1 + \dfrac{s}{0.0485}\right)\left(1 + \dfrac{s}{0.6248}\right)\left(1 + \dfrac{s}{4.5311}\right)}{\left(1 + \dfrac{s}{0.2207}\right)\left(1 + \dfrac{s}{1.6004}\right)\left(1 + \dfrac{s}{20.6273}\right)}$$

Box 8.

$$G_{MF}(s) = 0.1109 \, \frac{\left(1 + \dfrac{s}{0.0310}\right)\left(1 + \dfrac{s}{0.3327}\right)\left(1 + \dfrac{s}{1.4212}\right)\left(1 + \dfrac{s}{7.5707}\right)}{\left(1 + \dfrac{s}{0.1321}\right)\left(1 + \dfrac{s}{0.7036}\right)\left(1 + \dfrac{s}{3.0057}\right)\left(1 + \dfrac{s}{32.2790}\right)}$$

A Note on Comparison Between Methods of Approximation

The advantages and drawbacks of the recalled methods are now briefly discussed and synthesized.

The method by Maione is based on CFE and allows to compute the coefficients of the approximant by simple closed formulas that strictly depend on the fractional order v, see (14). Moreover, the approximation is formally proven to be characterized by minimum-phase zeros and stable poles that are interlaced along the negative real half-axis of the s-plane, as it was shown in Maione (2011, 28 August - 2 September). Finally, the obtained approximation shows a very reduced error with respect to the well-established method by Oustaloup, if we consider the phase diagram of the frequency response (See Figure 4 and Figure 6).

Instead, the method by Oustaloup is based on choosing the zeros and poles of the approximation by rules of thumb, and the stability, minimum-phase, and interlacing properties are *a priori* enforced by the choice of break-frequencies (17)-(19). By looking at the phase diagram, it is easily recognized that this method does not guarantee the same robustness degree, even if it is well-known and frequently used.

The method by Matsuda is very efficient and gives results very close to Maione's approximation. Both are based on CFE, that guarantee an efficient approximation. Namely, CFE converge much more rapidly than power series expansions, and in a much larger domain in the complex plane (Podlubny, 2002). Matsuda's method depends on choosing the sample points and the parameter λ, and it requires the recurrent computation of functions $m_k(\omega)$ and coefficients α_k.

PARTICLE SWARM OPTIMIZATION OF FRACTIONAL ORDER OPERATORS

Originally PSO was developed by Kennedy and Eberhart (1995). PSO is very robust in problems characterized by non-linearity, multiple optimum points, non-differentiability, and multidimensional search spaces (Clerc, 2002; Shi, 1998, 4-9 May). Fast and stable convergence, ease of implementation, reduced computational cost, very low sensitivity to population size and high sensitivity to values of its parameters (Shi, 1999; van den Bergh, 2006) are the main features of this technique. It is an evolutionary stochastic optimization based on the imitation of social behavior, movement and communication of ant colonies, bees, flocks of birds, swarms of insects, schools of fishes, and similar biological animal groups. Each particle in a swarm uses its own intelligence and a collective intelligence of the swarm, that is commonly not characterized by a leader. The main point is that each particle leaves and reinforces the information for the rest of the swarm that is useful to solve a problem, like finding the shortest path to food. Each particle moves, has memory of its successful movements, communicates them to other particles, and adapts its position and velocity. No particle is able to find the optimal solution, but each contributes to it thanks to numerous interactions based on simple rules that each particle obeys on the basis of local information. In this sense, a smart colony or swarm may have complex behaviors or solve complex problems that individual particles are not capable to approach, because none of them is intelligent enough but all together cooperate as a whole to the swarm solution. Also, the self-organizing group can quickly react to variations and disturbances.

PSO can be applied to function optimization, artificial neural network training, and control problems. In addition, a PSO algorithm is efficient, easy to implement, and it has few parameters to adjust. However, PSO may have convergence problems (Clerc, 2002; Shi, 1999), for example premature convergence or a slow convergence in the refined search stage. In this case, a weak local search ability implies a solution that is far from the optimum one.

Basically, two different PSO algorithms may be considered; the former is standard and is based on a weighting factor (PSOWF); the latter is an improved version based on a constriction factor (PSOCF) (Eberhart, 2000; Kennedy, 1995). Typically, a PSO algorithm is executed for a fixed number of iterations (N_{runs}); each iteration is characterized by a fixed number of generations (G_{max}); and each generation is defined by a fixed number of particles constituting the generated swarm population (N_{pop}). Obviously, these three numbers can be varied and properly tuned. In particular, the swarm size N_{pop} is a trade-off between the required computation time and the speed of convergence. Clearly, a big swarm implies a high computational effort, a small swarm may not or may slowly converge. The PSO algorithm stops the iterations when it achieves the optimum solution or when it reaches the maximum number of generations in the last iteration.

The PSO algorithm starts with a random initial population of particles. Each particle or solution is randomly defined in a search space properly chosen for the considered problem. A particle represents a different set of the unknown parameters (i.e. zeros and poles of $G(s)$) to be optimized. The solutions result from updating the population in successive generations. At every generation of a new population the algorithm updates two important variables: the "position", $p_j(i)$, and the "velocity", $v_j(i)$, of the particles, for $j = 1,..., N_{pop}$ and $i = 1,..., G_{max}$, in the trend towards the optimized solution inside the search space. Both algorithms update the position of each particle by using the past position and current velocity:

$$p_j(i) = p_j(i-1) + v_j(i). \qquad (23)$$

with initial zero velocity and initial random position taken from a uniform distribution between specified minimum and maximum values.

In the PSOWF (Shi, 1998, 25-27 March), the updated velocity of each particle is affected by three terms depending on three parameters: a weighting factor θ, with $0 < \theta < 1$, applied to the previous velocity, and two acceleration constants c_1, i.e. the individual learning rate ($0 < c_1 < 2$), and c_2, i.e. the social learning rate ($0 < c_2 < 2$):

$$v_j(i) = \theta\, v_j(i-1) + c_1\, r_1\, [p_{pbest}(i) - p_j(i-1)] + c_2\, r_2\, [p_{gbest}(i) - p_j(i-1)]. \qquad (24)$$

In the previous relation, $r_1 \sim U(0, 1)$ and $r_2 \sim U(0, 1)$ are two random numbers taken from a uniform distribution between 0 and 1.

The inertial weighting factor θ reduces the variation of velocity, to avoid missing of optimal solutions due to fast increments. A high value helps the global exploration, a low value favors the local search. Sometimes θ is linearly varied as $\theta(i) = \theta_{max} - i\,(\theta_{max} - \theta_{min})/G_{max}$, with $\theta_{min} = 0.4$ and $\theta_{max} = 0.9$.

The acceleration constants take into account, respectively, the difference of the previous position with respect to the best remembered individual particle position so far (i.e. the so-called "personal best" $p_{pbest}(i)$), and the difference of the previous position with respect to the best remembered swarm position of all particles until the current generation (i.e. the so-called "global best" $p_{gbest}(i)$). The personal best position is determined by the particle that achieves the minimum value of a properly defined objective function J, and the global best position results from $p_{gbest}(i) = \min\{p_{gbest}(i-1), p_{pbest}(i)\}$.

In the PSOCF, the whole velocity is affected by a constriction factor $\xi(c_1, c_2)$:

$$v_j(i) = \xi(c_1, c_2)\, \{v_j(i-1) + c_1\, r_1\, [p_{pbest}(i) - p_j(i-1)] + c_2\, r_2\, [p_{gbest}(i) - p_j(i-1)]\}. \qquad (25)$$

where $\xi(c_1, c_2)$ depends on c_1 and c_2 according to the following relation:

$$\xi(c_1, c_2) = \frac{2}{\left| 2 - c - \sqrt{c^2 - 4c} \right|} \qquad (26)$$

with $c = c_1 + c_2 > 4$.

The velocity parameters must be chosen to help the convergence of the solutions to values that are as close as possible to the ideal location of zeros and poles of $G(s)$ that would ensure null approximation error in a wide frequency range. In fact, this choice modifies the performance of the PSO algorithm.

Then, to apply a PSO algorithm, the first step is to choose the search interval of the solutions and then to fix the values of parameters that ensure good performance of the algorithm, in terms of the minimum achieved for the objective function and fast convergence. The search interval must include the range where the optimum solutions are thought to be. The search could have no particular restriction, even if, without constraint, the algorithm could lead to non-optimized solutions, with high approximation errors. Instead, as the approximation of fractional operator s^v is concerned, the search interval is restricted by enforcing the interlacing between zeros and poles. Moreover, the reference for each zero and pole of $G(s)$ is specified by the minimum and maximum values obtained by other approximation methods that gave the better results so far, in particular a recent approximation method that provided promising results (Maione, 2008). The gain of the approximating transfer function results from imposing unitary gain crossover angular frequency.

As other evolutionary procedures, the PSO algorithm minimizes an objective function J depending on the approximation error. Stated in another way, the goal of the PSO evolutionary process is to find poles and zeros of $G(s)$ such that the approximation error is as small as possible in the widest frequency range. The frequency response errors E_{mag} (measured in dB) and E_{pha} (measured in degrees) are between the magnitude and phase of the approximation $G(j\omega)$ and the magnitude and phase of the fractional operator $(j\omega)^v$. Then, it is important to define the objective function associated to the error behavior as in any optimization algorithm based on evolutionary process. In the literature, there are several types of error indices, and some of the most commonly used are the integral of absolute error (IAE), the integral of time absolute error (ITAE), and the integral of squared error (ISE). Here, the following index is considered:

$$J_1 = \sqrt{\sum_{i=1}^{\Omega} E_{mag,i}^2} + \sqrt{\sum_{i=1}^{\Omega} E_{pha,i}^2} \qquad (27)$$

where the errors are computed in a finite number Ω of frequency sample points ω_k, $k = 1, \ldots, \Omega$, inside the frequency range of interest. Another error index can be computed by the minimum of the sum of the maximum absolute errors:

$$J_2 = \min\{\max\{|E_{mag,i}|\} + \max\{|E_{pha,i}|\}\}. \qquad (28)$$

Moreover, a non-linear penalty function $P(\omega)$ can be introduced to give a higher weight $P(\omega) = 1$ to the approximation error in the frequency range of interest, and a lower weight $P(\omega) = e^{-\left|\log_{10}(\omega)\right|^2}$ outside this range. Thus, a penalized index J_3 can be defined in Box 9.

Box 9.

$$J_3 = \min\left\{\max\left\{\left|E_{mag,i}\right| \cdot P(\omega)\right\} + \max\left\{\left|E_{pha,i}\right| \cdot P(\omega)\right\}\right\} \qquad (29)$$

Obviously, choosing one of the defined indices gives different solutions and different values of the error index. In particular, the minimum value of the error index is computed in all iterations of the PSO algorithms; moreover, the average value of the index is computed in each iteration and then the lowest average is considered. After several simulation tests, the index that exhibited the smallest minimum values was J_3, that proved to guarantee the best approximation and it was therefore selected.

To sum up, each particle in a population defined by the PSO algorithm (i.e. each set of poles and zeros of an N-order rational transfer function) is characterized by a different frequency response, different errors, then a different value of J_3. The quality of the best solution after all generations in all iterations of the algorithms (minimum of J_3) is verified by plotting the frequency response of the obtained approximating transfer function:

$$G_{PSO}(s) = k_{PSO} \prod_{i=1}^{N} \frac{1 + \dfrac{s}{\overline{\omega}_{z_i}}}{1 + \dfrac{s}{\overline{\omega}_{p_i}}} \qquad (30)$$

SIMULATION RESULTS

Simulation is performed in the MATLAB® environment to test the efficiency of the PSO with respect to the other methods. The typical value $v = 0.5$ is selected. Then, the fractional order differential operator $s^{0.5}$ has a gain linearly varying with a slope of +10 dB/decade and a constant phase of +45°. More and different values for $0 < v < 1$ were not considered to save space and because they do not add further insight. The order of the different approximations was limited to $N = 3$, 4. Namely, higher values of N did not increase accuracy of the approximations significantly, but

required more complex realizations of filters. As previously remarked, the first and last frequency break points of all the approximations given by $G_{MA}(s)$, $G_{OU}(s)$, $G_{MF}(s)$, and $G_{PSO}(s)$ were nearly equal.

The PSOCF method is applied with $c_1 = c_2 = 2.05$, and consequently $\xi = 0.7298$. These particular values of c_1 and c_2 came after a trial and error method, by testing the PSOCF on the minimization of four nonlinear benchmark functions used for unconstrained global optimization, i.e. the Booth, Easom, Bohachecsky, and Ackley test functions (Hedar, 2006) and by comparing the performance with the results provided by the PSOWF to the same minimization problem, with the same fixed number of iterations, generations, and particles. In particular, the performance of the PSOCF algorithm becomes worst if c increases, then the best minimum values of c_1 and c_2 that satisfy $c > 4$ to obtain ξ were chosen.

To apply PSOCF to the approximation of fractional order operator, the parameters N_{runs}, G_{max}, and N_{pop} must be tuned. After several trials, good values for a trade-off between performance and fast convergence were found: $N_{runs} = 20$, $G_{max} = 30$, and $N_{pop} = 20$ are sufficient for $N = 3$; $N_{runs} = 50$, $G_{max} = 100$, and $N_{pop} = 50$, for $N = 4$. The search intervals for the zeros and poles (i.e. the associated break-away frequencies) are determined by considering the reference values provided by the three previously revisited methods. See Tables 1 and 2 for these reference values. More in details, to improve the performance of the PSOCF algorithm by restricting the search and to avoid exploration of solutions violating the interlacing property, the search intervals $[m, M]$ were determined by considering the minimum (*min*) and maximum (*max*) values given by the three reference methods, and then by adjusting these extreme values by trial and error. Furthermore, overlapping between adjacent intervals was avoided to guarantee interlacing. Finally, the penalty function $P(\omega) = 1$, for 0.1

Table 1. Values of break-away frequencies (N=3): [min, max] is defined by minimum and maximum values given by the reference approximation methods; [m, M] defines the search interval employed by the PSOCF method

Method	Maione	Oustaloup	Matsuda	[*min*, *max*]	PSOCF: [*m*, *M*]
z_1	0.0521	0.0518	0.0485	[0.0485, 0.0521]	[0.0480, 0.0610]
p_1	0.2319	0.1688	0.2207	[0.1688, 0.2319]	[0.1685, 0.2410]
z_2	0.6360	0.5509	0.6248	[0.5509, 0.6360]	[0.5505, 0.6710]
p_2	1.5724	1.7972	1.6004	[1.5724, 1.7972]	[1.5720, 1.7980]
z_3	4.3119	5.8634	4.5311	[4.3119, 5.8634]	[4.3115, 5.8640]
p_3	19.1957	19.1293	20.6273	[19.1293, 20.6273]	[17.6325, 20.6275]

Table 2. Values of break-away frequencies (N=4): [min, max] is defined by minimum and maximum values given by the reference approximation methods; [m, M] defines the search interval employed by the PSOCF method

Method	Maione	Oustaloup	Matsuda	[*min*, *max*]	PSOCF: [*m*, *M*]
z_1	0.0311	0.0311	0.0310	[0.0310, 0.0311]	[0.0290, 0.0315]
p_1	0.1325	0.0839	0.1321	[0.0839, 0.1325]	[0.0830, 0.1327]
z_2	0.3333	0.2261	0.3327	[0.2261, 0.3333]	[0.2260, 0.3335]
p_2	0.7041	0.6093	0.7036	[0.6093, 0.7041]	[0.6090, 0.7240]
z_3	1.4203	1.6419	1.4212	[1.4203, 1.6419]	[1.4201, 1.9570]
p_3	3.0000	4.4247	3.0057	[3.0000, 4.4247]	[2.9000, 5.0900]
z_4	7.5486	11.9237	7.5707	[7.5486, 11.9237]	[7.5400, 12.5900]
p_4	32.1634	32.1323	32.2790	[32.1323, 32.2790]	[32.1200, 45.0400]

rad/s $\leq \omega \leq$ 10 rad/s, was assumed for defining the frequency range of interest around the gain cross-over frequency ω_{gc} = 1 rad/s. Instead, $P(\omega) = e^{-\left|\log_{10}(\omega)\right|^2}$ for ω < 0.1 rad/s and for ω > 10 rad/s.

The PSOCF algorithm was simulated several times, to test efficiency of the method and quality of the achieved approximation. Each time the algorithm was iterated for N_{runs} simulation runs, and each time the results were slightly different because the initial population and parameters r_1 and r_2 in (25) were randomly generated. However, the algorithm always showed stability and fast convergence. Moreover, the results always guaranteed a good approximation to the fractional order operator, as testified by low values

of the approximation error index J_3. As an example, results are reported here of a test that yielded an accurate approximation of fractional slope of the amplitude diagram and fractional constant phase ($v\pi/2$). Table 3 and Table 4 report very low values of the error index.

Note that optimization by PSOCF was always performed with N_{pop} = 20, G_{max} = 30, and N_{runs} = 20 simulation runs when N = 3 and with N_{pop} = 50, G_{max} = 100, and N_{runs} = 50 simulation runs when N = 4. For each run the average value $J_{3r,avg}$ (r = 1,..., 50) of the objective function was computed over all the G_{max} generations. Then, the run $r*$ was determined giving the minimum among the computed average values. This run gives the best behavior of the objective function J_3 over

Figure 1. Approximation of order N = 3: convergence of J_3 in the run r* that shows the minimum average value of the error index: $J_{3r^*,min} = 1.4391$, $J_{3r^*,avg} = 1.5151$, $J_{3r^*,max} = 2.2996$, $J_{3r^*,std} = 1.5591 \cdot 10^{-1}$

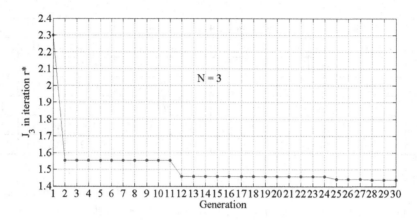

Figure 2. Approximation of order N = 4: convergence of J_3 in the run r* that shows the minimum average value of the error index: $J_{3r^*,min} = 5.8433 \cdot 10^{-1}$, $J_{3r^*,avg} = 6.1388 \cdot 10^{-1}$, $J_{3r^*,max} = 1.1470$, $J_{3r^*,std} = 8.5037 \cdot 10^{-2}$

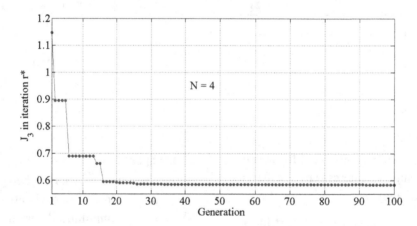

all simulation runs. For this selected iteration, the following indexes were computed: the minimum $J_{3r^*,min}$, the average $J_{3r^*,avg}$, the maximum $J_{3r^*,max}$, and the standard deviation $J_{3r^*,std}$. Figures 1 and 2 show the plot of J_{3r^*} for N = 3 and N = 4, respectively: it is easy to verify the convergence of the PSOCF algorithm.

Moreover, the best value of the objective function was determined, i.e. the minimum value $\bar{J}_3 = \min\{J_3\}$ provided by the generations in all the simulated runs. Tables 3 and 4 indicate the minimum value of J_3 in each run. $\bar{J}_3 = 1.3860$ for N = 3 and $\bar{J}_3 = 0.5838$ for N = 4, respectively (see bold values in Tables 3 and 4). Note that $\bar{J}_3$ is different from $J_{3r^*,min}$.

Therefore, the best values of zeros and poles of $G_{PSO}(s)$ are associated to $\bar{J}_3$ and determined by the corresponding generation and iteration. The obtained approximations are in Box 10 and Box 11.

Table 3. Minimum values of error index J_3 in 50 simulation runs for N = 3

Run	Minimum of J_3	Run	Minimum of J_3
1	1.9333	11	1.4176
2	1.8436	12	1.4423
3	1.5169	13	1.4255
4	1.4394	14	1.4411
5	1.4085	15	1.4504
6	1.4095	16	1.4391
7	1.4672	17	1.4382
8	1.4204	**18**	**1.3860**
9	1.4170	19	1.4341
10	1.4899	20	1.4349

Table 4. Minimum values of error index J_3 in 50 simulation runs for N = 4

Run	Minimum of J_3	Run	Minimum of J_3
1	0.8148	26	0.5860
2	0.6147	27	0.5878
3	0.6108	28	0.5873
4	0.6099	29	0.5868
5	0.6082	30	0.5886
6	0.6066	31	0.5884
7	0.6070	32	0.5870
8	0.6053	33	0.5871
9	0.6059	34	0.5862
10	0.5977	35	0.5860
11	0.5983	36	0.5868
12	0.5987	37	0.5852
13	0.6002	38	0.5844
14	0.5985	39	0.5839
15	0.5997	40	0.5847
16	0.5995	41	0.5845
17	0.5967	42	0.5856
18	0.5923	43	0.5844
19	0.5901	44	0.5878
20	0.5906	45	0.5875
21	0.5909	46	0.5857
22	0.5914	47	0.5842
23	0.5919	48	0.5843
24	0.5915	**49**	**0.5838**
25	0.5873	50	0.5840

Before comparing the approximations, let us note that all of them are characterized by stable poles and minimum-phase zeros that obey the interlacing property.

For comparison with $G_{PSO}(s)$, the approximations $G_{MA}(s)$, $G_{OU}(s)$, and $G_{MF}(s)$ are considered as developed in the previous section for $N = 3$ and $N = 4$.

Figures 3, 4, 5, and 6 show the frequency responses provided by the different approximation methods. As it can be verified, all the approximations exhibit a magnitude diagram quite close to the ideal linear one, especially in the two decades around the gain cross-over frequency, between 0.1 and 10 rad/s (see zoom in Figures 3b and 5b). The Oustaloup's approximation is a little bit closer.

But, if the phase diagram is considered, the PSOCF guarantees a much flatter plot in the frequency range of interest between 0.1 and 10 rad/s (see zoom in Figures 4b and 6b). This is achieved with respect to both the similar approximations by Maione and by Matsuda and the approximation by Oustaloup, that gives the worst results. In particular, for $N = 3$, even if the method by Maione gives a very flat phase diagram between 0.3 and 3 rad/s (see zoom in Figure 4b), the PSOCF gives better results if a wider range is considered. For $N = 4$, the PSOCF further improves the approximation achieved by the Maione's and other methods, except maybe for the interval between 0.4 and 2

Box 10.

$$G_{PSO}(s) = 6.5797 \; \frac{(s + 0.0554)\,(s + 0.6528)\,(s + 4.9188)}{(s + 0.2303)\,(s + 1.7631)\,(s + 19.8574)}$$

Box 11.

$$G_{PSO}(s) = 9.3375 \; \frac{(s + 0.0273)\,(s + 0.2859)\,(s + 1.6013)\,(s + 9.9258)}{(s + 0.1093)\,(s + 0.6802)\,(s + 3.8360)\,(s + 39.0441)}$$

Figure 3. Frequency response for N = 3: (a) Bode magnitude diagram; (b) Zoom in interval [0.1,10]

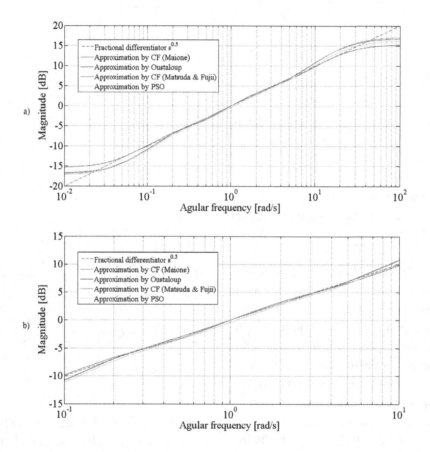

rad/s where Maione's approximation is slightly better (see zoom in Figure 6b).

Finally, note the similarity between the Maione's and Matsuda's approximations.

To conclude, the obtained results show that the PSOCF can guarantee a better approximation of s^{ν} than well-known methods, in a sufficiently wide frequency range and with a limited amount

Figure 4. Frequency response for N = 3: (a) Bode phase diagram; (b) Zoom in interval [0.1,10]

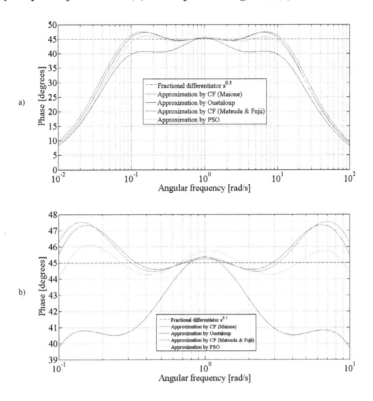

Figure 5. Frequency response for N = 4: (a) Bode magnitude diagram; (b) Zoom in interval [0.1,10]

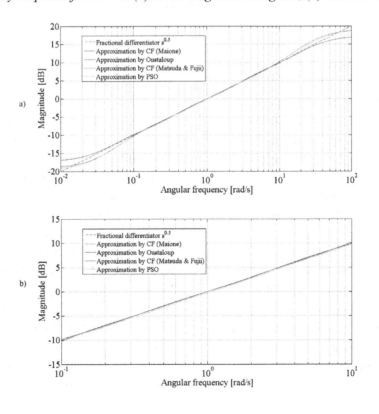

Figure 6. Frequency response for N = 4: (a) Bode phase diagram; (b) Zoom in interval [0.1,10]

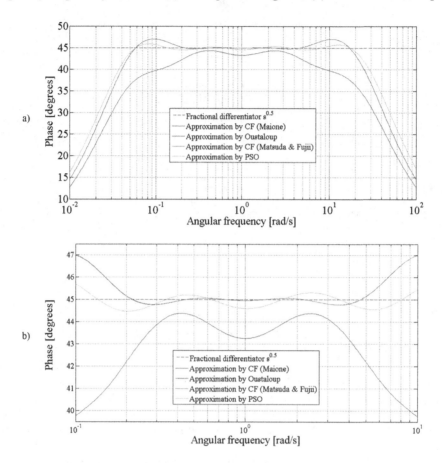

of pairs of zeros and poles of the approximating transfer function.

To conclude the simulation comparison, Figures 7-8 show the time response of the different transfer functions approximating the fractional differentiator $s^{0.5}$. To this aim, an input signal $x(t) = \sin(2\pi f t)$ is considered, with $f = 1$ Hz. The different approximate fractional derivatives of $x(t)$ are obtained by using the previously obtained transfer functions $G_{MA}(s)$, $G_{OU}(s)$, $G_{MF}(s)$, and $G_{PSO}(s)$ and are plotted together with the Grünwald-Letnikov approximation (G-L for short), i.e. $_{0}^{GL}D_{t}^{0.5}x(t)$, that is a well-known benchmark discrete time approximation and here is computed by using Equation 7 with a sampling period $h = 0.01$ s. Therefore, the input and other approximations are plotted with the same sampling period.

The errors in Figure 7b – Figure 8b are between the developed approximations and the G-L reference approximation.

Note that the approximation error is reduced by increasing the order N, and that the PSO approximation performs comparatively well. It is the best one for $N = 3$, and very close to the Maione's and Matsuda's curves for $N = 4$. However, it always improves the result achieved by the classical Oustaloup's approximation.

For sake of completeness, the step response is also simulated for the four approximations (see Figures 9-10). The typical indexes of the step responses, i.e. the 2% settling time (t_s) and the steady-state value (y_{ss}) are summarized in Table 5. The settling times are very similar in all cases but for the response by Oustaloup's approximation, that is the slower; for $N = 4$ the response obtained

Figure 7. Time responses to a sinusoidal input for N = 3: (a) Responses; (b) Approximation errors

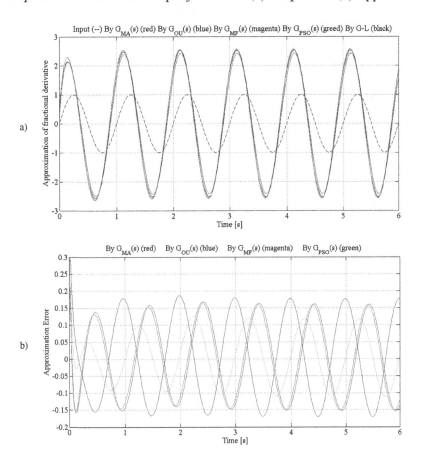

by PSO further speeds up. Similar considerations hold true when considering the rise times: note how prompt is the PSO response for $N = 4$. The same applies for steady-state values that tend to decrease for N increasing and for the PSO response.

CONCLUSION

This chapter illustrates an application of PSO to the approximation of fractional order operators that are the basic component of fractional order controllers. If compared to classical PID, these innovative controllers are capable to improve both closed-loop performance and robustness to gain and load variations. Since fractional order differential (or integral) operators s^v (or s^{-v}), with $0 < v < 1$, are irrational transfer functions, the main problem is to obtain an accurate and low-cost rational approximation of s^v. To this aim, many existing methods determine approximating rational transfer functions with a limited number of minimum-phase zeros and stable poles, because such features are very important for control purpose. Moreover, many popular approximations propose interlaced pairs of zeros and poles along the negative real half-axis of the s-plane. Among them, the well-known Oustaloup's recursive approximation receives application in many contexts, while the so-called Matsuda's method (Matsuda, 1993) and Maione's approach (Maione, 2008) employ continued fractions for providing very accurate approximations, especially in the phase diagram of the frequency response.

In this chapter, a new approximation technique based on PSO is compared to the three reference

Figure 8. Time responses to a sinusoidal input for N = 4: (a) Responses; (b) Approximation errors

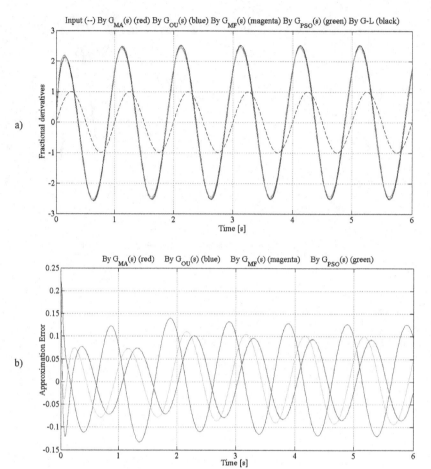

Figure 9. Step responses for N = 3

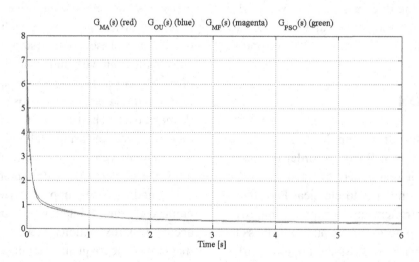

Figure 10. Step responses for N = 4

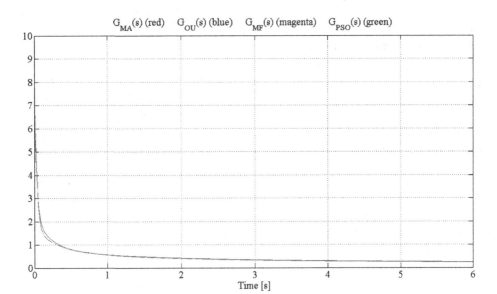

Table 5. Step response indexes for N = 3 and for N = 4

Approximation		t_s [s]	y_{ss}
N = 3	Maione	4.10	0.143
	Oustaloup	5.47	0.169
	Matsuda	4.11	0.137
	PSO	4.10	0.145
N = 4	Maione	3.81	0.111
	Oustaloup	4.55	0.138
	Matsuda	3.81	0.111
	PSO	3.61	0.104

approaches provided by Oustaloup, Matsuda, and Maione. The interlacing between zeros and poles is, however, enforced also by the proposed technique. Thanks to the optimization of this particular property, good results are obtained, even by low-order rational transfer functions. Namely, increasing the order N does not reduce errors noticeably, especially in the phase diagram of the frequency response. Then, to obtain a very precise approximation of the magnitude diagram of the fractional order operator, it is not worth-

while to use high values of N, as other authors do (Charef, 1992). More in details, the PSO-based approximation not only outperforms the classical Oustaloup's technique, but also improves the accuracy of the Matsuda's method and of other recent results by Maione (2008), as the frequency response diagrams testify in this chapter.

The final remark is that the interlacing property plays a crucial role, because the location of zeros and poles strongly affects the error and the width of the range where the approximation works. This is confirmed by the fact that Matsuda's and Maione's methods provide very similar results, even if they are apparently developed in a different way. They, indeed, lead to nearly coincident zero-pole distributions. To conclude, PSO helps to approach the optimum configuration of zeros and poles, that could be defined by a formal procedure and theoretical results on which future investigations can be made.

REFERENCES

Arena, P., Caponetto, R., Fortuna, L., & Porto, D. (2000). Nonlinear noninteger order circuits and systems - An introduction. In Chua, L. (Ed.), *World Scientific Series on Nonlinear Science, Series A* (*Vol. 38*). Singapore: World Scientific.

Åström, K. J., & Hagglund, T. (1995). *PID controllers: Theory, design, and tuning* (2nd ed.). Research Triangle Park, NC: Instrument Society of America.

Barbosa, R. S., & Tenreiro Machado, J. A. (2006). Implementation of discrete-time fractional-order controllers based on LS approximations. *Acta Polytechnica Hungarica*, *3*(4), 5–22.

Barbosa, R. S., Tenreiro Machado, J. A., & Ferreira, I. M. (2005, 4-8 July). Pole-zero approximations of digital fractional-order integrators and differentiators using signal modelling techniques. In *Proceedings of the 16th IFAC World Congress*, Prague, Czech Republic.

Barbosa, R. S., Tenreiro Machado, J. A., & Silva, M. F. (2006, October). Time domain design of fractional differintegrators using least squares. *Signal Processing*, *86*(10), 2567–2581. doi:10.1016/j.sigpro.2006.02.005

Bode, H. W. (1945). *Network analysis and feedback amplifier design*. New York, NY: Van Nostrand.

Bonabeau, E., Dorigo, M., & Theraulaz, G. (1999, August). *Swarm intelligence: From natural to artificial systems*. New York, NY: Oxford University Press.

Cao, J.-Y., & Cao, B.-G. (2006, December). Design of fractional order controller based on particle swarm optimization. *International Journal of Control, Automation, and Systems*, *4*(6), 775–781.

Carlson, G. E., & Halijak, C. A. (1964). Approximation of fractional capacitors $(1/s)^{1/n}$ by a regular Newton process. *IEEE Transactions on Circuit Theory*, *11*(2), 210–213.

Charef, A., Sun, H. H., Tsao, Y. Y., & Onaral, B. (1992, September). Fractal system as represented by singularity function. *IEEE Transactions on Automatic Control*, *37*(9), 1465–1470. doi:10.1109/9.159595

Chen, Y. Q., & Vinagre, B. M. (2003). A new IIR-type digital fractional order differentiator. *Signal Processing*, *83*(11), 2359–2365. doi:10.1016/S0165-1684(03)00188-9

Chen, Y. Q., Vinagre, B. M., & Podlubny, I. (2004). Continued fraction expansions approaches to discretizing fractional order derivatives. An expository review. *Nonlinear Dynamics*, *38*(1-2), 155–170. doi:10.1007/s11071-004-3752-x

Clerc, M., & Kennedy, J. (2002). The particle swarm - Explosion, stability, and convergence in a multidimensional complex space. *IEEE Transactions on Evolutionary Computation*, *6*(1), 58–73. doi:10.1109/4235.985692

Dorigo, M., & Colombetti, M. (1998). *Robot shaping: An experiment in behavior engineering*. Cambridge, MA: MIT Press/Bradford Books. doi:10.1177/105971239700500308

Dorigo, M., & Stützle, T. (2004, July). *Ant colony optimization*. Cambridge, MA: MIT Press/Bradford Books.

Dutta Roy, S. C. (1982). Rational approximation of some irrational functions through a flexible continued fraction expansion. *Proceedings of the IEEE*, *70*(1), 84–85. doi:10.1109/PROC.1982.12233

Eberhart, R. C., & Shi, Y. (2000, 16-19 July). Comparing inertia weights and constriction factors in particle swarm optimization. In *Proceedings of the IEEE International Congress on Evolutionary Computation* (Vol. 1, pp. 84-88). San Diego, CA, USA.

Groß, R., Bonani, M., Mondada, F., & Dorigo, M. (2006). Autonomous self-assembly in swarm-bots. *IEEE Transactions on Robotics*, *22*(6), 1115–1130. doi:10.1109/TRO.2006.882919

Hartley, T. T., Lorenzo, C. F., & Qammar, H. K. (1995). Chaos in fractional order Chua's system. *IEEE Transactions on Circuits & Systems – Part I*, *42*(8), 485–490. doi:10.1109/81.404062

Hedar, A.-R., & Fukushima, M. (2006). Tabu search directed by direct search methods for nonlinear global optimization. *European Journal of Operational Research*, *170*(2), 329–349. doi:10.1016/j.ejor.2004.05.033

Kennedy, J., & Eberhart, R. C. (1995, 27 November - 1 December). Particle swarm optimization. In *Proceedings of the IEEE International Conference on Neural Networks* (Vol. 4, pp. 1942-1948). Perth, Australia.

Khovanskii, A. N. (1963). *The application of continued fractions and their generalizations to problems in approximation theory*. Groningen, The Netherlands: P. Noordhoff N. V.

Khovanskii, A. N. (1965). Continued fractions. In L. A. Lyusternik & A. R. Yanpol'skii (Eds.), *Mathematical analysis: Functions, limits, series, continued fractions* (D. E. Brown, Trans.). International Series Monographs in Pure and Applied Mathematics. Oxford, UK: Pergamon Press.

Laakso, T. I., Välimäki, V., Karjalainen, M., & Laine, U. K. (1996). Splitting the unit delay: Tool for fractional delay filter design. *IEEE Signal Processing Magazine*, 30–60. doi:10.1109/79.482137

Lurie, B. J. (1994). Three-parameter tunable tilt-integral-derivative (TID) controller. US Patent US5371670. Alexandria, VA: United States Patent and Trademark Office (USPTO).

Lurie, B. J., & Enright, P. J. (2000). *Classical feedback control with Matlab. Control Engineering Series*. New York, NY: Marcel Dekker, Inc.

Maione, G. (2006). Concerning continued fractions representation of noninteger order digital differentiators. *IEEE Signal Processing Letters*, *13*(12), 725–728. doi:10.1109/LSP.2006.879866

Maione, G. (2008). Continued fractions approximation of the impulse response of fractional order dynamic systems. *IET Control Theory & Applications*, *2*(7), 564–572. doi:10.1049/iet-cta:20070205

Maione, G. (2011, March). High-speed digital realizations of fractional operators in the delta domain. *IEEE Transactions on Automatic Control*, *56*(3), 697–702. doi:10.1109/TAC.2010.2101134

Maione, G. (2011, 28 August - 2 September). Conditions for a class of rational approximants of fractional differentiators/integrators to enjoy the interlacing property. In S. Bittanti, A. Cenedese, & S. Zampieri (Eds.), *Preprints of the 18th IFAC World Congress* (pp. 13984-13989). Milan, Italy.

Manabe, S. (1961). The non-integer integral and its application to control systems. *Japanese Institute of Electrical Engineers Journal*, *6*(3/4), 83–87.

Matsuda, K., & Fujii, H. (1993). H∞ optimized wave-absorbing control: Analytical and experimental results. *Journal of Guidance, Control, and Dynamics*, *16*(6), 1146–1153. doi:10.2514/3.21139

Meng, L., & Xue, D. (2009, 9-11 December). Automatic loop shaping in fractional-order QFT controllers using particle swarm optimization. In *Proceedings of the IEEE International Conference on Control and Automation* (pp. 2182-2187). Christchurch, New Zealand.

Oustaloup, A., Cois, O., Lanusse, P., Melchior, P., Moreau, X., & Sabatier, J. (2006, 19-21 July). The CRONE approach: Theoretical developments and major applications. In *Proceedings of the 2nd IFAC Workshop on Fractional Differentiation and its Applications*, Porto, Portugal.

Oustaloup, A., Levron, F., Mathieu, B., & Nanot, F. M. (2000). Frequency band complex noninteger differentiator: Characterization and synthesis. *IEEE Transactions on Circuits and Systems I, Fundamental Theory and Applications, 47*(1), 25–39. doi:10.1109/81.817385

Oustaloup, A., Mathieu, B., & Lanusse, P. (1995). The CRONE control of resonant plants: Application to a flexible transmission. *European Journal of Control, 1*(2), 113–121.

Oustaloup, A., Moreau, X., & Nouillant, M. (1996). The CRONE suspension. *Control Engineering Practice, 4*(8), 1101–1108. doi:10.1016/0967-0661(96)00109-8

Oustaloup, A., Sabatier, J., Lanusse, P., Malti, R., Melchior, P., Moreau, X., & Moze, M. (2008, 6-11 July). An overview of the CRONE approach in system analysis, modeling and identification, observation and control. In *Proceedings of the 17th IFAC World Congress* (pp. 14254-14265). Seoul, Korea.

Podlubny, I. (1999, January). Fractional-order systems and $PI^\lambda D^\mu$-controllers. *IEEE Transactions on Automatic Control, 44*(1), 208–214. doi:10.1109/9.739144

Podlubny, I., Petráš, I., Vinagre, B. M., O'Leary, P., & Dorčák, L. (2002). Analogue realizations of fractional-order controllers. *Nonlinear Dynamics, 29*(1-4), 281–296. doi:10.1023/A:1016556604320

Shi, Y., & Eberhart, R. C. (1998, 4-9 May). A modified particle swarm optimizer. In *Proceedings of the IEEE World Congress on Computational Intelligence, The 1998 IEEE International Conference on Evolutionary Computation* (pp. 69-73). Anchorage, AK, USA.

Shi, Y., & Eberhart, R. C. (1998, 25-27 March). Parameter selection in particle swarm optimization. In V. W. Porto, N. Saravanan, D. E. Waagen, & A. E. Eiben (Eds.), *Proceedings of the 7th International Conference on Evolutionary Programming, Lecture Notes in Computer Science, Vol. 1447,* San Diego, CA, (pp. 591-600). Berlin, Germany: Springer-Verlag.

Shi, Y., & Eberhart, R. C. (1999, 6-9 July). Empirical study of particle swarm optimization. In *Proceedings of the IEEE Congress on Evolutionary Computation* (Vol. 3, pp. 1945-1950). Washington DC, USA.

Spanier, J., & Oldham, K. B. (1987). *An atlas of functions*. New York, NY: Hemisphere Publishing Co.

Tenreiro Machado, J. A. (1997). Analysis and design of fractional-order digital control systems. *Systems Analysis Modeling and Simulation, 27*(2-3), 107–122.

Tenreiro Machado, J. A. (2001). Discrete-time fractional-order controllers. *Fractional Calculus & Applied Analysis, 4*(1), 47–66.

Tenreiro Machado, J. A. (2010). Optimal tuning of fractional controllers using genetic algorithms. *Nonlinear Dynamics, 62*, 447–452. doi:10.1007/s11071-010-9731-5

Tenreiro Machado, J. A., & Galhano, A. M. (2009). Approximating fractional derivatives in the perspective of system control. *Nonlinear Dynamics, 56*(4), 401–407. doi:10.1007/s11071-008-9409-4

Tenreiro Machado, J. A., Galhano, A. M., Oliveira, A. M., & Tar, J. K. (2010, March). Optimal approximation of fractional derivatives through discrete-time fractions using genetic algorithms. *Communications in Nonlinear Science and Numerical Simulation, 15*(3), 482–490. doi:10.1016/j.cnsns.2009.04.030

Trianni, V., Nolfi, S., & Dorigo, M. (2006). Cooperative hole avoidance in a swarm-bot. *Robotics and Autonomous Systems, 54*(2), 97–103. doi:10.1016/j.robot.2005.09.018

Tricaud, C., & Chen, Y. Q. (2009, 10-12 June). Solution of fractional order optimal control problems using SVD-based rational approximations. In *Proceedings of the 2009 American Control Conference,* Hyatt Regency Riverfront, St. Louis, MO, USA (pp. 1430-1435).

Tseng, C. C. (2001). Design of fractional order digital FIR differentiators. *IEEE Signal Processing Letters, 8*(3), 77–79. doi:10.1109/97.905945

Tustin, A. (1958). The design of systems for automatic control of the position of massive objects. *The Proceedings of the Institution of Electrical Engineers, 105*(Part C, Suppl. No. 1), 1-57.

Välimäki, V., & Laakso, T. I. (2000, 5-9 June). Principles of fractional delay filters. In *Proceedings of the IEEE International Conference on Acoustics, Speech, and Signal Processing (ICASSP '00)* (Vol. 6, pp. 3870-3873). Istanbul, Turkey.

van Ast, J., Babuška, R., & De Schutter, B. (2008, 6-11 July). Particle swarms in optimization and control. In *Proceedings of the 17th IFAC World Congress* (pp. 5131-5136). Seoul, Korea.

van den Bergh, F., & Engelbrecht, A. P. (2006). A study of particle swarm optimization particle trajectories. *Information Sciences, 176,* 937–971. doi:10.1016/j.ins.2005.02.003

Villanova, R., & Visioli, A. (Eds.). (2012). *PID control in the third millennium – Lessons learned and new approaches.* London, UK: Springer. doi:10.1007/978-1-4471-2425-2

Vinagre, B. M., & Chen, Y. Q. (2002, 10-13 December). Fractional calculus applications in automatic control and robotics. In *Lecture Notes for Tutorial Workshop #2, 41st IEEE International Conference on Decision and Control* (pp. 1-310). Las Vegas, NV, USA.

Vinagre, B. M., Podlubny, I., Hernandez, A., & Feliu, V. (2000). Some approximations of fractional order operators used in control theory and applications. *Fractional Calculus and Applied Analysis, 3*(3), 231–248.

Westerlund, S., & Ekstam, L. (1994, October). Capacitor theory. *IEEE Transactions on Dielectrics and Electrical Insulation, 1*(5), 826–839. doi:10.1109/94.326654

Zamani, M., Karimi-Ghartemani, M., Sadati, N., & Parniani, M. (2009). Design of a fractional order PID controller for an AVR using particle swarm optimization. *Control Engineering Practice, 17*(12), 1380–1387. doi:10.1016/j.conengprac.2009.07.005

ADDITIONAL READING

Aguirre, C., Campos, D., Pascual, P., & Vázquez, L. (2006, 12-15 September). Computer simulations for a fractional calculus derived Internet traffic model. In B. H. V. Topping, G. Montero, & R. Montenegro (Eds.), *Proceedings of the Fifth International Conference on Engineering Computational Technology,* Las Palmas de Gran Canaria, Spain. Stirlingshire, UK: Civil-Comp Press.

Anastasio, T. J. (1994). The fractional-order dynamics of brainstem vestibulo-oculomotor neurons. *Biological Cybernetics, 72,* 69–79. doi:10.1007/BF00206239

Bagley, R. L., & Calico, R. A. (1991). Fractional-order state equations for the control of viscoelastic damped structures. *Journal of Guidance, Control, and Dynamics, 14*(2), 304–311. doi:10.2514/3.20641

Borredon, L., Henry, B., & Wearne, S. (1999). Differentiating the non-differentiable - Fractional calculus. *Parabola, 35*(2), 9–19.

Caponetto, R., Dongola, G., Fortuna, L., & Petráš, I. (2010). *Fractional order systems: Modeling and control applications.* Singapore: World Scientific.

Caputo, M. (1967). Linear models of dissipation whose Q is almost frequency independent – Part II. *Geophysical Journal of the Royal Astronomical Society, 13*(5), 529–539. doi:10.1111/j.1365-246X.1967.tb02303.x

Chen, Y. Q. (2006, 19-21 July). Ubiquitous fractional order controls? In *Proceedings of the 2nd IFAC Workshop on Fractional Differentiation and its Applications (FDA'06)* (Vol. 2, Part 1, Plenary Lecture), Porto, Portugal: The Institute of Engineering of Porto (ISEP).

Chen, Y. Q. (2009, 10-12 June). Fractional order control - A tutorial. In *Proceedings of the 2009 American Control Conference,* Hyatt Regency Riverfront, St. Louis, MO, USA (pp. 1397-1411).

Duarte, F. B., Tenreiro Machado, J. A., & Monteiro Duarte, G. (2010). Dynamics of the Dow Jones and the NASDAQ stock indexes. *Nonlinear Dynamics, 61*(4), 691–705. doi:10.1007/s11071-010-9680-z

Figueiredo, L., & Tenreiro Machado, J. A. (2007). Simulation and dynamics of freeway traffic. *Nonlinear Dynamics, 49*(4), 567–577. doi:10.1007/s11071-006-9115-z

Gorenflo, R., & Mainardi, F. (1988). Fractional calculus and stable probability distributions. *Archives of Mechanics, 50*(3), 377–388.

Hilfer, R. (2000). *Applications of fractional calculus in physics.* Singapore: World Scientific.

Koeller, R. C. (1984). Application of fractional calculus to the theory of viscoelasticity. *ASME Journal of Applied Mechanics, 51*(2), 299–307. doi:10.1115/1.3167616

Laskin, N. (2000). Fractional market dynamics. *Physica A, 287,* 482–492. doi:10.1016/S0378-4371(00)00387-3

Loverro, A. (2004). *Fractional calculus: History, definitions and applications for the engineer.* South Bend, IN: University of Notre Dame.

Lubich, C. (1986). Discretized fractional calculus. *SIAM Journal on Mathematical Analysis, 17*(3), 704–719. doi:10.1137/0517050

Magin, R. L. (2004). Fractional calculus in bioengineering, part 1. *Critical Reviews in Biomedical Engineering, 32*(1), 1–104. doi:10.1615/CritRevBiomedEng.v32.10

Magin, R. L. (2004). Fractional calculus in bioengineering, part 2. *Critical Reviews in Biomedical Engineering, 32*(2), 105–193. doi:10.1615/CritRevBiomedEng.v32.i2.10

Magin, R. L. (2004). Fractional calculus in bioengineering, part 3. *Critical Reviews in Biomedical Engineering, 32*(3-4), 195–377. doi:10.1615/CritRevBiomedEng.v32.i34.10

Mainardi, F. (1996). Fractional relaxation-oscillation and fractional diffusion wave phenomena. *Chaos, Solitons, and Fractals, 7*(9), 1461–1477. doi:10.1016/0960-0779(95)00125-5

Mainardi, F., & Bonetti, E. (1988). The application of real-order derivatives in linear viscoelasticity. *Rheologica Acta, 26,* 64–67.

Monje, C. A., Chen, Y. Q., Vinagre, B. M., Xue, D., & Feliu, V. (2010). *Fractional-order systems and controls: Fundamentals and applications.* London, UK: Springer-Verlag. doi:10.1007/978-1-84996-335-0

Monje, C. A., Vinagre, B. M., Feliu, V., & Chen, Y. Q. (2008). Tuning and auto-tuning of fractional order controllers for industry applications. *Control Engineering Practice, 16*, 798–812. doi:10.1016/j.conengprac.2007.08.006

Oldham, K. B. (2010, January). Fractional differential equations in electrochemistry. *Advances in Engineering Software, 41*(1), 9–12. doi:10.1016/j.advengsoft.2008.12.012

Oldham, K. B., & Spanier, J. (1974). *The fractional calculus: Integrations and differentiations of arbitrary order.* New York, NY: Academic Press.

Ortigueira, M. D. (2000, February). Introduction to fractional linear systems- Part 1: Continuous-time case. *IEE Proceedings. Vision Image and Signal Processing, 147*(1), 62–70. doi:10.1049/ip-vis:20000272

Ortigueira, M. D. (2000, February). Introduction to fractional linear systems- Part 2: Discrete-time systems. *IEE Proceedings. Vision Image and Signal Processing, 147*(1), 71–78. doi:10.1049/ip-vis:20000273

Ortigueira, M. D. (2008). An introduction to the fractional continuous-time linear systems: The 21st century systems. *IEEE Circuits and Systems Magazine, 8*(3), 19–26. doi:10.1109/MCAS.2008.928419

Ortigueira, M. D. (2011). *Fractional calculus for scientists and engineers. Lecture Notes in Electrical Engineering* (*Vol. 84*). Dordrecht, The Netherlands: Springer Verlag. doi:10.1007/978-94-007-0747-4

Oustaloup, A. (1983). *Systèmes asservis linéaires d'ordre fractionnaire: Théorie et pratique.* Paris, France: Masson.

Oustaloup, A. (1991). *La commande CRONE. Commande robuste d'ordre non entièr.* Paris, France: Editions Hermès.

Oustaloup, A. (1995, February). *La derivation non entière: Théorie, synthèse et applications.* Paris, France: Editions Hermès.

Podlubny, I. (1999). *Fractional differential equations.* San Diego, CA: Academic Press.

Saravanan, R., & McWilliams, J. C. (1998, February). Advective ocean-atmosphere interaction: An analytical stochastic model with implications for decadal variability. *Journal of Climate, 11*, 165–188. doi:10.1175/1520-0442(1998)011<0165:AO AIAA>2.0.CO;2

Tenreiro Machado, J. A. (2002). Special issue on fractional order calculus and its applications. *Nonlinear Dynamics, 29*(1-4), 1–385. doi:10.1023/A:1016508704745

Tenreiro Machado, J. A., Kiryakova, V., & Mainardi, F. (2011, March). Recent history of fractional calculus. *Communications in Nonlinear Science and Numerical Simulation, 16*(3), 1140–1153. doi:10.1016/j.cnsns.2010.05.027

Torvik, P. J., & Bagley, R. L. (1984). On the appearance of the fractional derivative in behaviour of real materials. *Journal of Applied Mechanics, Transactions of the ASME, 51*(2), 294-298.

Vinagre, B. M., Monje, C. A., Calderón, A. J., & Suárez, J. I. (2007, September). Fractional PID controllers for industry application: A brief introduction. *Journal of Vibration and Control, 13*(9-10), 1419–1429. doi:10.1177/1077546307077498

West, B. J., Bologna, M., & Grigolini, P. (2003). *Physics of fractal operators.* New York, NY: Springer-Verlag. doi:10.1007/978-0-387-21746-8

KEY TERMS AND DEFINITIONS

Approximation: In this context, a formal or empirical technique that is necessary to realize the non-integer (fractional) order differential/ integral operators.

Fractional Calculus: A mathematical theory in which calculus is based on non-standard definitions of integrals and derivatives of non-integer, e.g. fractional, order.

Fractional (Non-Integer) Order Differential/Integral Operators: Irrational operators that represent fractional (non-integer) order differentiation or integration in the frequency domain, related to the *s*-variable of the Laplace transform.

Fractional-Order Controllers: Innovative controllers that extend classical and standard PID controllers by replacing integer order with non-integer order integral and derivative control actions.

Interlacing: A property that is verified when zeros and poles of the approximating rational transfer function are alternated along the real axis of the *s*-plane, i.e. each zero follows a pole and vice versa.

Order of Approximation: The number of poles (and zeros) in the rational transfer function that approximates the irrational non-integer (fractional) order differential/integral operator.

ENDNOTES

[1] This work is supported by the Italian Ministry of University and Research (MIUR) under project *"Non integer order systems in modeling and control"*, grant no. 2009F4NZJP.

Chapter 11
Optimal Location of the Workpiece in a PKM–Based Machining Robotic Cell

E.J. Solteiro Pires
Universidade de Trás-os-Montes e Alto Douro, Portugal

J. A. Tenreiro Machado
Instituto Politécnico do Porto, Portugal

António M. Lopes
Universidade do Porto, Portugal

P. B. de Moura Oliveira
Universidade de Trás-os-Montes e Alto Douro, Portugal

ABSTRACT

Most machining tasks require high accuracy and are carried out by dedicated machine-tools. On the other hand, traditional robots are flexible and easy to program, but they are rather inaccurate for certain tasks. Parallel kinematic robots could combine the accuracy and flexibility that are usually needed in machining operations. Achieving this goal requires proper design of the parallel robot. In this chapter, a multi-objective particle swarm optimization algorithm is used to optimize the structure of a parallel robot according to specific criteria. Afterwards, for a chosen optimal structure, the best location of the workpiece with respect to the robot, in a machining robotic cell, is analyzed based on the power consumed by the manipulator during the machining process.

INTRODUCTION

A Parallel Kinematic Manipulator (PKM) is a complex closed-loop mechanism, made by two rigid bodies, or platforms, that are connected by at least two independent open kinematic chains. PKMs have considerable advantages over their serial-based counterparts, such as higher precision and stiffness, and better dynamic characteristics. In the last few years PKMs have been used in several areas, like robotics, motion simulators and machining-tools (Merlet and Gosselin, 1991, Lopes. 2009, Staicu and Zhang, 2008).

Machining has been regarded as a high-precision task, traditionally carried out by dedicated CNC machine-tools. However, there are many

DOI: 10.4018/978-1-4666-2666-9.ch011

tasks where the requirements for precision can be relaxed, such as machining of soft materials (e.g., polymers and composites), machining of large wood or stone workpieces, and end-machining of middle tolerance parts. In this kind of tasks, industrial (serial-based) robots started to be used as alternatives to CNC machine-tools, taking advantage of their flexibility and user-friendly programming interfaces (Olsson et al., 2010, Vosniakos and matsas, 2010, Mitsi et al., 2008, Zaghbani et al., 2011, Lopes and Pires, 2011). When properly designed, PKM-based robots can bring together both accuracy and flexibility that are usually needed in most machining tasks.

For a given PKM configuration, the geometric parameters that define the manipulator's kinematic structure must be determined. This is an extremely important question, since most performance characteristics, like structural stiffness and dexterity, depend on it.

Another important issue in a machining robotic cell is the positioning of the workpiece with respect to the robot. The best position depends on the trajectory and the forces that the robot has to exert on the workpiece during the machining process.

Both problems can be formulated as optimization problems where the objective functions are chosen in order to meet specific performance criteria.

The computational complexity in optimization problems appears in several robotics problems. In fact, we can find that problem in trajectory planning (Chen & Liao, 2011, Pires et al., 2007), robotic structure synthesis (Lopes et al, 2011), and manipulator control (Zi et al., 2008) among many others. Obtaining the optimal solution can result in a significant economic saving, or in the efficient use of scarce resources along the machinery lifetime. Lately, some bioinspired algorithms have been proposed to solve robotic problems. These algorithms represent a valid alternative since some problems are not differentiable and classical methods, such as gradient or derivative methods, can not be used. Another important factor is to solve

the problem while having a limited knowledge about all details. Moreover, a large part of the models used to solve the problems is simplified from real world. Therefore, the resulting solution leads to a poor performance, or is not adequate. The reduction of the objective number is a common practice. To overcome these shortcomings, a multi-objective PSO is used in this study.

Several works using bioinspired algorithms have been proposed in robotics. Barbosa et al. (2010) adopt a genetic algorithm to design a 6-dof parallel robotic manipulator. The dexterity of the robot is optimized using two algorithms: a genetic and a neuro-genetic algorithm. Chen and Pham (2012) propose a multi-objective genetic algorithm trajectory planner for a KPM. In this work is considered the dynamics, the trajectory duration and the required energy. Jamwal et al. (2009) propose a soft parallel robot for ankle joint rehabilitation. They use two genetic algorithms to determine an optimal robot design, adopting the Jacobian matrix as the performance index. Chen and Liao (2011) propose a hybrid scheme for determining the optimal time and the energy efficient trajectory of a parallel platform manipulator. The algorithm uses a combination of PSO and conjugate gradient methods. Lopes et al., (2011) use a MOPSO to optimize, simultaneously in one execution, the dexterity, the stiffness and its isotropy for a parallel robot. In addition, PSO algorithm has been also proposed to find similar ways of coordinating and controlling groups of robots (Sharkey, 2009, Atyabi and Powers, 2010).

Typically, the large number of optimization parameters, the complexity of the objective functions and the number of objectives that usually are involved, makes optimization, in this context, a difficult and time-consuming task. In such a case, evolutionary algorithms can be used as an effective option (Gosselin and Angeles, 1991, Pittens and Podhorodeski, 1993, Chen and Pham, 2012).

In this chapter a PKM-based machining robotic cell is proposed and the PKM manipulator designed according to specific criteria. Given the PKM configuration, the maximum stiffness and dexterity structure is firstly determined. Afterwards, for a chosen optimal structure, the best location of the workpiece with respect to the robot, in a machining robotic cell, is analyzed based on the power consumed by the manipulator during the machining process.

In this work, the PSO is adopted as optimization algorithm, because it is easier to implement and has a faster convergence than other bioinspired algorithms. In fact, the PSO does not includes the crossover and mutation operators, since only one particle (the best) shares information to the other solutions. Therefore, the evolution process search focuses primarily on the best fitness particle leading to a faster convergence of the algorithm.

Bearing these ideas in mind, the book chapter is organized as follows. Section 2 deals with the manipulator modeling. Section 3 presents the objective functions for structural optimization. In section 4, the optimization algorithm is presented. Structural optimization results are given in section 5. The optimal location of the workpiece is discussed in section 6. Finally, conclusions are drawn in section 7.

Describe the general perspective of the chapter. Toward the end, specifically state the objectives of the chapter.

PKM CONFIGURATION AND MODELLING

We adopt a 6-dof variant of the well-known Stewart platform PKM (Merlet and Gosselin, 1991). The mechanical structure includes a fixed (base) platform and a moving (payload) platform, connected by six independent, identical, open kinematic chains (Figure 1). Each chain has two links: the first link is a linear actuator that is always normal to the base and has variable length, l_i (i = 1, …, 6). One of its ends is fixed to the base and the other one is attached, by a universal joint, to the second link. The second link (arm) has a fixed

length, L, and is attached to the payload platform by a spherical joint.

Two frames, {P} and {B}, are attached to the centre of mass of the moving and base platforms, respectively. The generalized position (pose) of frame {P} relative to frame {B} may be represented by the vector:

$$^{B}\mathbf{x}_{P}\big|_{B|E} = \begin{bmatrix} ^{B}\mathbf{x}_{P(pos)}^{T}\big|_{B} & ^{B}\mathbf{x}_{P(o)}^{T}\big|_{E} \end{bmatrix}^{T} \quad (1)$$

where $^{B}\mathbf{x}_{P(pos)}\big|_{B} = \begin{bmatrix} x_{P} & y_{P} & z_{P} \end{bmatrix}^{T}$ is the position of the origin of frame {P} relative to frame {B}, and $^{B}\mathbf{x}_{P(o)}\big|_{E} = \begin{bmatrix} \psi_{P} & \theta_{P} & \phi_{P} \end{bmatrix}^{T}$ defines an Euler angles system representing the orientation of frame {P} relative to {B}. The rotation matrix is related to the Euler angles and is given by:

$$^{B}\mathbf{R}_{P} = \begin{bmatrix} C\psi_{P}C\theta_{P} & C\psi_{P}S\theta_{P}S\phi_{P} - S\psi_{P}C\phi_{P} & C\psi_{P}S\theta_{P}C\phi_{P} + S\psi_{P}S\phi_{P} \\ S\psi_{P}C\theta_{P} & S\psi_{P}S\theta_{P}S\phi_{P} + C\psi_{P}C\phi_{P} & S\psi_{P}S\theta_{P}C\phi_{P} - C\psi_{P}S\phi_{P} \\ -S\theta_{P} & C\theta_{P}S\phi_{P} & C\theta_{P}C\phi_{P} \end{bmatrix}$$
$$(2)$$

$S(\bullet)$ and $C(\bullet)$ correspond to the sine and cosine functions, respectively.

Figure 1. PKM configuration

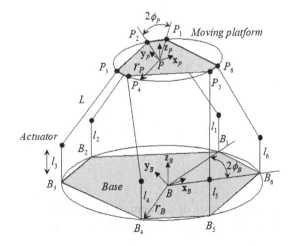

Figure 2. Schematic representation of a kinematic chain

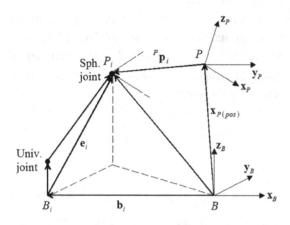

The inverse kinematic jacobian, $\mathbf{J}_C$, relates the velocities of the joints and the velocities of the payload platform (linear and angular):

$$\mathbf{\dot{l}} = \mathbf{J}_C \cdot {}^B\dot{\mathbf{x}}_P \big|_B \tag{3}$$

Vector $\mathbf{\dot{l}} = [\dot{l}_1 \quad \dot{l}_2 \quad \cdots \quad \dot{l}_6]^T$ represents the joints velocities, and vector $ {}^B\dot{\mathbf{x}}_P \big|_B = [{}^B\dot{\mathbf{x}}_{P(pos)}^T \big|_B \quad {}^B\acute{\mathbf{E}}_P^T \big|_B]^T$ represents the Cartesian-space velocities.

The inverse jacobian matrix can be computed using vector algebra and is given by Equation 4.

$$\mathbf{J}_C = \begin{bmatrix} \dfrac{\left(\mathbf{e}_1 - l_1\mathbf{z}_B\right)^T}{\left(\mathbf{z}_B^T\mathbf{e}_1 - l_1\right)} & \dfrac{\left({}^P\mathbf{p}_1\big|_B \times \left(\mathbf{e}_1 - l_1\mathbf{z}_B\right)^T\right)}{\left(\mathbf{z}_B^T\mathbf{e}_1 - l_1\right)} \\ \vdots & \vdots \\ \dfrac{\left(\mathbf{e}_6 - l_6\mathbf{z}_B\right)^T}{\left(\mathbf{z}_B^T\mathbf{e}_6 - l_6\right)} & \dfrac{\left({}^P\mathbf{p}_6\big|_B \times \left(\mathbf{e}_6 - l_6\mathbf{z}_B\right)^T\right)}{\left(\mathbf{z}_B^T\mathbf{e}_6 - l_6\right)} \end{bmatrix} \tag{4}$$

The analysis of a kinematic chain (Figure 2) allows the computation of all vectors that are involved.

Vector $\mathbf{e}_i$ is given by:

$$\mathbf{e}_i = {}^B\mathbf{x}_{P(pos)}\big|_B - \mathbf{b}_i + {}^P\mathbf{p}_i\big|_B \tag{5}$$

where $\mathbf{b}_i$ and $ {}^P\mathbf{p}_i\big|_B$ represent the positions of points B_i and P_i with reference to the base and payload platforms, respectively. Both vectors are expressed in the base frame. The scalar l_i is the displacement of each actuator, given by Equation 6.

$$l_i = e_{iz} - \sqrt{L^2 - e_{ix}^2 - e_{iy}^2} \tag{6}$$

As the angular velocity and the time derivatives of the Euler angles are related by

$$ {}^B\acute{\mathbf{E}}_P\big|_B = \mathbf{J}_A \cdot {}^B\dot{\mathbf{x}}_{P(o)}\big|_E \tag{7}$$

The following equation can be written:

$$\mathbf{\dot{l}} = \mathbf{J}_E \cdot {}^B\dot{\mathbf{x}}_P\big|_{B|E} \tag{8}$$

where $\mathbf{J}_E$ is the Euler angles inverse jacobian matrix, and $\mathbf{J}_A$ is a matrix given by

$$\mathbf{J}_A = \begin{bmatrix} 0 & -S\psi_P & C\theta_P C\psi_P \\ 0 & C\psi_P & C\theta_P S\psi_P \\ 1 & 0 & -S\theta_P \end{bmatrix} \tag{9}$$

The PKM dynamics can be computed using the Lagrange-Euler equation (LE). Using the moving platform pose, $ {}^B\mathbf{x}_P\big|_{B|E}$, as generalized coordinates, The LE equation becomes:

$$\frac{d}{dt}\left(\frac{\partial K\left({}^B\mathbf{x}_P\big|_{B|E}, {}^B\dot{\mathbf{x}}_P\big|_{B|E}\right)}{\partial {}^B\dot{\mathbf{x}}_P\big|_{B|E}}\right) - \frac{\partial K\left({}^B\mathbf{x}_P\big|_{B|E}, {}^B\dot{\mathbf{x}}_P\big|_{B|E}\right)}{\partial {}^B\mathbf{x}_P\big|_{B|E}} + \frac{\partial P\left({}^B\mathbf{x}_P\big|_{B|E}\right)}{\partial {}^B\mathbf{x}_P\big|_{B|E}} = {}^P\mathbf{f}\big|_{B|E} \tag{10}$$

where K and P are the total kinetic and potential energies, and represents the generalized force applied on the centre of mass of the moving platform.

Using the Euler angles inverse jacobian, the actuators forces, $\boldsymbol{\tau}$, may be computed using the statics equation:

$$\ddot{\mathbf{A}} = \mathbf{J}_E^{-T} \cdot {}^P\mathbf{f}\Big|_{B|E} \tag{11}$$

STRUCTURAL OPTIMIZATION

An important requirement in robotic machining is structural stiffness, which means the manipulator's ability to resist deformation due to the external forces' action. Lower stiffness affects precision, limits maximum velocities and induces manipulator vibrations.

At any pose the stiffness of the robot can be characterized by the stiffness matrix, which relates the Cartesian forces and torques, applied on the end-effector, to the corresponding linear and angular displacements.

Giving the Equation 3 we may write

$$\Delta \mathbf{l} \approx \mathbf{J}_C \cdot {}^B\Delta\mathbf{x}_P\Big|_B \tag{12}$$

where $\Delta \mathbf{l}$ represents an infinitesimal displacement of the actuators and ${}^B\Delta\mathbf{x}_P\big|_B$ represents the corresponding displacement in the Cartesian space. Moreover, using the duality between differential kinematics and static we get

$$ {}^P\mathbf{f}\Big|_B = \mathbf{J}_C^T \cdot \ddot{\mathbf{A}} \tag{13}$$

Figure 3. Standard PSO algorithm

```
t = 0
Initialize swarm X(t)
for each particle
  Evaluate fitness function of xᵢ(t)
  if fitness(xᵢ(t)) > fitness(pᵢ)
          pᵢ = xᵢ(t)
  end if
end for
choose the particle with the best fitness value of all the particles as the pₘ


while ( !(termination criteria) )
  for each particle i (and parameter d)
    evaluate particle velocity according equation (20)
    update particle position according equation (21)
  end for

  for each particle
    Evaluate fitness function of xᵢ(t)
    if fitness(xᵢ(t)) > fitness(pᵢ)
            pᵢ = xᵢ(t)
    end if
  end for
  choose the particle with the best fitness value of all the particles as the pₘ
end while
```

Figure 4. Standard MOPSO algorithm

$t = 0$
initialize swarm $X(t)$
evaluate $X(t)$
generate archive $A(t)$
while(!(termination criterion))
    $t = t+1$
    select guiding particles from archive
    % generate $X(t+1)$
    update particles velocity
    update particles position
    update $p_i(t)$ for all particles
    evaluate $X(t+1)$
    update $A(t)$
end while

where $^{P}\mathbf{f}\big|_{B}$ represents the force and torque applied on the payload platform, and $\boldsymbol{\tau}$ represents the forces on the actuators.

Adopting k_i as the stiffness of each actuator, the corresponding force, τ_i, and displacement, Δl_i, are related by $\tau_i = k_i \cdot \Delta l_i$. Considering that all actuators are identical, the following equation can be written:

$$\boldsymbol{\tau} = \mathbf{K} \cdot \Delta \mathbf{l} \qquad (14)$$

with $\mathbf{K} = \mathrm{diag}\left(\begin{bmatrix} k_1 & k_2 & \cdots & k_6 \end{bmatrix}^T\right)$ representing the stiffness matrix expressed in the space of the actuators.

Using Equations 12 to 14, results in:

$$^{P}\mathbf{f}\big|_{B} = \mathbf{J}_C^T \cdot \mathbf{K} \cdot \mathbf{J}_C \cdot \Delta^{B}\mathbf{x}_P\big|_{B} \qquad (15)$$

where $\mathbf{K}\big|_{B} = \mathbf{J}_C^T \cdot \mathbf{K} \cdot \mathbf{J}_C$ is the Cartesian-space stiffness matrix.

In the particular case of identical actuators, then $k_i = k$ and

$$^{P}\mathbf{f}\big|_{B} = k \cdot \mathbf{J}_C^T \cdot \mathbf{J}_C \cdot \Delta^{B}\mathbf{x}_P\big|_{B} \qquad (16)$$

$$\mathbf{K}\big|_{B} = k \cdot \mathbf{J}_C^T \cdot \mathbf{J}_C \qquad (17)$$

Mathematically, the objective function that is adopted to quantify the stiffness of the PKM is given by the trace of matrix $\mathbf{K}\big|_{B}$, as expressed by Equation 18.

$$v = \mathrm{trace}\left\{k \cdot \mathbf{J}_C^T \cdot \mathbf{J}_C\right\} \qquad (18)$$

Another important objective is dexterity, which measures the manipulator's ability to exert force and execute movements in any direction. Mathematically, the condition number of $\mathbf{J}_C$ is used to quantify dexterity as given by:

$$\kappa = \frac{\sigma_{\max}\left(\mathbf{J}_C\right)}{\sigma_{\min}\left(\mathbf{J}_C\right)} \qquad (19)$$

where $\sigma_{\max}\left(\mathbf{J}_C\right)$ and $\sigma_{\min}\left(\mathbf{J}_C\right)$ represent the maximum and minimum singular values of $\mathbf{J}_C$.

In this work a multi-criteria optimization problem is formulated. Accordingly, the manipulator will be designed for maximum dexterity and maximum stiffness. This means that we have to maximize the objective function given by Equation 18 and minimize the one given by Equation 19.

Firstly, in order to obtain a dimensionally homogeneous $\mathbf{J}_C$, the manipulator payload platform radius, r_p, is used as a characteristic length. Thus, the Jacobian results dependent upon ten variables, four of them are manipulator kinematic parameters: the position and orientation of the payload platform; the base radius (r_B); the separation angles on the payload platform (ϕ_p); the separation angles on the base (ϕ_B); and the arm length (L).

We will consider a particular manipulator pose, corresponding to the centre of the manipulator workspace *i.e.*, $(0\ 0\ 2\ 0\ 0\ 0)^T$ (units in r_p and degrees, respectively). Thus, for this pose, $\mathbf{J}_C$ will be a function of the four kinematic parameters.

PARTICLE SWARM OPTIMIZATION

Particle swarm optimization (PSO) is a popular technique used in a large number of optimization and search problems. Its operating principle is inspired in social behaviors found among bird flocks or fish schools. PSO is considered a metaheuristic that was developed by Kennedy and Eberhart (1995).

The PSO popularity, like evolutionary computation, is due essentially to work in several problems where it makes few or no assumptions about the problem to solve. Moreover, it is not required any prior knowledge of the search space. The PSO algorithm is a population based technique where a set of potential solutions evolves in order to approach to better regions of the search space during a predefined number of iterations. The swarm evolution is governed by a few simple equations (Equations 20-21), guided by its best position, p_i, found so far, and the best particle position, p_g, of the set. These guidelines represent the personal and social experience of each particle. The parameters c_1 and c_2 represent the stochastic weight terms for local best and global best, respectively. The variables φ_1 and φ_2 are random numbers in the range (0, 1). And the inertial weight, w, introduced later in the algorithm is used to prevent the swarm global exploration. Equations 20-21 are applied to all d parameters of the problem.

$$v_{id} = w \cdot v_{id} + c_1 \cdot \phi_1 \left(p_{id} - x_{id} \right) + c_2 \cdot \phi_2 \left(p_{gd} - x_{id} \right) \tag{20}$$

$$x_{id} = x_{id} + v_{id} \tag{21}$$

At the end of the algorithm, it is expected to find the optimal or at least a good solution. However, the algorithm does not guarantee that the optimal solution is found.

The simple PSO algorithm is illustrated in Figure 3.

In the same manner as Genetic Algorithms (GA) was applied to multi-objective problems (MOGA), the PSO has also been applied successfully in this type of problems (Reyes-Sierra and Coello, Pires et al.). Besides that, the basic principles used in the MOGA were successful transposed to PSO.

A standard multi-objective PSO (MOPSO) follows the algorithm presented in Figure 4. The swarm, X, can be initialized in several ways using a priori knowledge or totally randomly.

The MOPSO reported here has two minor changes when compared to the originally PSO algorithm proposed for single-objective problems. The first is the introduction of a population archive that holds the best solutions found so far, that describes the Pareto front. The aim of the archive

Figure 5. Swarm and archive evolution over one iteration

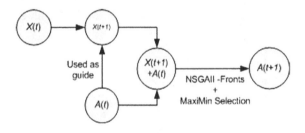

Figure 6. Achieve update

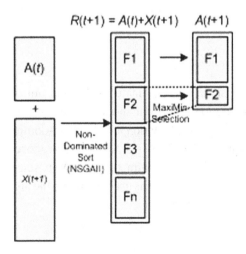

Figure 7. Maximin selection

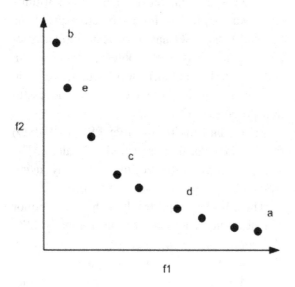

Figure 8. Selection of the global guide particle

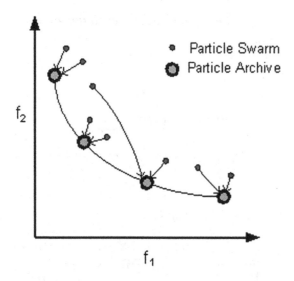

Therefore, the MOPSO uses an archive, A, in which the best non-dominated swarm solutions are kept in every iteration, according to a sorted ranking scheme. In each iteration, t, the swarm $X(t+1)$, and archive $A(t)$, are merged into a set $R(t+1)$. The solutions are evaluated in terms of Pareto domination and sorted according to a non-dominated front ranking scheme. The new archive, $A(t+1)$ is obtained selecting solutions from $R(t+1)$, using both the non-dominated front ranking (Deb, 2001) as well as its diversity evaluated using the MaxiMin sorting scheme proposed by Solteiro Pires et al. (2005) (See Figure 6). The overall objective of MaxiMin is to achieve a non-dominated Pareto front as wide as possible and uniformly distributed.

The main idea behind the MaxiMin selection is to choose the solutions in order to decrease the large gap areas existing in the already selected population. To illustrate the algorithm, consider the non-dominated solutions in Figure 7. At the algorithm begin, the extreme solutions, of each objective, are selected $S \equiv \{a, b\}$. Then, solution c is selected because it has the greater distance to the set S. After that, solutions d and e are selected into the set $S \equiv \{a, b, c\}$, for the same reason. The process is repeated until the S set is completed.

The second change introduced has to do with the choice of the guide for each solution. While in the uni-objective PSO, the global best particle of the swarm (or neighborhood) is trivial to identify and can be unique for all the swarm, this is not the case in the MOPSO.

In the MOPSO used here (Oliveira et al., 2009) each global archive particle can be selected by a particle as its global guide a predefined maximum number of times. For instance, if the swarm size is 100 and the archive size is 50 the maximum number of times that each archive member can be selected as guide is defined as 2. Therefore, the two swarm particles that are closest to each archive particle "follow" it. The remaining swarm particles, having no guide, select randomly a guide that has not the

is to store non-dominated solutions uniformly spread along the front and a wide front. For this purpose, an adaptation of the Non-dominated Sorting GA, (NSGA-II) technique proposed by Deb et al. (2001) with the MaxiMin selection (Pires et al., 2005) was used to update the archive A (See Figure 5).

total number of followers. This methodology has the advantage to force some particles to select as global guiding particles which are far away than others already selected by neighbors.

In the example shown in Figure 8, applying this technique will force one of the particles to choose a guide which are further away from it. This motivates the exploration of the search space in order to achieve a wide Pareto front and with the MaxiMin algorithm, uniformly spread.

STRUCTURAL OPTIMIZATION RESULTS

This section presents the experimental results when the κ and v optimization functions (equations (18) and (19)) are simultaneously considered. Four parameters $\{r_B, \phi_P, \phi_B, L\}$ are used in the MPSO particles. These parameters can vary in the intervals given by Table 1.

Figure 9 shows the Pareto front obtained at the end of 500 MPSO iterations. The front is basically obtained by varying the parameter r_B in the range [1.771 r_P, 2.303 r_P]. The other parameters practically remain the same around the values $\phi_P = 0$, $\phi_B = 0$, and L = 2.067 r_P].

WORKPIECE OPTIMAL LOCATION

The mechanical power dissipated by the robot actuators, P_{act}, along the machining trajectory is adopted as one objective function. Mathematically, it is given by (22):

$$P_{act} = \sqrt{\sum_{k=1}^{K}\sum_{i=1}^{6}\left[\tau_i(k) \cdot \dot{l}_i(k)\right]^2} \qquad (22)$$

where K is the total number of points of the discretized trajectory, which depends on the discretizing period and the trajectory time length.

Table 1. Optimization parameters range

Parameter	Minimum	Maximum
r_B	1 r_P	2.5 r_P
ϕ_P	0	20°
ϕ_B	0	20°
L	2 r_P	4.5 r_P

Figure 9. Pareto optimal front for $\kappa \times v$ optimization

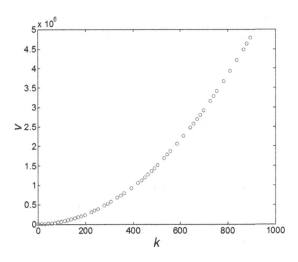

Figure 10. Pareto front of optimal solutions

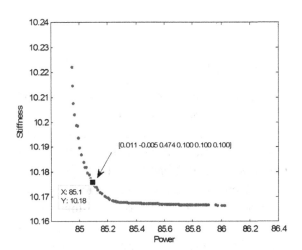

Table 2. Kinematic and dynamic parameters of the robot

r_P	0.2500 m	m_P	5.9555 kg
r_B	0.5000 m	m_L	1.4851 kg
ϕ_P	0°	m_A	1.3161 kg
ϕ_B	0°	IP_{xx}	0.0859 kg·m2
L	0.5168 m	IP_{yy}	0.0801 kg·m2
b_{cm}	0.3000 m	IP_{zz}	0.1583 kg·m2

Table 3. Optimization parameters range

Parameter	Minimum	Maximum
x_W	-0.1 m	0.1 m
y_W	-0.1 m	0.1 m
z_W	0.4 m	0.6 m
ψ_W	-0.6°	0.6°
θ_W	-0.3°	0.3°
φ_W	-0.3°	0.3°

A second objective is the manipulator's stiffness along the trajectory. This means that the objective function given by Equations 18 and 22 have to be maximized and minimized, respectively.

The robot has to follow a trajectory on the surface of a workpiece, exerting a constant normal force, F_N.

The coordinates of the workpiece are given by the vector

$$^B\mathbf{x}_W\big|_{B|E} = \begin{bmatrix} x_W & y_W & z_W & \psi_W & \theta_W & \phi_W \end{bmatrix}^T,$$

with reference to the base of the robot, and constitute the optimization parameters. The trajectory of the robot is specified in terms of the position, velocity and acceleration with respect to the workpiece's local frame {W}. The trajectory is then computed with respect to the base of the robot and the optimization algorithm is run.

We illustrate the approach with a simple example. The considered workpiece is a cone and the manipulator has to follow a trajectory on its

surface, with $F_N = 200$N. The parametric equations of the trajectory with reference to {W} are:

$$
\begin{aligned}
x(t) &= a \cdot t \cdot \cos(b \cdot t) \\
y(t) &= a \cdot t \cdot \sin(b \cdot t) \\
z(t) &= b \cdot t \\
\psi(t) &= 0 \\
\theta(t) &= 0 \\
\phi(t) &= 0
\end{aligned}
\tag{23}
$$

with the constants $a = 0.001$ and $b = 0.01$.

The corresponding homogeneous matrix is

$$^W\mathbf{T}\big|_P = \begin{bmatrix} \mathbf{I} & {}^W\mathbf{x}\big|_P \\ \mathbf{0} & 1 \end{bmatrix} \tag{24}$$

where $^W\mathbf{x}\big|_P = \begin{bmatrix} x(t) & y(t) & z(t) \end{bmatrix}^T$.

The trajectory of the robot with respect to the base frame, {B}, expressed in homogeneous coordinates will be:

$$^B\mathbf{T}\big|_P = {}^B\mathbf{T}\big|_W \cdot {}^W\mathbf{T}\big|_P \tag{25}$$

where,

$$^B\mathbf{T}\big|_W = \begin{bmatrix} {}^B\mathbf{R}\big|_W & {}^B\mathbf{x}\big|_W \\ \mathbf{0} & 1 \end{bmatrix} \tag{26}$$

$$^B\mathbf{R}_W = \begin{bmatrix} C\psi_W C\theta_W & C\psi_W S\theta_W S\phi_W - S\psi_W C\phi_W & C\psi_W S\theta_W C\phi_W + S\psi_W S\phi_W \\ S\psi_W C\theta_W & S\psi_W S\theta_W S\phi_W + C\psi_W C\phi_W & S\psi_W S\theta_W C\phi_W - C\psi_W S\phi_W \\ -S\theta_W & C\theta_W S\phi_W & C\theta_W C\phi_W \end{bmatrix} \tag{27}$$

$$^B\mathbf{x}\big|_W = \begin{bmatrix} x_W & y_W & z_W \end{bmatrix}^T \tag{28}$$

$^B\mathbf{x}\big|_W$ and $^B\mathbf{R}_W$ are the position and the orientation of the frame {W} with respect to {B}, respectively.

The kinematic and dynamic parameters of the robot are shown in Table 2. Where, b_{cm} represents the location of the centre of mass of the arms, m_p, m_L and m_A are the masses of the payload platform, arms and actuators, respectively, and I_{Pxx}, I_{Pyy} and I_{Pzz} represent the moments of inertia of the payload platform.

The MOPSO optimization algorithm described in Section 4 is used. The search is carried out by a 200 and 100 elements for the population and for the archive, respectively, during 500 iterations. The project parameters can change in the range shown by the Table 3.

Several independent optimizations were performed and the algorithm always converged to the same nondominated front. Figure 10 illustrates one of the experiments carried one, underlining one possible solution and the respective values of the objective functions. As can be seen, the two objectives are quarrelsome and several alternative solutions can be chosen as the desired workpiece position.

CONCLUSION

PKM-based robots can perform tasks that require high accuracy and are very flexible. However, the design and optimization is a very complex task, especially where it is required the definition of the geometric parameters of the kinematic structure and the operational trajectory. In this paper two MOPSO are used in a PKM robot in order to define the structural parameters of the parallel robot and to find out the operational trajectory. In both optimizations two objectives were used and both problems have revealed quarrelsome. Using a multi-objective optimization allows the decision maker to decide at a higher-level. Thus, the decision maker is aware of the final performance that will get in each of the objectives. On the other hand, if a weighted uni-objective was used, some optimization experiments had to be performed to get the desired trade-off between the conflicting

objectives. The described algorithm was generic and can be used with other optimization criteria.

REFERENCES

Atyabi, A., & Powers, D. M. W. (2010). The use of area extended particle swarm optimization (AEPSO) in swarm robotics. *11th International Conference on Control Automation Robotics & Vision (ICARCV 2010)*, (pp. 591–596).

Barbosa, M. R., Solteiro Pires, E. J., & Lopes, A. M. (2010). Optimization of parallel manipulators using evolutionary algorithms. *Soft Computing Models in Industrial and Environmental Applications, 5th International Workshop* (SOCO 2010), Guimarães, Portugal, (pp. 79-86).

Chen, C., & Pham, H. (2012). Trajectory planning in parallel kinematic manipulators using a constrained multi-objective evolutionary algorithm. *Nonlinear Dynamics, 67*, 1669–1681. doi:10.1007/s11071-011-0095-2

Chen, C. T., & Liao, T. T. (2011). A hybrid strategy for the time- and energy-efficient trajectory planning of parallel platform manipulators. *Robotics and Computer-integrated Manufacturing, 27*(1), 72–81. doi:10.1016/j.rcim.2010.06.012

de Moura Oliveira, P. B., Solteiro Pires, E. J., Cunha, J. B., & Vrancic, D. (2009, June). *Multiobjective particle swarm optimization design of PID controllers. 10th International Work-Conference on Artificial Neural Networks, IWANN 2009 Workshops*, Salamanca, Spain, (pp. 1222-1230).

Deb, K. (2001). *Multi-objective optimization using evolutionary algorithms*. Wiley-Interscience Series in Systems and Optimization.

Gosselin, C., & Angeles, J. (1991). A global performance index for the kinematic optimization of robotic manipulators. *ASME Journal of Mechanical Design, 13*, 220–226. doi:10.1115/1.2912772

Jamwal, P. K., Xie, S., & Aw, K. C. (2009). Kinematic design optimization of a parallel ankle rehabilitation robot using modified genetic algorithm. *Robotics and Autonomous Systems, 57*(10), 11018–11027. doi:10.1016/j.robot.2009.07.017

Kennedy, J., & Eberhart, R. C. (1995). Particle swarm optimization. *IEEE International Conference on Neural Networks,* Perth, Australia, (pp. 1942-1948).

Lopes, A. M. (2009). Dynamic modeling of a Stewart platform using the generalized momentum approach. *Communications in Nonlinear Science and Numerical Simulation, 14,* 3389–3401. doi:10.1016/j.cnsns.2009.01.001

Lopes, A. M., Freire, H., de Moura Oliveira, P. B., & Solteiro Pires, E. J. (2011). *Multi-objective optimization of a parallel robot using particle swarm algorithm.* Paper presented at the meeting of the 7th European Nonlinear Dynamics Conference (ENOC 2011), Rome, Italy, 2011.

Lopes, A. M., & Solteiro Pires, E. J. (2011). Optimization of the workpiece location in a machining robotic cell. *International Journal of Advanced Robotic Systems, 8,* 37–46.

Merlet, J.-P., & Gosselin, C. (1991). Nouvelle architecture pour un manipulateur parallele a six degres de liberte. *Mechanism and Machine Theory, 26,* 77–90. doi:10.1016/0094-114X(91)90023-W

Mitsi, S., Bouzakis, K. D., Sagris, D., & Mansour, G. (2008). Determination of optimum robot base location considering discrete end-effector positions by means of hybrid genetic algorithm. *Robotics and Computer-integrated Manufacturing, 24,* 50–59. doi:10.1016/j.rcim.2006.08.003

Olsson, T., Haage, M., Kihlman, H., Johansson, R., Nilsson, K., & Robertsson, A. (2010). Cost-efficient drilling using industrial robots with high-bandwidth force feedback. *Robotics and Computer-integrated Manufacturing, 26,* 24–38. doi:10.1016/j.rcim.2009.01.002

Pittens, K., & Podhorodeski, R. (1993). A family of stewart platforms with optimal dexterity. *Journal of Robotic Systems, 10,* 463–479. doi:10.1002/rob.4620100405

Reyes-Sierra, M., & Coello, C. A. C. (2006). Multi-objective particle swarm optimizers: A survey of the state-of-the-art. *International Journal of Computer Intelligence Research, 2,* 287–30.

Sharkey, A. J. C. (2009). Swarm robotics. In Rabunal, J. R., Dorado, J., & Pazos, A. (Eds.), *Encyclopedia of artificial intelligence* (pp. 1537–1542).

Solteiro Pires, E. J., de Moura Oliveira, P. B., & Tenreiro Machado, J. A. (2005, March). Multi objective MaxiMin sorting scheme. *Evolutionary Multi-Criterion Optimization,* (pp. 165-175). Guanajuato, Mexico.

Solteiro Pires, E. J., de Moura Oliveira, P. B., & Tenreiro Machado, J. A. (2007). Manipulator trajectory planning using a MOEA original research article. *Applied Soft Computing, 7*(3), 659–667. doi:10.1016/j.asoc.2005.06.009

Staicu, S., & Zhang, D. (2008). A novel dynamic modelling approach for parallel mechanisms analysis. *Robotics and Computer-integrated Manufacturing, 24,* 167–172. doi:10.1016/j.rcim.2006.09.001

Vosniakos, G., & Matsas, E. (2010). Improving feasibility of robotic milling through robot placement optimization. *Robotics and Computer-integrated Manufacturing, 26,* 517–525. doi:10.1016/j.rcim.2010.04.001

Zaghbani, I., Lamraoui, M., Songmene, V., Thomas, M., & El Badaoui, M. (2011). Robotic high speed machining of aluminium alloys. *Advanced Materials Research, 188,* 584–589. doi:10.4028/www.scientific.net/AMR.188.584

Zi, B., Duan, B. Y., Du, J. L., & Bao, H. (2008). Dynamic modeling and active control of a cable-suspended parallel robot. *Mechatronics, 18*(1), 1–12. doi:10.1016/j.mechatronics.2007.09.004

ADDITIONAL READING

Baumgartner, U., Magele, C., & Renhart, W. (2004). Pareto optimality and particle swarm optimization. *IEEE Transactions on Magnetics, 40*(2), 1172–1175. doi:10.1109/TMAG.2004.825430

Boudreau, R., & Gosselin, C. M. (1999). The synthesis of planar parallel manipulators with a genetic algorithm. *ASME Journal of Mechanical Design, 121*, 533–537. doi:10.1115/1.2829494

Chablat, D., Wenger, P., Majou, F., & Merlet, J.-P. (2004). An interval analysis based study for the design and the comparison of three-degrees-of-freedom parallel kinematic machines. *The International Journal of Robotics Research, 23*, 615–624. doi:10.1177/0278364904044079

Coello Coello, C. A., & Salazar Lechuga, M. (2002). MOPSO: A proposal for multiple objective particle swarm optimization. In *Congress on Evolutionary Computation (CEC'2002)*, Vol. 2, (pp. 1051-1056). Piscataway, New Jersey.

CoelloCoello. C. A., Lamont, G. B., & van Veldhuizen, D. A. (2007). *Evolutionary algorithms for solving multi-objective problems*, 2nd ed. Springer.

Gao, Z., Zhang, D., & Ge, Y. (2010). Design optimization of a spatial six degree-of-freedom parallel manipulator based on artificial intelligence approaches. *Robotics and Computer-integrated Manufacturing, 26*(2), 180–189. doi:10.1016/j.rcim.2009.07.002

Gosselin, C. M., & Guillot, M. (1991). The synthesis of manipulators with prescribed workspace. *ASME Journal of Mechanical Design, 113*, 451–455. doi:10.1115/1.2912804

Hay, A. M., & Snyman, J. A. (2004). Methodologies for the optimal design of parallel manipulators. *International Journal for Numerical Methods in Engineering, 59*, 131–152. doi:10.1002/nme.871

Hay, A. M., & Snyman, J. A. (2006). Optimal synthesis for a continuous prescribed dexterity interval of a 3-dof parallel planar manipulator for different prescribed output workspaces. *International Journal for Numerical Methods in Engineering, 68*, 1–12. doi:10.1002/nme.1691

Hu, X., & Eberhart, R. (2002). Multiobjective optimization using dynamic neighborhood particle swarm optimization. *Congress on Evolutionary Computation (CEC'2002)*, Vol. 2, (pp. 1677-1681). Piscataway, New Jersey.

Kennedy, J., & Eberhart, R. C. (2001). *Swarm intelligence*. San Francisco, CA: Morgan Kaufmann Publishers Inc.

Laribi, M. A., Romdhane, L., & Zeghloul, S. (2007). Analysis and dimensional synthesis of the DELTA robot for a prescribed workspace. *Mechanism and Machine Theory, 42*, 859–870. doi:10.1016/j.mechmachtheory.2006.06.012

Liu, X.-J. (2006). Optimal kinematic design of a three translational DoFs parallel manipulator. *Robotica, 24*, 239–250. doi:10.1017/S0263574705002079

Lopes, A., Freire, H., & de Moura Oliveira, P. B. Solteiro Pires & E. J., Reis, C. (2010). *Multi-objective optimization of parallel manipulators using a particle swarm algorithm*. In The 10th WSEAS International Conference on Applied Informatics and Communications (AIC '10), Taipei, Taiwan.

Lou, Y., Liu, G., Chen, N., & Li, Z. (2005). Optimal design of parallel manipulators for maximum effective regular workspace. *Proceedings of the IEEE/RSJ International Conference on Intelligent Robots Systems*, Alberta, (pp. 795-800).

Merlet, J.-P. (1997). Designing a parallel manipulator for a specific workspace. *The International Journal of Robotics Research, 16*(4), 545–556. doi:10.1177/027836499701600407

Merlet, J.-P. (2006). *Parallel robots* (2nd ed.). Springer Solid Mechanics and its Applications, Vol. 128. The Netherlands.

Miller, K. (2004). Optimal design and modeling of spatial parallel manipulators. *The International Journal of Robotics Research, 23,* 127–140. doi:10.1177/0278364904041322

Poli, R., Kennedy, J., & Blackwell, T. (2007). Particle swarm optimization. *Swarm Intelligence, 1*(1), 33–57. doi:10.1007/s11721-007-0002-0

Reyes-Sierra, M., & Coello Coello, C. A. (2006). Multi-objective particle swarm optimizers: A survey of the state-of-the-art. *International Journal of Computational Intelligence Research, 2*(3), 287–308.

Shi, Y., & Eberhart, R. C. (1998). Parameter selection in particle swarm optimization. *International Conference on Evolutionary Programming,* (pp. 591-600).

Stamper, R. E., Tsai, L.-W., & Walsh, G. C. (1997). Optimization of a three DOF translational platform for well-conditioned workspace. *Proceedings of the IEEE International Conference on Robotic Automation,* Albuquerque, (pp. 3250-3255).

Steward, D. (1965). A platform with six degrees of freedom. *Proceedings - Institution of Mechanical Engineers, 180*(5), 371–386. doi:10.1243/PIME_PROC_1965_180_029_02

Stock, M., & Miller, K. (2003). Optimal design of spatial parallel manipulators: Application to linear DELTA robot. *ASME Journal of Mechanical Design, 125,* 292–301. doi:10.1115/1.1563632

Toscano Pulido, G., & Coello Coello, C. A. (2004). Using clustering techniques to improve the performance of a particle swarm optimizer. *GECCO 2004 - Proceedings of the Genetic and Evolutionary Computation Conference* (pp. 225-237). Seattle, Washington, USA.

Wan, Y. H., & Wang, G. (2008). A y-star robot: Optimal design for the specified workspace. *Proceedings of the IEEE International Conference on Computer Science and Software Engineering,* Wuhan, China.

Zhao, J.-S., Zhang, S.-L., Dong, J.-X., Feng, Z.-J., & Zhou, K. (2005). Optimizing the kinematic chains for a spatial parallel manipulator via searching the desired dexterous workspace. *Robotics and Computer-integrated Manufacturing, 23,* 38–46. doi:10.1016/j.rcim.2005.09.003

KEY TERMS AND DEFINITIONS

Modelling: Modelling is the process of generating abstract, conceptual, graphical or mathematical representations of physical entities.

Multiobjective Optimization: Multiobjective Optimization is the method of simultaneously optimizing two or more conflicting objectives subject to constraints.

Parallel Robots: Parallel Robots are closed-loop mechanisms, made by two rigid bodies, or platforms, that are connected by at least two independent open kinematic chains. Each chain comprises active and passive joints.

Particle Swarm Optimization: Particle Swarm Optimization is a population-based stochastic approach for solving optimization problems. It is inspired in social behaviors found among flocks of birds or schools of fishes.

Robot Design: Robot Design is the method of defining the mechanical structure of the robot.

Stewart Platform PKM: Stewart platform PKM is a six degree-of-freedom parallel robot that comprises six identical kinematic chains connecting two planar platforms. Each kinematic chain has an active linear joint, a passive universal joint and a passive spherical joint.

Workpiece Optimization: Workpiece Optimization is the process of finding the best location for the workpiece in the robot workspace. Typically involves optimizing several conflicting objectives subject to constraints.

Chapter 12
The Generalized Particle Swarm Optimization Algorithm:
Idea, Analysis, and Engineering Applications

Željko S. Kanović
University of Novi Sad, Serbia

Milan R. Rapaić
University of Novi Sad, Serbia

Zoran D. Jeličić
University of Novi Sad, Serbia

ABSTRACT

A generalization of the popular and widely used Particle Swarm Optimization (PSO) algorithm is presented in this chapter. This novel optimizer, named Generalized PSO (GPSO), is inspired by linear control theory. It enables direct control over the key aspects of particle dynamics during the optimization process, overcoming some typical flaws of classical PSO. The basic idea of this algorithm with its detailed theoretical and empirical analysis is presented, and parameter-tuning schemes are proposed. GPSO is also compared to the classical PSO and Genetic Algorithm (GA) on a set of benchmark problems. The results clearly demonstrate the effectiveness of the proposed algorithm. Finally, two practical engineering applications of the GPSO algorithm are described, in the area of electrical machines fault detection and classification, and in optimal control of water distribution systems.

DOI: 10.4018/978-1-4666-2666-9.ch012

INTRODUCTION

Successful optimizers are often inspired by natural processes and phenomena. The field of global optimization has prospered much from these nature-inspired techniques, such as Genetic Algorithms (GAs) (Michalewitz, 1999), Simulated Annealing (SA) (Kirkpatrick et al, 1983), Ant Colony Optimization (ACO) (Dorigo & Blum, 2005) and others. Among these search strategies, the Particle Swarm Optimization (PSO) algorithm is relatively novel, yet well studied and proven optimizer based on the social behavior of animals moving in large groups (particularly birds) (Kennedy & Eberhart, 1995). PSO uses a set of particles called swarm to investigate the search space, imitating the movement of birds in a flock. The position of each particle is a potential solution, characterized by the value of the optimality criterion. During the search process, particles move through the search space and eventually discover the location(s) which provide the best value of the optimality criterion.

Compared to other evolutionary techniques, PSO has only a few adjustable parameters and it is computationally inexpensive and very easy to implement (Ratnaweera et al, 2004; Schutte & Groenwold, 2005). Thus, it provides high calculation speed, compared to other evolutionary algorithms. However, this algorithm also has some flaws, such as premature convergence tendency and the inability to independently control various aspects of the search - oscillation frequency and the impact of personal and global knowledge.

In this chapter a modification of the original PSO algorithm will be presented, named Generalized Particle Swarm Optimization (GPSO) (Kanović et al, 2011). The idea of this modification is to address PSO in a new and conceptually different fashion, i.e., to consider each particle within the swarm as a second-order linear stochastic system. Such systems are extensively studied in engineering literature. The authors found and explained that the stability and response proper-

ties of such a system can be directly related to its performance as an optimizer, i.e., its explorative and exploitative properties.

Based on formal analysis of response properties and stability of the control system, which represents a particle and simulates its dynamics, a new algorithm formulation with a new set of parameters is proposed, which enables more direct and independent control over key aspects of the search process. This way, one can overcome the mentioned inherent flaws of the PSO, keeping at the same time all advantages of the original version of the algorithm, concerning its simplicity, small number of parameters and easy practical implementation.

Using the stability criterions well-known from the control theory, some recommendations for parameter tuning are also proposed, resulting in desired properties of the algorithm, concerning its convergence, exploration and exploitation properties.

Extensive experimental analysis of the proposed algorithm is also presented in this chapter. The algorithm is tested on a set of standard benchmark functions and the results are compared to some earlier versions of the PSO algorithm, including the modification with time-varying coefficients. The obtained results are discussed and some crucial conclusions are emphasized regarding performance of algorithms used in comparison.

This chapter will also present two engineering applications of GPSO algorithm. Firstly, the application of this algorithm in induction motor fault detection and classification procedure based on vibration analysis will be described. GPSO is used in a cross-validation procedure of a Support Vector Machines classifier in order to provide a more reliable and accurate fault detection and classification process. Some of the most common faults are considered, such as damaged bearings, broken rotor bar and static rotor eccentricity. This application is tested and already implemented in real industrial system, demonstrating the potential of GPSO algorithm for practical applications. Secondly, the application in water distribution systems will be presented, where GPSO is used to determine the optimal variable-speed pump control strategy. This solution is implemented as a part of SCADA system of water distribution facility and provides better, more reliable and more efficient water distribution with minimal energy costs.

BACKGROUND

In PSO, each particle in the swarm is described by its position (x) and velocity (v). The position of each particle is a potential solution, and the best position that each particle achieved during the entire optimization process is memorized (p). The swarm as a whole memorizes the best position ever achieved by any of its particles (g). The position and the velocity of each particle in the k-th iteration are updated as

$$
\begin{aligned}
v[k+1] = & w \cdot v[k] + cp \cdot rp[k] \cdot (p[k] - x[k]) \\
& + cg \cdot rg[k] \cdot (g[k] - x[k]) \\
x[k+1] = & x[k] + v[k+1]
\end{aligned}
\tag{1}
$$

Acceleration factors cp and cg control the relative impact of the personal (local) and common (global) knowledge on the movement of each particle. In original version of the algorithm, these coefficients have constant value during the search process. Inertia factor w, which was introduced for the first time in (Shi & Eberhart, 1999), keeps the swarm together and prevents it from diversifying excessively and therefore diminishing PSO into a pure random search. Random numbers rp and rg are mutually independent and uniformly distributed on the range [0, 1]. Many improved versions of PSO emerged so far, based on velocity clamping (Shahzad et al, 2009), constriction factor approach (Clerk & Kennedy, 2002), variable inertia factor (Eberhart & Shi, 2000), time-

varying acceleration coefficients (Ratnaweera et al, 2004), particle position extrapolation (Senthil Arumugam et al, 2009), sub-swarm (Jiang et al, 2007) or dynamical multi-swarm (Liang & Suganthan, 2005) strategies, knowledge-based cooperation (Jie et al, 2008) and many other, which all improve the performance of the algorithm. In recent literature, several other modifications of the original PSO also emerged; these incorporate mutation-like operators (Senthil Arumugam &Rao, 2007; Cui et al, 2008), modify the topology of the swarm (Van der Bergh, 2001; Niu et al, 2006; DeBao & ChunXia, 2009; Jiang et al, 2010) or utilize hybridization with other techniques (Chen et al, 2010; Thangaraj et al, 2011). However, these modifications imply more complicated calculation procedure and extended calculation time and therefore reduce the main advantage of PSO – its simplicity and high calculation speed, so they cannot be compared with previously mentioned algorithms.

Numerous studies have been published addressing PSO both empirically and theoretically (Shi & Eberhart, 1999; Van der Berg, 2001). Over the years, the effectiveness of the algorithm has been proven for various engineering problems (Dong et al, 2005, Dimopoulos, 2007; He & Wang, 2007; Olamaei et al, 2008). In electrical engineering, which is the main focus of this book, PSO has been extensively used for practical application in many areas, such as energy conversion and power distribution (Chatterjee et al, 2011; Ramesh et al, 2012; Mallipeddi et al, In press), controller design and optimization (Fang et al, 2011; Bouallegue et al, 2012; Menhas et al, 2012), system identification (Alireza, 2011; Majhi & Panda, 2011), fault detection (Samanta & Nataraj, 2009) and many other.

However, the theoretical justification of the PSO procedure remained long an open question. The first formal theoretical analyses were undertaken by Ozcan and Mohan (1998). They addressed the dynamics of a simplified, one-dimensional, deterministic PSO model. Clerc and Kennedy (2002) also analyzed PSO by focusing on swarm stability and the so-called "explosion" phenomenon. Jiang et al. (2007) were the first to analyze the stochastic nature of the algorithm. Rapaić and Kanović (2009) explicitly addressed PSO with time-varying parameters, which is the most common case in practical implementations. In (Kanović et al, 2011), the idea and comprehensive analysis of a new, improved algorithm modification is presented, which overcomes in great deal the flaws of the original PSO mentioned earlier. This algorithm modification and its implementation in some engineering applications is the subject of this chapter.

GENERALIZED PARTICLE SWARM OPTIMIZATION

The Idea of the Algorithm

Most theoretical analyses reported in the literature are conducted on the basis of the second-order PSO model

$$x[k+1] - (1 + w - cp \cdot rp[k] - cg \cdot rg[k])x[k]$$
$$+ w \cdot x[k-1] = cp \cdot rp[k] \cdot p[k] + cg \cdot rg[k] \cdot g[k]$$

$$(2)$$

Equivalent to the model described by Equation 1. This equation can also be interpreted as a difference equation describing the motion of a stochastic, second-order, discrete-time, linear system with two external inputs. The output of such a system is the current position of the particle (x), while its inputs are personal and global best positions (p and g, respectively). In general, a second-order discrete-time system can be modeled by the recurrent relation

$$x[k+1] + a_1 \cdot x[k] + a_0 \cdot x[k-1] = b_p \cdot p[k]$$
$$+ b_g \cdot g[k],$$

$$(3)$$

where some or all of the parameters a_1, a_0, b_p and b_g are stochastic variables with appropriate, possibly time-varying probability distributions.

In order to make Equation 3 a successful optimizer, several restrictions should be imposed on these parameters. First, the system should be stable (in the sense of control theory (Åström & Wittenmark, 1997)), with stability margins growing during the optimization process. The particles should initially be allowed to move more freely in order to explore the search space. As the search process approaches its final stages, the good solutions that were previously found should be exploited, and the effort of the swarm should be concentrated in the vicinity of known solutions. Second, the response of the system to the perturbed initial conditions should be oscillatory in order for the particles to overshoot or fly over their respective attractor points. Also, if both external inputs approach the same limit value as k grows, the particle position x should also converge to this limit. Convergence is understood in the mean square sense, as will be elaborated later. Finally, in the early stages of the search, the system should primarily be governed by the cognitive input p, allowing particles to behave more independently in these stages. In later stages, the social input g should be dominant because the swarm should become more centralized, and global knowledge of the swarm as a whole should dominate the local knowledge of each individual particle. All of these requirements are formulated in a sequel using notions from control theory.

The characteristic polynomial of Equation 3 is $f(z) = z^2 + a_1 z + a_0$. The particles' dynamics is determined by roots of this polynomial, which are also known as the eigenvalues of the system. The system is stable if and only if the modulus ρ of the eigenvalues is less than 1. In order for the system to be able to oscillate, the roots of the characteristic polynomial must not be positive real numbers. The argument φ of the eigenvalues determines the discrete frequency of the characteristic oscillations of the system. Argument values close to π result in more frequent oscillations.

The requirement that particle positions tend to the global best position when personal best and global best are equal is equivalent to

$$1 + a_1 + a_0 = b_p + b_g. \qquad (4)$$

It is also clear that an increase in b_p favors the cognitive component of the search, while an increase in b_g favors the social component. All of the requirements can easily be satisfied if system Equation 3 is rewritten in the following canonical form, often used in control theory (Åström & Wittenmark, 1997):

$$x[k+1] - 2\zeta\rho x[k] + \rho^2 x[k-1] = \\ (1 - 2\zeta\rho + \rho^2)(c \cdot p[k] + (1-c) \cdot g[k]) \qquad (5)$$

In Equation 5, ρ is the eigenvalues module, and ζ is the cosine of their arguments. Parameter c is introduced to replace both b_p and b_g. Clearly, requirement of Equation 4 is satisfied by Equation 5. The primary idea of GPSO algorithm is to use Equation 5, instead of Equations 1 or 2, in optimizer implementation. The parameters in this equation allow a more direct and independent control of the various aspects of the search procedure. Figure 1 depicts particle trajectories governed by Equation 5 with different parameter values in a two-dimensional search space. Both attractor points (i.e., global and personal best) are assumed to be equal to zero, for simplicity. Note that lower values of parameter ρ lead to faster convergence, while higher values result in less stable motion and slower convergence. Thus, higher values of ρ would enable the swarm to cover a wider portion of the search space, which improves exploration abilities of the algorithm. This approach is usually used in early stages of

Figure 1. GPSO particle trajectories with different sets of parameters. The search space is assumed to be two-dimensional, and both attractors (personal and global best) are assumed to be zero. The starting point is assumed to be (1,1).

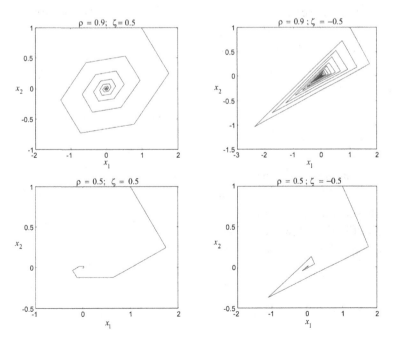

the search. On the other hand, lower values of ρ are beneficial in the later stages of the search, when search space is extensively investigated and faster convergence is preferable. Parameter ζ determines the way particles oscillate over attractor points. For ζ equal to 1, the particle motion would be non-oscillatory. For ζ equal to -1, a particle would erratically oscillate from one side of the attractor to another. Discrete frequency of oscillations, determined by ζ, has not been previously considered as a parameter in PSO dynamics. Conducting numerous numerical experiments, some of which are presented in this chapter, the authors noticed that by adjusting ζ to a value between -1 and 1, one can control oscillatory nature of particles' motion and obtain a desired behavior, in the sense of exploration properties of the algorithm. Exploitation properties, however, are mainly controlled adjusting the value of parameter c. Higher value of c favors personal (local) knowledge, while lower value of this parameter enable better exploitation of global swarm knowledge.

Convergence Analysis

There are several ways to define convergence for stochastic sequences. The notion of mean square convergence is adopted here. A stochastic sequence $x[k]$ is said to converge in mean square to the limit point a if and only if

$$\lim_{k \to \infty} \mathbf{E}(x[k] - a)^2 = 0, \tag{6}$$

where $\mathbf{E}$ denotes the mathematical expectation operator (Papoulis, 1965). The investigation of convergence of a stochastic sequence can therefore be replaced by the investigation of two deterministic sequences, since Equation 6 is equivalent to

$$\lim_{k \to \infty} \mathbf{E}x[k] = a \tag{7}$$

and

$$\lim_{k \to \infty} \mathbf{E}x^2[k] = a^2. \tag{8}$$

In the sequel, $\overline{(\cdot)}$ is used instead of $\mathbf{E}(\cdot)$ to simplify notation and make it more compact.

If we assume that $p[k]$ and $g[k]$ both converge to their respective values p and g, GPSO's convergence is equivalent to the stability of the dynamical system (Equation 5). It has already been mentioned that p and g can be considered as inputs, which do not affect stability, as is well known from control theory (Åström & Wittenmark, 1997). Let us introduce $\mu = c \cdot p + (1 - c) \cdot g$ and $y[k] = x[k] - \mu$. The following equation is a direct consequence of Equation 5:

$$y[k + 1] - 2\zeta\rho y[k] + \rho^2 y[k - 1] = 0. \tag{9}$$

Because the inertia factor w of the classical PSO is closely related to the eigenvalue modulus, ρ is considered a deterministic parameter. The ζ factor, on the other hand, is assumed to be stochastic and independent of particle position.

Applying the expectation operator to Equation 9 one obtains

$$\overline{y[k + 1]} - 2\zeta\rho\overline{y[k]} + \rho^2\overline{y[k - 1]} = 0 \tag{10}$$

The eigenvalues of this system are

$$p_{1,2} = \rho\zeta \pm j\rho\sqrt{1 - \zeta^2}, \tag{11}$$

where j is the imaginary unit, and the stability conditions are given by

$$0 < \rho < 1, \tag{12}$$

$$|\zeta| \leq 1. \tag{13}$$

If System 10 converges, its limit point is zero, and therefore, the mathematical expectation of any particle's position tends to μ. If personal best and global best positions are asymptotically equal, which is true for the particle achieving the global best position, then $\mu = p = g$.

The variance of $x[k]$ is addressed next. From Equation 9, it is readily obtained that

$$\overline{y^2[k+1]} = 4\overline{\zeta^2}\rho^2\overline{y^2[k]} + \rho^4\overline{y^2[k-1]}$$
$$- 4\overline{\zeta}\rho^3\overline{y[k]y[k-1]}, \qquad (14)$$

$$\overline{y[k+1]y[k]} = 2\overline{\zeta}\rho\overline{y^2[k]} - \rho^2\overline{y[k]y[k-1]}. \qquad (15)$$

Equations 14 and 15 can be considered a model of a deterministic linear system. State variables can be designated as $\alpha_1[k] = \overline{y^2[k]}$, $\alpha_2[k] = \overline{y^2[k-1]}$ and $\alpha_3[k] = \overline{y[k]y[k-1]}$. The system can be rewritten in state-space form as

$$\boldsymbol{\alpha}[k+1] = \mathbf{A}\boldsymbol{\alpha}[k], \qquad (16)$$

with state vector $\boldsymbol{\alpha} = [\alpha_1 \ \alpha_2 \ \alpha_3]^T$ and the matrix $\mathbf{A}$ defined as

$$\mathbf{A} = \begin{bmatrix} 0 & 1 & 0 \\ \rho^4 & 4\overline{\zeta^2}\rho^2 & -4\overline{\zeta}\rho^3 \\ 0 & 2\overline{\zeta}\rho & -\rho^2 \end{bmatrix}. \qquad (17)$$

The System 16 is stable if and only if all eigenvalues of $\mathbf{A}$ are less than 1 in modulus. The eigenvalues of a matrix are roots of its characteristic polynomial. The characteristic polynomial of the matrix $\mathbf{A}$ is

$$f(z) = z^3 + z^2\rho^2(1 - 4\overline{\zeta^2})$$
$$- z\rho^4(4\overline{\zeta^2} - 8(\overline{\zeta})^2 + 1) - \rho^6. \qquad (18)$$

Stability can be investigated by various methods. Jury's criterion (Åström & Wittenmark, 1997) is utilized in the sequel. For a general third-order polynomial $f(z) = z^3 + a_2z^2 + a_1z + a_0$, the conditions under which all zeros are less than one in modulus are

$$f(1) > 0, \ f(-1) < 0, \ |a_0| < 1,$$
$$|a_0^2 - 1| > |a_0a_2 - a_1|. \qquad (19)$$

Applying these conditions to Equation 18 yields

$$(1 + \rho^2)(1 - \rho^4) > 4\rho^2(\overline{\zeta^2} + \rho^2(\overline{\zeta^2} - 2(\overline{\zeta})^2)), \qquad (20)$$

$$(1 - \rho^2)(1 - \rho^4) > 4\rho^2(-\overline{\zeta^2} + \rho^2(\overline{\zeta^2} - 2(\overline{\zeta})^2)), \qquad (21)$$

$$|\rho^6| < 1, \qquad (22)$$

$$1 - \rho^{12} > \rho^4\left|1 - \rho^4 + 4\overline{\zeta^2}(1 + \rho^4) - 8(\overline{\zeta})^2\right|. \qquad (23)$$

Due to Equation 12, both Equations 21 and 22 are identically satisfied; that is, right-hand side of Equation 21 is always negative, whereas its left-hand side is always positive. The stability conditions are therefore Equations 20 and 23. Unfortunately, these conditions are not as nearly as compact as Equations 12 and 13, and they provide no direct insight into the influence of parameters ρ and ζ on stability. However, some clarification can be obtained by the introduction of the helper polynomial

$$f_1(q) = \frac{f(\rho^2 q)}{\rho^6}$$
$$= q^3 + q^2(1 - 4\overline{\zeta^2}) - q(4\overline{\zeta^2} - 8(\overline{\zeta})^2 + 1) - 1 \qquad (24)$$

and investigation of its roots by means of Equation 19:

$$f_1(1) = 8((\bar{\zeta})^2 - \overline{\zeta^2}) > 0, \tag{25}$$

$$f_1(-1) = -8(\bar{\zeta})^2 < 0, \tag{26}$$

$$|1| < 1, \tag{27}$$

$$0 > 8\left|\overline{\zeta^2} - (\bar{\zeta})^2\right| \tag{28}$$

Condition 25 is not satisfied because the variance of ζ ($\operatorname{var}\zeta = \overline{\zeta^2} - (\bar{\zeta})^2$) is non-negative. Conditions 27 and 28 are also violated. It is therefore clear that f_1 has at least one root outside the unit circle or on its boundary at best. Since the roots of f are ρ^2 times the roots of f_1, it can be concluded that at least one root of f is greater than or equal to ρ^2. Thus, it can be concluded that an increase in ρ has a negative influence on the convergence of both the mean value and the variance of particle positions. It is also clear that by increasing the variance of ζ, $f_1(1)$ decreases, and the right-hand side of Condition 28 increases; therefore, the increase in the variance of ζ impedes the convergence of the algorithm.

The limit point of $\bar{\alpha}$ is $[0\ \ 0\ \ 0]^T$, and therefore, $\overline{y^2[k]}$ is zero in limit. Because $\overline{y^2[k]}$ is asymptotically equal to the variance of x, it follows that the variance of the position of any particle is asymptotically equal to zero. This concludes the proof that under Conditions 12, 13, 20 and 23, GPSO System 5 exhibits mean square convergence.

Empirical Analysis

It is known that proper parameter selection is crucial for the performance of PSO. The relationships between adjustable factors within classical PSO and GPSO are

$$\rho = \sqrt{w}, \tag{29}$$

$$\zeta = \frac{1 - w - cp \cdot rp - cg \cdot rg}{2\sqrt{w}}, \tag{30}$$

$$c = \frac{cp \cdot rp}{cp \cdot rp + cg \cdot rg}. \tag{31}$$

The inertia factor w is the "glue" of the swarm; its responsibility is to keep the particles together and to prevent them from wandering too far away. Since its introduction by Shi and Eberhart (1999), it was noted that it is beneficial for the performance of the search to gradually decrease its value. A common choice is to select a value close to 0.9 at the beginning of the search and a value close to 0.4 at the end. Based on Equation 29, it would be reasonable to designate ρ as decreasing from about 0.95 to about 0.6. Regarding the acceleration factors, $cp = cg = 2$ is the choice used in the original variant of the PSO algorithm (Kennedy & Eberhart, 1995). Other commonly used schemes include $cp = cg = 0.5$ as well as $cp = 2.8$ and $cg = 1.3$. It is generally noted that the choice of acceleration factors is not critical, but they should be chosen so that $cp + cg < 4$ (Schutte & Groenwold, 2005). In fact, if the last condition is violated, PSO does not converge, regardless of the inertia value (Jiang et al, 2007; Rapaić & Kanović, 2009). Ratnaweera et al. (2004) introduced the time-varying acceleration coefficients PSO (TVAC-PSO) and demonstrated that it outperforms other commonly used PSO schemes. They suggested that cp should linearly decrease from 2.5 to 0.5, while cg should simultaneously linearly increase from 0.5 to 2.5. This would correspond to changing c from about 0.8 to approximately 0.2.

It is important to note that the novelty of the GPSO algorithm does not reduce to a change in coefficient expressions. Indeed, if the coefficients of GPSO are fixed, then according to Formulas 29-31, one can find an equivalent set of classical PSO parameters. However, if parameters vary in time, which is the case in most practical applications, linear variations in GPSO parameters are equivalent to nonlinear variations in inertia w and acceleration coefficients cp and cg. It is this nonlinearity that accounts for the performance of GPSO, as the effects of the new parameterization cannot be achieved by the linear alteration of the classical PSO parameters. One could utilize the classical PSO with a highly nonlinear parameter adaptation law to achieve this, but this would be impractical. In GPSO, nonlinearity has been hidden within the algorithm definition, thus allowing the same effect to be achieved by linear alteration of the newly proposed parameters. These parameters have a well-defined interpretation in terms of particle dynamics, and they allow independent control of the particles' dynamic properties, such as stability, characteristic frequency and relative impact of local to global knowledge.

Based on recommendations stated earlier, authors have proposed two parameters adjustment schemes, with ρ linearly decreasing from 0.95 to 0.6 and c linearly decreasing from 0.8 to 0.2. The proper selection of ζ proved to be the most difficult task, as this factor has no direct analogy to any of the parameters of the classical PSO. The ζ parameter equals the cosine of particle eigenvalues in the dynamic model. Because this is a discrete system, the eigenvalue argument equals the discrete circular characteristic frequency of particle motion. Thus, ζ is directly related to the ability of particles to oscillate around attractor points (i.e., global and personal best) and, therefore, has a crucial impact on the exploitative abilities of the algorithm. If ζ equals 1, this would prevent particles from flying over the attractor points; however, if ζ is close to -1, this would

result in the desultory movement of the particles. In both cases, particles cannot explore the vicinity of the attractor points. Based on thorough empirical analysis and numerical experiments conducted by the authors, it is shown that the most appropriate and robust choice is to select ζ uniformly distributed across a wider range of values. The two most successful schemes are presented further. In the first GPSO scheme (GPSO1), ζ was adopted as a stochastic parameter with uniform distribution ranging from -0.9 to 0.2, while in the second scheme (GPSO2), ζ was uniformly distributed in the range $[-0.9, 0.6]$.

The empirical analysis of various GPSO parameter adjustment strategies has been performed using several well-known benchmark functions presented in Table 1, all of which have a minimum value of zero.

The proposed GPSO schemes are compared to TVAC-PSO developed by Ratnaweera et al. (2004), which will be denoted simply as PSO in the sequel. Among many existing PSO modifications, this algorithm is chosen for comparison because it alters the value of all three coefficients during the search, keeping at the same time the original form and implementation simplicity of PSO algorithm, which makes it similar to proposed GPSO schemes. A comparison is also made with respect to GA with linear ranking, stochastic universal sampling, uniform mutation and uniform crossover (Michalewitz, 1999), since GA is extensively used in engineering applications.

In the performed experiment, the search was conducted within a 5-dimensional search space using 100 iterations with 30 particles in the swarm. The population was initialized within a hypercube of edge 10 units. The center of the initial hypercube is displaced from the global optimal point of a particular benchmark by 100 units in each direction. The results of the experiment are presented in Table 2; values are shown for the mean, median and standard deviation of the obtained minimum after 100 consecutive runs.

Table 1. Account of benchmark functions used for comparison

Dixon-Price	$f(\mathbf{x}) = (x_1 - 1)^2 + \sum_{i=2}^{n} i(2x_i^2 - x_{i-1})^2$
Rosenbrock	$f(\mathbf{x}) = \sum_{i=1}^{n-1} \left[100(x_i^2 - x_{i+1})^2 + (x_i - 1)^2 \right]$
Zakharov	$f(\mathbf{x}) = \sum_{i=1}^{n} x_i^2 + \left(\sum_{i=1}^{n} 0.5 i x_i \right)^2 + \left(\sum_{i=1}^{n} 0.5 i x_i \right)^4$
Griewank	$f(\mathbf{x}) = \sum_{i=1}^{n} \frac{x_i^2}{4000} - \prod_{i=1}^{n} \cos\left(\frac{x_i}{\sqrt{i}} \right) + 1$
Rastrigin	$f(\mathbf{x}) = 10n + \sum_{i=1}^{n} (x_i^2 - 10\cos(2\pi x_i))$
Ackley	$f(\mathbf{x}) = 20 + e - 20\exp(-\frac{1}{5}\sqrt{\frac{1}{n}\sum_{i=1}^{n} x_i^2}) - \exp(-\frac{1}{n}\sum_{i=1}^{n}\cos(2\pi x_i))$
Michalewitz	$f(\mathbf{x}) = 5.2778 - \sum_{i=2}^{n} \sin(x_i) \left[\sin\left(\frac{i x_i^2}{\pi} \right) \right]^{2m}, \ m = 10$
Perm	$f(\mathbf{x}) = \sum_{k=1}^{n} \left[\sum_{i=1}^{n} \left(i^k + 0.5 \right) \left(\left(\frac{x_i}{i} \right)^k - 1 \right) \right]^2$
Spherical	$f(\mathbf{x}) = \sum_{i=1}^{n} x_i^2$

It is clear that both newly proposed schemes perform better than either PSO or GA in the majority of cases; the exception is Michalewitz's benchmark, where they are consistently outperformed by GA and slightly outperformed by PSO. Note also that in many of the considered cases, both GPSO variants show results that are several orders of magnitudes better than the results obtained by the other two optimizers.

Figure 2 depicts changes in the objective value of the best, mean and worst particles within the PSO and GPSO1 swarms.

One can notice that the best particle converges very fast in both settings. However, in the GPSO swarm, other particles do not follow so rapidly, effectively keeping the diversity of the population sufficiently large. To illustrate further, Figure 3 shows the maximum distance between two particles in the swarm during the optimization process. It is clear that the GPSO swarm becomes

Table 2. Results of the experiment: mean, median and standard deviation of the minimal obtained values

		PSO	GPSO1	GPSO2	GA
Dixon-Price	mean	$9.29 \cdot 10^{-1}$	$2.60 \cdot 10^{-1}$	$2.34 \cdot 10^{-1}$	$1.06 \cdot 10^{3}$
	median	$2.50 \cdot 10^{-2}$	$\mathbf{8.25 \cdot 10^{-15}}$	$1.47 \cdot 10^{-14}$	$2.94 \cdot 10^{2}$
	std. dev.	$1.82 \cdot 10^{1}$	$3.22 \cdot 10^{-1}$	$3.17 \cdot 10^{-1}$	$1.710 \cdot 10^{3}$
Rosenbrock	mean	$3.167 \cdot 10^{2}$	$1.29 \cdot 10^{2}$	$1.94 \cdot 10^{2}$	$1.81 \cdot 10^{3}$
	median	$9.17 \cdot 10^{1}$	$\mathbf{2.59 \cdot 10^{1}}$	$7.62 \cdot 10^{1}$	$2.29 \cdot 10^{2}$
	std. dev.	$6.127 \cdot 10^{2}$	$2.64 \cdot 10^{2}$	$3.19 \cdot 10^{2}$	$3.77 \cdot 10^{3}$
Zakharov	mean	$7.60 \cdot 10^{2}$	$2.21 \cdot 10^{-11}$	$4.85 \cdot 10^{-11}$	$1.76 \cdot 10^{1}$
	median	$2.19 \cdot 10^{1}$	$\mathbf{6.88 \cdot 10^{-17}}$	$1.38 \cdot 10^{-14}$	2.86
	std. dev.	$2.11 \cdot 10^{3}$	$3.62 \cdot 10^{-10}$	$3.15 \cdot 10^{-10}$	$4.46 \cdot 10^{1}$
Griewank	mean	$8.06 \cdot 10^{-1}$	$7.50 \cdot 10^{-2}$	$8.39 \cdot 10^{-2}$	$8.94 \cdot 10^{-2}$
	median	$3.32 \cdot 10^{-1}$	$\mathbf{6.64 \cdot 10^{-2}}$	$7.02 \cdot 10^{-2}$	$7.23 \cdot 10^{-2}$
	std. dev.	$9.25 \cdot 10^{-1}$	$4.68 \cdot 10^{-2}$	$5.25 \cdot 10^{-2}$	$6.46 \cdot 10^{-2}$
Rastrigin	mean	4.12	2.87	3.50	7.39
	median	3.08	$\mathbf{1.98}$	2.98	7.27
	std. dev.	3.25	1.88	2.30	2.82
Ackley	mean	$2.00 \cdot 10^{1}$	$2.00 \cdot 10^{1}$	$2.00 \cdot 10^{1}$	$2.00 \cdot 10^{1}$
	median	$\mathbf{2.00 \cdot 10^{1}}$	$\mathbf{2.00 \cdot 10^{1}}$	$\mathbf{2.00 \cdot 10^{1}}$	$\mathbf{2.00 \cdot 10^{1}}$
	std. dev.	$2.00 \cdot 10^{-2}$	$2.99 \cdot 10^{-2}$	$4.10 \cdot 10^{-2}$	$5.40 \cdot 10^{-3}$
Michalewitz	mean	2.05	1.99	2.17	$9.94 \cdot 10^{-1}$
	median	2.06	2.07	2.17	$\mathbf{9.65 \cdot 10^{-1}}$
	std. dev.	$4.49 \cdot 10^{-1}$	$5.32 \cdot 10^{-1}$	$4.35 \cdot 10^{-1}$	$2.38 \cdot 10^{-1}$
Perm	mean	$5.50 \cdot 10^{14}$	$2.08 \cdot 10^{1}$	$2.72 \cdot 10^{1}$	$2.46 \cdot 10^{12}$
	median	$1.43 \cdot 10^{3}$	$\mathbf{4.79}$	8.49	$2.78 \cdot 10^{3}$
	std. dev.	$2.85 \cdot 10^{15}$	$9.85 \cdot 10^{1}$	$6.33 \cdot 10^{1}$	$1.96 \cdot 10^{13}$
Spherical	mean	$2.45 \cdot 10^{-6}$	$4.48 \cdot 10^{-19}$	$3.44 \cdot 10^{-20}$	$1.22 \cdot 10^{-1}$
	median	$4.43 \cdot 10^{-7}$	$\mathbf{2.29 \cdot 10^{-23}}$	$6.30 \cdot 10^{-21}$	$9.85 \cdot 10^{-2}$
	std. dev.	$1.50 \cdot 10^{-5}$	$1.39 \cdot 10^{-17}$	$8.65 \cdot 10^{-20}$	$1.18 \cdot 10^{-1}$

extremely diverse at certain points, spreading across the vast area of the search space, which is the main and most important effect of the newly proposed parameterization. This phenomenon is likely the explanation of the superior performance that GPSO exhibits in most of the analyzed cases. We must note that Figure 2 and 3 depict data obtained optimizing Griewank's function, but similar behavior can be noticed for all considered test-functions.

Figure 4 illustrates one of the most important advantages of PSO algorithm and its modifications over other evolution algorithms – its high calculation speed, as a consequence of algorithm simplicity. Calculation of minimum of Ackley's function is depicted on Figure 4 as an example, but the calculation time is similar for all other benchmarks. One can notice that both PSO and GPSO1 algorithms finish the calculation in less than 0.1 seconds, while GA needs about 1.3 seconds to complete calculations for same swarm

Figure 2. Objective values of the best, the mean and the worst particle within a swarm during 100 iterations- a) PSO swarm; b) GPSO1 swarm

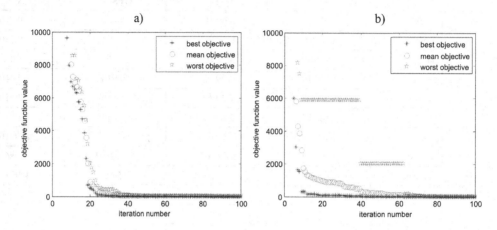

Figure 3. Maximum distance between two particles during the optimization process for GPSO1 and PSO

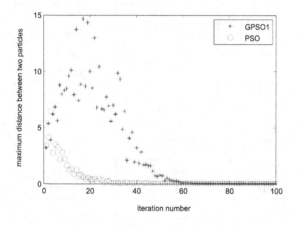

Figure 4. Calculation time for PSO, GPSO1 and GA

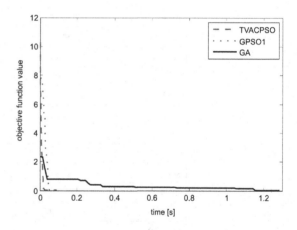

size and same number of iterations. This property of PSO is very promising and implies its application in real-time optimization problems. We need to remark that specified calculation times are related to relatively simple test-functions (Ackley' benchmark in this particular case). When applying described optimization algorithms to some real scenarios, total calculation time is extended, due to complexity of cost function. This, however, does not depend on the type of the algorithm chosen for optimization, but must be taken into consideration, especially in real-time applications.

Two other experiments were also conducted, one with same swarm size and iteration number, but with initial hypercube centered on global optimal point for each benchmark, and the other with the same swarm size but with 1000 iterations. Both these experiments showed results that are consistent with previously described experiment (see (Kanović et al, 2011) for more details).

Application Example 1: GPSO in Fault Detection

This section describes an application of GPSO to induction motor fault detection. GPSO is used within the classification procedure for optimization of Support Vector Machines parameters.

The experiment was conducted in a sunflower oil processing and production plant during a maintenance period. Vibration analysis was used, which is a popular technique due to its easy measurability, high accuracy and reliability (Widodo et al, 2007). This technique uses a motor vibration signal, i.e. some of its parameters, to determine the presence and origin of potential faults. Two typical faults were considered, static eccentricity and bearing wear. For each of them, a classifier was constructed based on acquired vibration signals that detects whether a fault on the observed motor is present. Classifier parameters were optimized using GPSO, resulting in better performance, efficiency and accuracy.

Fault classification based on vibration analysis consists of following phases:

- Data collection and processing
- Characteristic feature set calculation and dimensional reduction
- Fault detection and classification

In this experiment, vibration signals from horizontal and vertical vibration sensors mounted on ten induction motors were acquired. High sensitivity (100 mV/g) ceramic shear accelerometers with magnetic mounting were used. Two different types of motors were considered, five of each type. The first type is a 5.5 kW motor with one pair of poles, a nominal speed of 2925 rpm and a nominal current of 10.4 A. These motors drive the screw conveyors in the process plant. Two motors of this type were healthy; one motor had wear on both the inner and outer race of the bearing, one motor had a static eccentricity level of 30%, and one motor had both faults present at the same time (i.e., inner and outer race wear and 50% static eccentricity). The second type is a 15 kW motor driving the crushed oilseed conditioners with one pair of poles, a nominal speed of 2940 rpm and a nominal current of 26.5 A. Two of these motors were healthy, two had static eccentricity levels of 30% and 50%, and one had defects on the bearing ball, inner and outer race.

Multiple vibration measurements were conducted on each motor. Each signal was collected for 2 seconds with a sampling frequency of 25.6 kHz. A total of 200 signals were collected, with 78 healthy signals, 58 signals representing only static eccentricity, 44 signals representing only bearing wear and 20 signals with both faults present at the same time.

Collected signals were analyzed in both time and frequency domain. In both these domains, some characteristic parameters of signal were calculated. These parameters are usually called characteristic features of the vibration signal. In time domain, the signal is represented as a series of values that show the amplitude of motor vibration versus time. Characteristic features in this domain are some typical statistical parameters of the signal. Nine statistical characteristic features were used in this experiment, including arithmetic mean value, root mean square value, square mean root value, skewness index, kurtosis index, C factor, L factor, S factor and I factor. These parameters can provide information about the presence of considered faults (Stepanić et al, 2009; Samanta & Nataraj, 2009). In the frequency domain, signal is observed using its spectrum, calculated by Fast Fourier Transformation – FFT. If some fault is present, it will cause amplified vibration amplitude on some frequencies in signal spectrum. These frequencies, typical for every fault type, are called characteristic frequencies. They are used as characteristic features for fault detection in frequency domain. Eight characteristic features were used in this experiment, which represent the sum of amplitudes of the spectrum in the region around the characteristic frequencies. The summation is performed in the band $\left[f_c - 3Hz, f_c + 3Hz\right]$, where f_c is the characteristic frequency. First three features represent the sum around twice the supply frequency ($f_s = 50Hz$) and its sidebands ($2f_s \pm f_r$, where f_r is rotor frequency), which are the indicators of static eccentricity in an induction motor (Dorell et al, 1997). The next five features are related to the bearing conditions. They represent the sum of the amplitudes of the power spectrum around the bearing characteristic frequencies, which are:

Outer race fault frequency:

$$f_{rpfo} = f_r \times \frac{N}{2}\left(1 - \frac{d}{D}\cos\phi\right) \tag{32}$$

Inner race fault frequency:

$$f_{bpfi} = f_r \times \frac{N}{2}\left(1 + \frac{d}{D}\cos\phi\right) \tag{33}$$

Rotation frequency of the rolling element:

$$f_{bsf} = f_r \times \frac{D}{2d}\left[1 - \left(\frac{d}{D}\right)^2 \cos^2\phi\right] \tag{34}$$

Rolling element fault frequency:

$$f_{bff} = 2 \times f_{bsf} \tag{35}$$

Cage fault frequency:

$$f_{ftf} = f_r \times \frac{1}{2}\left(1 - \frac{d}{D}\cos\phi\right) \tag{36}$$

Note that N is the number of rolling elements in the bearing, ϕ is the contact angle of the rolling element, d is the rolling element diameter,

and D is the diameter of the bearing shell (Stepanić et al, 2009). Data on bearing geometry and expected characteristic frequencies are shown in details in (Kanović et al, 2011).

The feature set of totally seventeen features for every recorded signal was calculated. However, not all of the features are essential for fault detection. Some of them are sometimes irrelevant and can cause redundancy, so the original feature set must be dimensionally reduced. Dimensional reduction in this experiment was performed using Principal Component Analysis. This technique transforms the original feature set in the set containing new features, called principal components, which are uncorrelated. The transformed feature set has the same number of components (dimensions) as the original set, but these components are sorted by variance (Widodo et al, 2007; He et al, 2009). The components with higher variance usually carry more information about the presence of the fault and its nature, so the dimensional reduction is conducted by keeping only certain number of principal components. Only first six principal components of the original feature set were used in this experiment, resulting in a more comprehensive and less redundant feature set that contains sufficient information for successful classification and fault detection. This modified feature set was then divided into two subsets, the training set and the test set, containing features for 146 and 54 signals, respectively. These sets were formed using signals of different motors to avoid overfitting at the classifier training stage.

Classification was performed using Support Vector Machines (SVM), a kind of learning machine based on statistical learning theory. A variant of SVM with a soft margin (penalty parameter C) and a Gaussian RBF kernel function (parameter σ) was used (Wang, 2005; Kankar et al, 2011). For both the static eccentricity and bearing wear faults, a classifier was applied. Classifiers were trained using a training feature set,

Figure 5. The diagram of the SVM parameters optimization process

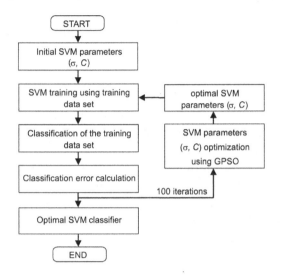

and the classification error on this data set, i.e., the number of false classifications, was used as the optimality criterion for SVM parameter optimization. This criterion is optimized using GPSO algorithm and optimal values of parameters C and σ were determined. GPSO parameters were set according to the GPSO1 settings, with 100 iterations and 30 particles. The diagram of the SVM parameter optimization process is shown in Figure 5.

The trained classifiers were tested using the test set of 54 signal samples: 18 samples were from healthy motors, 4 samples represented only bearing wear and 32 samples were from motors with both faults (static eccentricity and bearing wear) present at the same time. Both classifiers, applied on described test set, had only one false classification, i.e. classification error for both of them was only 1.85%. These classification results demonstrate the efficiency of classifiers and the effect of SVM parameter optimization. Described experimental results proved that GPSO can be successfully applied in fault detection, providing accurate, efficient and reliable fault classifiers applicable in real industrial systems.

Application Example 2: GPSO in Water Distribution System Control Optimization

In this section, an application of GPSO to determine the optimal control strategy of pump operation in a water distribution system is presented. The goal of the research was to determine optimal flow, i.e. the optimal pump rotation speed in specified periods during one day. The energy cost of the pump drive is used as optimization criterion. It is defined using quasy steady-state model of water distribution network, consisting of two reservoirs, two variable speed pumps and a distribution pipe network. At the same time, all operational parameters of the system (pressure in all nodes, level in a reservoir and its daily trajectory) must be in assigned range, which represents stiff constraints in optimization process. These constraints are implemented in optimization criterion using penalty functions with adjustable weights, as it will be explained in the sequel.

The model represents a part of water distribution network of Novi Sad. The start point in the network is the reservoir 1 located on 132 m altitude. Beside this reservoir, a pump station is located, which pumps water to main pipeline. This station has two 45 kW variable speed pumps controlled by frequency controllers. The main pipeline transports water from reservoir 1 to reservoir 2 which is located on 150 m altitude and has accumulative function. This pipeline also supplies consumers located between these two reservoirs. Detailed system description can be seen in (Kanović et al, 2008).

Using real system data, such as pipe length, diameter and material, estimated consumption in nodes and input variables, i.e. e. pump speed in every simulation step, quasy steady-state model is formed based on mass and energy conservation laws. As a result of this model, pressures in nodes and flows in pipes in entire system are calculated (Sanks, 1998). Duration of the simulation period is 24 hours, and the simulation step is adopted to be 30 minutes. Based on simulation results, total energy costs are calculated:

$$C = \sum_i C_i E_i \Delta t_i \qquad (37)$$

where C represents total costs, C_i is price of kWh in appropriate simulation step, Δt_i is simulation step duration (30 minutes), i is simulation step and E_i is electrical power consumed by pump, obtained from:

$$E_i = k \gamma Q_i H p_i / \eta \qquad (38)$$

where k is conversion factor, γ is fluid density in N/m³, Q_i is pump flow in l/s, Hp_i is pump head in meters and η is total efficiency factor (Sanks, 1998).

Following constraints have been imposed to the system:

- Water level in reservoir 2 must not exceed minimal or maximal level.
- Pressure in nodes must not exceed minimal or maximal allowed pressure.
- Water level at the start and at the end of the simulation must be equal, to close daily cycle.

These constraints are embedded in cost function using penalty functions. For every parameter an appropriate penalty is formed, as follows:

$$f = \begin{cases} (P - P_{max}) / (P_{max} - P_{min}) , P > P_{max} \\ 0 , P_{max} > P > P_{min} \\ (P_{min} - P) / (P_{max} - P_{min}) , P < P_{min} \end{cases}$$
$$(39)$$

where f is the penalty for the appropriate parameter (water level in reservoir, pressure in nodes and water level at the end of simulation), P is parameter value and P_{min} and P_{max} are minimal and maximal

allowed parameter values. Finally, cost function is defined as follows:

$$F = C + \sum_i \sum_j \alpha_j f_{ij} \qquad (40)$$

where F is cost function value, C is total energy cost obtained from (37), f_{ij} is penalty for appropriate parameter, α_j is appropriate penalty weight factor, i is simulation step and j is parameter index.

Calculating minimum of Cost Function 40, the optimal pump speed in every simulation step, i.e. the optimal pump control strategy is determined, satisfying all physical constraints in the system. The GPSO optimization was conducted in 200 iterations, with 100 particles in the swarm. The energy cost was set to 0.1 money units in lower rate (from 11 pm to 7 am) and to 0.3 money units in higher rate (from 7 am to 11 pm). Pressure in nodes must be in range from 2 bar to 6.5 bar, and the water level in reservoir 2 must be in the range from 1 to 3.5 meters. GPSO parameters were set according to scheme GPSO1.

Figure 6 a) shows optimal pump flow and total consumption in the network. The period from 11 pm to 6 am, which is in lower energy rate, is used for filling up the reservoir 2 (pump flow is higher than consumption). In the period from 4 pm to 7 pm the pump flow is also higher than consumption, to prevent reservoir discharge. In the rest of the day, the pump flow is lower than consumption, and consumer are additionally supplied by draining the reservoir 2.

Figure 6 b) shows water level trajectory in reservoir 2. The reservoir is drained in higher energy rate, and lower rate is used for filling. Also, one can notice that the reservoir is once filled and drained during the day, which prevents water aging and ensures good microbiological water quality.

The obtained optimal pump flow is used as a system control strategy. Implemented in SCADA of water distribution system, it provides more efficient and more reliable water distribution with optimal energy cost.

Figure 6. a) Pump flow and total consumption in system; b) Water level in reservoir 2

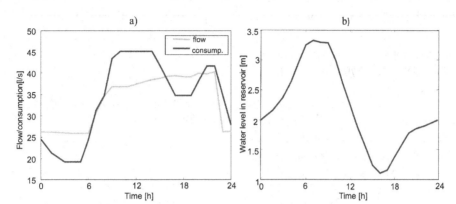

FUTURE RESEARCH DIRECTIONS

There are several possibilities for further research and development in the area of PSO and GPSO modification. In particular, more complex models than Equation 5, whether linear or even non-linear, can be derived. Further research should also address the topology of the swarm and exploit various patterns of communication among the particles. Different swarm topologies have already been proven to be beneficial for the performance of classical PSO (Niu et al, 2006). Hybridization with evolutionary algorithms is also known to improve particle swarm optimizers (Senthil Arumungam & Rao, 2007). The incorporation of evolutionary operators, such are crossover and mutation, should also improve the performance of GPSO.

CONCLUSION

Finding the global minimum of a function is generally an ill-posed problem (Schute & Groenwold, 2005). However, swarm-based methods in general and PSO in particular provide us with powerful and robust tools for tackling the global optimization problems in science and engineering. In this chapter, the idea of PSO algorithm modification, named GPSO, is presented. This modification incorporates requirements concerning explora-

tion and exploitation properties in a formalism usually connected to the linear control theory. Convergence conditions of this algorithm have been derived, and the influence of parameters on particle dynamics and optimizer's performance has been investigated. A broad empirical study based on various benchmark problems was also conducted. Based on an analysis of the obtained results, the parameter adjusting schemes are recommended. In most of the examples, the proposed GPSO schemes perform better in comparison to some other algorithms, such as TVAC-PSO and GA. Finally, two application examples of GPSO are presented, demonstrating the potential of this algorithm for practical engineering problems.

REFERENCES

Alireza, A. (2011). PSO with adaptive mutation and inertia weight and its application in parameter estimation of dynamic systems. *Acta Automatica Sinica, 37*(5), 541–548.

Bouallegue, S., Haggege, J., Ayadi, M., & Benrejeb, M. (2012). PID-type fuzzy logic controller tuning based on particle swarm optimization. *Engineering Applications of Artificial Intelligence, 25*, 484–493. doi:10.1016/j.engappai.2011.09.018

Chatterjee, A., Ghoshal, S. P., & Mukherjee, V. (2011). Craziness-based PSO with wavelet mutation for transient performance augmentation of thermal system connected to grid. *Expert Systems with Applications, 38*, 7784–7794. doi:10.1016/j.eswa.2010.12.128

Chen, M. R., Li, X., Zhang, X., & Lu, Y. Z. (2010). A novel particle swarm optimizer hybridized with extremal optimization. *Applied Soft Computing, 10*, 367–373. doi:10.1016/j.asoc.2009.08.014

Clerc, M., & Kennedy, J. (2002). The particle swarm - Explosion, stability and convergence in a multidimensional complex space. *IEEE Transactions on Evolutionary Computation, 6*(1), 58–73. doi:10.1109/4235.985692

Cui, Z., Cai, X., Zeng, J., & Sun, G. (2008). Particle swarm optimization with FUSS and RWS for high dimensional functions. *Applied Mathematics and Computation, 205*, 98–108. doi:10.1016/j.amc.2008.05.147

DeBao, C., & ChunXia, Z. (2009). Particle swarm optimization with adaptive population size and its application. *Applied Soft Computing, 9*, 39–48. doi:10.1016/j.asoc.2008.03.001

Dimopoulos, G. G. (2007). Mixed-variable engineering optimization based on evolutionary and social metaphors. *Computer Methods in Applied Mathematics and Engineering, 196*, 803–817. doi:10.1016/j.cma.2006.06.010

Dong, Y. (2005). An application of swarm optimization to nonlinear programming. *Computers & Mathematics with Applications (Oxford, England), 49*, 1655–1668. doi:10.1016/j.camwa.2005.02.006

Dorell, D. G., Thomson, W. D., & Roach, S. (1997). Analysis of air gap flux, current, and vibration signals as a function of the combination of static and dynamic air gap eccentricity in 3-phase induction motors. *IEEE Transactions on Industry Applications, 33*(1), 24–34. doi:10.1109/28.567073

Dorigo, M., & Blum, C. (2005). Ant colony optimization theory: A survey. *Theoretical Computer Science, 344*, 243–278. doi:10.1016/j.tcs.2005.05.020

Eberhart, R. C., & Shi, Y. (2000). Comparing inertia weights and constriction factors in particle swarm optimization. *Proceedings of the Congress on Evolutionary Computation* (pp. 84-88). La Jolla, California.

Fang, H., Chen, L., & Shen, Z. (2011). Application of an improved PSO algorithm to optimal tuning of PID gains for water turbine governor. *Energy Conversion and Management, 52*, 1763–1770. doi:10.1016/j.enconman.2010.11.005

He, Q., & Wang, L. (2007). An effective co-evolutionary particle swarm optimization for constrained engineering design problems. *Engineering Applications of Artificial Intelligence, 20*, 89–99. doi:10.1016/j.engappai.2006.03.003

He, Q., Yan, R., Kong, F., & Du, R. (2009). Machine condition monitoring using principal component representations. *Mechanical Systems and Signal Processing, 23*, 446–466. doi:10.1016/j.ymssp.2008.03.010

Jiang, M., Luo, Y. P., & Yang, S. Y. (2007). Stochastic convergence analysis and parameter selection of standard particle swarm optimization algorithm. *Information Processing Letters, 102*, 8–16. doi:10.1016/j.ipl.2006.10.005

Jiang, Y., Hu, T., Huang, C. C., & Wu, X. (2007). An improved particle swarm optimization algorithm. *Applied Mathematics and Computation, 193*, 231–239. doi:10.1016/j.amc.2007.03.047

Jiang, Y., Liu, C., Huang, C., & Wu, X. (2010). Improved particle swarm algorithm for hydrological parameter optimization. *Applied Mathematics and Computation, 217*, 3207–3215. doi:10.1016/j.amc.2010.08.053

Jie, J., Zeng, J., Han, C., & Wang, Q. (2008). Knowledge-based cooperative particle swarm optimization. *Applied Mathematics and Computation*, *205*, 861–873. doi:10.1016/j.amc.2008.05.100

Kankar, P. K., Sharma, S. C., & Harsha, S. P. (2011). Fault diagnosis of ball bearings using machine learning methods. *Expert Systems with Applications*, *33*, 1876–1886. doi:10.1016/j.eswa.2010.07.119

Kanović, Ž., Rapaić, M., & Erdeljan, A. (2008). Generalized PSO algorithm in optimization of water distribution. *Proceedings of Planning and Management of Water Resources Systems*, (pp. 203-210). Novi Sad.

Kanović, Ž., Rapaić, M., & Jeličić, Z. (2011). Generalized particle swarm optimization algorithm - Theoretical and empirical analysis with application in fault detection. *Applied Mathematics and Computation*, *217*, 10175–10186. doi:10.1016/j.amc.2011.05.013

Kennedy, J., & Eberhart, R. C. (1995). Particle swarm optimization. *Proceedings of IEEE International Conference on Neural Networks* (pp. 1942-1948). Perth, Australia.

Kirkpatrick, S., Gellat, C. D., & Vecchi, M. P. (1983). Optimization by simulated annealing. *Science*, *220*(4598), 671–680. doi:10.1126/science.220.4598.671

Liang, J. J., & Suganthan, P. N. (2005). Dynamic multi-swarm particle swarm optimizer. *Proceedings of IEEE Swarm Intelligence Symposium* (pp. 124-129). Pasadena, California.

Majhi, B., & Panda, G. (2011). Robust identification of nonlinear complex systems using low complexity ANN and particle swarm optimization technique. *Expert Systems with Applications*, *38*, 321–333. doi:10.1016/j.eswa.2010.06.070

Mallipeddi, R., Jeyadevi, S., Suganthan, P. N., & Baskar, S. (2012). Efficient constraint handling for optimal reactive power dispatch problems. *Swarm and Evolutionary Computation*, *5*, 28–36. doi:10.1016/j.swevo.2012.03.001

Menhas, M. I., Wang, L., Fei, M., & Pan, H. (2012). Comparative performance analysis of various binary coded PSO algorithms in multi-variable PID controller design. *Expert Systems with Applications*, *39*, 4390–4401. doi:10.1016/j.eswa.2011.09.152

Michalewicz, Z. (1999). *Genetic algorithms + data structures = Evolution programming* (3rd ed.). Berlin, Germany: Springer.

Niu, B., Zhu, Y., He, X., & Wu, H. (2006). MCPSO: A multi-swarm cooperative particle swarm optimizer. *Applied Mathematics and Computation*, *185*, 1050–1062. doi:10.1016/j.amc.2006.07.026

Olamaei, Y., Niknam, T., & Gharehpetian, G. (2008). Application of particle swarm optimization for distribution feeder reconfiguration considering distributed generators. *Applied Mathematics and Computation*, *201*, 575–586. doi:10.1016/j.amc.2007.12.053

Ozcan, E., & Mohan, C. K. (1998). Analysis of simple particle swarm optimization system. *Intelligent Engineering Systems through Artificial. Neural Networks*, *8*, 253–258.

Papoulis, A. (1965). *Probability, random variables and stochastic processes*. New York, NY: McGraw-Hill.

Ramesh, L., Chakraborthy, N., Chowdhury, S. P., & Chowdhury, S. (2012). Intelligent DE algorithm for measurement location and PSO for bus voltage estimation in power distribution system. *Electrical Power and Energy Systems*, *39*, 1–8. doi:10.1016/j.ijepes.2011.10.009

Rapaić, M. R., & Kanović, Ž. (2009). Time-varying PSO - Convergence analysis, convergence related parameterization and new parameter adjustment schemes. *Information Processing Letters, 109*, 548–552. doi:10.1016/j.ipl.2009.01.021

Ratnaweera, A., Saman, K. H., & Watson, H. C. (2004). Self-organizing hierarchical particle swarm optimizer with time-varying acceleration coefficients. *IEEE Transactions on Evolutionary Computation, 8*(3), 240–255. doi:10.1109/TEVC.2004.826071

Samanta, B., & Nataraj, C. (2009). Use of particle swarm optimization for machinery fault detection. *Engineering Applications of Artificial Intelligence, 22*, 308–316. doi:10.1016/j.engappai.2008.07.006

Sanks, R. L. (Ed.). (1998). *Pumping station design.* Woburn, MA: Butterworth – Heinemann.

Schutte, J. C., & Groenwold, A. A. (2005). A study of global optimization using particle swarms. *Journal of Global Optimization, 31*, 93–108. doi:10.1007/s10898-003-6454-x

Senthil Arumugam, M., & Rao, M. V. C. (2007). On the improved performances of the particle swarm optimization algorithms with adaptive parameters, cross-over operators and root mean square (RMS) variants for computing optimal control of a class of hybrid systems. *Applied Soft Computing, 8*, 324–336. doi:10.1016/j.asoc.2007.01.010

Senthil Arumugam, M., Rao, M. V. C., & Tan, A. W. C. (2009). A novel and effective particle swarm optimization like algorithm with extrapolation technique. *Applied Soft Computing, 9*(1), 308–320. doi:10.1016/j.asoc.2008.04.016

Shahzad, F., Rauf Baig, A., Masood, S., Kamran, M., & Naveed, N. (2009). Opposition-based particle swarm optimization wit velocity clamping (OVCPSO) . In Yu, W., & Sanchez, E. N. (Eds.), *Advances in computational intelligence* (pp. 339–348). Berlin, Germany: Springer. doi:10.1007/978-3-642-03156-4_34

Shi, Y., & Eberhart, R. C. (1999). Empirical study of particle swarm optimization. *Proceedings of IEEE International Congress on Evolutionary Computation,* Vol. 3, (pp. 101-106). Washington, DC.

Stepanić, P., Latinović, I., & Đurović, Ž. (2009). A new approach to detection of defects in rolling element bearings based on statistical pattern recognition. *International Journal of Advanced Manufacturing Technology, 45*, 91–100. doi:10.1007/s00170-009-1953-7

Åström, K. J., & Wittenmark, B. (1997). *Computer-controlled systems - Theory and design,* 3rd ed. New Jersey: Prentice Hall.

Thangaraj, R., Pant, M., Abraham, A., & Bouvry, P. (2011). Particle swarm optimization: Hybridization perspectives and experimental illustrations. *Applied Mathematics and Computation, 217*, 5208–5226. doi:10.1016/j.amc.2010.12.053

Van der Bergh, F. (2001). *An analysis of particle swarm optimizers.* PhD Thesis, University of Pretoria, Pretoria.

Wang, L. (Ed.). (2005). *Support vector machines: Theory and applications.* Berlin, Germany: Springer.

Widodo, A., Yang, B., & Han, T. (2007). Combination of independent component analysis and support vector machines for intelligent faults diagnosis of induction motors. *Expert Systems with Applications, 32*, 299–312. doi:10.1016/j.eswa.2005.11.031

ADDITIONAL READING

Andrade, F. A., Esat, I., & Badi, M. N. M. (2001). A new approach to time-domain vibration condition monitoring: gear tooth fatigue crack detection and identification by the Kolmogorov-Smirnov test. *Journal of Sound and Vibration, 240*(5), 909–919. doi:10.1006/jsvi.2000.3290

Chow, M.-Y. (2000). Guest editorial special section on motor fault detection and diagnosis. *IEEE Transactions on Industrial Electronics*, *47*(5), 982–983. doi:10.1109/TIE.2000.873205

Corne, D., Dorigo, M., & Glover, F. (Eds.). (1999). *New ideas in optimization*. New York, NY: McGraw Hill.

Delac, K., Grgic, M., & Grgic, S. (2005). Independent comparative study of PCA, ICA, and LDA on the FERET data set. *International Journal of Imaging Systems and Technology*, *15*, 225–260. doi:10.1002/ima.20059

Filippetti, F., Franceschini, G., Tassoni, C., & Vas, P. (1998). AI techniques in induction machines diagnosis including the speed ripple effect. *IEEE Transactions on Industry Applications*, *34*(1), 98–108. doi:10.1109/28.658729

Floudas, C. A. (1999). *Handbook of test problems in local and global optimization*. New York, NY: Springer-Verlag.

Fukuyama, H. Y. Y., Takayama, K. K. S., & Nakanishi, Y. (2001). A particle swarm optimization for reactive power and voltage control considering voltage security assessment. *IEEE Transactions on Power Systems*, *15*(4), 1232–1239.

He, Q., Yan, R., Kong, F., & Du, R. (2009). Machine condition monitoring using principal component representations. *Mechanical Systems and Signal Processing*, *23*, 446–466. doi:10.1016/j.ymssp.2008.03.010

Jack, L. B., & Nandi, A. K. (2002). Fault detection using support vector machines and artificial neural networks augmented by genetic algorithms. *Mechanical Systems and Signal Processing*, *16*(2–3), 373–390. doi:10.1006/mssp.2001.1454

Jolliffe, I. J. (1986). *Principal component analysis*. New York, NY: Springer.

Kanović, Ž., Rapaić, M., & Erdeljan, A. (2007). PSO algorithm in optimization of water distribution system operation. *Proceedings of 9th International Symposium Interdisciplinary Regional Research ISIRR 2007*, Novi Sad.

Kecman, V. (2001). *Learning and soft computing-Support vector machines, neural networks, and fuzzy logic models*. Cambridge, MA: MIT Press.

Obaid, R. R. (2002). *Detection of rotating mechanical asymmetries in small induction machines*. Ph.D. Dissertation, Georgia Institute of Technology, USA.

Rapaić, M., Kanović, Ž., & Jeličić, Z. (2008). Discrete particle swarm optimization algorithm for solving optimal sensor deployment problem. *Journal of Automatic Control*, *18*(1), 9–14. doi:10.2298/JAC0801009R

Rapaić, M., Kanović, Ž., & Jeličić, Z. (2009). A theoretical and empirical analysis of convergence related particle swarm optimization. *WSEAS Transactions on Systems and Control*, *4*(11), 541–550.

Rapaić, M. R., Kanović, Ž., Jeličić, Z. D., & Petrovački, D. (2008). Generalized PSO algorithm – An application to Lorenz system identification by means of neural-networks. *Proceedings of 9th Symposium on Neural Network Applications in Electrical Engineering NEUREL-2008* (pp. 31-35). Belgrade.

Tandon, N., & Choudhury, A. (1999). A review of vibration and acoustic measurement methods for the detection of defects in rolling element bearings. *Tribology International*, *32*, 469–480. doi:10.1016/S0301-679X(99)00077-8

Van den Bergh, F., & Engelbrecht, A. P. (2002). A new locally convergent particle swarm optimiser. *Proceedings of the IEEE International Conference on Systems, Man and Cybernetics*, Tunisia.

Vapnik, V. N. (1995). *The nature of statistical learning theory*. New York, NY: Springer.

Williams, T. (2001). Rolling element bearing diagnostics in run-to-failure lifetime testing. *Mechanical Systems and Signal Processing, 15*(5), 979–993. doi:10.1006/mssp.2001.1418

Yang, H., Mathew, J., & Ma, L. (2003). Vibration feature extraction techniques for fault diagnosis of rotating machinery: A literature survey. *Proceedings of Asia-Pacific Vibration Conference*. Gold Coast, Australia.

KEY TERMS AND DEFINITIONS

Control Theory: A science that deals with monitoring and controlling processes and studying their properties.

Evolutionary Computation: Subfield of artificial intelligence (more particularly computational intelligence) that involves combinatorial optimization problems.

Fault Detection: A set of methods and techniques for equipment monitoring and fault determination and diagnosis.

Genetic Algorithms: Global optimization algorithms based on concepts of Charles Darwin's natural evolution theory.

Global Optimization: Branch of applied mathematics that deals with the optimization of the objective function or a set of objective functions with multiple local optima.

Particle Swarm Optimization: Global optimization algorithm inspired by social behavior of animals living and moving in large groups (birds, fish, insects…).

Swarm Intelligence: Collective behavior of decentralized, self-organized systems, natural or artificial; a kind of artificial intelligence that aims to simulate the behavior of swarms.

Section 4
Scheduling and Diagnosis

Nowadays, many scientific experiments are conducted through complex and distributed scientific computations that are represented and structured as scientific workflows. Moreover, nondestructive testing and imaging techniques are widely used in many industrial and research applications. This section reports some recent studies of PSO to solve hard problem of scheduling, task allocation and diagnosis. In detail, the first chapter illustrates the implementation of PSO to determine all the changes needed in the electric transmission system infrastructure allowing the balance between the projected demand and the power supply, at minimum investment and operational costs. The second chapter proposes the application of PSO to solve the problem of short-term hydrothermal generation scheduling problem which consists of minimizing the fuel cost for thermal plants under the constraints of the water available for hydro generation in a given time period. The third chapter illustrates the ant colony optimization method in the framework of the nondestructive analysis of dielectric targets by using electromagnetic approaches based on inverse scattering. Finally, the forth chapter provides a survey of some common evolutionary algorithms used in electroencephalogram studies.

Chapter 13
Transmission Expansion Planning by using DC and AC Models and Particle Swarm Optimization

Santiago P. Torres
University of Campinas (UNICAMP), Brazil

Carlos A. Castro
University of Campinas (UNICAMP), Brazil

Marcos J. Rider
São Paulo State University (UNESP), Brazil

ABSTRACT

The Transmission Expansion Planning (TEP) entails to determine all the changes needed in the electric transmission system infrastructure in order to allow the balance between the projected demand and the power supply, at minimum investment and operational costs. In some type of TEP studies, the DC model is used for the medium and long term time frame, while the AC model is used for the short term. This chapter proposes a load shedding based TEP formulation using the DC and AC model, and four Particle Swarm Optimization (PSO) based algorithms applied to the TEP problem: Global PSO, Local PSO, Evolutionary PSO, and Adaptive PSO. Comparisons among these PSO variants in terms of robustness, quality of the solution, and number of function evaluations are carried out. Tests, detailed analysis, guidelines, and particularities are shown in order to apply the PSO techniques for realistic systems.

DOI: 10.4018/978-1-4666-2666-9.ch013

INTRODUCTION

The Transmission Expansion Planning (TEP) problem consists of determining all the changes needed in the transmission system infrastructure, i.e. additions, modifications and/or replacements of obsolete transmission facilities, in order to allow the balance between the projected demand and the power supply, at minimum investment and operational costs. The TEP problem is a large scale, mixed–integer, non-linear, non-convex and combinatorial problem. It has been largely discussed in the specialized literature and this is still considered a very complex problem where better algorithms are needed. The publications, according to the optimization technique, are classified into mathematical, heuristic, and meta-heuristic approaches. Techniques such as dynamic programming (Dusonchet, 1973), linear programming (Oliveira, 2007), non-linear programming (Youssef, 1989), mixed-integer programming (Alguacil, 2009), Benders decomposition (Akbari, 2011), Hierarchical decomposition (Romero, 1994), and Branch and Bound (Choi, 2007) method have been used and are categorized as mathematical based approaches. These techniques demand large computing time, due to the dimensionality curse of this kind of problem. Heuristic methods emerged as an alternative to classical optimization methods; their use has been very attractive since they were able to find good feasible solutions, demanding a small computational effort. Some heuristic approaches have been proposed using constructive heuristic algorithms (Romero, 2005), and the forward – backward approach (Seifi, 2007). Meta-heuristic methods emerged as an alternative to the two previous approaches, producing high quality solutions with moderate computing time. Genetic algorithms (Gallego, 2007; Rodriguez, 2009), Differential evolution (Georgilakis, 2011), Greedy Randomized Adaptive Search Procedure (Binato, 2001), Harmony

search algorithm (Verma, 2010), Tabu search (Da Silva, 2001) have been used to solve the TEP problem among other metaheuristic optimization techniques. It is important to point out that they cannot guarantee the global optimal solution to the TEP problem.

In the last years, several novel meta-heuristic techniques have been proposed. In particular, Particle Swarm Optimization (PSO) has been successful in tackling power systems related problems, and constitute a serious option when one has to solve complex optimization problems (Del Valle, 2008; Torres, 2011).

The TEP problem involves several hierarchical levels of power system analysis studies and requires a large amount of expertise of the system planners. In each stage of the planning horizon, some alternatives for the TEP are selected and used as a starting point for the next planning stage. This kind of study can be performed for three time frames, namely long term, medium term, and short term. Each stage presents some particularities, which requires the utilization of an adequate power system model. The DC model is used for medium and long term studies, while the AC Model has been recently proposed to be used for short term ones (Rider, 2007).

There are several works reported in the literature regarding the use of the DC Model in TEP (for instance Romero, 2002; LaTorre, 2003; Romero, 2005). On the other hand, only very few works have been developed using the AC network model (Rider, 2007; Rider, 2007; Rodriguez, 2008; Gallego, 2009; Rodriguez, 2009; Rahmani, 2010).

One of the most common approaches to deal with the TEP problem is by modeling the load shedding, which has been proposed only using the DC model. The load shedding approach is not only useful in quantifying this variable in the obtained expansion plans, but also it is important to penalize the objective function, in case of a constraint violation, in an easy way. In this research work, the main contribution is to extend the load shedding formulation to solve the Transmission Expansion Planning using the AC network model. Additionally, it is presented and compared four PSO variations to solve the TEP problem. Section II presents a brief explanation of the different perspectives of the Transmission Expansion Planning problem, the state of art of the application of PSO in TEP, and the formulation of the TEP problem using both DC and AC electric network models. Section III shows the theoretical key points of the following PSO variants used in this work: Global Particle Swarm Optimization (GPSO), Local Particle Swarm Optimization (LPSO), Evolutionary Particle Swarm Optimization (EPSO), and Adaptive Particle Swarm Optimization (APSO). The implementation issues regarding the application of the four PSO variants to the TEP problem are discussed in Section IV. In order to test the PSO algorithms applied to the TEP problem, four test systems will be used, and the results are shown in Section V. Some conclusions and future work perspectives are discussed in Section VI.

Extensive simulations have been performed by using some small, medium, and large-sized test systems. Also, detailed analyses are shown in order to allow the implementation of PSO for the TEP problem. This work contributes with a formulation based on load shedding by using the AC Model integrating real–reactive power planning, and the application of four PSO variants to solve the TEP problem using both DC and AC models.

THE TRANSMISSION EXPANSION PLANNING PROBLEM USING DC AND AC MODELS

As mentioned in the introductory section, the main goal of the TEP problem is to determine, in the most economical way, the changes needed in the transmission system infrastructure due to the load increase and the future power generation plants

added to the electrical network. This topic can be studied from different points of view, and the TEP problem can be classified according to the: a) regulatory structure, b) uncertainties involved in the problem, and c) time frame structure.

Regulatory Structure: The Transmission Planning in a vertically integrated electricity sector is centralized, where a unique organization performs the transmission expansion planning process. On the other hand, under a market structure, there are many agents (regulatory agency, utility agents, and external agents) where each one works in their own plans. Some work that takes the regulatory structure into account can be found in references (Bugyi, 2004; De la Torre, 2008).

Uncertainties: There are some uncertainties involved in the TEP problem, specialty those related to the load increasing and the power generation planning. From this point of view, the problem can be classified as deterministic or non-deterministic the uncertainty could be taken into account using a simply scenario based approach or a more sophisticated technique considering the load (or generation) as a variable inside the optimization process (Enamorado, 1999; Silva, 2006).

Time Frame: The TEP problem can be classified as either static (Romero, 2002; Romero, 2005), where the expansion plan includes the determination of where and what additions the electric power system needs, or dynamic (Escobar 2004), where the expansion plan must also include the best schedule for such additions. The multistage or dynamic transmission expansion planning problem determines not only optimal locations in order to satisfy the continuing growth of the demand and generation. In this case, the planning horizon is divided into several stages and the circuits must be added to each stage of the planning horizon; investment is carried out at the beginning of each stage. The objective is to minimize the present value of the sum of all the investments carried out throughout the years corresponding to the simulated periods.

In this work, the problem is considered as deterministic and static without taking into account regulatory aspects. Commonly, this type of study is performed in three stages according to the time horizon (Rodriguez, 2008):

- Long-term, with a time horizon up to 20 years, in which the large transmission interconnections, closely associated with new energy sources and power demand, are defined. In this stage, the study is usually performed using simplified models of the network such as the DC or the transportation model.
- Medium-term, up to 10 years, in which more details of the interconnections are determined (voltage level, etc.) and alternatives for the regional systems are provided.
- Short-term, up to 5 years, where the final adjustments are made regarding to alternatives previously chosen, such as reactive compensation, dynamic behavior (voltage stability, angle stability, etc), reliability, short circuits, etc., and information regarding to the system operation are added to the planning process. In the short-term horizon, detailed studies of the few alternatives left are conducted. This kind of study uses the full AC power flow formulation in order to assess the power losses and the reactive compensation requirements in an accurate way, for both the base case and contingencies.

The application of the Particle Swarm Optimization (PSO) technique as an optimization tool is not new in the power systems area (Del Valle, 2008). In the transmission expansion planning area, most research works in the literature propose solving the TEP by using some basic *global best* based PSO variants, and evaluate them for very small test systems (Dong-Xiao, 2006; Kavitha, 2006; Pringles, 2007; Ren, 2005; Sensarma, 2002; Shayegui 2009; Torres, 2011; Verma, 2009; Yi-Xiong, 2005; Yi-Xiong, 2006). All these works

use the DC electric network model only, and there is no work related to the application of PSO using the AC model.

The use of a heuristic approach based on a sensitivity index to add lines, transformers or capacitors, in order to solve the short-term transmission expansion planning problem was proposed by (Rider, Gallego, 2007). An improved version of this work is presented in (Rider, Garcia, 2007). The disadvantage of using a sensitivity index, as pointed out in the references themselves, is that the solutions obtained could not be near optimal, specialty for larger power networks. In (Rider, 2007; Gallego, 2009; Rodriguez, 2009; Rahmani, 2010) some optimization tools based on genetic algorithms are used to solve the TEP problem by using the AC model.

PARTICLE SWARM OPTIMIZATION VARIANTS

In this section, the Particle Swarm based optimization tools used in this work are described, mainly based on references (Kennedy, 2001; Clerc, 2006; Miranda, 2005; Miranda, 2007; Zhan, 2009).

Particle Swarm Optimization Basics

Swarm intelligence is a branch of artificial intelligence that studies the collective behavior of complex, self-organized, decentralized systems with social structure. Such systems consist of simple interacting agents organized in small societies (swarms). The aggregated behavior of the whole swarm exhibits traits of intelligence. In the algorithms based on swarm intelligence, there are five basic principles:

1. **Proximity:** Ability to perform space and time computations.
2. **Quality:** Population ability to respond to environmental quality factors.
3. **Diverse Response:** Ability to produce a wide set of different responses.

4. **Stability:** Ability to retain robust behaviors under mild environmental changes.
5. **Adaptability:** Ability to change behavior when it is dictated by external factors.

PSO, a swarm intelligence technique developed by Kennedy and Eberhart (Kennedy, 2001), is a stochastic optimization algorithm based on social simulation models. The development of PSO was based on concepts that govern socially organized populations in nature, such as bird flocks, fish schools, and animal herds. It employs a population of search points that moves stochastically in the search space. The best experience or position achieved by each individual is retained, and then communicated to part or the whole population. The communication scheme is determined by a fixed or adaptive social network that plays a crucial role on the convergence properties of the algorithm.

Global Version of Particle Swarm Optimization

Two variants of PSO were initially developed to take advantage of properties of exploration/exploitation, namely, the global (GPSO), where the entire swarm is considered as the neighborhood of each particle, and the local (LPSO), where neighborhoods are strictly smaller. In a mathematical framework, let, $A \subset R^n$, be the search space, and, $f : A \to Y \subseteq R^n$, be the objective function, the population is called the *swarm* and its individuals are called the *particles*. The swarm is defined as a set $S = (x_{i1}, x_{i2}, ..., x_{im})^T$ of N particles, defined as $x_i = (x_{i1}, x_{i2}, ..., x_{im})^T \epsilon A$ where N is a

user-defined parameter dependent on the problem. Each particle can be an m-component vector, which also defines the dimension of the problem. The objective function, $f(x)$, is assumed to be available for all points in A. Therefore, each particle has a unique function value $f_i = f(x_i) \epsilon Y$. The particles are assumed to move within the search space A iteratively, adjusting their position shift, called *velocity*, and denoted as $v_1 = (v_{i1}, v_{i2}, ..., v_{im})^T$, $i = 1, 2, ..., N$. Velocity is updated based on information obtained in previous steps of the algorithm. This is implemented in terms of a memory, where each particle can store the *best position* it has ever visited during its search. Also, let $P = \{p_1, p_2, ..., p_m\}$, be the *memory* set which contains the best positions $p_i = (p_{i1}, p_{i2}, ..., p_{im})^T \epsilon A$, $i=1, 2, ..., N$, ever visited by each particle. Assuming a minimization problem, let g be the index of the best position with the lowest function value in P at a given iteration t, then the global version of PSO is defined in Equations 1 and 2 (see Box 1) where $v_{ij}(t+1)$ is the velocity of each particle; $x_{ij}(t+1)$ is the position of each particle; $i=1, 2, ..., N$, $j=1, 2, ..., m$; t represents the iteration counter; R_1 and R_2 are random variables distributed uniformly within [0, 1], c_1 and c_2 are weighting factors called the cognitive and social parameters respectively. χ is a parameter called the constriction factor introduced by Clerc and Kennedy (Clerc, 2006) in order to enhance the PSO performance.

At each iteration, after the update and evaluation of particles, the best positions (memory) are also updated. The new determination of index g for the updated best positions completes an iteration of PSO.

Box 1

$$v_{ij}(t+1) = \chi \left[v_{ij}(t) + c_1 R_1 \left(p_{ij}(t) - x_{ij}(t) \right) + c_2 R_2 \left(p_{gj}(t) - x_{ij}(t) \right) \right] \qquad (1)$$

$$x_{ij}(t+1) = x_{ij}(t) + v_{ij}(t+1) \qquad (2)$$

Local Version of Particle Swarm Optimization

The concept of neighborhood comes into play in order to reduce the global information exchange scheme to a local one, where information is diffused only in small parts of the swarm at each iteration. Each particle assumes a set of other particles to be its neighbors and, at each iteration, it communicates its best position only to these particles, instead of to the whole swarm. Thus, information regarding the overall best position is initially communicated only to the neighborhood of the best particle, and successively to the rest through their neighbors. In general, a neighborhood of a particle x_i belonging to a swarm $S = \{x_1, x_2, ..., x_N\}$, is defined as a set $NB_i = \{x_{n1}, x_{n2}, ..., x_{ns}\}$, where, $\{n1, n2, ..., ns\} \subseteq \{1, 2, ..., N\}$ is the set of indices of its neighbors. The global variant converges faster towards the overall best position than the local variant. Therefore, the former stands out for its exploitation ability. The local variant (LPSO) has better exploration abilities (Parsopoulus 2010) since information regarding the best position of each particle is gradually communicated to the other particles through their neighbors. The precise neighbor's definition adopted here is as following:

$$NB_i = \{x_{i-r}, x_{i-r+1}, ..., x_{i-1}, x_i, x_{i+1}, ..., x_{i+r-1}, x_{i+r}\},$$

where parameter r determines the neighborhood size, and is known as *neighborhood radius (NR)*. The local version of Particle Swarm Optimization, used in this work tries to take advantage of exploration properties. The velocity and position of the swarm in the local version of PSO can be described by Equations 3 and 4 respectively (See Box 2) where l represents the index of the best position P_i in the neighborhood of x_i.

Evolutionary Particle Swarm Optimization

Evolutionary Particle Swarm Optimization (EPSO) is a novel optimization meta-heuristic algorithm. It combines the concepts of Evolutionary Strategies (ES) and Particle Swarm Optimization. Under the name of Evolutionary Strategies an important number of models have been developed. Evolutionary algorithms have been inspired in the biological evolution of species; these rely on Darwinist selection to promote progress toward the optimal. The algorithm EPSO relies on a set of particles that evolve in the search space trying to find the optimal point in this space. Unlike PSO, the evolution not only looks at the behavior of particles but also in the weights that affect their movement when they move forward in the search space. One of the most important features of EPSO is that it is a self-adaptive method, which automatically tunes its parameters or behaviors in order to produce an adequate rate of progress towards the optimum (Miranda, 2005; Miranda, 2007; Lee, 2008, Miranda, 2009). EPSO has two mechanisms (evolutionary and self-adaptive) acting in sequence, each one with its own probability of producing not only better individuals, but also a better average group. At a given iteration, each particle is defined by a position in the search space $x_i(t)$ and a velocity $v_i(t)$. At a given moment in time t, there is at least one particle that has the best position in the search space. The population of particles recognizes such position b_g,

Box 2.

$$v_{ij}(t+1) = \chi \left[v_{ij}(t) + c_1 R_1 \left(p_{ij}(t) - x_{ij}(t) \right) + c_2 R_2 \left(p_{lj}(t) - x_{ij}(t) \right) \right] \tag{3}$$

$$x_{ij}(t+1) = x_{ij}(t) + v_{ij}(t+1) \tag{4}$$

then the particles tend to move in that direction, furthermore each particle is also attracted to his previous best position b_i.

In the course of each generation (iteration), the particles will reproduce and evolve according to the following steps:

Replication: Each particle is replicated r times, giving rise to identical new particles.

Mutation: Strategic parameters (w_i) which affect the particles movement are mutated.

Reproduction: Each mutated particle generates an offspring according to the particle movement rule.

Evaluation: Each offspring is evaluated with an objective function.

Selection: By stochastic tournament or other selection procedure, the best particles survive to form a new generation.

The movement or reproduction rule of the particles is the same as represented by Equation 2. In EPSO, the velocity of the particle in Equation 1 is replaced in Equation 5 (see Box 3) where b_i is the best point found by particle i in its past life up to the current generation; b_g is the best global point found by the swarm of particles in their past life up to the current generation, $x_{ij}(t)$ is the location of particle i, dimension j, at generation t; $v_{ij}(t)$ is the velocity of particle i at generation k, w_{i1} is the weight of inertia term, w_{i2} is the weight of memory term, w_{i3} is the weight of cooperation term, P is the communication factor.

The particle velocity $v_{ij}(t)$ is made up by three terms: the first term of the summation represents inertia, because the particle keeps moving in the direction it had previously moved; the second

term represents memory, the particle is attracted to the best point in its past trajectory, finally the third term represents the cooperation between particles of the swarm, i.e. the particle is attracted to the best point found by all particles. The particle movement rule is illustrated in Figure 1.

In the movement rule, symbol * indicates that these parameters present an evolution as a result of the mutation process. This is an important difference with PSO where the weights are fixed in the optimization process. The mutation rule is

$$w_{ik}^* = w_{ik} \left[\log N(0,1) \right]^\tau \tag{6}$$

where $\log N(0,1)$ is a random variable with log-normal distribution obtained from the Gaussian distribution with mean 0 and variance 1, i.e. $N(0,1)$; τ is a learning parameter, fixed exogenously, which controls the amplitude of the mutation.

Commonly, another form used to mutate the weight is

$$w_{ik}^* = w_{ik} \left[1 + \tau N(0,1) \right] \tag{7}$$

where $N(0,1)$ is a random variable with Gaussian distribution of mean 0 and variance 1. The previous two equations are equivalent provided that τ is small, so that the negative weights are rejected. Moreover, the global best solution b_g is randomly disturbed, which is expressed by

$$b_g^* = b_g + w_{i4}^* N(0,1) \tag{8}$$

where w_{i4} is the fourth strategic parameter (weight) associated with particle i. This parameter controls

Box 3.

$$v_{ij}(t+1) = w_{i1}^* v_{ij}(t) + w_{i2}^* \left(b_i - x_{ij}(t) \right) + w_{i3}^* \left(b_g^* - x_{ij}(t) \right) P \tag{5}$$

Figure 1. Movement of particles in EPSO algorithm

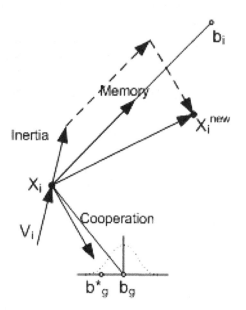

the size of the neighborhood of b_g where it is more likely to find the global best solution or, at least, a better solution than the current b_g. This weight w_{i4} is mutated according to the general mutation rule shown above.

The communication factor P introduces a stochastic draw for the communication between particles. P is a diagonal matrix that affects all particles, contains binary variables of value 1 with probability p and value 0 with probability $(1-p)$. The value of p is an exogenous parameter and it controls the communication of information within the particle swarm. In the classical formulation, p is considered equal to 1. This parameter allows a more individual search by each particle, avoiding premature convergence to local optima.

Adaptive Particle Swarm Optimization

Adaptive Particle Swarm Optimization (APSO) is one of the most advanced variations of Particle Swarm Optimization (Zhan, 2009) so far. It is based on two main strategies. First, by evaluating the population distribution and particle fitness, a

real-time evolutionary state estimation procedure is performed to identify one of the following four defined evolutionary states including exploration, exploitation, convergence, and jumping out in each generation. By doing so, an automatic control of some parameters such as the inertia weight, acceleration coefficients, and other algorithmic parameters is possible at run time to improve the search efficiency and convergence speed. After that, an Elitist Learning Strategy (ELS) is performed when the evolutionary state is classified as a converged state. The strategy is supposed to act on the globally best particle to jump out of the likely local optima. The whole procedure is known as Evolutionary State Estimation (ESE), which is the core of APSO, and can be explained in a summarized way as follows.

1. At the current position, calculate the mean distance of each particle i to all the other particles. This can be calculated using an Euclidean metric:

$$d_i = \frac{1}{N-1} \sum_{j=1,j\neq i}^{N} \sqrt{\sum_{k=1}^{D} \left(x_i^k - x_j^k\right)^2} \qquad (9)$$

2. Let d_g the global best particle. Compare all d_i's obtained in the previous step, and determine the maximum and minimum distances d_{max} and d_{min}. Compute an "evolutionary factor" f as defined by:

$$f = \frac{d_g - d_{min}}{d_{max} - d_{min}} \in \left[0,1\right] \qquad (10)$$

3. Classify f into one of the four sets S_1, S_2, S_3, and S_4, which represents the states of exploration, exploitation, converge and jumping out, respectively. For better results, according to the original work (Zhan, 2009), it is better to define them using fuzzy classification. For space reasons, the fuzzy classification

is not presented here and can be consulted in the same reference above.

4. Control of acceleration coefficients. The acceleration coefficients (c_1, c_2) can be controlled in an adaptive way. Parameter c_1 represents the "self-cognition" that pulls the particle to its own historical best position, helping to explore local regions and maintaining the swarm diversity. Parameter c_2 represents the "social influence" that helps the swarm to fast converge to the current globally best region. The acceleration coefficients are both initialized to 2.0 and adaptively controlled to the evolutionary state, using some strategies as follows.

   a. Exploration (increasing c_1 and decreasing c_2): this strategy helps particles to explore individually the search space and achieve their own historical best positions.

   b. Exploitation (increasing c_1 slightly and decreasing c_2 slightly): increasing c_1 slowly and maintaining a relatively large value can emphasize the search and exploitation around their own historical best position. At the same time, decreasing c_2 slowly and maintaining a small value can avoid to get a local optimum.

   c. Convergence (increasing c_1 slightly and increasing c_2 slightly): in the convergence state, the swarm seems to find the globally optimal region, and, hence, the influence of c_2 should be emphasized to lead the other particles to the probable global optimal region; therefore, the value of c_2 should be increased. On the other hand, to avoid a premature convergence, which is harmful if the current best region is a local optimum, the value of c_1 is also slightly increased.

   d. Jumping out (decreasing c_1 and increasing c_2): when the globally best particle is jumping out of local optimum toward a better optimum, it is likely to be far away from the many particles of the swarm. As soon as this new region is found by a particle, which becomes the new leader, other should follow it and fly to this new region as fast as possible.

The above adjustments should not be too irruptive; therefore the maximum increment or decrement between two generations is bounded by

$$\left| c_i(g+1) - c_i(g) \right| \leq \delta, \qquad i = 1, 2 \qquad (11)$$

where δ is the acceleration rate, and the best results are obtained by using an uniformly generated random value of δ in the interval [0.05, 0.1]. It is recommended to use 0.5 δ for strategies 4.2 and 4.3, where slightly changes are needed. In addition, the interval [1.5, 2.5] was chosen in order to clamp both parameters c_1 and c_2. If the sum is larger than 4.0, then both c_1 and c_2 are normalized as follows:

$$c_i = \frac{c_i}{c_1 + c_2} 4.0, \qquad i = 1, 2 \qquad (12)$$

5. If the optimization process is in the convergence state, it is possible to perform the ELS. The ELS randomly choses one dimension of the global historical best position, which is denoted by P^d for the dth dimension. Similar to simulated annealing, the mutation operation in evolutionary programming or in evolution strategies, elitist learning is performed through a Gaussian perturbation.

$$P^d = P^d + (x_{max}^d - x_{min}^d) \cdot Gaussian(\mu, \sigma^2)$$
$$(13)$$

where the search range $[x^d_{\min}, x^d_{\max}]$ is the same as the lower and upper bounds of the problem. The *Gaussian*(μ,σ^2) is a random number of a Gaussian distribution with zero mean μ and a standard deviation σ. It is recommended that σ be linearly decreased according the generation number:

$$\sigma = \sigma_{\max} - (\sigma_{\max} - \sigma_{\min})\frac{g}{G} \qquad (14)$$

where $\sigma_{\max}$ and $\sigma_{\min}$ are the upper and lower bounds of σ, which represents the learning scale to reach a new region. It is recommended to use [0.1, 1.0] as bounds for σ.

3. The inertia weight w is adaptively controlled according to the evolutionary state of f using a sigmoid mapping in the following way.

$$w(f) = \frac{1}{1 + 1.5e^{-2.6f}} \in [0.4, 0.9] \qquad \forall f \in [0,1] \qquad (15)$$

In this work, the ESE approach is used together with the local PSO variation, which constitutes the APSO. This section summarized the key points to implement APSO based on reference (Zhan, 2009), where more detailed information can be found.

Mathematical Modeling: DC Model

The mathematical modeling used in this work corresponds to the exact DC model (Romero, 2002) as shown below.

$$\min v = \sum_{(k,l)\in\Omega} c_{kl} n_{kl} + w \qquad (16)$$

$$\min w = \sum_{k\in\Lambda} \alpha_1 r_{Pk} \qquad (17)$$

Subject to

$$Sf + g + r_P = d \qquad (18)$$

$$f_{kl} - \gamma_{kl}\left(n^o_{kl} + n_{kl}\right)\left(\theta_k + \theta_l\right) = 0 \qquad (19)$$

$$\left|f_{kl}\right| \leq \left(n^o_{kl} + n_{kl}\right)\overline{f}_{kl} \qquad (20)$$

$$0 \leq g \leq \overline{g} \qquad (21)$$

$$\underline{r}_P \leq r_P \leq \overline{r}_P \qquad (22)$$

$$0 \leq n_{kl} \leq \overline{n}_{kl} \qquad (23)$$

n integer; θ unbounded

$(k,l) \in \Omega$

Where equation (16) corresponds to the objective function to deal with the expansion planning problem, and equation (17) handle the operational problem. Variable v is the investment due to the addition of new circuits. Variable w is the load shedding when some of the operational constraints (Equations 18-22) are violated. $c_{kl}, \gamma_{kl}, n_{kl}, n^o_{kl}, f_{kl}, \overline{f}_{kl}$

correspond respectively to the cost of a circuit that can be added to the right of way k-l, the susceptance of that circuit k-l, the number of circuits in the right of way k-l, the number of circuits in the base case, the total power flow and the corresponding maximum power flow by circuit in the right of way k-l. S is the branch-node incidence transposed matrix of the power system, f is a vector with f_{kl} elements, g is a vector of active power generation whose maximum value is $\overline{g}$, r_P is the active load shedding modeled by artificial generators whose maximum value is $\overline{r}_P$; α_l is the cost of the load shedding, d is the demand vector, $\overline{n}_{kl}$ is the maximum number of circuits that can be added to the right of way k-l, θ_l is the phase angle of bus l, and $_\Omega$ is the set of all rights of way. The constraint (18) represents the power balance for each node. This constraint models Kirchhoff's current law in the equivalent DC network. Constraint 19 corresponds to applying Ohm's law for the equivalent DC network. Thus, Kirchhoff's voltage law is implicitly taken into account, and these constraints are nonlinear constraints. Constraint 20 represents the power flow limit of transmission lines. Constraints 21 and 22 are the power generation limit of existing and artificial generators, respectively. Constraint 23 corresponds to the maximum number circuits allowed in the right of way *k-l*. The decision variables for this problem are $[n_{kl}^1, n_{kl}^2, ..., n_{kl}^{nr}]$ where *nr* is the number of right-of-ways available to add circuits.

The operational problem is a common optimal power flow solved by one of the MATPOWER solvers using an interior point method (Zimmerman, 2011).

Mathematical Modeling: AC Model

The objective function used here to cope with the expansion problem is

$$\min v = \sum_{(k,l)\in\Omega} c_{kl} n_{kl} + \sum_{k\in\Lambda} c_k q_k + w \tag{24}$$

where *v* is the investment due to the addition of new circuits and capacitor banks to the network, c_{kl} corresponds to the cost of a circuit that can be added to the right of way *k-l*, n_{kl} the number of circuits added in the right of way *k-l*, c_k the cost of a capacitor added to a PQ node, q_k the number of capacitor banks added to bus *k*, and *w* the load shedding. Ω is the set of all rights of way, and Λ is the set of all load buses.

The operation problem is handled by using the AC model. As in the DC Model, this is done by using an optimal power flow with some operational constraints; in this case, the objective function is the load shedding, which is modeled by adding artificial generators to the PQ nodes in such a way that the loading shedding is minimized.

$$\min w = \sum_{k\in\Lambda} (\alpha_1 r_{Pk} + \alpha_2 r_{Qk}) \tag{25}$$

Subject to

$$P(V,\theta,n,q) - P_G + P_D - r_P = 0 \tag{26}$$

$$Q(V,\theta,n,q) - Q_G + Q_D - r_Q = 0 \tag{27}$$

$$\underline{P}_G \leq P_G \leq \bar{P}_G \tag{28}$$

$$\underline{Q}_G \leq Q_G \leq \bar{Q}_G \tag{29}$$

$$\underline{r}_P \leq r_P \leq \bar{r}_P \tag{30}$$

$$\underline{r}_Q \leq r_Q \leq \bar{r}_Q \tag{31}$$

$$\underline{V} \leq V \leq \bar{V} \tag{32}$$

$$\underline{q} \leq q \leq \bar{q} \tag{33}$$

$$(N + N^0)S^{from} \leq (N + N^0)\bar{S} \tag{34}$$

$$(N + N^0)S^{to} \leq (N + N^0)\bar{S} \tag{35}$$

$$0 \leq n \leq \bar{n} \tag{36}$$

n integer; θ unbounded

Where *c* and *n* represent the circuit cost vector and the added lines vector, respectively; *N* and N^0 are diagonal matrices containing vector *n* and the existing circuits in the base case configuration respectively; $\bar{n}$ is a vector containing the maximum number of circuits in the base configuration. θ is the phase angles vector; P_G and Q_G are the existing real and reactive power generation vectors; P_D and Q_D are the real and reactive power demand vectors; *V* is the voltage magnitudes vector; r_P and r_Q are the active and reactive load shedding; α_1 and α_2 are the costs of the load shedding; $\overline{P}_G, \overline{Q}_G, \underline{V}$ are the vectors of maximum real and reactive power generation limits and voltage magnitudes, respectively. In this work, the maximum and minimum voltage magnitude limits are set to 95 e 105% of the nominal value. $S^{from}, S^{to}, \overline{S}$ are the apparent power flow vectors (MVA) through the branches in both terminals and their limits. The decision variables for this problem are $[n_{kl}^1, n_{kl}^2, ..., n_{kl}^{nr}; q_k^1, q_k^2, ..., q_k^{nl}]$ where nr is the number of right-of-way available to add circuits, and nl is the number of load buses to add shunt compensation. In this formulation, shunt compensation is also considered as a discrete variable which is a more realistic hypothesis.

Also, in this case, the operational problem is an AC optimal power flow solved by one of the MATPOWER solvers using an interior point method (Zimmerman, 2011).

IMPLEMENTATION ISSUES

This section describes the pseudo code of PSO implementation in the TEP problem, and the settings to test the performance of the algorithm by using either DC or AC power network models.

Pseudocode of PSO

1. Prepare electric data network.
2. Set parameters of PSO (i.e. swarm size, number of neighbors, maximum number of iterations, c_1, c_2, χ, initial iteration, etc).
3. Initialize particles positions and velocities randomly.
4. Evaluate the objective function by using either the optimal DC or AC power flow.
5. Update the best overall, individual, and local particle positions.
6. While the stopping criterion is not met, do

a. Increase iterations counter.
b. For each particle update velocities by using correspondingly equations to the different PSO variations.
c. Check velocity limits.
d. Update swarm.
e. Check swarm limits.
f. Evaluate the objective function using the optimal DC or AC power flow.
g. Update the best overall, individual, and local particle positions.

7. End

Network Data

Normally, the data used for the TEP implementation are those needed for either the DC or AC model. The dimension of the problem is given by the number of right of ways where it is possible to add circuits; the number of load buses is also taken into account, in case the short term transmission expansion planning is performed, where shunt compensation is added to the system.

Parameter Settings

Some parameters are very important in order to assure the convergence of PSO algorithms. The number of particles N is a problem dependent parameter, which should be chosen according to the dimension of the problem; however tests performed show that, with the formulation used in this problem, using between 20 - 120 particles leads to reasonable results. There is no formal procedure to select such parameter, which is mainly determined by a trial and error process. In the case of PSO, constants c_1 and c_2 are set to 2.05; therefore, factor χ is 0.729 (Clerc, 2006). In general, the neighborhood radius NR can take values less than the number of particles, in this work NR was chosen to be 1.

Swarm and Velocities Initialization

The technique of random uniform initialization is used in this work to generate the initial swarm. It is the most popular in evolutionary computation since it allows an equal exploration of the search space. In practice, it only consists of a random vector defined within [0, 1] with uniform probability distribution. The produced value is then scaled in the corresponding search space, so that the particles and their corresponding velocities lie strictly within their bounds.

Swarm and Velocities Bounds

In order to limit the search space, the minimum and maximum bounds of each particle x has to be defined. In this problem, it corresponds to the number of circuits in each right of way in the original electric network (x_{min}) and the maximum number of circuits (x_{max}) allowed in each right of way of the future network, respectively. In addition, the velocity components are also checked in order to clamp it within its limits, $[-v_{max}, v_{max}]$, where v_{max} was defined as $x_{max}/2$.

Update Velocities and Swarm

The PSO techniques previously presented are suited for continuous variables. The TEP problem formulated in this paper works with integer variables. Therefore, Equation 2 has been slightly modified, in order to cope with integer numbers in the following way

$$x_{ij}(t+1) = round[x_{ij}(t) + v_{ij}(t+1)] \qquad (37)$$

So each time x_{ij} is updated with the movement equation, then it is rounded to the nearest integer.

Stopping Criteria

In this work, two criteria have been used to stop the algorithm and get a solution. The first one is the maximum allowed number of iterations, which also limits the number of function evaluations. When optimal solutions were known, the second criterion is related to the convergence towards function known values. In practice, search stagnation could also have been used as an alternative stopping criterion. In addition, the number of function evaluations was used as measure of performance. There is no formal procedure to select the maximum number of iterations; in this work, it was chosen by a trial and error process taking into account the dimension of the problem.

Search Space

Suppose the number of candidate corridors to be d (which is also the dimension of the problem) and x_{max} the maximum number of candidate circuits to be feasible in each corridor, then the search space is given by the number of possible topologies (x_{max} +1) ^{d}; evidently, there are included feasible and unfeasible network topologies in this calculation.

TEST RESULTS

In this work, four test systems have been used, namely 1) Garver 6-bus, 2) IEEE 24-bus, 3) Southern Brazilian equivalent 46-bus, and 4) Northeastern Brazilian equivalent 87-bus systems. These systems are well known in the transmission planning literature as benchmark systems. The complete data for the DC and AC network model can be found in (Romero, 2002; Romero, 2005; Rider, 2006). The algorithms were implemented in MATLAB, running on an Intel i7, 2.80 GHz, 8GB RAM, hardware platform. The open source tool MATPOWER (Zimmerman, 2011), a well-known tool for power systems analysis, was also used, and can manage the function evaluation process by using whichever optimal power flow models presented in the previous sections. Since the TEP is considered a not very time constrained task, which can last days or weeks, the criteria for comparison to choose the best algorithms are the following: 1) Robustness, 2) Quality of the final solution, and 3) Number of function evaluations employed, in that order. The tests were performed allowing generation rescheduling.

Garver's 6-Bus System

This system has 6 buses, 15 candidate branches, with a total load of 760MW and 1100MW of maximum power generation. The maximum allowed lines per right-of-way is five and the number of feasible and unfeasible topologies is $(5+1)^{15}=6^{15}$.

DC Model

By using this model, the best known solution was obtained (Romero, 2002), with line additions in $n_{3-5}=1$, $n_{4-6}=3$. The cost of this expansion plan is US\$ 110. In order to compare the performance of the four proposed algorithms (EPSO, APSO, GPSO, LPSO), 10 tests were carried on using 20, 40, 60, 80, 100 and 120 particles. The results obtained can be seen on Figures 2-4.

The success rate for the different algorithms is shown in Figure 2. It can be seen that LPSO and APSO are the most robust algorithms getting 100% of success in 10 tries (this means that the algorithm got the optimal solution) in most of the tests performed. EPSO and APSO do not obtain the optimal solution in most of the tests, thus they are not very reliable. Figure 3 shows that the lowest cost solutions are also obtained using LPSO and APSO. Regarding the number of function evaluations, it was taken into account only those tests where the algorithm reached the optimal solution. In general, as shown in Figure 4, GPSO is the fastest in reaching the optimal solution but its robustness is very poor; for this reason, it is possible to say that LPSO performs much better in general. More function evaluations are needed for APSO and EPSO in order to get the optimal solution. In general, the choice of 60 – 100 particles could be a good option for this test system, where it is possible to work with a very robust algorithm in a fast way. The average calculation time to finish the test using 60 particles was 308.5 sg, and 52.3 sg in order to get the optimal solution. All the tests used 150 iterations as a stopping criterion.

Figure 2. Success rate using different swarm size

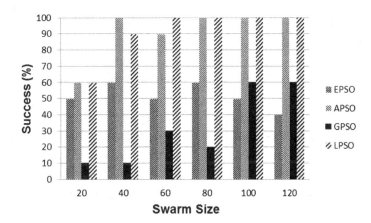

Figure 3. Average cost for different swarm size

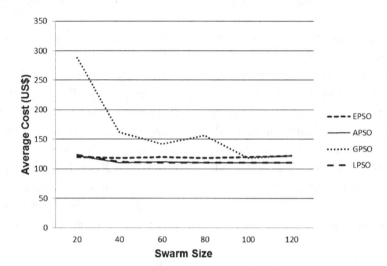

Figure 4. Average number of function evaluations for different swarm size

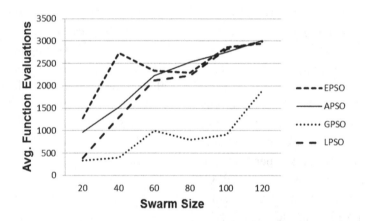

AC Model

Three tests were performed using this model, one without shunt compensation, and two others with the use of fixed capacitors with different costs. Shunt compensation was allowed at buses 1, 2, 3, 4, and 5. Results of simulations are presented below.

Without shunt compensation, the line additions were $n_{2\text{-}6}=2$, $n_{3\text{-}5}=2$, $n_{4\text{-}6}=2$. In this case, the cost associated to the additions was of US\$160. This agrees with the scarce literature on this topic (Rahmani, 2010; Rider, 2007). The algorithm LPSO

was able to obtain the same solution mentioned earlier in 100% of the tries (5), using an average of 19 iterations and 1140 optimal power flows. Regarding the PSO parameters, 60 particles with 100 iterations were allowed, so the total number of function evaluations was 6000. The latter confirm the robustness of the local PSO variation, which was also demonstrated by using the DC Model.

By allowing the allocation of fixed capacitors with the cost of US\$ 1 per Mvar, the line additions were $n_{2\text{-}6}=2$, $n_{3\text{-}5}=1$, $n_{4\text{-}6}=2$, and 9Mvar of shunt compensation at bus 5. This means US\$ 140 corresponding to line additions and US\$ 9

Figure 5. Garver System optimal plan for AC model using Shunt compensation

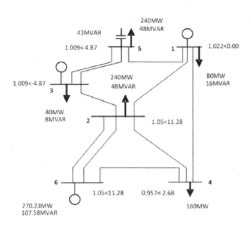

corresponding to shunt compensation. Therefore, by using shunt compensation (at this price), it is not necessary to add a line in the right-of-way *3-5*. In this case, it was also used LPSO to perform the test, using 60 particles and 100 iterations.

By using fixed capacitors with the cost of US$ 0.5 per Mvar, the line additions were $n_{2-6}=2$, $n_{3-5}=1$, $n_{4-6}=1$, and 43Mvar of shunt compensation at bus 5. This means US$ 110 corresponding to line additions and US$ 21.5 corresponding to shunt compensation. The line additions correspond exactly to the reported in (Rider, 2007) as shown in Figure 5. For this case, several tests were also done using the algorithms studied in this work. Five tries were performed for each algorithm using 20, 40, 60, 80, 100 and 120 particles. The results in Table 1 exhibit the same trends as in the case of the DC Model, therefore, it is demonstrated

that, for this problem, LPSO outperforms the other PSO variants studied in this research work.

IEEE 24-Bus System

The system consists of 24 buses and 41 rights of way. The original topology is shown in Figure 6. The maximum allowed number of circuits per right-of-way is five. In this case, the number of total topologies is $(5+1)^{41}=6^{41}$.

DC Model

In this case, the optimal solution was also obtained. The line additions were $n_{6-10}=1$, $n_{7-8}=2$, $n_{10-12}=1$, $n_{14-16}=1$ with a total cost of US$ 152,000. For this test, the optimal solution was obtained in 100% of 10 tries by using LPSO with 80 and 120 particles with 250 iterations allowed. Five tries for each algorithm were performed using from 20 to 120 particles. Using 80 particles and 250 iterations, the average time to finish one test is 1061sg, and 298.7sg was necessary in order to get the optimal solution. In this case, it was confirmed again that LPSO outperforms the other PSO variants, not only in robustness and quality of solution, but also in the smaller number of function evaluations employed. This can be seen in Table 2.

AC Model

Two tests were performed, namely a) without and b) with shunt compensation. The candidate buses for compensation were all load buses. Sixty

Table 1. Results of test performed for a) EPSO, b) APSO, c) GPSO, and d) LPSO using the AC model

PARTICLES	20				40				60				80				100				120			
PSO VARIANT	a	b	c	d	a	b	c	d	a	b	c	d	a	b	c	d	a	b	c	d	a	b	c	d
TRIALS SUCC. (%)	80	100	20	80	100	100	20	100	100	100	40	100	100	100	80	100	100	100	20	100	100	100	80	100
AVERAGE ITER.	49	32	12	40	26	36	12	46	19	36	10	31	24	29	12	32	49	36	10	31	23	28	13	29
AVER.FUNC.EVAL.	1920	649	240	805	2056	1458	480	1856	2220	2156	570	1836	3824	2323	960	2576	9620	3626	1000	3280	5400	3338	1590	3504
MIN	110	110	110	110	110	110	110	110	110	110	110	110	110	110	110	110	110	110	110	110	110	110	110	110
AVERAGE	133	110	182	114	110	110	158	110	110	110	159	110	110	110	114	110	110	110	150	110	110	110	122	110
MAX	223	110	170	130	110	110	200	110	110	110	250	110	110	110	130	110	110	110	190	110	110	110	170	110

Figure 6. IEEE 24-bus test system

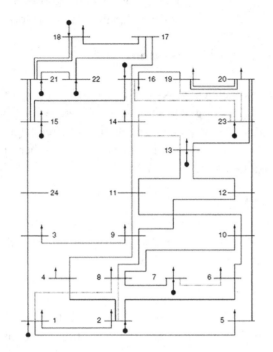

particles were used and a maximum of 300 iterations was allowed for this test system. The results obtained in this case are described below.

- Without using shunt compensation, the line additions were $n_{1-3}=1$, $n_{2-4}=1$, $n_{3-24}=2$, $n_{6-10}=2$, $n_{7-8}=2$, $n_{9-11}=1$, $n_{15-24}=1$. The cost associated to the additions was of US\$ 374,000.
- By using unlimited fixed reactive power sources at zero cost, the line additions were $n_{6-10}=1$ and $n_{14-16}=1$. The buses where shunt compensation was added were $q_3=366$, $q_8=29$, $q_9=568$, $q_{10}=431$, and $q_{19}=143$ Mvar. The cost of the new lines is US\$ 70,000. This means, when compared to the expansion using the DC model, that the line in the right-of-way *10-12* is not necessary, due to the reactive power sources.

Table 2. Results of test performed for a) EPSO, b) APSO, c) GPSO, and d) LPSO using the DC model

PARTICLES	20				40				60			
PSO VARIANT	a	b	c	d	a	b	c	d	a	b	c	d
TRIALS SUCC. (%)	20	20	0	0	10	60	0	30	30	80	0	70
AVERAGE ITER.	143	99	250	250	42	115	250	114	124	88	250	104
AVER.FUNC.EVAL.	5680	2042	5000	5000	3320	4649	10000	4547	14860	5338	15000	6266
MIN	152000	152000	692000	160000	152000	152000	296000	152000	152000	152000	465000	152000
AVERAGE	271200	236800	1222900	325000	231300	169000	732500	179600	207600	159200	756400	157200
MAX	415000	390000	1893000	560000	318000	260000	1319000	242000	313000	202000	1009000	184000
PARTICLES	80				100				120			
PSO VARIANT	a	b	c	d	a	b	c	d	a	b	c	d
TRIALS SUCC. (%)	20	80	0	100	20	90	0	90	40	100	0	100
AVERAGE ITER.	139	111	250	108	33	98	250	112	105	108	250	91
AVER.FUNC.EVAL.	22080	8972	20000	8608	6400	9903	25000	11178	24960	13003	30000	10968
MIN	152000	152000	557000	152000	152000	152000	324000	152000	152000	152000	432000	152000
AVERAGE	253100	158600	788600	152000	186300	154200	553800	152800	203200	152000	667100	152000
MAX	373000	196000	1385000	152000	265000	174000	945000	160000	248000	152000	954000	152000

Brazilian 46-Bus Equivalent System

This is a realistic system that represents the southern part of the Brazilian interconnected system, with 46 buses and 79 rights of way, where it is possible to add new circuits. The total demand for this system is 6,880MW. The maximum number of lines allowed per right of way is five and, in this case, the number of topologies that could be explored is $(5+1)^{79}=6^{79}$. The Mvar limit for shunt compensation was 1000Mvar, and its cost was US$ 10,000 per Mvar.

DC Model

In this case, by using LPSO, it was obtained the best known solution so far (US$ 72,870,000). For this case, the line additions were $n_{2-5}=1$, $n_{5-6}=2$, $n_{13-20}=1$, $n_{20-21}=2$, $n_{20-23}=1$, $n_{42-43}=1$, $n_{46-06}=1$. This result was achieved using 60 particles and 700 maximum iterations. The total time of this test was 84min approximately, and the optimal solution was obtained in 51min approximately.

AC Model

By using the AC model with shunt compensation, the circuits added were: $n_{4-9}=4$, $n_{5-9}=4$, $n_{2-5}=3$, $n_{20-23}=3$, $n_{26-27}=3$, $n_{33-34}=1$, $n_{34-35}=2$, $n_{40-42}=4$, $n_{20-21}=1$, $n_{14-15}=1$, $n_{5-11}=3$, $n_{21-25}=1$, $n_{31-32}=1$, $n_{27-29}=1$, $n_{28-41}=1$. Shunt compensation was added as follows: $q_2=25$, $q_4=77$, $q_{13}=88$, $q_{20}=376$, $q_{23}=356$, $q_{33}=21$, $q_{38}=254$, $q_{40}=20$, $q_{42}=8$, and $q_{44}=109$, corresponding to 1,334Mvar. The cost obtained for this system was US$ 325,342,000 and US$ 13,340,000 for lines and shunt compensation respectively. The test was performed using 60 particles and allowing 800 iterations, thus 42,000 optimal power flows were run in approximately 102 min.

Brazilian 87-Bus Equivalent System

This is a very challenging system as far as transmission expansion planning is concerned. This system is a reduced version of the Brazilian northeastern network. The system has 87 buses, 183 right-of-ways for the addition of new circuits, and a total demand of 20,316MW. The maximum number of lines allowed per right-of-way is 15 in this case. In this case, the search space is composed by 16^{183} maximum transmission lines topologies. When used the AC Model, all the load buses were allowed to receive reactive power compensation.

DC Model

The best solution obtained for this test system was achieved using LPSO with the following line additions: $n_{4-68}=1$, $n_{5-56}=1$, $n_{5-58}=2$, $n_{5-68}=2$, $n_{13-14}=2$, $n_{14-59}=1$, $n_{16-44}=5$, $n_{17-18}=3$, $n_{18-50}=15$, $n_{19-20}=1$, $n_{22-37}=2$, $n_{24-25}=1$, $n_{25-55}=15$, $n_{26-54}=1$, $n_{27-28}=1$, $n_{29-30}=1$, $n_{30-63}=1$, $n_{35-47}=6$, $n_{36-46}=5$, $n_{39-42}=4$, $n_{40-46}=1$, $n_{44-46}=1$, $n_{47-48}=6$, $n_{48-49}=3$, $n_{51-52}=1$, $n_{54-63}=1$, $n_{60-66}=15$, $n_{61-85}=2$, $n_{63-64}=1$. The total cost for these additions is US$ 1,066,112,000. This solution was obtained with 100 particles and 1500 iterations. The time employed to get this solution was approximately 310min. The convergence process for this system is shown in Figure 7.

Figure 7. Convergence process for Brazilian 87-bus system using the DC model

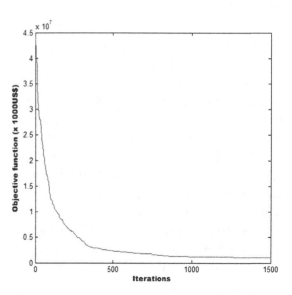

AC Model

For this test, the solution was obtained using 100 particles and 1000 iterations with the LPSO algorithm in 322 min. Also, in this test, there were used unlimited fixed reactive power sources at zero cost. The following line additions were obtained: $n_{3-83}=1$, $n_{3-87}=1$, $n_{4-32}=2$, $n_{5-6}=14$, $n_{4-32}=15$, $n_{4-69}=1$, $n_{5-38}=13$, $n_{5-56}=1$, $n_{5-58}=1$, $n_{5-80}=1$, $n_{6-37}=14$, $n_{6-75}=2$, $n_{7-62}=1$, $n_{10-11}=3$, $n_{12-13}=4$, $n_{13-14}=14$, $n_{13-17}=2$, $n_{15-46}=13$, $n_{16-44}=13$, $n_{16-77}=1$, $n_{18-50}=14$, $n_{19-20}=4$, $n_{22-58}=15$, $n_{25-55}=3$, $n_{26-27}=15$, $n_{26-29}=5$, $n_{26-54}=15$, $n_{27-28}=14$, $n_{29-30}=15$, $n_{30-31}=15$, $n_{30-63}=15$, $n_{33-67}=1$, $n_{34-41}=13$, $n_{35-51}=4$, $n_{36-46}=13$, $n_{39-86}=1$, $n_{42-44}=3$, $n_{42-85}=1$, $n_{44-46}=1$, $n_{48-50}=15$, $n_{48-51}=15$, $n_{49-50}=15$, $n_{58-78}=2$, $n_{60-66}=13$, $n_{61-64}=2$, $n_{61-85}=2$, $n_{62-72}=2$, $n_{65-87}=1$, $n_{67-71}=1$, $n_{68-83}=1$, $n_{70-82}=1$, $n_{73-74}=1$, $n_{81-83}=15$. In this case, the Shunt compensation constraints was left: $q_1=559$, $q_{20}=701$, $q_{22}=24$, $q_{24}=17$, $q_{25}=100$, $q_{26}=1000$, $q_{27}=716$, $q_{28}=177$, $q_{29}=15$, $q_{30}=19$, $q_{34}=39$, $q_{36}=205$, $q_{40}=1000$, $q_{41}=243$, $q_{42}=41$, $q_{44}=808$, $q_{46}=1000$, $q_{48}=48$, $q_{49}=485$, $q_{50}=1000$, $q_{51}=30$, $q_{52}=988$, $q_{85}=400$, corresponding to 9615Mvar. The total cost for these additions is US\$ 3,555,203,000.

FUTURE RESEARCH DIRECTIONS

Solving the TEP problem is not an easy task, and many efforts have been employed to find efficient algorithms to deal with this complex combinatorial problem. Heuristic and meta-heuristic optimization methods seem to be very promising to solve the TEP problem. It is possible to extend this work combining heuristic methods to obtain good quality initial solutions in order to help PSO methods to converge fast. Also, as the results of this work suggest, local variants of PSO should be developed in order to tackle this kind of combinatorial problem. On the other hand, this approach should be extended to deal with the multistage transmission expansion problem. The formulation used in this work allows the flexibility to incorporate different kinds of constraints and sophisticated models of transmission equipment such as FACTS devices.

CONCLUSION

This work shows the use of some PSO variations applied to the Transmission Expansion Planning in electrical networks using both DC and AC power network models. Also, it represents a new contribution to the few existing works that use the AC model so far, and presents a novel formulation using a load shedding scheme with reactive power planning in the same framework. In addition, extensive simulations were performed for small, medium and large size systems. The results obtained with the proposed approach were usually comparable than the ones that can be found in the literature. The results of this work suggest that the local version of PSO (LPSO) outperforms other PSO variants such as GPSO, EPSO and APSO, not only in robustness and quality of solutions, but also in the capacity to rapidly converge to optimal solutions. LPSO showed to be a very robust algorithm obtaining excellent solutions in small and medium size systems, and also good quality solutions in larger power networks. This makes LPSO a good option in order to solve the TEP problem using both AC and DC models in realistic power systems.

ACKNOWLEDGMENT

This work was supported in part by the Brazilian funding agency Fundação de Amparo a Pesquisa do Estado de São Paulo/FAPESP (www.fapesp. br), grant no. 2010/01014-7.

REFERENCES

Akbari, T., Rahimikian, A., & Kazemi, A. (2011). A multi-stage stochastic transmission expansion planning method. *Energy Conversion and Management, 52,* 2844–2853. doi:10.1016/j.enconman.2011.02.023

Alguacil, N., Carrión, M., & Arroyo, J. M. (2009). Transmission network expansion planning under deliberate outages. *International Journal of Electrical Power & Energy Systems, 31,* 553–561. doi:10.1016/j.ijepes.2009.02.001

Binato, S., de Oliveira, G. C., & de Araujo, J. L. (2001). A greedy randomized adaptive search procedure for transmission expansion planning. *IEEE Transactions on Power Systems, 16*(2), 247–253. doi:10.1109/59.918294

Buygi, M., Balzer, G., Shanechi, H., & Shahidehpour, M. (2004). Market-based transmission expansion planning. *IEEE Transactions on Power Systems, 19*(4), 2060–2067. doi:10.1109/TPWRS.2004.836252

Cedeño, E. B., & Arora, S. (2011). Performance comparison of transmission network expansion planning under deterministic and uncertain conditions. *International Journal of Electrical Power & Energy Systems, 33,* 1288–1295. doi:10.1016/j.ijepes.2011.05.005

Choi, J., Mount, T., & Thomas, R. (2007). Transmission expansion planning using contingency criteria. *IEEE Transactions on Power Systems, 22,* 2249–2261. doi:10.1109/TPWRS.2007.908478

Clerc, M. (2006). *Particle swarm optimization.* doi:10.1002/9780470612163

Da Silva, E. L., Ortiz, J. M. A., De Oliveira, G. C., & Binato, S. (2001). Transmission network expansion planning under a Tabu search approach. *IEEE Transactions on Power Systems, 16*(1), 62–68. doi:10.1109/59.910782

De la Torre, S., Conejo, A. J., & Contreras, J. (2008). Transmission expansion planning in electricity markets. *IEEE Transactions on Power Systems, 23*(1), 238–248. doi:10.1109/TPWRS.2007.913717

De Salvo, C. A., & Smith, H. L. (1965). Automatic transmission planning with ac load flow and incremental transmission loss evaluation. *IEEE Transactions on Power Apparatus and Systems, 84,* 156–163. doi:10.1109/TPAS.1965.4766166

Del Valle, Y. (2008). Particle swarm optimization: Basic concepts, variants and applications in power systems. *IEEE Transactions on Evolutionary Computation, 12*(2). doi:10.1109/TEVC.2007.896686

Dong-Xiao, N., Yun-Peng, L., Zhao, Q., & Qing-Ying, Z. (2006). An improved particle swarm optimization method based on borderline search strategy for transmission network expansion planning. *Proceedings of the Fifth International Conference on Machine Learning and Cybernetics,* (pp. 2846-2850). Dalian, 13-16 August.

Dusonchet, Y. P., & El-Abiad, A. H. (1973). Transmission planning using discrete dynamic optimization. *IEEE Transactions on Power Apparatus and Systems, 92,* 1358–1371. doi:10.1109/TPAS.1973.293543

Enamorado, J. C., Gómez, T., & Ramos, A. (1999). Multi-area regional interconnection planning under uncertainty. *Proceedings of the 13th Power Systems Computer Conference* Trondheim, 1999, (pp. 599–606).

Escobar, A. H., Gallego, R. A., & Romero, R. (2004). Multistage and coordinated planning of the expansion of transmission systems. *IEEE Transactions on Power Systems, 19*(2), 735–744. doi:10.1109/TPWRS.2004.825920

Gallego, L. A., Rider, M. J., Romero, R., & Garcia, A. V. (2009). *A specialized genetic algorithm to solve the short term transmission expansion planning*. IEEE Bucharest Power Tech Conference, June 28th – July 2nd 2009, Bucharest, Romania.

Georgilakis, P. S., Korres, G. N., & Hatziargyriou, N. D. (2011). Transmission expansion planning by enhanced differential evolution. *16th International Conference on Intelligent System Application to Power Systems (ISAP)*, 25 – 28 Sept.

Kavitha, D., & Swarup, K. S. (2006). *Transmission expansion planning using LP-based particle swarm optimization*. Power Indian Conference, IEEE, India.

Kennedy, J., & Eberhart, R. (2001). *Swarm intelligence*. Academic Press.

Latorre, G., Cruz, R., Areiza, J., & Villegas, A. (2003). Classification of publications and models on transmission expansion planning. *IEEE Transactions on Power Systems*, *18*(2). doi:10.1109/TPWRS.2003.811168

Lee, K. Y., & El-Sharkawi, M. A. (2008). *Modern heuristic optimization techniques*. IEEE Press. doi:10.1002/9780470225868

Miranda, V. (2005). Evolutionary algorithms with particle swarm movements. *Proceedings of the 13th International Conference on Intelligent Systems Application to Power Systems*, (pp. 6-21).

Miranda, V., De Magalhaes, L., da Rosa, M. A., Da Silva, A. M. L., & Singh, C. (2009). Improving power system reliability calculation efficiency with EPSO variants. *IEEE Transactions on Power Systems*, *24*, 1772–1779. doi:10.1109/TPWRS.2009.2030397

Miranda, V., Keko, H., & Jaramillo, A. (2007). EPSO: Evolutionary particle swarms. *Evolutionary Computing for System Design. Series* [Springer.]. *Studies in Computational Intelligence*, *66*, 139–168. doi:10.1007/978-3-540-72377-6_6

Oliveira, G., Binato, S., & Pereira, M. (2007). Value-based transmission expansion planning of hydrothermal systems under uncertainty. *IEEE Transactions on Power Systems*, *22*, 1429–1435. doi:10.1109/TPWRS.2007.907161

Parsopoulus, K., & Vrahatis, M. (2010). *Particle swarm optimization and intelligence: Advances and applications*. Information Science Reference. doi:10.4018/978-1-61520-666-7

Pringles, R., Miranda, V., & Garcés, F. (2007). Expansión optima del sistema de transmisión utilizando EPSO. *VII Latin American Congress on Electricity Generation & Transmission*, October 24-27.

Rahmani, M., Rashididejad, M., Carreno, E. M., & Romero, R. (2010). Efficient method for AC transmission network expansion planning. *Electric Power Systems Research*, 1056–1064. doi:10.1016/j.epsr.2010.01.012

Ren, P., Gao, L., Li, N., Li, Y., & Lin, Z. (2005). Transmission network optimal planning using the particle swarm optimization method. *Proceedings of the Fourth International Conference on Machine Learning and Cybernetics, Guangzhou*, 16-21 August.

Rider, M. (2006). *Planejamento da expansão de sistemas de transmissão usando os modelos CC – CA e técnicas de programação não – Linear*. Ph.D Thesis in Portuguese, University of Campinas, Brazil.

Rider, M. J., Gallego, L. A., Romero, R., & Garcia, A. V. (2007). Heuristic algorithm to solve the short term transmission expansion planning. *Proceedings of the IEEE General Meeting*.

Rider, M. J., Garcia, A. V., & Romero, R. (2007). Power system transmission network expansion planning using AC model. *Proceedings of IET Generation, Transmission, and Distribution*, *1*(5), 731–742. doi:10.1049/iet-gtd:20060465

Rodriguez, J. I., Falcão, D. M., & Taranto, G. N. (2008). *Short-term transmission expansion planning with AC network model and security constraints.* 16th PSCC, Glasgow, Scotland, July 14-18.

Rodriguez, J. I., Falcão, D. M., Taranto, G. N., & Almeida, H. L. S. (2009). *Short-term transmission expansion planning by a combined genetic algorithm and hill-climbing technique.* 15th International Conference on Intelligent System Applications to Power Systems, ISAP '09.

Romero, R., & Monticelli, A. (1994). A hierarchical decomposition approach for transmission network expansion planning. *IEEE Transactions on Power Systems, 9*, 373–380. doi:10.1109/59.317588

Romero, R., Monticelli, A., García, A., & Haffner, S. (2002). Test systems and mathematical models for transmission network expansion planning. *IEE Proceedings. Generation, Transmission and Distribution, 149*(1), 27–36. doi:10.1049/ip-gtd:20020026

Romero, R., Rocha, C., Mantovani, J. R. S., & Sanchez, I. G. (2005). Constructive heuristic algorithm for the DC model in network transmission expansion planning. *IEE Proceedings. Generation, Transmission and Distribution, 152*(2), 277–282. doi:10.1049/ip-gtd:20041196

Seifi, H., Sepasian, M. S., Haghighat, H., Foroud, A. A., Yousefi, G. R., & Rae, S. (2007). Multi-voltage approach to long-term network expansion planning. *IET Generation. Transmission & Distribution, 1*(5), 826–835. doi:10.1049/iet-gtd:20070092

Sensarma, P. S., Rahmani, M., & Carvalho, M. (2002). A comprehensive method for optimal expansion planning using particle swarm optimization. *IEEE Power Engineering Society Winter Meeting*, Vol. 2, (pp. 1317–1322).

Shayegui, H., Mahdavi, M., & Bagheri, A. (2009). Discrete PSO algorithm based optimization of transmission lines loading in TNEP problem. [Elsevier.]. *Energy Conversion and Management, ▪▪▪*, 112–121.

Silva, I. J., Rider, M. J., Romero, R., & Murari, C. A. F. (2006). Transmission network expansion planning considering uncertainty in demand. *IEEE Transactions on Power Systems, 21*(4), 1565–1573. doi:10.1109/TPWRS.2006.881159

Torres, S. P., Castro, C. A., Pringles, R., & Guaman, W. (2011). Comparison of particle swarm based meta-heuristics for the electric transmission network expansion planning problem. *Proceedings of the IEEE General Meeting*, Detroit, 24-29 July 2011, USA.

Verma, A., Panigrahi, B. K., & Bijwe, P. R. (2009). Transmission network expansion planning with adaptive particle swarm optimization. *World Congress on Nature & Biologically Inspired Computing (NaBIC 2009)*, (pp. 1099–1104).

Verma, A., Panigrahi, B. K., & Bijwe, P. R. (2010). Harmony search algorithm for transmission expansion planning. *IET Generation. Transmission & Distribution, 4*(6), 663–673. doi:10.1049/iet-gtd.2009.0611

Yi-Xiong, J., Hao-Zhong, G., Jian-Yong, Y., & Li, Z. (2005). *Local optimum embranchment based convergence guarantee particle swarm optimization and its application in transmission network planning.* IEEE/PES Transmission and Distribution Conference & Exhibition: Asia and Pacific, Dalian, China.

Yi-Xiong, J., Hao-Zhong, G., Jian-Yong, Y., & Li, Z. (2006). New discrete method for particle swarm optimization and its application in transmission network expansion planning. [Elsevier.]. *Electric Power Systems Research*, 227–233.

Youssef, H. K., & Hackam, R. (1989). New transmission planning model. *IEEE Transactions on Power Systems*, *4*, 9–18. doi:10.1109/59.32451

Zhan, Z. H., Zhang, J., Li, Y., & Chung, H. S. (2009). Adaptive particle swarm optimization. *IEEE Transactions on Systems, Man, and Cybernetics. Part B, Cybernetics*, *39*(6).

Zimmerman, R. D., Murillo-Sánchez, C. E., & Thomas, R. J. (2011). MATPOWER: Steady-state operations, planning, and analysis tools for power systems research and education. *IEEE Transactions on Power Systems*, *26*(1). doi:10.1109/TPWRS.2010.2051168

ADDITIONAL READING

Bahiense, L., Oliveira, G. C., Pereira, M., & Granville, S. (2001). A mixed integer disjunctive model for transmission network expansion. *IEEE Transactions on Power Systems*, *16*, 560–565. doi:10.1109/59.932295

Baldick, R., & Kahn, E. (1993). Transmission planning issues in a competitive economic environment. *IEEE Transactions on Power Systems*, *8*, 1497–1503. doi:10.1109/59.260951

Bent, R., Toole, G. L., & Berscheid, A. (2012). Transmission network expansion planning with complex power flow models. *IEEE Transactions on Power Systems*, *27*(2), 904–2912. doi:10.1109/TPWRS.2011.2169994

Fang, R., & Hill, D. J. (2003). A new strategy for transmission expansion in competitive electricity markets. *IEEE Transactions on Power Systems*, *18*(1), 374–380. doi:10.1109/TPWRS.2002.807083

Farmani, R., & Wright, J. A. (2003). Self-adaptive fitness formulation for constrained optimization. *IEEE Transactions on Evolutionary Computation*, *7*(5). doi:10.1109/TEVC.2003.817236

Garver, L. L. (1989). Transmission network estimation using linear programming. *IEEE Transactions in PAS*, *7*, 1688–1697.

Haffner, S., Monticelli, A., Garcia, A., Mantovani, J., & Romero, R. (2000). Branch and bound algorithm for transmission system expansion planning using a transportation model. *Proceedings of the Institute of Electrical Engineers—Generation, Transmission, and Distribution*, *147*, 149–156.

Leite da Silva, A. M., Rezende, L. S., Honorio, L. M., & Manso, L. A. F. (2011). Performance comparison of metaheuristics to solve the multi-stage transmission expansion planning problem. *IET Generation, Transmission, and Distribution*, *5*(3), 360–367. doi:10.1049/iet-gtd.2010.0497

Maghouli, P., Hosseini, S. H., Buygi, M. O., & Shahidehpour, M. (2011). A scenario-based multi-objective model for multi-stage transmission expansion planning. *IEEE Transactions on Power Systems*, *26*(1). doi:10.1109/TPWRS.2010.2048930

Manoharan, P. S., Kannan, P. S., Baskar, S., & Iruthayarajan, M. W. (2008). Penalty parameter-less constraint handling scheme based evolutionary algorithm solutions to economic dispatch. *IET Generation, Transmission, and Distribution*, *2*(4), 478–490. doi:10.1049/iet-gtd:20070423

Miranda, V., & Proença, L. M. (1998). Why risk analysis outperforms probabilistic choice as the effective decision support paradigm for power system planning. *IEEE Transactions on Power Systems*, *13*, 643–648. doi:10.1109/59.667394

Monticelli, A., Santos, A., Pereira, M. V. F., Cunha, S. H. F., Parker, B. J., & Praça, J. C. G. (1982). Interactive transmission network planning using a least-effort criterion. *IEEE Transactions on Power Apparatus and Systems*, *101*, 3919–3925. doi:10.1109/TPAS.1982.317043

Nakamura, S. (1995). *Numerical analysis and graphic visualization with MATLAB*. Englewood Cliffs, NJ: Prentice-Hall.

Pereira, M. V., & Pinto, L. M. V. G. (1985). Application of sensitivity analysis of load supplying capability to interactive transmission expansion planning. *IEEE Transactions on Power Apparatus and Systems*, *104*, 381–389. doi:10.1109/TPAS.1985.319053

Romero, R., Gallego, R. A., & Monticelli, A. (1990). Transmission network expansion planning by simulated annealing. *IEEE Transactions on Power Systems*, *11*(1), 364–369. doi:10.1109/59.486119

Romero, R., Rider, M. J., & Silva, I. J. (2007). A metaheuristic to solve the transmission expansion planning. *IEEE Transactions on Power Systems*, *22*(4), 2289–2291. doi:10.1109/TPWRS.2007.907592

Rudnick, H., & Quinteros, R. (1998). Power system planning in the southamerica electric market restructuring. *Proceedings of the VI Symposium on Specialists in Electrical Operations and Expansion Planning*, Bahia, Brazil.

Santos, A., França, P. M., & Said, A. (1989). An optimization model for long range transmission expansion planning. *IEEE Transactions on Power Systems*, *4*, 94–101. doi:10.1109/59.32462

Seifi, H., & Sepasian, M. (2011). *Electric power system planning: Issues, algorithms and solutions*. Berlin, Germany: Springer-Verlag. doi:10.1007/978-3-642-17989-1

Silva, I. D. J., Rider, M. J., Romero, R., Garcia, A. V., & Murari, C. A. (2005). Transmission network expansion planning with security constraints. *IEE Proceedings. Generation, Transmission and Distribution*, *152*(6), 828–836. doi:10.1049/ip-gtd:20045217

Sum-Im, T., Taylor, G. A., Irving, M. R., & Song, Y. H. (2009). Differential evolution algorithm for static and multistage transmission expansion planning. *IET Generation. Transmission & Distribution*, *3*(4), 365–384. doi:10.1049/iet-gtd.2008.0446

Tessema, B., & Yen, G. G. (2009). An adaptive penalty formulation for constrained evolutionary optimization. *IEEE Transactions on Systems, Man, and Cybernetics. Part A, Systems and Humans*, *39*(3). doi:10.1109/TSMCA.2009.2013333

Ting, T. O., Rao, M. V. C., & Loo, C. K. (2006). A novel approach for unit commitment problem via an effective hybrid particle swarm optimization. *IEEE Transactions on Power Systems*, *21*(1). doi:10.1109/TPWRS.2005.860907

Villasana, R., Garver, L. L., & Salon, S. L. (1985). Transmission network planning using linear programming. *IEEE Transactions on Power Apparatus and Systems*, *104*, 349–356. doi:10.1109/TPAS.1985.319049

Zimmerman, R. D., Murillo-Sánchez, C. E., & Thomas, R. J. (2009). *MATPOWER's extensible optimal power flow architecture*. IEEE Power & Energy Society General Meeting, PES '09.

KEY TERMS AND DEFINITIONS

AC Model: Alternate current component models for electric power networks.

Adaptive Particle Swarm Optimization (APSO): The PSO variant that adapts its parameters according to some rules based on observation and experience to improve exploration and exploitation properties.

DC Model: Direct current component models for electric power networks.

Evolutionary Particle Swarm Optimization (EPSO): The PSO version in which the parameters are adjusted automatically in an evolutionary way.

Global Particle Swarm Optimization (GPSO): The earliest version of PSO which uses the global best particle position to take advantage of exploitation properties to get a fast convergence.

Local Particle Swarm Optimization (LPSO): The PSO variation which uses the concept of neighborhoods to take advantage of exploration properties to get good quality solutions.

Particle Swarm Optimization (PSO): Meta-heuristic optimization technique based on social behavior of some animals such as fish, birds, etc.

Transmission Expansion Planning (TEP): The process carried out in electric utilities to determine the future electric transmission assets to meet planned generation and future demand.

Chapter 14

Short–Term Generation Scheduling Solved with a Particle Swarm Optimizer

Víctor Hugo Hinojosa Mateus
Universidad Técnica Federico Santa María, Chile

Cristhoper Leyton Rojas
Universidad Técnica Federico Santa María, Chile

ABSTRACT

In this chapter, a particle swarm optimizer is applied to solve the problem of short-term Hydrothermal Generation Scheduling Problem – one day to one week in advance. The optimization problems have been formulated taking into account binary and real variables (water discharge rates and thermal states of the units). This proposal is based on a strategy to generate and keep the decision variables on feasible space through the correction operators, which were applied to each constraint. Such operators not only improve the quality of the final solutions, but also significantly improve the convergence of the search process due to the use of feasible solutions. The results and effectiveness of the proposed technique are compared to those previously discussed in the literature such as PSO, GA, and DP, among others.

INTRODUCTION

One of the most important priorities in all economic environments is the efficient use of variables, with the objective of minimizing production costs to obtain a product or final service. The electricity

generating companies invest significant amounts of money in fuel (diesel, coal, gas, etc.). For example, a thermal unit with a level of power of 10 000 MW, may spend close to US$7.784 million in fuel per year (Wood, 1996) with the fuel cost of 82,81 US$/bbl. So, in the electric industry, the

DOI: 10.4018/978-1-4666-2666-9.ch014

most important priorities for the responsible agents are cost reduction and safe operation of the electric power system, considering the efficient use of the various types of fuels to generate electric power.

Customer load demands in electric power systems are subject to changes because human activities follow daily, weekly and monthly cycles. Sufficient generating units must be committed in order to satisfy the load in electric power systems. By committing enough generating units to withstand the peak load demand and by keeping these units on at times, we provide a solution for the generation scheduling. However, turning units off when they are not required can have a great impact and save a lot of money. Therefore, the way hydro, fuel, wind, geothermal, and solar are scheduled is an important topic in electrical power systems. We need to satisfy the load while operating the power system economically.

The hydrothermal generation scheduling problem depends on determining a secure operation strategy subject to a variety of constraints. The limitations in the water storage capacities in the reservoirs, together with the stochastic nature of their readiness, present a complex challenge. A profitable schedule of generation in a hydrothermal system involves sharing it among the thermal and hydraulic units in such a way that the total operating cost of the thermal units is minimized.

The efficient scheduling of available energy resources for satisfying load has become an important task and has been extensively studied because of its significant economic impact. The problem of finding the energy production of every power plant (hydro, nuclear, thermal and renewables) in all sub-periods of a given planning period is subject to a variety of technical constraints. Decisions to be made are coupled in time; for instance, future reservoir storage depends on the previous operation of the system. Generations of plants must be coordinated: not only the necessary constraints, such as load power balance and reserve, but also the characteristics of the power system (for instance

hydro plants in cascade). In addition, uncertainties of both load and hydrological conditions have to be managed at the same time.

GENERATION PLANNING MODELS USED IN POWER SYSTEMS

Planners need a set of basic elements for the problem formulation.

Hydrothermal Generation Scheduling Problem (HGSP)

In hydrothermal generation power systems, the well-timed allocation of hydro energy resources is a very complex task that requires probability analysis and long-term consideration, because if water is used now, it will not be available in the future, thus increasing the future operation costs of the power systems.

The problem of minimizing the operational cost of a hydrothermal system can be reduced essentially to that of minimizing the fuel cost for thermal plants under the constraints of the water available for hydro generation in a given period of time.

In thermal generation power systems, the decision variables are modeled as integers, i.e., ON and OFF. For an exhaustive overview of short-term thermal scheduling until 2011, the reader may review the survey (Inostroza, 2011).

In this study, the HGSP is decomposed into three subproblems, namely, the hydrothermal coordination problem (HCP), the unit commitment problem (UCP), and the economic dispatch problem (EDP). In this way, the HGSP involves three main decision stages, usually separated using a time hierarchical decomposition, which is shown in Figure 1.

In hydrothermal generation systems, the cyclical nature of water flows and load, as well as the validity of model assumptions, suggest splitting

Figure 1. Time hierarchical decomposition for the HGSP

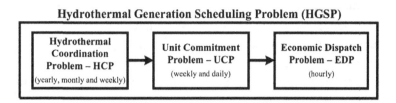

the hydro problem into long and short periods, so the HGSP can be broken down into smaller problems in order to solve it (Wood, 1996).

1. The long-term problem considers the yearly cyclical nature of reservoir, water inflows, seasonal load, and scheduling period; for instance, a period of one year is used. The solution of a long-range problem considers the dynamics of head variations through the water flow continuity equation.

2. In the short-term problem, the load on the power system exhibits cyclical variation over a day or a week, and correspondingly the scheduling interval for short-term is either a day or a week. As the scheduling interval of short-term problem is small, the solution of the problem can be assumed as constant, so the amount of water to be utilized for the short-term scheduling problem is known from the solution of the long-term scheduling problem.

Therefore, each hydro plant is constrained by the amount of water available for draw-down in the scheduling period, so the coordination between the hydrothermal coordination problem and the unit commitment problem can be done by setting targets in order to guarantee a proper system optimization.

The HGSP is a non-explicit objective function based on a nonlinear optimization problem with high dimensionality, continuous and discrete variables, unit forbidden operation zones, and equality and inequality constraints including hy-

draulic coupling, storage, and released flow limits of the reservoirs. The forbidden zones cause a non-continuous operation of the generating units, making the solution of the problem more difficult due to the associated combinatorial nature.

Hydrothermal Coordination Problem (HCP)

HCP is the first stage in the solution of the HGSP. This optimization problem consists of determining the optimal amounts of hydro and thermal generation to be used during a scheduling period. The HCP is broken down into long, medium, and short-term models (Pereira, 1983; Gil, 2007), depending on the reservoirs' storage capacity. According to the output, HCP approaches can be classified into two main categories:

1. Fixed reservoir storage level for each stage: the use of the water in each stage is determined strictly by the desired volume to be discharged by each reservoir over the scheduling period under study. The final target for each reservoir is a result from the long/medium-term generation scheduling.

2. Future cost functions: a function that shows the future or opportunity cost of the water used during the present stage versus the storage level at the end of the scheduling period.

Hydro generation has an opportunity cost associated to the thermal generation displaced. If a larger amount of water is used during the present stage, the immediate cost decreases and the water available in the future, i.e., the final water storage decreases. Hence, if less water will be available, the future costs will increase. The future cost of thermal units –FCF– can be obtained by calculating recursively the system operation costs in the future, starting from the end of the period under analysis, and considering different starting values of water storage. The FCF allows the long/medium-term coordination activity to be separated from the short-term hydrothermal coordination activity.

Unit Commitment Problem (UCP)

Once the hydroelectric generation is determined for each hour, thermal units must meet the load not covered by hydroelectric generation. The UCP deals with the decision of which of the thermal units will be started up or shutdown during each hour of the scheduling period. The committed units must be able to meet the system load at minimum operating cost, subject to a variety of technical constraints. The UCP is an important optimizing task in daily operational planning of power systems, which can be mathematically formulated as a large-scale nonlinear mixed-integer minimization problem for which there is no exact solution technique. The solution to the problem can be obtained only by complete enumeration, often at a prohibitive computation time requirement for realistic power systems. The solution space of the UCP is $[2^{G\_T}-1]^T$, where G_T is the number of thermal units and T is the scheduling period. For example, if G_T=10 and T=24, the total number of combinations are 1.7259 e72 "NP-Hard".

Economic Dispatch Problem (EDP)

Once the running units have been determined by the solution of the UCP, it is necessary to distribute the load by solving the EDP. The EDP consists of finding the optimal allocation of power among the running thermal units, satisfying the power balance equations and the unit's operation constraints at the minimum operating cost. The EDP is done through the lambda-iteration. The procedure converges very rapidly for this particular type of optimization problem.

Lambda-Iteration Algorithm

The algorithm can approach the solution by considering the Bisection Method (Wood, 1996). The algorithm starts with values of lambda below and above the optimal value, corresponding to too much and too little total output, and then iteratively brackets the optimal value. In the iterative method, the stopping rules must be established. Two general forms of stopping rules seem appropriate for this application: 1) finding the proper operating point within a specified tolerance; and 2) counting the number of times through the iterative loop and stopping when a maximum number is exceeded. In this proposal, the relative tolerance used is 1e-6.

Generation Scheduling Approach

In a mathematical optimization model, the decision variables are a set of variables that the planner has considered as the solution to the problem. The HGSP can be solved using the cost of the thermal generation. This fuel cost can be obtained in two ways:

1. By using a standard unit commitment and economic dispatch technique to find the optimal operation cost of the thermal generators, so the model handles the different sub-problems of the HGSP -HCP, UCP and EDP- simultaneously. The objective function in the optimization problem must minimize the future cost of the hydro reservoirs, the fuel cost, the starting up and shutting down costs from thermoelectric units. It must satisfy the

operating constraints of both the electrical system's (spinning reserve and load), and of each hydro and thermal unit's (continuity equation for the reservoir network, the desired water volume of each reservoir at the end of the period under study, physical limitations on reservoirs storage volumes and discharge rates, minimum up/down times, minimum and maximum power capacity, and load ramps, among others).

2. By assuming the thermal generation is represented by an equivalent plant, whose characteristics can be determined as described in Wood and Wollenberg (1996).

When the generation system is hydrothermal, the quantity of decision variables is too large, thereby causing the solution space to increase dramatically, and therefore, too, the degree of difficulty to find solutions. In this chapter, the HGSP is solved with this approach.

In this formulation, the HGSP can be solved in two stages. In the first stage, the HCP is solved considering continuous/real variables in the optimization problem, such as decision variables; i.e., the HGSP is managed by establishing the levels of hydro generation. In the second stage, the UCP is solved considering binary variables such as decision variables. In the optimization problem, the load balance constraint must consider the levels of hydro power generation determined previously, so the new load is equal to the initial load minus the obtained power generation. Considering this approach to model the optimization problem, the mathematical model is based on a real optimization problem for the first stage and on a binary optimization problem for the second stage.

In the first stage, the objective is to maximize the energy production from hydro resources through the fuel cost of the equivalent thermal plant. The second task decides the unit commitment obtained by minimizing the cost of

fuel, starting up and shutting down costs from thermoelectric units. At last, the EDP is solved using the lambda-iteration algorithm –continuous optimization problem–.

In the state of the art, some authors consider that the thermal generation can be represented by an equivalent plant where the fuel cost is considered as a quadratic function of the power from the composite thermal plants (Bernholtz, 1960; Wong, 1994; Yang, 1996; Orero, 1998; Hota, 1999; Sinha, 2003, Lakshminarasimman, 2006; Mandal, 2008; Hota, 2009).

Generation Planning Methodologies: Literature Review

The problem has been largely discussed in the specialized literature: transactions, magazines, and conference proceedings from the IEEE, Institution of Electrical Engineers, Electric Power Systems Research, and International Journal of Electrical Power and Energy Systems, among others journals, since 1960 (Bernholtz, 1960) to 2011 (Sahin, 2011; Inostroza, 2011; Constantinescu, 2011). In short-term HGSP different optimization algorithms have been applied to the optimization problem, which are based on mathematical and heuristic methodologies.

Mathematical Optimization Methods

These methodologies make use of mathematical formulation of the problem for finding optimal scheduling. Many classical methods have been successfully employed to solve this problem. However, the fuel cost curves of thermal plants and the input-output curves of hydro plants are usually represented as nonlinear and nonconvex ones with prohibited operation regions. Hence, most of the conventional gradient-based methods encounter difficulties due to the nonconvex feasible region of the optimization problems. Among these methods, only dynamic programming is able to handle them without strict requirements on linearity, convexity, and even continuity of the objective function and various constraints. However, it suffers from a computational overburden when it is applied to large-scale power systems. Due to the impossibility of considering all aspects and constraints, the obtained solution is optimum only under simplifications.

According to the reviewed literature, the most important approaches in this category are the following (Wood, 1996): methods based on Lagrangian relaxation, linear programming, dynamic programming, nonlinear programming, and mixed-integer linear programming.

The problem considering the equivalent thermal plant has been solved using the extended differential continuation method (Calderon, 1987), Lagrangian relaxation (Soares, 1980), dynamic programming (Tang, 1995), and progressive optimality algorithm (Nanda, 1981).

The success of the lagrangian relaxation approach lies in the fact that it allows relaxing the hard constraints. The large-scale problem is broken down into unit-wise thermal and river-wise hydro sub-problems that can be easily solved with conventional optimization techniques such as dynamic programming. However, on account of its own nature, the solution attained for the dual problem is almost always non-practical. It requires further complex adjustments to meet every constraint of the primary problem, especially for inter-temporal constrains, such as the multi-period water balance of hydropower schemes (Sifuentes, 2007). Therefore, the solution of hydrothermal systems is extremely difficult to deal with using Lagrangian relaxation approaches. The head-dependent water-power conversion adds other dimension of difficulty since the objective functions of hydro sub-problems are no longer stage-wise additive in comparison with water discharge. Therefore, there are two major disadvantages: 1) conver-

gence of the commonly employed sub-gradient algorithms to the dual maximum is very slow and the solution to the sub-problems may be very sensitive to variations of the prices, and 2) due to the non-convexity of the problem search space, the values of the multipliers that maximize the dual function do not guarantee feasibility of the primal problem.

Dynamic programming is flexible and can handle the constraints in a straightforward way. However, it needs to discretize reservoir levels. For systems with multiple and cascaded reservoirs, and discrete operating states, the state space expands exponentially with problem size, causing the "curse of dimensionality" for practical applications.

Evolutionary Approaches

The solution for such a highly nonlinear and nonconvex problem demands more robust and versatile techniques. Heuristic methods are the current alternative to the mathematical optimization models. They are very efficient in solving the highly nonlinear problems since they do not place any restrictions on the shape of cost curves and other nonlinearities in model representation.

Heuristic methods entail step-by-step generating, evaluating, and selecting options from the solutions region, with or without the user's help. To accomplish this task, the heuristic models perform searches with the guidance of logical or empirical rules. These rules are used to generate and classify the options during the search. The heuristic process is carried out until the algorithm is not able to find a better solution considering the assessment criteria that were agreed upon.

Heuristic techniques do not have any limitations regarding applications in problems where the objective functions are very complex or when the solution spaces are non-convex. This characteristic is a very important incentive for applying heuristic methods to face hydrothermal generation scheduling. Although these heuristic methods do not always guarantee the globally optimal solution, they will provide a reasonable solution or a suboptimal/optimal solution. One of the first heuristic approaches attempting to solve the HGSP problem was proposed by Chang and Cheng (1996).

The problem considering the equivalent thermal plant has been solved using neural networks (Naresh, 1999), evolutionary strategy (Werner, 1999), evolutionary programming (Yang, 1996-2; Hota, 1999; Sinha, 2003), simulated annealing (Wong, 1994; Yang, 1996-2), genetic algorithm (Yang, 1996-2, Orero, 1998), particle swarm optimization (Hota, 2009), and differential evolution (Mandal, 2008; Lakshminarasimman, 2006).

As a conclusion on the literature review, there is no strong formulation available for application to the short-term HGSP. Therefore, the contributions of this study are as follows: 1) a different and efficient repair algorithm is developed to model the hydro constraints, so the algorithm handles the constraints using corrector operators; 2) a very robust algorithm, the standard deviation observed in the test cases was close to zero. The methodology is applied to standard systems and the results are compared to the ones previously reported in the literature. The methodology can produce better solutions (minimal fuel cost of the thermal units) than the reference techniques such as dynamic programming, PSO, and genetic algorithms.

Short-Term Hydrothermal Generation Scheduling Formulation

A model of medium-term must deliver the amount of water to use in a short-term model for completing the weekly and daily scheduling. These water resources, associated with the generation of electrical energy, can be delivered to a short-term model in two ways: 1) by providing the future cost curves for each reservoir as a function of the final volume of each reservoir; and 2) by determining the volume that each reservoir must have in the system at the end of the scheduling period. The authors have considered the second case to model the optimization problem, so the future cost curves for each reservoir are not modeled in the objective function.

The optimizing schedule for hydrothermal power systems is modeled as a constrained optimization problem with a nonlinear objective function and a set of linear, nonlinear and dynamic constraints. Mathematically, the problem can be formulated as shown in Equation 1 (see Box 1).

The fuel cost can be defined using the expression $CC_i = a_i + b_i * Pt_{i(t)} + c_i * Pt^2_{i(t)}$; where, a_i, b_i and c_i are fuel cost coefficients of the i-th thermal unit. The startup and shutdown costs can be considered independent of time, and in the problem formulation, it is usually considered that the shutdown costs are much less than the startup costs, so it is disregarded.

The hydro and thermal units are connected to the same bus supplying the load, so the transmission losses and the network constraints are not

Box 1.

$$z = \min \left[\sum_{t=1}^{T} \left(\min \sum_{i=1}^{G_r} CC_i \left(Pt_{i(t)} \cdot U_{i(t)} \right) \right) + \left(\sum_{i=1}^{G_r} \left(c_i^{up} \cdot \left(U_{i(t)} - U_{i(t-1)} \right) \right) + \left(C_i^{down} \cdot \left(U_{i(t-1)} - U_{i(t)} \right) \right) \right) \right] \tag{1}$$

considered in the model. The independent system operator –ISO– knows the number of hours that each thermal unit has been startup or shutdown at the beginning of the scheduling period. In addition, the ISO must forecast the load system and the inflows for each hydro plant. Each generation company must inform the following: hydraulic dynamic model of each reservoir considering filtrations, evaporation and spillage, hydraulically coupled models, water storage capacity limits, water discharge rates, reservoir volume for each plant, and power generation function, among other technical aspects.

The optimization problem is subject to:

Load balance: The total active power generated by the units must be enough to satisfy the load forecast.

$$\sum_{i=1}^{G_r} Pt_{i(t)} \cdot U_{i(t)} + \sum_{j=1}^{G_H} Ph_{j(t)} + \sum_{i=1}^{GROR} Ph_{j(t)} = Dem_{(t)} \tag{2}$$

Generation limits: The active power generation of the units must satisfy minimum and maximum bounds.

$$Pt_i^{\min} \leq Pt_{i(t)} \leq Pt_i^{\max} \\ Ph_i^{\min} \leq Ph_{i(t)} \leq Ph_i^{\max} \tag{3}$$

Minimum up/down times: The constraints are modeled with the following equations.

$$s_{i(t)}^{up} \geq MIN_i^{up} \cdot \left(U_{i(t)} - U_{i(t-1)} \right) \\ \left| s_{i(t)}^{down} \right| \geq MIN_i^{down} \cdot \left(U_{i(t-1)} - U_{i(t)} \right) \tag{4}$$

Physical limitations on reservoir storage volumes: This constraint represents the limitations on the storage capacity for each reservoir.

$$Vol_j^{\min} \leq Vol_{j(t)} \leq Vol_j^{\max} \tag{5}$$

Physical limitations on water discharge rates: This constraint represents the limitations on the water discharge for each reservoir.

$$Q_j^{\min} \leq Q_{j(t)} \leq Q_j^{\max} \tag{6}$$

Hydraulic dynamic of each reservoir: Each hydro generation system has its own particular hydraulic restrictions, depending mainly on geographical and hydrological conditions. Sometimes, water discharge from one reservoir can affect availability in another reservoir, the so-called hydraulically coupled units.

$$Vol_{j(t+1)} = Vol_{j(t)} + infl_j - \\ \left(Q_{j(t)} + filt_t + ev_j + spil_j \right) + \sum_{j=1}^{R_j^{up}} Q_{j(t)} \tag{7}$$

Target volume to be discharged by each reservoir over the scheduling period: If the formulation does not consider the FCF, the mathematical formulation must include this restriction to guarantee the future availability of water for hydro plants.

$$Vol_{j(t=0)} \leq Vol_j^{begin} \\ Vol_{j(t=T)} \leq Vol_j^{\max} \tag{8}$$

Hydraulic power generation: In general, the hydro generator power output is a function of the reservoir volume and the rate of water discharge, where the reservoir volume is a function of the net hydraulic head of the reservoir. The model can also be written in terms of reservoir volume instead of the reservoir net head.

$$Ph_{j(t)} = C_1 \cdot Vol_{j(t)}^2 + C_2 \cdot Q_{j(t)}^2 + C_3 \cdot Vol_{j(T)} \cdot Q_{j(t)} \\ + C_4 \cdot Vol_{j(T)} + C_5 \cdot Q_{j(t)} + C_6 \tag{9}$$

EPSO Algorithm Applied to HGSP Problem

In 1995, Kennedy and Eberhart introduced the Particle Swarm Optimizer (PSO), a modern meta-heuristic optimization technique based on population. The PSO is based on two fundamental disciplines: social science and computer science. In addition, the PSO uses the swarm intelligence concept, which is the property of a system whereby the collective behaviors of unsophisticated agents that are interacting locally with their environment create coherent global functional patterns. The information available for each individual is based on its own experience, i.e., the decisions that it has made so far, the success of each decision, and the knowledge of the performance of other individuals in its neighborhood. Since the relative importance of these two factors can vary from one decision to another, it is reasonable to apply random weights to each part. There are three important aspects: 1) inertia that has the individual in his motion; 2) the individual memory that has come from positions; and 3) cooperation with the group's leader, who guides the movements in a certain proportion.

The optimization process is carried out with four basic operations: initialization, constraint handling, updating, evaluation, and selection. Figure 2 shows its flowchart.

The swarm size is an algorithm control parameter selected by the user, and each particle of the swarm is a vector that contains as many elements as the problem decision variables. The initial population is chosen randomly in order to cover the entire searching region uniformly, considering that the decision variables are real and binary, respectively. A uniform probability distribution for the decision variables is considered.

The success of the evolutionary algorithms like PSO is heavily dependent on the setting of control parameters and the constraint handling technique. It is important to discuss these two issues in depth before applying the proposed algorithm to the problem under consideration.

Control Parameters

When implementing the PSO, several considerations must be taken into account to facilitate the convergence and prevent an "explosion" of the algorithm. The performance of the evolutionary algorithms does not depend only on a good design and implementation, but it is also strongly influenced by the parameter values that determine its behavior. Determining the most appropriate parameters for an arbitrary problem is a complex task, since such parameters interact with each other in a highly nonlinear manner.

Two main types of methods have been proposed for setting up the parameter values of an evolutionary algorithm: off-line (called tuning) and on-line strategies (self-adapt parameters). Miranda *et. al.* introduced self-adaptation capabilities to the particle swarm by modifying the concept of a particle to include, in addition to the objective

Figure 2. PSO structure

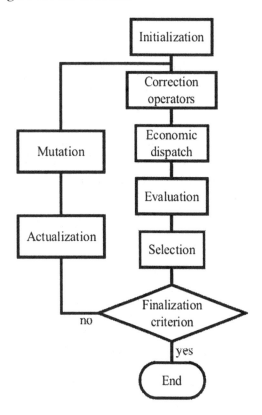

parameters, a set of strategic parameters: inertia and acceleration constants. This hybrid variant of PSO is called evolutionary programming and PSO –EPSO–, which defines these parameters as the genotype of a moving solution. Several publications on EPSO have shown better behavior than other meta-heuristics (Miranda, 2002; Miranda, 2002-2; Miranda, 2005; Miranda, 2009).

The EPSO rules generate new individuals as a weighted combination of parents: individual, its best ancestor, and the best ancestor of the present generation. The new individual is formed from a weighted mix of ancestors, and this weighted mix may vary in each space dimension. The general EPSO scheme can be summarized as follows.

- **Replication:** Each particle is replicated r times. In this study $r=1$.
- **Mutation:** Each particle has its weights mutated. The parameters of each individual are mutated according to (10):

$$w^*_{ij} = w_{ij} + \log N(0,1)^\tau \quad (10)$$

where, w_{ij} are the strategic parameters, $\log N(0,1)$ is a random variable which follows a Lognormal distribution from a Gaussian with zero mean and unit variance, and τ is a learning dispersion parameter that controls the amplitude of mutations. Small values of τ lead to a higher probability of having values close to pivot value in real applications, for instance in (Miranda, 2005) $\tau =0.1$

and in (Miranda, 2009) $\tau =0.3$. In this study, the optimal value is $\tau =0.5$.

- **Reproduction:** In order to update the real and binary variables, the position and the velocity for each particle in the swarm should be updated in different ways considering the real and binary spaces. The binary PSO also introduced by Kennedy and Eberhart, enables the PSO to operate in binary spaces. It is implemented by defining an intermediate variable called a sigmoid limiting transformation. Inostroza and Hinojosa (Inostroza, 2011) propose a new way to update the position. A particle decreases its speed as it converges toward the solution; thus, a great speed in absolute value implies that the particle is far away, and its position state must be changed and vice versa. The results improve substantially when the way to update the particle is changed. The mixed transformation process is given in Equation 11 (see Exhibit 1) where $G^* = G + w^*_{i4}$. $N(0,1)$ is the particle in the neighborhood of G, $N(0,1)$ is a random variable which follows a normal distribution.

The HCP has been formulated taking into account the water discharge rates as a decision variable via a continuous EPSO where the position $x_{i(k)}$ is the water discharge for each j-th reservoir,

Exhibit 1.

$$v_{i(k)} = w^*_{i1}v_{i(k-1)} + w^*_{i2}(p_i - x_{i(k-1)}) + w^*_{i3}(G^* - x_{i(k-1)})$$
$$\text{Real spaces}\left(x_{i(k)} \epsilon R^n\right)$$
$$x_{i(k)} = x_{i(k-1)} + v_{i(k)}$$

$$\text{Binary spaces}\left(X_{i(k)} \text{ is 0 or 1}\right)$$
$$X_{i(k)} = 1 - X_{i(k-1)} \quad \text{if } v_{i(k)} > \mu_i$$
$$X_{i(k)} = X_{i(k-1)} \quad \text{otherwise}$$

(11)

$Q_{j(t)}$, where j runs from 1 to G_H and t from 1 to T. The maximum values of each particle are always forced in the algorithm, so the physical limitations on water discharge rates are never violated. With $Q_{j(t)}$ the volume for each j-th reservoir $Vol_{j(t)}$ is computed using Equation 7, and finally so is the hydro power generation with Equation 9.

The UCP has been formulated taking into account the thermal states of the units via a binary EPSO where the position $X_{i(k)}$ is the status of the thermal units $U_{i(t)}$, where t runs from 1 to G_T.

In the solution of the HGSP, the algorithm should find the velocity of each particle using $v_{i(k)}$ $[v_{i(k)} \in R^n]$ and the position will be determined for continuous space by $x_{i(k)}$ and for binary space by $X_{i(k)}$.

- **Evaluation:** Each particle is evaluated according to its current position $x_{i(k)}$ and $X_{i(k)}$.
- **Selection:** A stochastic tournament is carried out in order to select the best particle to survive to the next generation. The fitness or objective function of an offspring particle is compared to that of its parent. The particle is replaced by its offspring if the offspring is more fit than its parent.
- **Stop Criterion:** The optimization process is repeated for several iterations in order to improve their fitness, while exploring the solution space.

Constraint Handling Technique

The constraint handling can be accomplished with several kinds of methods: methods based on preserving feasible solutions known as correction operators or heuristic rules; methods based on penalty functions; methods which make a clear distinction between feasible and infeasible solutions; and hybrid methods. Penalty functions are the most common technique used in evolutionary algorithms to incorporate restrictions on the fitness function. However, in the HGSP formulation,

an optimal penalty value for each constraint is a very complex task.

In evolutionary algorithms like genetic algorithms and PSO, the process is solved randomly; thereby the constraints may be frequently violated. In this study, the modeling of the constraints to preserve solution feasibility was accomplished by correction operators. Such operators not only improve the quality of the final solutions but also significantly improve the convergence of the search process due to work with feasible solutions. The correction operators are used after the initialization and the updating stage.

In the entire specialized literature review, the correction operators have been applied to short-term thermal generation scheduling (Inostroza, 2011), to hydrothermal generation scheduling (Gil, 2003) and to the optimal power flow problem (Oñate, 2009).

The purpose of the "correction operators" stage is to force the constraints in the HCP and EDP problems: 1) In the HCP problem, the volume of water to be discharged by each reservoir over the scheduling period and the physical limitations on reservoir storage capacity; and 2) in the EDP problem, the minimum up/down times and load balance constraint.

1. Volume of water to be discharged by each reservoir over the scheduling period: This operator is responsible for forcing the constraint imposed by the ISO on the volume at the end of the period for each reservoir. The pseudo-code is detailed in Exhibit 2.
2. Physical limitations on reservoir storage capacity: The previous operator fixed the final volume for each reservoir, but the algorithm does not verify the volume in each stage. So it is very important to fulfill the physical limitations on the storage capacity for each reservoir during each hour t. The pseudo-code is detailed in Exhibit 3.

Exhibit 2.

```
if Vol_{j(T)} ≠ Vol^{max}_{j(T)} {
    E= Vol_{j(T)} - Vol^{max}_{j(T)}
    if E > 0 {
        t_d ← [t| Vol^{min}_j < Vol_{j(t)} ≤ Vol^{max}_j]; t ∈ [1, 24]
        while E ≠ 0 {
            n ← rand (t_d)
            Q_{j(n)} = max (Q_j^{min}, Q_{j(n)} - E )
            update → [t|Vol_{j(t)}]; t ∈ [n,24]
            E= Vol_{j(T)} - Vol^{max}_{j(T)}
            t_d(n)=[] // delete position n from t_d
        }
    }
    else {
        t_u ← [t| Vol^{min}_j ≤ Vol_{j(t)} < Vol^{max}_j]; t ∈ [1,24]
        while E ≠ 0 {
            n ← rand (t_u)
            Q_{j(n)} = min (Q_j^{max}, Q_{j(n)} - E )
            update → [t|Vol_{j(t)}]; t ∈ [n,24]
            E= Vol_{j(T)} - Vol^{max}_{j(T)}
            t_u(n)=[] // delete position n from t_u
        }
    }
}
```

Exhibit 3.

```
if Vol_{j(t*)} > Vol^{max}_j
{
    E= Vol_{j(t)} - Vol^{max}_j
    if E>0 {
        t_d ← [t| Q_{j(t)} < Q^{max}_j ∩ V_{j(t+1)} > Vol^{min}_j]; t ∈ [1,t*-1]
        aux=E
        while E ≠ 0 { // decrease water discharge
            n ← rand (t_d)
            Q_{j(n)} = max (Q_j^{min}, Q_{j(n)} - E )
            update → [t|Vol_{j(t)}]; t ∈ [n,24]
            E= Vol_{j(t)} - Vol^{max}_j
            t_d(n)=[]
        }
    }
    t_u ← [t| Q_{j(t)} >Q^{min}_j ∩ V_{j(t+1)} > Vol^{max}_j]; t ∈ [t*+1,24]
    aux=E
    while E ≠ 0 { // increase water discharge
        n ← rand (t_u)
        Q_{j(n)} = min (Q_j^{max}, Q_{j(n)} - E )
        update → [t|Vol_{j(t)}]; t ∈ [n,24]
        E= Vol_{j(t)} - Vol^{min}_j
        t_u(n)=[]
    }
}
```

The increase in the water discharge at time *t* must be compensated for a decrease in the slack reservoir after the time *t**, because each reservoir meets the volume over the final period. The selection of the slack reservoir is accomplished randomly. In the case that there are no hours to decrease or increase water, i.e., set *t_d* or *t_u* is empty, the algorithm must undo the last change made and change the discharges to an upstream reservoir considering the same pseudo-code shown. In addition, the constraint $Vol_{j(t*)} < Vol^{min}_j$ has a similar approach as the algorithm previously executed.

3. Minimum up/down times and load balance constraint: The subroutine begins by analyzing if the state of the generating unit had a change of state. If there is no change, it retains its previous state. Otherwise it is analyzed whether the change has violated the minimum operating time restriction.

In addition, two situations can occur: 1) the maximum power that all units can deliver is excessive, so the algorithm can shutdown any feasible unit and still meet the load balance constraint; or 2) the power is not sufficient for the load, so the algorithm must startup feasible units randomly only if there are units. In the case that there are no units, depending on a random value, a unit can be started up in a previous time. However, when the unit changes its state, the constraint of spinning reserve changes from that hour until the analyzed hour, and therefore the process must be restarted (minimum up/down times).

If there is an invalid change, one of two corrections is performed, depending on a random value. One possibility is to preserve the previous state, i.e., if it is startup or shutdown, it remains with its state. The other correction changes the previous status of the preceding hours that are different from the current status.

Finally, the production costs for each thermal unit over the scheduling period are calculated, so the value of fuel cost in the first brackets of the objective function can be computed. Also, the startup cost and the future cost of the thermal units can be obtained in order to be able to evaluate the fitness function of each particle.

SIMULATIONS AND RESULTS

The statistical estimators considered in the simulation are the following: the best solution obtained, the average of the solutions, the worst solution,

Table 1. River transport delay between reservoirs

Plant	1	2	3	4
R_u	0	0	2	1
t_d	2	3	4	0

Table 2. Total load of the power systems (MW)

hour	load	hour	load	hour	load
1	1 370	9	2 240	17	2 130
2	1 390	10	2 320	18	2 140
3	1 360	11	2 230	19	2 240
4	1 290	12	2 310	20	2 280
5	1 290	13	2 230	21	2 240
6	1 410	14	2 200	22	2 120
7	1 650	15	2 130	23	1 850
8	2 000	16	2 070	24	1 590

Table 3. Hydro power generation coefficients

Plant	C1	C2	C3	C4	C5	C6
1	-0.0042	-0.42	0.03	0.9	10	-50
2	-0.004	-0.3	0.015	1.14	9.5	-70
3	-0.0016	-0.3	0.014	0.55	5.5	-40
4	-0.003	-0.31	0.027	1.44	14	-90

Figure 3. Hydraulic system network

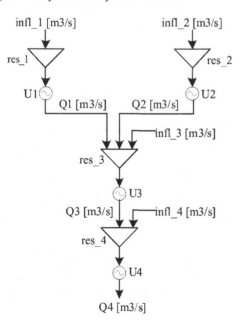

Table 4. Reservoir inflows (10^4 m^3)

hour	Reservoir				hour	Reservoir			
	1	2	3	4		1	2	3	4
1	10	8	8.1	2.8	13	11	8	4	0
2	9	8	8.2	2.4	14	12	9	3	0
3	8	9	4	1.6	15	11	9	3	0
4	7	9	2	0	16	10	8	2	0
5	6	8	3	0	17	9	7	2	0
6	7	7	4	0	18	8	6	2	0
7	8	6	3	0	19	7	7	1	0
8	9	7	2	0	20	6	8	1	0
9	10	8	1	0	21	7	9	2	0
10	11	9	1	0	22	8	9	2	0
11	12	9	1	0	23	9	8	1	0
12	10	8	2	0	24	10	8	0	0

Table 5. Technical data and restrictions for the reservoirs and generation limits

Plant	V_{min}	V_{max}	$V_{initial}$	V_{final}	Q_{min}	Q_{max}	Ph_{min}	Ph_{max}
1	80	150	100	120	5	15	0	500
2	60	120	80	70	6	15	0	500
3	100	240	170	170	10	30	0	500
4	70	160	140	140	13	25	0	500

Table 6. Technical parameters for the thermal units

Unit	1	2	3	4	5	6	7	8	9	10
P_{max} MW	592	592	169	169	211	104	114	72	72	72
P_{min} MW	150	150	20	20	25	20	20	10	10	10
a	1 000	970	700	680	450	370	480	660	665	670
b	16.19	17.26	16.6	16.5	19.7	22.26	27.74	25.92	27.27	27.79
c	0.00048	0.00031	0.002	0.00211	0.00398	0.00712	0.00079	0.00413	0.00222	0.00173
t_{min_up} h	8	8	5	5	6	3	3	1	1	1
t_{min_down} h	8	8	5	5	6	3	3	1	1	1
C_{hot_up} \$	4 500	5 000	550	560	900	170	260	30	30	30
C_{cold_up} \$	9 000	10 000	1 100	1 120	1 800	340	520	60	60	60
t_{cold_up} h	5	5	4	4	4	2	2	0	0	0
Initial state h	8	8	-5	-5	-6	-3	-3	-1	-1	-1

Table 7. Sensitivity analysis of τ

τ	0.05	0.1	0.2	0.3	0.4	0.5	0.6	0.7
best	927 307	926 537	926 223	926 643	928 116	**922 386**	923 224	929 449
average	930 335	933 523	926 765	926 659	930 733	**924 200**	927 984	934 709

and the coefficient of variation which is defined as CV = 100 * standard deviation/average solution.

The proposed test system consists of a multi-chain cascade of four hydro plants and ten thermal units.

- The multi-chain cascade of four hydro plants is shown in Figure 3 and Table 1. The hydraulic subsystem is characterized by the following: 1) a multi-chain cascade flow network, with all the plants on one stream; 2) river transport delay between successive reservoirs; 3) variable head hydro plants; and 4) variable natural inflow rates into each reservoir.

Where: I_j is the natural inflow to reservoir i, Q_i is the discharge of the plant i, R_u is the number of upstream plants, and T_d is the time delay to immediate downstream plant. This hydraulic test network models most of the complexities encountered in practical hydro networks.

The load system, hydro unit power generation coefficients, river inflows and reservoir limits for

Table 8. Optimal hourly plant discharges

hour	Reservoir 1	Reservoir 2	Reservoir 3	Reservoir 4
1	7.2632	6.3857	30.0000	13.0000
2	8.4878	6.9691	30.0000	13.0000
3	10.5477	6.7844	30.0000	13.1993
4	9.4040	6.3857	30.0000	13.0000
5	7.8537	6.0000	30.0000	13.0000
6	8.5682	6.1286	14.9313	13.0000
7	7.0521	7.5602	13.8210	13.0000
8	8.6620	7.9005	14.8998	13.0000
9	7.9484	10.1209	14.5670	13.1781
10	8.8240	7.4200	15.2631	14.3539
11	9.6841	8.4375	15.5280	14.2633
12	9.3521	7.6949	16.3367	14.8483
13	7.5952	6.6730	15.2595	15.5170
14	9.0703	8.3186	16.4639	15.1293
15	7.9228	10.4977	17.0279	15.3823
16	7.2312	8.2738	16.8548	15.3367
17	8.8568	8.0192	15.2664	15.3188
18	7.4835	9.4295	15.2843	16.5257
19	9.2566	10.4782	14.5060	16.8821
20	8.7143	11.5546	12.6028	17.1024
21	7.9681	10.9207	10.0125	17.8312
22	6.6261	10.5127	11.0000	18.7954
23	5.6277	9.5132	10.0000	19.0580
24	5.0000	10.0212	10.0000	21.6910

Table 9. Hourly hydro reservoir storage volumes

Vol [1e4 m3]	Reservoir 1	Reservoir 2	Reservoir 3	Reservoir 4
initial	100	80	170	120
1	102.74	81.61	148.10	109.80
2	103.25	82.65	126.30	99.20
3	100.70	84.86	107.56	87.60
4	98.30	87.48	94.44	74.60
5	96.44	89.48	84.95	91.60
6	94.88	90.35	90.21	108.60
7	95.82	88.79	93.63	125.60
8	96.16	87.89	95.30	142.60
9	98.21	85.76	94.91	159.42
10	100.39	87.34	96.87	160.00
11	102.70	87.91	98.19	159.56
12	103.35	88.21	102.80	159.61
13	106.76	89.54	108.64	158.66
14	109.69	90.22	112.97	158.79
15	112.76	88.72	114.23	158.94
16	115.53	88.45	115.12	159.94
17	115.68	87.43	118.10	159.88
18	116.19	84.00	122.54	159.82
19	113.94	80.52	126.16	159.96
20	111.22	76.97	130.06	159.72
21	110.25	75.05	140.74	157.15
22	111.63	73.53	151.93	153.64
23	115.00	72.02	162.45	149.09
final	120.000	70.000	170.000	140.000

the test network are given in Table 2, Table 3, Table 4, and Table 5, respectively (Orero, 1998).

The modeling of the thermal generation system is given in Table 6.

In this first analysis, it is assumed that the cost functions of the 10 thermal units can be combined into an equivalent function. This function reduces the complexity of the optimization problem, so the modeling allows a reduction in running time, without losing too much precision (Wood, 1996). The thermal generation system is represented by an equivalent thermal plant where the fuel cost is considered as a quadratic function, with a=5 000, b= 19.2, c=0.002, and $500 \leq Pt_{i(t)} \leq 2\ 500$. The scheduling period is 24 h with a 1h time interval.

Clerc *et. al.* proposes that the swarm size s can be computed by a formula based on the dimension d, s=10+(2*sqrt(d)). Some researchers use s=[40,60] as a suggested value and others, for instance Oñate *et. al.* (2009), use s=100. However, a large population is usually employed to obtain a good solution, and it is quite evident that a large population is not suitable for real power systems due to the long CPU computation time.

The authors must conduct a sensibility analysis in order to obtain the optimal size of the swarm.

Table 10. Hydro plant power outputs (MW) and total thermal generation (MW)

hour	Plant 1	Plant 2	Plant 3	Plant 4	Thermal
1	71	52	0	206	1 041
2	79	57	0	194	1 060
3	89	57	0	182	1 032
4	83	55	0	165	987
5	73	54	0	168	995
6	77	55	29	189	1 060
7	67	65	34	208	1 276
8	78	66	33	226	1 597
9	74	77	34	243	1 812
10	80	62	33	262	1 883
11	85	69	33	262	1 781
12	84	65	32	267	1 862
13	74	59	37	272	1 788
14	84	70	36	269	1 741
15	77	81	36	271	1 665
16	73	69	37	271	1 620
17	84	67	41	271	1 667
18	75	73	43	282	1 667
19	86	76	46	284	1 748
20	83	78	48	286	1 785
21	78	74	49	290	1 749
22	68	71	53	294	1 634
23	60	65	54	291	1 380
24	55	67	55	298	1 115

Table 11. Comparison of optimal cost for the hydrothermal power system

Method	solution	Cost [$]
EPSO-CO	**best**	**922 386**
	average	924 200
	worst	924 989
MDE	best	922 555
NLP	best	924 249
DP	best	928 919
GA	best	926 707
IPSO	best	923 443
PSO	best	925 384
	average	926 353
	worst	927 240
IFEP	best	930 130
	average	930 290

different values of τ, ranging from 0.05 to 0.7. Table 7 shows the results for each case.

It can be seen that the best parameter is $\tau = 0.5$. If the value is very small, the motion will be restricted. On the other hand, if the value is large, the motion of the swarm is chaotic in the simulations.

The average solution, in terms of the fitness function, is usually taken as the optimal solution.

In Table 8, the optimal hydro discharges solution obtained by the proposed algorithm is shown.

In addition to the turbine discharges, it is also useful to provide as an output quantities such as reservoir storage levels, total thermal generation and hydro unit power outputs, during each time interval. These quantities are computed over the scheduling period using the water discharge rates, the hourly river inflows, water transport delay and the load at each time interval.

In Table 9, the volume trajectories for each reservoir are shown.

It is clearly evident that the final volume of water for each reservoir satisfies the constraint. Finally, the hourly unit power outputs are given in Table 10.

The initial and final swarms consider 40 particles and 60 particles, respectively. The search is performed with 2500 iterations and 20 simulations. The selected parameters for the algorithm is s=50 based on a compromise solution between the computation time and the objective function.

Learning dispersion parameter τ: The original EPSO proposes that τ is fixed externally, and it is constant in the searching process. This parameter controls the speed and accuracy of the evolution strategy. For this case, a sensitivity analysis is conducted and 20 simulations are performed with

As a final point, the hourly hydrothermal generation scheduling for the power system is shown in Figure 4.

Table 11 provides comparisons of the optimal system costs obtained by different authors using evolutionary algorithms: modified differential evolution MDE (Lakshminarasimman, 2006), nonlinear programming NLP (Lakshminarasimman, 2006), dynamic programming (Lakshminarasimman, 2006), genetic algorithm GA (Orero, 1998), improved particle swarm optimization IPSO (Hota, 2009), particle swarm optimization PSO (Yu, 2007), and improved evolutionary programming IFEP (Sinha, 2003). The proposed approach yields better results than other methodologies while satisfying the reservoir end-volume constraints. In addition, there is no clarity on which statistical estimators are presented in some studies, so it is assumed that the solution is the best solution.

Lastly, the UCP and EDP must be solved, considering that the hydro generation is discounted from total load. The problem is a thermal scheduling, which is successfully solved by Inostroza and Hinojosa (2011) using the PSO-CO.

The optimal parameters of the learning coefficients and the inertia used by the PSO-CO are the following: the number of the particles is 50, 100 iterations, the maximal velocity is +/- 4, ϕ_1 is 2.8 and ϕ_2 is 1.2, and the inertia varies from 0.9 to 0.0.

The average fuel cost for 20 simulations is [\$] 707 015.93 and the average time is 27.90s, and the CV=0%. This means that the algorithm can reach the same optimum irrespective of the initial operating point. This is the better performance index of the PSO-CO algorithm. The results obtained in the simulations of the proposed algorithm demonstrate the feasibility of applying the PSO-CO as a solution to the unit commitment problem, permitting solutions to be obtained near the global optimum, mainly because of i) correction operators, which supports working with feasible solutions to the problem; ii) the mutation stage in the best solution, in order to avoid easily failing into local optimum; and iii) a new way of updating solutions.

In Table 12, the best solution obtained for the thermal power system is described.

Figure 4. Load forecast with the hydro and equivalent thermal generation

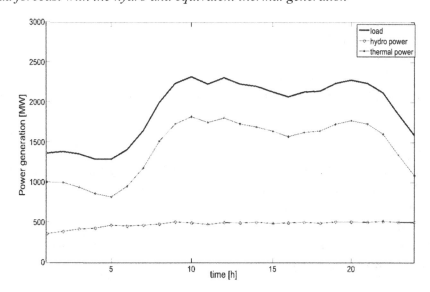

Table 12. Best solution obtained for the thermal system

time	Hydro power [MW]				Thermal power [MW]									
	U1	U2	U3	U4	U1	U2	U3	U4	U5	U6	U7	U8	U9	U10
1	71	52	0	206	592	449	0	0	0	0	0	0	0	0
2	79	57	0	194	592	468	0	0	0	0	0	0	0	0
3	89	57	0	182	592	440	0	0	0	0	0	0	0	0
4	83	55	0	165	592	395	0	0	0	0	0	0	0	0
5	73	54	0	168	592	403	0	0	0	0	0	0	0	0
6	77	55	29	189	592	468	0	0	0	0	0	0	0	0
7	67	65	34	208	592	515	0	169	0	0	0	0	0	0
8	78	66	33	226	592	592	169	169	75	0	0	0	0	0
9	74	77	34	243	592	592	169	169	211	79	0	0	0	0
10	80	62	33	262	592	592	169	169	211	104	0	46	0	0
11	85	69	33	262	592	592	169	169	211	48	0	0	0	0
12	84	65	32	267	592	592	169	169	211	104	0	25	0	0
13	74	59	37	272	592	592	169	169	211	55	0	0	0	0
14	84	70	36	269	592	592	169	169	199	20	0	0	0	0
15	77	81	36	271	592	592	169	169	143	0	0	0	0	0
16	73	69	37	271	592	592	169	169	98	0	0	0	0	0
17	84	67	41	271	592	592	169	169	145	0	0	0	0	0
18	75	73	43	282	592	592	169	169	145	0	0	0	0	0
19	86	76	46	284	592	592	169	169	206	20	0	0	0	0
20	83	78	48	286	592	592	169	169	211	52	0	0	0	0
21	78	74	49	290	592	592	169	169	207	20	0	0	0	0
22	68	71	52	294	592	592	169	169	112	0	0	0	0	0
23	60	65	54	291	592	592	0	0	196	0	0	0	0	0
24	55	67	55	298	592	523	0	0	0	0		0	0	0

FUTURE RESEARCH DIRECTIONS

Although extensively investigated, the hydro-thermal problem still attracts the attention of researchers because of the strong needs for lower cost operating schedules. To evaluate the performance of the proposed algorithm, the authors are studying the application of the methodology to a reduced version of the Chilean Central Interconnected System (SIC). It consists of six water reservoirs with 11 hydro units that are not hydraulically coupled and ten thermal units, which are the most representative. The initial tests and results in the modeling of the generation system are very promising.

A large population in the PSO is usually employed to obtain a good solution, but the CPU computation time will be very long. In contrast, an algorithm with a small population will be welcomed by power system operators and planners. Instead of using hundreds of particles, several particles (five to ten particles) will be enough to find the optimal solution. Furthermore, the transmission network with the equivalent thermal plant will render erroneous solutions because the constraint related to transmission limits could not be fulfilled due to the hydro and thermal units not being located in the network properly. Therefore, it is possible to extend this work considering the modeling of the transmission constraint in the optimization problem via a small population Particle Swarm Optimizer.

Currently, the authors are studying the modeling of uncertainty considering the water inflows and the load forecasting, and the inclusion of reliability and security constraints. In addition, with the advent of parallel and distributed computing, the implementation of this methodology on parallel machines will be investigated in the future.

CONCLUSION

The presented research has proposed a new approach to model the scheduling problem of short-term generation in hydrothermal power systems, which is characterized by three problems: the hydrothermal coordination problem, the unit commitment problem, and the economic dispatch problem, which were modeled in a nested and recursive optimization process.

This chapter has presented a novel approach based on the particle swarm optimizer. The model was implemented and exhaustively validated. In addition, in the algorithm a robust constraint handling based on correction operators/heuristic rules is introduced: volume of water to be discharged by each reservoir, physical limitations on reservoir storage capacity, minimum up/down times and the load balance. Such operators not only improve the quality of the final solutions but also significantly improve the convergence of the search process because of the use of feasible solutions. The results achieved show many advantages compared to penalty functions: 1) penalty functions do not work on a broad search space full of infeasible solutions, and 2) the problem of choosing penalties of different natures for each of the constraints disappears. The advantage of the approach is that it can obtain a reasonably good solution with acceptable computation time.

Another important aspect to be considered is that the proposed method does not easily fall into local optimum. This can be seen in the variance of the results, where all optimized values are bound in small ranges with several trials; the standard deviation/coefficient of variation in the unit commitment of the thermal system is zero.

Finally, it is necessary to emphasize that the evolutionary particle swarm optimizer algorithm is applicable to any nonlinear optimization problem.

ACKNOWLEDGMENT

This study was supported by the Centro Científico y Tecnológico de Valparaíso – CCTVal/Conicyt, Chile, under project FB0821; and by the Universidad Técnica Federico Santa María, Chile, under project No. USM 22.11.20.

REFERENCES

Bernholtz, B., & Graham, L. J. (1960). Hydrothermal economic scheduling part 1. Solution by incremental dynamic programing. *Transactions in AIEE Power Apparatus and Systems, 141*(5), 497–506.

Calderon, L. R., & Galiana, F. D. (1987). Continuous solution simulation in the short-term hydrothermal coordination problem. *IEEE Transactions on Power Systems, 2*(3), 737–743. doi:10.1109/TPWRS.1987.4335203

Chang, H. C., & Cheng, P. H. (1996). Genetic aided scheduling of hydraulically coupled plants in hydrothermal coordination. *Proceedings of IET, Generation, Transmission, and Distribution, 11*(2), 975–981.

Constantinescu, E. M., Zavala, V. M., Rocklin, M., Lee, S., & Anitescu, M. (2011). A computational framework for uncertainty quantification and stochastic optimization in unit commitment with wind power generation. *IEEE Transactions on Power Systems, 26*, 431–441. doi:10.1109/TPWRS.2010.2048133

Gil, E., Bustos, J., & Rudnick, H. (1983). Short-term hydrothermal generation scheduling model using a genetic algorithm. *IEEE Transactions on Power Systems, 18*(4), 1256–1264. doi:10.1109/TPWRS.2003.819877

Hota, P. K., Chakrabarti, R., & Chattopadhyay, P. K. (1999). Short-term hydrothermal scheduling through evolutionary programming technique. *Electric Power Systems Research, 52*, 189–196. doi:10.1016/S0378-7796(99)00021-8

Hota, P. K., Barisal, A. K., & Chakrabarti, R. (2009). An improved PSO technique for short-term optimal hydrothermal scheduling. *Electric Power Systems Research, 79*(7), 1047–1053. doi:10.1016/j.epsr.2009.01.001

Inostroza, J. C., & Hinojosa, V. H. (2011). Short-term scheduling solved with a particle swarm optimizer. *IET Generation, Transmission, and Distribution, 5*(11), 1091–1104. doi:10.1049/iet-gtd.2011.0117

Lakshminarasimman, L., & Subramanian, S. (2006). Short-term scheduling of hydrothermal power system with cascaded reservoirs by using modified differential evolution. *IEE Proceedings. Generation, Transmission and Distribution, 153*(6), 693–700. doi:10.1049/ip-gtd:20050407

Mandal, K. K., & Chakraborty, N. (2008). Differential evolution technique-based short-term economic generation scheduling of hydrothermal systems. *Electric Power Systems Research, 78*, 1972–1979. doi:10.1016/j.epsr.2008.04.006

Miranda, V., & Fonseca, N. (2002). *EPSO best of two worlds meta-heuristic applied to power system problems.* Paper presented at IEEE Congress on Evolutionary Computing.

Miranda, V., & Fonseca, N. (2002-2). *EPSO evolutionary particle swarm optimization, a new algorithm with applications in power systems.* Paper presented at IEEE/PES Transmission and Distribution Conf. Exhibition Asia Pacific.

Miranda, V. (2005). *Evolutionary algorithms with particle swarm movements.* Paper presented at 13th International Conference on Intelligent Systems Application to Power Systems.

Miranda, V., de Magalhaes, L., daRosa, M. A., daSilv, A. M. L., & Singh, C. (2009). Improving power system reliability calculation efficiency with EPSO variants. *IEEE Transactions on Power Systems, 24*, 1772–1779. doi:10.1109/TPWRS.2009.2030397

Nanda, J., & Bijwe, P. R. (1981). Optimal hydro-thermal scheduling with cascaded plants using progressive optimality algorithm. *IEEE Transactions on Power Apparatus and Systems, 100*(4), 2093–2099. doi:10.1109/TPAS.1981.316486

Naresh, R., & Sharma, J. (1999). Two-phase neural network based solution technique for short term hydrothermal scheduling. *IEE Proceedings. Generation, Transmission and Distribution, 146*(6), 657–663. doi:10.1049/ip-gtd:19990855

Oñate, P., Ramirez, J., & Coello, C. (2009). Optimal power flow subject to security constraints solved with a particle swarm optimizer. *IEEE Transactions on Power Systems, 23*, 96–104.

Orero, S. O., & Irving, M. R. (1998). A genetic algorithm modelling framework and solution technique for short term optimal hydrothermal scheduling. *IEEE Transactions on Power Systems, 13*(2), 501–518. doi:10.1109/59.667375

Pereira, M. V. F., & Pinto, L. M. V. G. (1983). Application of decomposition techniques to the mid-and short-term scheduling of hydrothermal systems. *IEEE Transactions on Power Apparatus and Systems*, *102*, 3611–3618. doi:10.1109/TPAS.1983.317709

Sahin, C., Zuyi, L., Shahidehpour, M., & Erkmen, I. (2011). Impact of natural gas system on risk-constrained midterm hydrothermal scheduling. *IEEE Transactions on Power Systems*, *26*, 520–531. doi:10.1109/TPWRS.2010.2052838

Sifuentes, W. S., & Vargas, A. (2007). Hydrothermal scheduling using benders decomposition: Accelerating techniques. *IEEE Transactions on Power Systems*, *22*(3), 1351–1359. doi:10.1109/TPWRS.2007.901751

Sinha, N., Chakrabarti, R., & Chattopadhyay, P. K. (2003). Fast evolutionary programming techniques for short-term hydrothermal scheduling. *IEEE Transactions on Power Systems*, *18*(1), 214–220. doi:10.1109/TPWRS.2002.807053

Soares, S., Lyra, C., & Tavares, H. (1980). Optimal generation scheduling of hydro-thermal power system. *IEEE Transactions on Power Apparatus and Systems*, *PAS-99*(3), 1107–1115. doi:10.1109/TPAS.1980.319741

Tang, J., & Luh, P. B. (1995). Hydrothermal scheduling via extended differential dynamic programming and mixed coordination. *IEEE Transactions on Power Systems*, *10*(4), 2021–2028. doi:10.1109/59.476071

Werner, T. G., & Verstege, J. F. (1999). An evolution strategy for short-term operation planning of hydrothermal power systems. *IEEE Transactions on Power Systems*, *14*(4), 1362–1368. doi:10.1109/59.801897

Wong, K. P., & Wong, Y. W. (1994). Short-term hydrothermal scheduling - Part-I and Part-II: Simulated annealing approach. *IEE Proc. Gen. Transactions Dist., 141 (5)*, 497-506.

Wood, A. J., & Wollenberg, B. F. (1996). *Power generation, operation and control*. John Wiley & Sons.

Yang, J. S., & Chen, N. (1989). Short-term hydrothermal generation scheduling model using a genetic algorithm. *IEEE Transactions on Power Systems*, *4*(3), 1050–1056.

Yang, P. C., Yang, H. T., & Huang, C. L. (1996). Scheduling short-term hydrothermal generation using evolutionary programming techniques. *Proceedings of IET, Generation, Transmission, and Distribution*, *143*(4), 371–376. doi:10.1049/ip-gtd:19960463

ADDITIONAL READING

Arroyo, J. M., & Conejo, A. J. (2002). A parallel repair genetic algorithm to solve the unit commitment problem. *IEEE Transactions on Power Systems*, *17*(4), 1216–1224. doi:10.1109/TPWRS.2002.804953

Houzhong, Y., Luh, P. B., Guan, X., & Rogan, P. M. (1993). Scheduling of hydrothermal power systems. *IEEE Transactions on Power Systems*, *8*(3), 1358–1365. doi:10.1109/59.260857

Pereira, M. V. F., & Pinto, L. M. V. G. (1991). Multi stage stochastic Optimization applied to energy planning. *Mathematical Programming*, *52*, 359–375. doi:10.1007/BF01582895

Yu, B. H., Yuan, X. H., & Wang, J. W. (2007). Short-term hydro-thermal scheduling using particle swarm optimization method. *Energy Conversion and Management*, *48*(7), 1902–1908. doi:10.1016/j.enconman.2007.01.034

Zoumas, C. E., Bakirtzis, A. G., Theocharis, J. B., & Petridis, V. (2004). A genetic algorithm solution approach to the hydrothermal coordination problem. *IEEE Transactions on Power Systems*, *19*(2), 1356–1364. doi:10.1109/TPWRS.2004.825896

KEY TERMS AND DEFINITIONS

Correction Operators: These operators are based on preserving feasibility of solutions, i.e., the operators are strategies to generate and keep control variables on feasible space.

Economic Dispatch: The economic dispatch is an optimization problem that seeks to assign the power levels that will supply each unit at each hour of the scheduling period, to minimize costs and satisfy the power load, while respecting the operation constraints.

Hydrothermal Generation Scheduling: The hydrothermal generation scheduling consists of determining the optimal amounts of hydro and thermal generation to be used during a scheduling period.

Particle Swarm: The particle swarm optimization is a modern metaheuristic optimization technique based on populations. It consists of a stochastic iterative process that operates on a particle swarm, where the position of each particle represents a potential solution of the problem simultaneously satisfying the operating constraints, for both the electrical system (spinning reserve and load) and for each unit (minimum up/down times, minimum and maximum power, and load ramps).

Unit Commitment: The unit commitment consists of deciding what units are in operation at each stage (1 hour) to minimize the fuel cost, and starting up and shutting down costs from thermoelectric units, and simultaneously satisfying the operating constraints, for both the electrical system (spinning reserve and load) and for each unit (minimum up/down times, minimum and maximum power, and load ramps).

APPENDIX: NOMENCLATURE

z: total system operation cost

T: number of periods (24 h)

G_T: number of thermal units

G_H: number of hydro units with reservoirs

$R^{up}_{(j)}$: number of the immediate upstream reservoirs of reservoir j

G_{ROR}: number of run-of-river units

$Pt_{i(t)}$: power supplied by thermal unit i in period t

$Ph_{j(t)}$: power supplied by hydro unit j in period t

Pt^{min}_i: minimum power of the thermal unit i

Ph^{min}_j: minimum power of the hydro unit j

Pt^{max}_i: maximum power of the thermal unit i

Ph^{max}_j: maximum power of the hydro unit j

CC_i: fuel cost of thermal units during period t

$U_{i(t)}$: status of the thermal unit i in period t (1 if it is startup or 0 if shutdown)

$Vol_{j(t)}$: volume for reservoir j in period t

$Vol_{j(T)}$: volume for reservoir j at the final period

Vol^{min}_j: minimum volume for reservoir j.

Vol^{max}_j: maximum volume for reservoir j

$Vol^{max}_{j(T)}$: maximum volume for reservoir j at the final period

$Q_{j(t)}$: water discharge for reservoir j in period t

Q^{min}_j: minimum discharge for reservoir j

Q^{max}_j: maximum discharge for reservoir j

$infl_j$: inflows forecasting for reservoir j

$filt_j$: filtrations for reservoir j

$spil_j$: spillage for reservoir j

ev_j: evaporation for reservoir j

$C_1, ..., C_6$: hydro power generation coefficients

$Dem_{(t)}$: load forecast for the system in period t

$C^{up}_{i(t)}$: startup cost of the thermal unit i in period t

$C^{down}_{i(t)}$: shutdown cost in period t

$S_{i(t)}$: operating time of unit i in period t

$S_{i(t)}^{up}$: minimal uptime of unit i in period t

$S_{i(t)}^{down}$: minimal downtime of unit i in period t

C^{hot}_i: hot cost of the unit i in period t

C^{cold}_i: cold cost of the unit i in period t

MIN^{up}_i: minimum uptime of thermal unit i

MIN^{down}_i: minimum downtime of thermal unit i

S^{Ccold}_i: cold critical time of thermal unit i

Chapter 15
Nondestructive Analysis of Dielectric Bodies by Means of an Ant Colony Optimization Method

Matteo Pastorino
University of Genoa, Italy

Andrea Randazzo
University of Genoa, Italy

ABSTRACT

Electromagnetic approaches based on inverse scattering are very important in the field of nondestructive analysis of dielectric targets. In most cases, the inverse scattering problem related to the reconstruction of the dielectric properties of unknown targets starting from measured field values can be recast as an optimization problem. Due to the ill-posedness of this inverse problem, the application of global optimization techniques seems to be a very suitable choice. In this chapter, the authors review the use of the Ant Colony Optimization method, which is a stochastic optimization algorithm that has been found to provide very good results in a plethora of applications in the area of electromagnetics as well as in other fields of electrical engineering.

INTRODUCTION

Global optimization algorithms are now very common in several areas of electromagnetic engineering and, in particular, in nondestructive testing and imaging. Among them, the most successful ones are the genetic algorithm (GA) (Haupt, 1995), the differential evolution method (DEM) (Price, 1999), and the particle swarm optimization (PSO) (Robinson & Rahmat-Samii, 2004). In the scientific literature, several different implementations have been proposed for such

DOI: 10.4018/978-1-4666-2666-9.ch015

algorithms. Sometimes, these algorithms are hybridized with deterministic methods in order to improve the convergence speed, which is usually rather slow (Pastorino, 2010).

The Ant Colony Optimization (ACO) method is a relatively new global stochastic optimization approach that has been inspired by the way ants find the optimal path from their nest to the food. It has been introduced by Dorigo in (Dorigo, Maniezzo, & Colorni, 1996) and successively discussed in (Dorigo & Gambardella, 1997; Socha & Dorigo, 2008). The ACO has been proven to outperform other optimization algorithms in some case, e.g., in the detection of dielectric objects inspected by interrogating electromagnetic waves (Pastorino, 2010). This latter application will be considered in this chapter and the result reported in the mentioned book justify the need for further assessing the use of ACO-based inverse scattering methods for electromagnetic nondestructive testing. However, the ACO has been successfully applied with different implementations in several other applicative fields. Some examples are reported in the following for illustrative purposes.

In (S. Xu, Bing, Lina, Shanshan, & Lianru, 2010), for example, the ACO has been used for image clustering applications. In particular, it has been combined with the so-called K-means algorithm in order "to solve the problem of misclassification of K-means and slow convergence of ACO". The results of K-means algorithm are used as elicitation information of ACO. In the same paper, the K-means-ACO algorithm has been proven to "improve the clustering accuracy for adjusting the misclassification" of the K-means approach (both by using simulations and real data). The proposed strategy is essentially an hybrid approach, which has been considered in several other implementations of the method. In electromagnetic imaging, which is the topic dealt with in this chapter, the ACO has been combined with the Linear Sampling Method (LSM) (Colton, Haddar, & Piana, 2003), which is a very fast method able to retrieve the shape of an unknown target starting

by measurements of the field it scatters when it is illuminated by a series of incident waves. In (Brignone, Bozza, Randazzo, Piana, & Pastorino, 2008) and (Brignone, Bozza, Randazzo, Aramini, et al., 2008), the ACO has been combined with the LSM in order to retrieve positions, shapes, and dielectric parameters (electric conductivity and dielectric permittivity) of homogeneous and inhomogeneous targets, respectively.

In (Chan, Chao, Jheng, Hsin, & Wu, 2010), the ACO has been applied for optimizing the performances of Networks-on-Chip (NoC), which are "generally dominated by traffic distribution and routing". By properly exploiting the precise network information for the path selection by using *pheromone*, the considered ACO-based adaptive routing has been shown to overcome the unbalance and unpredictable traffic load (Chan et al., 2010). In this case, too, the ACO has been combined with a Cascaded Adaptive Routing (CAR) in a hybrid solution.

Another hybrid approach has been proposed in the area of energy generation (Olamei, Niknam, Arefi, & Mazinan, 2011), where the ACO has been combined with the Simulated Annealing (SA), which is one of the first adopted stochastic optimization methods (Caorsi, Gragnani, Medicina, Pastorino, & Zunino, 1991). In particular, the ACO-SA strategy has been used in (Olamei et al., 2011) for the distribution feeder reconfiguration considering distributed generators (DGs), in which a cost-based compensation method is applied to "encourage DGs in active and reactive power generation." The approach has been tested on a real distribution feeder and the robustness of the approach has been verified (Olamei et al., 2011).

The routing problem has also been addressed by using of ACO, for example, for the ground service scheduling of an airport (Du, Zhang, & Chen, 2008). In this case, a *model* with multiple objectives has been considered in order to minimize the number of vehicles used, the total start time of serving flights, and the total flow time of vehicles. An efficient heuristic strategy (which has been called "the earliest due date (EDD) first") has been *incorporated* into the ACO original scheme (Du et al., 2008).

In military applications, too, hybrid approaches based on ACO has been successfully proposed. In (Lu, Wang, Chen, & Su, 2011), for example, a hybrid algorithm based on artificial immune systems (AIS) and ACO has been proposed for multiple uninhabited combat aerial vehicles (multi-UCAV) cooperative path planning. The effectiveness of the proposed cooperative planning model has been assessed by means of numerical simulations. Furthermore, the ACO method has been exploited for solving the so-called Target Assignment Problem (TAP) for an air defense command and control system of surface to air missile tactical units (Wang, Gao, Zhu, & Wang, 2010).

Other electromagnetic applications of ACO have been reported concerning the design of antennas and microwave components. In (Tenglong, Xiaoying, Jian, & Yihan, 2011), a modified version of ACO has been used for the optimization of antenna layouts. The main modification of the considered implementation of ACO with respect to the original version of the algorithm concerned the use of a strategy indicated by the authors as "discordant encoding to variable and optimization in vicinity". For the considered application, the new approach seems to outperform the original implementation in avoiding that the solution be trapped in local optima.

ACO has been also used for antenna design purposes in applications based on the use of Radio Frequency Identification Devices (RFIDs). In (Weis, Lewis, Randall, & Thiel, 2010), for example, an implementation allowing a fine tuning of the search procedure has been considered.

In (Vilovic, Burum, Sipus, & Nad, 2007), the ACO has been exploited for the optimal allocation of WLAN base stations, and, for this application, a comparison with results obtained by using the Particle Swarm Optimization (PSO) method have been reported in the same paper.

In addition, ACO has been proposed for optimizing the coverage mechanism in sensor networks by considering an energy-efficient strategy (R. Huang, Zhu, & Xu, 2007). The proposed approach "guarantee that each target should be covered by at least one active sensor at the minimum cost" (R. Huang et al., 2007).

ACO-based algorithms have been also used in telecommunications (C. Xu, El-Hajjar, Maunder, Yang, & Hanzo, 2010; Chowdhury, Baker, & Choudhury, 2008), and remote sensing, including spectral mixture analysis by hyperspectral sensors (Zhang, Sun, Gao, & Yang, 2011) and maritime applications, e.g., ship detection starting from synthetic aperture radar (SAR) images (Xie, Li, Bo, & Zhang, 2009).

However, it must be recognized that, although ACO has been widely used, relatively few theoretical studies are available. One of them has been reported in (H. Huang, Wu, & Hao, 2009), where, on the bases of the "absorbing Markov chain model", the ACO convergence time has been analyzed. The study was based on the relationship between the convergence time and the pheromone rate. It has been concluded that the pheromone rate and its deviation determine the expected convergence time. The reported results has been also verified by means of numerical simulations.

Beside theoretical studies, it is also important to have proper comparison results among different "alternative" methods of the same class. Comparing stochastic optimization methods is not an easy task, due to the large number of triggering parameters, which can be suitably chosen only for specific applications. However, some comparative studies have been recently reported in the scientific literature. For example, ACO, GA, and PSO algorithms have been compared in the framework of the Fractal Image Compression (FIC) (Uma, Palanisamy, & Poornachandran, 2011). The new ACO-based approach proposed in (Uma et al., 2011) has been found to be robust against outliers in the image and able to effectively reduce the encoding time.

In electromagnetic imaging, ACO has been compared in (Pastorino, 2007) with the GA and DEM, in the case of the inspection of a two-layer circular dielectric cylinder (the forward scattering data – used as input data for the reconstruction procedure - have been analytically computed and corrupted by Gaussian noise). A multi-illumination/multiview imaging approach (see Section 2) has been used, in which sources and probes (uniformly distributed on a circumference) have been positioned to form a tomographic set up. The reconstruction results reported in (Pastorino, 2007) showed that, at least in the considered case, the ACO "reaches a more accurate reconstruction than those obtained by the other two methods" and "requires a lower number of function evaluations". It has been concluded in (Pastorino, 2007) that ACO is "a good candidate for RF and microwave imaging applications."

As previously mentioned, in this chapter, the ACO is considered in the framework of the nondestructive analysis of dielectric targets, which is a research area that has received a significant attention in the last years (Albanese, Rubinacci, Takagi, & Udpa, 1998; Chang, Chou, & Lee, 1995; Giakos et al., 1999; Kharkovsky & Zoughi, 2007; Lesselier & Razek, 1999; Mudanyal, Yldz, Semerci, Yapar, & Akduman, 2008). In particular, the aim is to get information about a dielectric structure under test by using interrogating wave radiations. In most cases, such information is represented by the position, shape, and (possibly) dielectric properties of cracks and defects inside a known structure. When microwaves are used for illuminating the structure, an inverse scattering problem must be solved, in which the scattered field data (collected by probes outside the object under test) are related to geometrical/physical properties of the inspected target (usually, by integral equations).

The present chapter is organized as follows. In Section 2 the formulation of the considered inverse scattering problem is reported. In particular, the basic equations relating the target to be inspected to the measured values of the scattered electromagnetic field are briefly reviewed. Moreover, Section 3 describes the implemented ACO method, with details about the initialization phase, the iterative construction of the solution, the so-called pheromone update, and the adopted stopping criterion. Finally, some numerical results concerning the reconstruction of dielectric cylinders are provided in Section 4, whereas some conclusions are drawn in Section 5.

2. MATHEMATICAL FORMULATION

Approaches based on inverse scattering are aimed at retrieving the information about an object under test starting from measurements of the electric field scattered by the target when illuminated by a given source (Garnero, Franchois, Hugonin, Pichot, & Joachimowicz, 1991; Gragnani, 2002; Pastorino, 2010). Usually, a multi-illumination configuration is considered, i.e., the target is successively illuminated by waves impinging from different directions. Let $\mathbf{E}_s^{inc}(\mathbf{r})$, $s = 1,...,S$, be the electric field vector associated to the sth illuminating wave. If the target under test has an elongated shape (e.g., along the $\hat{\mathbf{z}}$ direction) and the incident waves are linearly polarized with the electric field vectors directed along the target major axis (i.e., $\mathbf{E}_s^{inc}(\mathbf{r}) = \Psi_s^{inc}(\mathbf{r})\hat{\mathbf{z}}$), the scattering problem can be approximated as a scalar two-dimensional problem, in which the only interest is the retrieval of the transversal section of the body (cross section).

In the case of nonmagnetic lossy dielectric materials and under harmonic time dependence, the relationship among the measured quantities and the unknown parameters to be retrieved is the Lippmann-Schwinger integral equation, which can be written as (Balanis, 1989)

$$\Psi_s^{scatt}\left(x,y\right) = \iint_D \gamma\left(x',y'\right)$$
$$\left(\Psi_s^{scatt}\left(x',y'\right) + \Psi_s^{inc}\left(x',y'\right)\right)\Gamma\left(x,y,x',y'\right)dx'dy'$$

$$(1)$$

where $\Psi_{m,s}^{scatt}$ denotes the z-component of the electric field vector $\mathbf{E}_s^{scatt}$, which is scattered by the object when illuminated by the sth incident electric field; Γ is the kernel of the integral equation and relates the current integration points with the measurement points of the sth view (i.e., Γ is the 2D Green's function for the assumed imaging configuration). D is the cross section of the target (if it is a known quantity) or a fixed investigation area (containing, by hypothesis, the cross section of the target). Finally γ is the contrast function, which is related to the values of the dielectric parameters by the following relation

$$\gamma\left(x,y\right) = -K^2\left(\epsilon\left(x,y\right) - \epsilon_0 - j\frac{\sigma\left(x,y\right)}{\omega}\right)$$

$$(2)$$

where ϵ and σ are the dielectric permittivity [F/m] and the electric conductivity [S/m] of the body, respectively. Both these parameters are in general inhomogeneous. Moreover, ω is the angular frequency [rad/s], ϵ_0 is the dielectric permittivity of vacuum, and K is the wavenumber of the propagation medium.

In order to solve Equation 1, it is necessary to discretize the involved quantities. The most straightforward way is to use pixel-based representations of the unknowns. In this case, the investigation area is subdivided into N square subdomains with centers $\left(x_n^{(D)},y_n^{(D)}\right)$, $n=1,...,N$, and side d. The contrast function and the electric field distributions are assumed to be constant in each subdomain (i.e., a piecewise-constant approximation is adopted). Moreover, for any view, the scattered electric field is col-

lected in a discrete set of M measurement points $\left(x_{m,s}^{(O)},y_{m,s}^{(O)}\right)$, $m=1,...,M$, $s=1,...,S$. Consequently, $I=M\times S$ measurement samples are acquired. From a practical point of view, such scattered field values can be obtained by moving a single probe or by using an array of probes.

By means of the previous discretization procedure, Equation 1 can be written in a discrete form as

$$\Psi_s^{scatt} = \Gamma_s^{(O)}\mathrm{diag}\left(\gamma\right)\Psi_s^D, \; s=1,...,S \qquad (3)$$

where

$$\Psi_s^{scatt} = \left[\Psi_s^{scatt}\left(x_{1,s}^{(O)},y_{1,s}^{(O)}\right),...,\Psi_s^{scatt}\left(x_{M,s}^{(O)},y_{M,s}^{(O)}\right)\right]^t$$

(the superscript t denotes transposition), is an array containing the scattered electric field in the M measurement points for the sth view,
$$\gamma = \left[\gamma\left(x_1^{(D)},y_1^{(D)}\right),...,\gamma\left(x_N^{(D)},y_N^{(D)}\right)\right]^t, \; \mathrm{diag}\left(\cdot\right) \text{ is}$$
a diagonal matrix whose elements are the values contained in its argument,
$\Psi_s^D = \Psi_s^{inc,D} + \Psi_s^{scatt,D}$,
being
$$\Psi_s^{inc,D} = \left[\Psi_s^{inc}\left(x_1^{(D)},y_1^{(D)}\right),...,\Psi_s^{inc}\left(x_N^{(D)},y_N^{(D)}\right)\right]^t$$
and
$$\Psi_s^{scatt,D} = \left[\Psi_s^{scatt}\left(x_1^{(D)},y_1^{(D)}\right),...,\Psi_s^{scatt}\left(x_N^{(D)},y_N^{(D)}\right)\right]^t$$
the arrays containing the incident and scattered electric field values in the N subdomains for the sth view, and $\Gamma_s^{(O)}$ is a matrix whose elements are the integrals of the Green's function over the subdomains. It is worth noting that the array Ψ_s^D (in Equation 3) is an unknown quantity, since it represents the total electric field inside the investigation area. Thus, a second equation is needed in order to solve the inverse scattering problem. In particular, a Lippmann-Schwinger integral equation similar to Equation 3 (but applied to the electric field in the investigation domain) can be

considered. In a discrete setting, such equation can be written as

$$\Psi_s^D = \Psi_s^{inc,D} + \Gamma^{(D)}\text{diag}\left(\gamma\right)\Psi^D, \ s = 1,...,S \tag{4}$$

where $\Gamma^{(D)}$ is a $N \times N$ matrix, which depends on the integrals of the Green's function.

The pair of Equations 3 and 4 can be solved in different ways. In particular, it is possible to directly consider a system of two equations. This approach has been followed, for example, in in (Bozza, Estatico, Pastorino, & Randazzo, 2006). Otherwise, the two equations can be manipulated in order to obtain a single nonlinear equation. In the former case, since it can be proven that equation 4 is well conditioned, the internal electric field can be formally obtained as

$$\Psi_s^D = \left(\mathbf{I} - \Gamma^{(D)}\text{diag}\left(\gamma\right)\right)^{-1}\Psi_s^{inc,D}, \ s = 1,...,S \tag{5}$$

where $\mathbf{I}$ is the identity matrix. Consequently, by substituting Equation 5 into Equation 3, one obtains the following nonlinear relationship

$$\Psi_s^{scatt} = \Gamma_s^{(O)}\text{diag}\left(\gamma\right) \\ \left(\mathbf{I} - \Gamma^{(D)}\text{diag}\left(\gamma\right)\right)^{-1}\Psi_s^{inc,D}, \ s = 1,...,S \tag{6}$$

It is worth noting that Equation 5 can be efficiently solved by means of iterative methods based on the fast Fourier transform (e.g., the BiCGStab-FFT (X. Xu, Liu, & Zhang, 2002)), resulting in a computational complexity of $CN\log N$ (being C the number of iterations needed to solve the forward problem, usually with $C \ll N$).

In order to retrieve the parameters of the cross section (contained in the array γ) starting from the I measurements, Relationship 6 must be

inverted. An approach followed when using global methods is to recast the solution of Equation 6 as the iterative minimization of a suitable functional representing a measure of the "residual" between the measured values and those predicted by the considered trial solution. A commonly adopted functional is the following

$$f\left(\gamma\right) = \frac{1}{\sum_{s=1}^{S}\left\|\Psi_s^{scatt}\right\|^2}\sum_{s=1}^{S} \\ \left\|\Psi_s^{scatt} - \Gamma_s^{(O)}\text{diag}\left(\gamma\right)\left(\mathbf{I} - \Gamma^{(D)}\text{diag}\left(\gamma\right)\right)^{-1}\Psi_s^{inc,D}\right\|^2 \tag{7}$$

where $\left\|\mathbf{x}\right\|^2 = \sum_{m=1}^{M}\left|x_m\right|^2$ is the standard l^2 norm, being $\mathbf{x}$ a generic array of size M and x_n its mth component. A regularization term can also be considered, e.g., by adding or multiplying F with a function $\mathcal{R}\left(\gamma\right)$ aimed at reducing the ill-conditioning of the problem.

It is worth noting that it is usually necessary to use large N values in order to obtain good spatial resolutions. Consequently, Equation 7 must be minimized with respect to a very large number of unknowns. In order to speed-up the reconstruction process, it is possible to use a different representation of the contrast function. In this case, the problem of retrieving the array γ of N (possibly complex) unknowns is recast as the retrieval of an auxiliary array $\mathbf{a}$ of size $L \ll N$. A possible choice is a representation of the contrast function in terms of B-spline (Dierckx, 1996), which are usually able to provide fairly good approximations even with a reduced number of unknowns. Following the formulation detailed in (Baussard, Miller, & Prémel, 2004), the array γ is expressed as

$$\gamma = \mathbf{Ba} \tag{8}$$

where $\mathbf{a} = \left[a_1, \ldots, a_L \right]^t$ is an array containing the unknown expansion coefficients and $\mathbf{B}$ is a (sparse) matrix containing the cubic B-spline functions (Dierckx, 1996). In this case, Equation 6 becomes

$$\Psi_s^{scatt} = \Gamma_s^{(O)} \mathrm{diag}\left(\mathbf{Ba} \right)$$
$$\left(\mathbf{I} - \Gamma^{(D)} \mathrm{diag}\left(\mathbf{Ba} \right) \right)^{-1} \Psi_s^{inc,D}, \ s = 1, \ldots, S$$

$$(9)$$

and consequently, the cost function is given by

$$f\left(\mathbf{a} \right) = \frac{1}{\sum_{s=1}^{S} \left\| \Psi_s^{scatt} \right\|^2} \sum_{s=1}^{S}$$
$$\left\| \Psi_s^{scatt} - \Gamma_s^{(O)} \mathrm{diag}\left(\mathbf{Ba} \right) \left(\mathbf{I} - \Gamma^{(D)} \mathrm{diag}\left(\mathbf{Ba} \right) \right)^{-1} \Psi_s^{inc,D} \right\|^2$$

$$(10)$$

In both cases (pixel- or spline-based representations), the final functional equation can be solved by using both deterministic (e.g., Gauss-Newton methods) and stochastic methods (e.g., GA, DEM, PSO, ACO). As mentioned in the Introduction, ACO is a very good candidate to perform this task.

3. ANT COLONY OPTIMIZATION

Ant Colony Optimization is a recently developed stochastic method for functional optimization. In its first versions, it was designed for imitating the behavior of ants, and, in particular, how they find the optimal path for reaching the food starting from their nest (Dorigo et al., 1996). Such an approach turned out to be very effective in solving hard combinatorial problems, such as the Traveling Salesman Problem (Dorigo & Gambardella, 1997). In the last few years, various different versions of the original algorithm have been proposed. In particular, several works concerned the extension

of the ACO to continuous domains (Bilchev & Parmee, 1995; Dréo & Siarry, 2002; Monmarché, Venturini, & Slimane, 2000).

In the present Chapter, the $\mathrm{ACO}_{\mathbb{R}}$ version (Socha & Dorigo, 2008) is considered and applied for solving the inverse scattering problem described in the previous section. A description of the method is reported in the following.

Let us consider a cost function $f : S \subseteq \mathbb{R}^G \to [0, +\infty)$ to be minimized. The unknown array $\mathbf{x} = \left[x_1, x_2, \ldots, x_G \right]^t \in S \subseteq \mathbb{R}^G$ can be subjected to arbitrary constraints. In the following, the case of bound constraints (i.e., $l_g \leq x_g \leq u_g$, being l_g and u_g the lower and upper bounds for the g th component of $\mathbf{x}$) is considered. As other swarm-based methods, the ACO algorithm is an iterative method that continuously update a population of P trial solutions until some convergence criteria are fulfilled. In the following, the population at the k th iteration is denotes as $\wp_k = \left\{ \mathbf{x}_{k,p}, p = 1, \ldots, P \right\}$, $k = 1, \ldots, K_{opt}$ (being K_{opt} the number of the iteration at which the method is stopped). For computational convenience, the population is stored in an ordered archive, i.e., $f\left(\mathbf{x}_{k,1} \right) \leq f\left(\mathbf{x}_{k,2} \right) \leq \cdots \leq f\left(\mathbf{x}_{k,P} \right)$.

A flow chart of the $\mathrm{ACO}_{\mathbb{R}}$ algorithm is shown in Figure 1.

The blocks shown in Figure 1 perform the following tasks.

Initialization

The initial population $\wp_0 = \left\{ \mathbf{x}_{0,p}, p = 1, \ldots, P \right\}$ is created by generating P random trial solutions $\mathbf{x}_{0,p} = \left[x_{0,p,1}, \ldots, x_{0,p,G} \right]^t$. In particular, in the case of boundary constraints, the g th components of the p th trial solution is generated as

Figure 1. Flow chart of the ACO algorithm

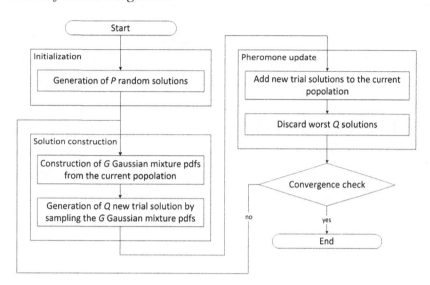

$$x_{0,p,g} = l_g + \left(u_g - l_g\right)U\left(0,1\right),$$
$$g = 1,...,G, p = 1,...,P \tag{11}$$

where $U\left(0,1\right)$ is a function returning a random value uniformly distributed in the range $\left[0,1\right]$. If additional a-priori information is available, it can be included in the initialization procedure (e.g., if the target of the inspection process is the reconstruction of one or more localized objects, it is possible to generate random shapes and then use their pixel- or spline-based representations as starting random solutions).

Solution Construction

At each iteration, Q new random solutions $\tilde{\mathbf{x}}_{k,q}$, $q = 1,...,Q$, are generated. In particular, each component $\tilde{x}_{k,q,g}$ of $\tilde{\mathbf{x}}_{k,q}$ is drawn from a Gaussian mixture probability density function, defined as

$$G_{k,g}\left(x\right) = \sum_{p=1}^{P} \omega_p \frac{1}{s_{k,p,g}\sqrt{2\pi}} e^{-\frac{(x-m_{k,p,g})^2}{2s_{k,p,g}^2}} \tag{12}$$

where w_p, $p = 1,...,P$, are weighting parameters given by

$$w_p = \frac{1}{\rho P\sqrt{2\pi}} e^{-\frac{(p-1)^2}{2\rho^2 P^2}}, \ p = 1,...,P \tag{13}$$

and the mean and standard deviation of the Gaussian kernels are computed as

$$m_{k,p,g} = x_{k,p,g}, \ p = 1,...,P, \ g = 1,...,G \tag{14}$$

$$s_{k,p,g} = \xi\sum_{i=1}^{P} \frac{\left|x_{k,p,g} - x_{k,i,g}\right|}{P-1}, \tag{15}$$
$$p = 1,...,P, \ g = 1,...,G$$

The quantities ρ and ξ, which is usually called the *pheromone evaporation rate*, are key parameters of the ACO algorithm (Socha & Dorigo, 2008), whose best choice depends on the specific application.

Pheromone Update

The new population $\wp_{k+1}$ at the $(k+1)$th iteration is obtained by performing the following two steps:

- **Positive Update:** The Q newly created trial solutions are added to the archive containing the population. In this way, we obtain a pool of $P+Q$ arrays.

$$\tilde{\wp}_{k+1} = \left\{ \mathbf{x}_{k,p}, p = 1, \ldots, P \right\} \cup \left\{ \tilde{\mathbf{x}}_{k,q}, q = 1, \ldots, Q \right\}$$

- **Negative Update:** The worst Q elements (i.e., those having the highest values of the cost function) of $\tilde{\wp}_{k+1}$ are discarded.

Convergence Check

The algorithm is stopped when some predefined stopping criteria are fulfilled. In particular, the stopping criteria can be composed by several conditions. Some of the most used are the following

- **Maximum Number of Iterations:** The method is stopped when a given number of iteration k_{max} is reached;
- **Cost Function Threshold:** The method is stopped when the value of cost function of the best trial solution falls below a given threshold f_{th};
- **Cost Function Improvement Threshold:** The method is stopped if the improvement of the cost function of the best individual is below a fixed threshold, Δf_{th}, after k_{th}. iterations.

Figure 2. (a) Actual configuration and (b) reconstructed distribution of the dielectric permittivity. (c) cost function and (d) relative reconstruction errors for the best individual of the population versus the iteration number. Two adjacent cylinders. Pixel-based procedure.

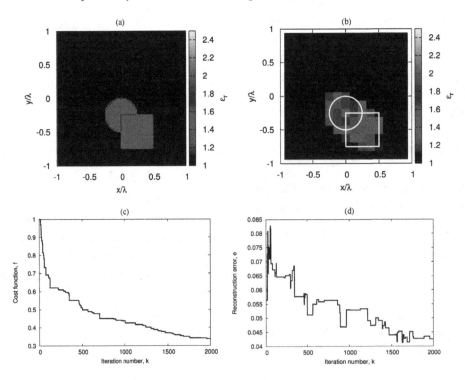

Figure 3. (a) Reconstructed distribution of the dielectric permittivity. (b) cost function and (c) relative reconstruction errors for the best individual of the population versus the iteration number. Two adjacent cylinders. Spline-based procedure.

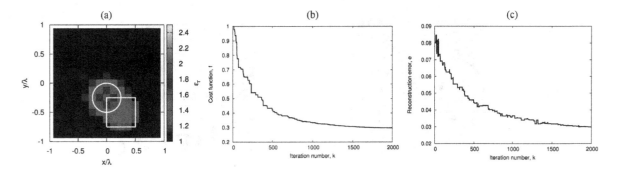

4. NUMERICAL RESULTS

Some results obtained by applying the previously described ACO-based inversion algorithm to the reconstruction of lossless dielectric objects are reported in the following. In all cases, the investigation domain is a square area of side $W = 2\lambda$ (being λ the wavelength of the incident radiation in the background medium). The incident electric fields are generated by means of $S = 8$ line-current sources located on a circumference of radius $R^{(I)} = 1.5\lambda$ at angular positions $\theta_s^{(I)} = 2\pi(s-1)/S$, $s = 1,\ldots,S$. For each view, the scattered electric field is collected at $M = 51$ points $\left(x_{m,s}^{(O)}, y_{m,s}^{(O)}\right)$, located on a circumference of radius $R^{(O)} = 1.5\lambda$, at angular positions $\theta_{m,s}^{(O)} = 0.25\pi + \theta_s^{(I)} + 1.5\pi(m-1)/(M-1)$, $s = 1,\ldots,S$, $m = 1,\ldots,M$. The electric field data used as input data for the inversion procedure are obtained by using a numerical simulator based on the method of moments (Harrington, 1993) with pulse basis and Dirac's delta weighting functions. The number of pixel, N_{fw}, used to discretize the investigation area in the forward problem is chosen to be equal to $N_{fw} = 3969$. Moreover,

the computed data are corrupted by a Gaussian noise with zero mean value and variance corresponding to a signal-to-noise ratio of 25 dB. The assumed value of SNR is in agreement with the values achievable by using (for example) the prototype of a microwave tomograph reported in (Salvade, Pastorino, Monleone, Bozza, & Randazzo, 2009).

In the inverse problem, both the discretization procedures introduced in Section 2 are considered. For the pixel-based approach, the investigation area is discretized into $N = 256$ square subdomains. In this case, since lossless object are considered, the number of unknowns (representing the real part of the contrast function) is $G = N$. For the spline-based approach, $T = 9$ inner knots for each direction (i.e., along the x and y directions) are used. Consequently, the number of unknowns is $G = (T+4)^2 = 169$. Cubic B-spline are used (Dierckx, 1996). In both cases, the values of the parameters of the ACO solver have been chosen on the basis of the suggestions available in the literature (Socha & Dorigo, 2008), i.e., $\rho = 0.1$, $\xi = 0.85$, $P = G$, and $Q = G/10$. The algorithm is stopped when a maximum number of iteration, $k_{max} = 2000$, is reached.

Figure 4. (a) Actual configuration and (b) reconstructed distribution of the dielectric permittivity. (c) cost function and (d) relative reconstruction errors for the best individual of the population versus the iteration number. Two separate cylinders. Pixel-based procedure.

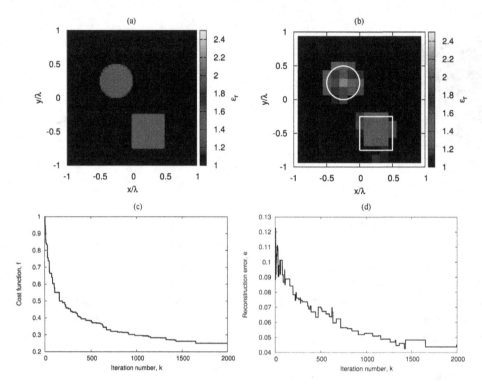

Two target configurations are considered. The first one is a composition of two adjacent objects (Figure 2(a)): A square cylinder with side 0.5λ, a center located at $\left(0.25\lambda, -0.5\lambda\right)$, and with a relative dielectric permittivity equal to 1.5; a sector of circular cylinder with radius 0.25λ, a center located at $\left(0, -0.25\lambda\right)$, and with a relative dielectric permittivity equal to 2.0.

The final reconstruction obtained by employing the pixel-based procedure is shown in Figure 2(b). As can be seen, the object is correctly localized and shaped. Moreover, the values of the relative dielectric permittivity of the two parts of the scattering configuration are correctly identified. Figure 2(c) reports the values of the normalized cost function for the best individual of the population versus the iteration number. As expected, the plot has a steep descent for the first

iterations, and then its slope decreases. The same behavior is confirmed in Figure 2(d), which reports the mean relative reconstruction error, which is defined as

$$e = \sum_{n=1}^{N} \frac{\left|\epsilon_n - \tilde{\epsilon}_n\right|}{\epsilon_n} \qquad (16)$$

being ϵ_n and $\tilde{\epsilon}_n$ the actual and reconstructed values of the relative dielectric permittivity of the n th subdomain of the investigation area) for the best trial solution versus the iteration number.

The reconstruction obtained by using the spline-based approach is shown in Figure 3(a) (the pixel image is produced by sampling the spline approximation on a regular grid of 63×63 pixels). In this case, although the number of unknowns is quite limited, a good reconstruction is

Figure 5. (a) Reconstructed distribution of the dielectric permittivity. (b) cost function and (c) relative reconstruction errors for the best individual of the population versus the iteration number. Two separate cylinders. Spline-based procedure.

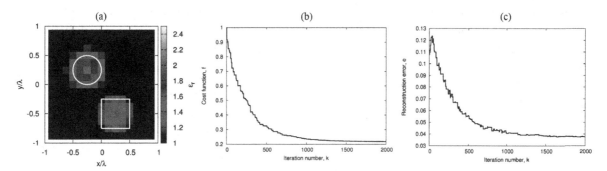

obtained. Figure 3(b) reports the values of the cost function of the best individual versus the iteration number. The overall behavior of the plot is similar to that of the pixel-based approach. However, as expected, the method converges slightly faster, thanks to the lower number of unknowns to be retrieved. Finally, Figure 3(c) shows the mean reconstruction error of the best individual versus the iteration number. As can be seen, the final reconstruction error is lower than the one obtained by using the pixel-based approach.

In the second considered configuration, two separate targets are assumed (Figure 4(a)): The first one is a circular cylinder with radius 0.25λ, a center located at $\left(-0.25\lambda, 0.25\lambda\right)$, and with a relative dielectric permittivity equal to 2.0. The second one is a square cylinder of side 0.5λ, a center located at $\left(0.25\lambda, -0.5\lambda\right)$, and with a relative dielectric permittivity equal to 1.5. The reconstructed distribution obtained by using the pixel-based inversion algorithm is shown in Figure 4(b). Again, the method is able to correctly reconstruct the two object, both in term of geometrical and dielectric properties. Figure 4(c) and Figure 4(d) show the values of the cost function and of the mean relative reconstruction errors for the best individual of the population, versus the iteration number. As expected, the behaviors of the two curves are similar to those obtained for

the first configuration. The results provided by the spline-based approach are shown in Figure 5. In particular, Figure 5(a) reports the distribution of the relative dielectric permittivity, whereas Figure 5(b) and Figure 5(c) show the behavior of the cost function and of the mean relative errors for the best trial solution versus the iteration number. In this case, too, the ACO-based inversion procedure is able to correctly reconstruct the investigated scene. Moreover, as for the first test case, the spline-based approach yield a lower final reconstruction error and a faster convergence.

5. CONCLUSION

In this chapter, the Ant Colony Optimization method has been adopted to solve the inverse scattering problem related to the nondestructive analysis of targets made of dielectric materials. Considering a tomographic approach, dielectric targets with cylindrical shapes have been inspected. In particular, the capabilities of the approach in retrieving the cross sections of quite complex targets (composed by the intersection of canonical scatterers and separate cylinders) have been proven. The Ant Colony Optimization method, which is a stochastic optimization technique widely used (as mentioned in the Introduction) in several areas of electrical engineering, has been found to be quite efficient for the considered diagnostic application. In particular, the reported results seem to indicate the possibility of further steps toward the development of fast, efficient and accurate inspection tools based on radiofrequency and microwave signals.

REFERENCES

Albanese, R., Rubinacci, G., Takagi, T., & Udpa, S. S. (Eds.). (1998). *Electromagnetic nondestructive evaluation (II). Studies in Applied Electromagnetics and Mechanics.* Amsterdam, The Netherlands: IOS Press.

Balanis, C. A. (1989). *Advanced engineering electromagnetics.* New York, NY: Wiley.

Baussard, A., Miller, E. L., & Prémel, D. (2004). Adaptive B-spline scheme for solving an inverse scattering problem. *Inverse Problems, 20*(2), 347–365. doi:10.1088/0266-5611/20/2/003

Bilchev, G., & Parmee, I. C. (1995). The ant colony metaphor for searching continuous design spaces. *Selected Papers from AISB Workshop on Evolutionary Computing* (pp. 25–39). London, UK: Springer-Verlag.

Bozza, G., Estatico, C., Pastorino, M., & Randazzo, A. (2006). An inexact Newton method for microwave reconstruction of strong scatterers. *Antennas and Wireless Propagation Letters, 5*(1), 61–64. doi:10.1109/LAWP.2006.870360

Brignone, M., Bozza, G., Randazzo, A., Aramini, R., Piana, M., & Pastorino, M. (2008). Hybrid approach to the inverse scattering problem by using ant colony optimization and no-sampling linear sampling. *Proceedings of the 2008 IEEE Antennas and Propagation Society International Symposium (APS2008)* (pp. 1–4). San Diego, CA: IEEE. doi:10.1109/APS.2008.4619941

Brignone, M., Bozza, G., Randazzo, A., Piana, M., & Pastorino, M. (2008). A hybrid approach to 3D microwave imaging by using linear sampling and ACO. *IEEE Transactions on Antennas and Propagation, 56*(10), 3224–3232. doi:10.1109/TAP.2008.929504

Caorsi, S., Gragnani, G. L., Medicina, S., Pastorino, M., & Zunino, G. (1991). Microwave imaging method using a simulated annealing approach. *IEEE Microwave and Guided Wave Letters, 1*(11), 331–333. doi:10.1109/75.93902

Chan, E.-J., Chao, C.-H., Jheng, K.-Y., Hsin, H.-K., & Wu, A.-Y. (2010). ACO-based cascaded adaptive routing for traffic balancing in NoC systems. *Proceedings of the 2010 International Conference on Green Circuits and Systems (ICGCS)* (pp. 317–322). Shanghai, China: IEEE. doi:10.1109/ICGCS.2010.5543045

Chang, M., Chou, P., & Lee, H. (1995). Tomographic microwave imaging for nondestructive evaluation and object recognition of civil structures and materials. *Proceedings of the 29th Asilomar Conference on Signals, Systems and Computers,* ASILOMAR '95 (Vol. 2, pp. 1061–1065). Washington, DC: IEEE Computer Society.

Chowdhury, N. M., Baker, S. M., & Choudhury, E. H. (2008). A new adaptive routing approach based on ant colony optimization (ACO) for ad hoc wireless networks. *Proceedings of the 11th International Conference on Computer and Information Technology (ICCIT2008)* (pp. 51–56). Khulna, Bangladesh: IEEE. doi:10.1109/ICCITECHN.2008.4803126

Colton, D., Haddar, H., & Piana, M. (2003). The linear sampling method in inverse electromagnetic scattering theory. *Inverse Problems, 19*(6), S105–S137. doi:10.1088/0266-5611/19/6/057

Dierckx, P. (1996). *Curve and surface fitting splines.* Oxford, UK: Clarendon Press.

Dorigo, M., & Gambardella, L. M. (1997). Ant colony system: A cooperative learning approach to the traveling salesman problem. *IEEE Transactions on Evolutionary Computation, 1*(1), 53–66. doi:10.1109/4235.585892

Dorigo, M., Maniezzo, V., & Colorni, A. (1996). Ant system: optimization by a colony of cooperating agents. *IEEE Transactions on Systems, Man, and Cybernetics. Part B, Cybernetics, 26*(1), 29–41. doi:10.1109/3477.484436

Dréo, J., & Siarry, P. (2002). A new ant colony algorithm using the heterarchical concept aimed at optimization of multiminima continuous functions. *Proceedings of the Third International Workshop on Ant Algorithms (ANTS'02)* (pp. 216–221). London, UK: Springer-Verlag.

Du, Y., Zhang, Q., & Chen, Q. (2008). ACO-IH: An improved ant colony optimization algorithm for airport ground service scheduling. *Proceedings of the 2008 IEEE International Conference on Industrial Technology (ICIT2008)* (pp. 1–6). Chengdu, China: IEEE. doi:10.1109/ICIT.2008.4608674

Garnero, L., Franchois, A., Hugonin, J.-P., Pichot, C., & Joachimowicz, N. (1991). Microwave imaging-complex permittivity reconstruction-by simulated annealing. *IEEE Transactions on Microwave Theory and Techniques, 39*(11), 1801–1807. doi:10.1109/22.97480

Giakos, G. C., Pastorino, M., Russo, F., Chowdhury, S., Shah, N., & Davros, W. (1999). Noninvasive imaging for the new century. *IEEE Instrumentation & Measurement Magazine, 2*(2), 32–35. doi:10.1109/5289.765967

Gragnani, G. L. (2002). *Two-dimensional imaging of dielectric scatterers based on Markov random field models: A short review. Microwave Nondestructive Evaluation and Imaging* (pp. 121–145). Trivandrum, India: Research Signpost.

Harrington, R. (1993). *Field computation by moment methods.* Piscataway, NJ: IEEE Press. doi:10.1109/9780470544631

Haupt, R. L. (1995). An introduction to genetic algorithms for electromagnetics. *IEEE Antennas and Propagation Magazine, 37*(2), 7–15. doi:10.1109/74.382334

Huang, H., Wu, C.-G., & Hao, Z.-F. (2009). A pheromone-rate-based analysis on the convergence time of ACO algorithm. *IEEE Transactions on Systems, Man, and Cybernetics. Part B, Cybernetics, 39*(4), 910–923. doi:10.1109/TSMCB.2009.2012867

Huang, R., Zhu, J., & Xu, G. (2007). An optimization coverage mechanism in sensor networks based on ACO. *Proceedings of the 2007 International Conference on Microwave and Millimeter Wave Technology (ICMMT'07)* (pp. 1–4). Gulin, China: IEEE. doi:10.1109/ICMMT.2007.381508

Kharkovsky, S., & Zoughi, R. (2007). Microwave and millimeter wave nondestructive testing and evaluation - Overview and recent advances. *IEEE Instrumentation & Measurement Magazine, 10*(2), 26–38. doi:10.1109/MIM.2007.364985

Lesselier, D., & Razek, A. (Eds.). (1999). *Electromagnetic nondestructive evaluation (III). Studies in Applied Electromagnetics and Mechanics.* Amsterdam, The Netherlands: IOS Press.

Lu, J., Wang, N., Chen, J., & Su, F. (2011). Cooperative path planning for multiple UCAVs using an AIS-ACO hybrid approach. *Proceedings of the 2011 International Conference on Electronic and Mechanical Engineering and Information Technology (EMEIT)* (pp. 4301–4305). Harbin, China: IEEE. doi:10.1109/EMEIT.2011.6023113

Monmarché, N., Venturini, G., & Slimane, M. (2000). On how Pachycondyla apicalis ants suggest a new search algorithm. *Future Generation Computer Systems, 16*, 937–946. doi:10.1016/S0167-739X(00)00047-9

Mudanyal, O., Yldz, S., Semerci, O., Yapar, A., & Akduman, I. (2008). A microwave tomographic approach for nondestructive testing of dielectric coated metallic surfaces. *IEEE Geoscience and Remote Sensing Letters, 5*(2), 180–184. doi:10.1109/LGRS.2008.915602

Olamei, J., Niknam, T., Arefi, A., & Mazinan, A. H. (2011). A novel hybrid evolutionary algorithm based on ACO and SA for distribution feeder reconfiguration with regard to DGs. *Proceedings of the 2011 IEEE GCC Conference and Exhibition (GCC)* (pp. 259–262). Dubai, UAE: IEEE. doi:10.1109/IEEEGCC.2011.5752495

Pastorino, M. (2007). Stochastic optimization methods applied to microwave imaging: A review. *IEEE Transactions on Antennas and Propagation, 55*(3), 538–548. doi:10.1109/TAP.2007.891568

Pastorino, M. (2010). *Microwave imaging.* Hoboken, NJ: John Wiley. doi:10.1002/9780470602492

Price, K. V. (1999). *An introduction to differential evolution* (pp. 79–108). Maidenhead, UK: McGraw-Hill.

Robinson, J., & Rahmat-Samii, Y. (2004). Particle swarm optimization in electromagnetics. *IEEE Transactions on Antennas and Propagation, 52*(2), 397–407. doi:10.1109/TAP.2004.823969

Salvade, A., Pastorino, M., Monleone, R., Bozza, G., & Randazzo, A. (2009). A new microwave axial tomograph for the inspection of dielectric materials. *IEEE Transactions on Instrumentation and Measurement, 58*(7), 2072–2079. doi:10.1109/TIM.2009.2015521

Socha, K., & Dorigo, M. (2008). Ant colony optimization for continuous domains. *European Journal of Operational Research, 185*(3), 1155–1173. doi:10.1016/j.ejor.2006.06.046

Tenglong, K., Xiaoying, Z., Jian, W., & Yihan, D. (2011). A modified ACO algorithm for the optimization of antenna layout. *Proceedings of the 2011 International Conference on Electrical and Control Engineering (ICECE)* (pp. 4269–4272). Yichang, China: IEEE. doi:10.1109/ICECENG.2011.6057613

Uma, K., Palanisamy, P. G., & Poornachandran, P. G. (2011). Comparison of image compression using GA, ACO and PSO techniques. *Proceedings of the 2011 International Conference on Recent Trends in Information Technology (ICRTIT)* (pp. 815–820). Chennai, India: IEEE. doi:10.1109/ICRTIT.2011.5972298

Vilovic, I., Burum, N., Sipus, Z., & Nad, R. (2007). PSO and ACO algorithms applied to location optimization of the WLAN base station. *Proceedings of the 19th International Conference on Applied Electromagnetics and Communications (ICECom2007)* (pp. 1–5). Dubrovnik, Croatia: IEEE. doi:10.1109/ICECOM.2007.4544491

Wang, J., Gao, X.-G., Zhu, Y.-W., & Wang, H. (2010). A solving algorithm for target assignment optimization model based on ACO. *Proceedings of the 2010 Sixth International Conference on Natural Computation (ICNC)* (pp. 3753–3757). Yantai, China: IEEE. doi:10.1109/ICNC.2010.5583099

Weis, G., Lewis, A., Randall, M., & Thiel, D. (2010). Pheromone pre-seeding for the construction of RFID antenna structures using ACO. *Proceedings of the 2010 IEEE Sixth International Conference on e-Science (e-Science)* (pp. 161–167). Brisbane, Australia: IEEE. doi:10.1109/eScience.2010.39

Xie, H., Li, L.-L., Bo, H., & Zhang, Y.-N. (2009). A novel method for ship detection based on NSCT and ACO. *Proceedings of the 2nd International Congress on Image and Signal Processing (CISP '09)* (pp. 1–4). Tianjin, China: IEEE. doi:10.1109/CISP.2009.5304472

Xu, C., El-Hajjar, M., Maunder, R. G., Yang, L.-L., & Hanzo, L. (2010). Performance of the space-time block coded DS-CDMA uplink employing soft-output ACO-aided multiuser space-time detection and iterative decoding. *Proceedings of the 2010 IEEE 71st Vehicular Technology Conference (VTC 2010-Spring)* (pp. 1–5). Taipei, Taiwan: IEEE. doi:10.1109/VETECS.2010.5493768

Xu, S., Bing, Z., Lina, Y., Shanshan, L., & Lianru, G. (2010). Hyperspectral image clustering using ant colony optimization (ACO) improved by K-means algorithm. *Proceedings of the 2010 3rd International Conference on Advanced Computer Theory and Engineering (ICACTE)* (Vol. 2, pp. V2–474–V2–478). IEEE. doi:10.1109/ICACTE.2010.5579337

Xu, X., Liu, Q. H., & Zhang, Z. W. (2002). The stabilized biconjugate gradient fast Fourier transform method for electromagnetic scattering. *Proceedings of the 2002 IEEE Antennas and Propagation Society International Symposium* (Vol. 2, pp. 614–617). San Antonio, TX: IEEE. doi:10.1109/APS.2002.1016722

Zhang, B., Sun, X., Gao, L., & Yang, L. (2011). Endmember extraction of hyperspectral remote sensing images based on the ant colony optimization (ACO) algorithm. *IEEE Transactions on Geoscience and Remote Sensing, 49*(7), 2635–2646. doi:10.1109/TGRS.2011.2108305

ADDITIONAL READING

Bertero, M. (1998). *Introduction to inverse problems in imaging*. Bristol, UK: Institute of Physics. doi:10.1887/0750304359

Bond, E. J., Xu Li, Hagness, S. C., & Van Veen, B. D. (2003). Microwave imaging via space-time beamforming for early detection of breast cancer. *IEEE Transactions on Antennas and Propagation, 51*(8), 1690–1705. doi:10.1109/TAP.2003.815446

Caorsi, S., Costa, A., & Pastorino, M. (2001). Microwave imaging within the second-order Born approximation: Stochastic optimization by a genetic algorithm. *IEEE Transactions on Antennas and Propagation, 49*(1), 22–31. doi:10.1109/8.910525

Caorsi, S., Massa, A., & Pastorino, M. (1999). Genetic algorithms as applied to the numerical computation of electromagnetic scattering by weakly nonlinear dielectric cylinders. *IEEE Transactions on Antennas and Propagation, 47*(9), 1421–1428. doi:10.1109/8.793322

Caorsi, S., Massa, A., Pastorino, M., Raffetto, M., & Randazzo, A. (2003). Detection of buried inhomogeneous elliptic cylinders by a memetic algorithm. *IEEE Transactions on Antennas and Propagation, 51*(10), 2878–2884. doi:10.1109/TAP.2003.817984

Caorsi, S., Massa, A., Pastorino, M., & Randazzo, A. (2005). Optimization of the difference patterns for monopulse antennas by a hybrid real/integer-coded differential evolution method. *IEEE Transactions on Antennas and Propagation, 53*(1), 372–376. doi:10.1109/TAP.2004.838788

Caorsi, S., Massa, A., Pastorino, M., Randazzo, A., & Rosani, A. (2004). A reconstruction procedure for microwave nondestructive evaluation based on a numerically computed Green's function. *IEEE Transactions on Instrumentation and Measurement, 53*(4), 987–992. doi:10.1109/TIM.2004.831446

Chen, W.-T., & Chiu, C.-C. (2000). Electromagnetic imaging for an imperfectly conducting cylinder by the genetic algorithm [medical application]. *IEEE Transactions on Microwave Theory and Techniques, 48*(11), 1901–1905. doi:10.1109/22.883869

Chew, W. C. (1999). *Waves and fields in inhomogeneous media.* IEEE Press. doi:10.1109/9780470547052

Colton, D. L., & Kress, R. (1998). *Inverse acoustic and electromagnetic scattering theory.* New York, NY: Springer.

De Zaeytijd, J., Franchois, A., Eyraud, C., & Geffrin, J.-M. (2007). Full-wave three-dimensional microwave imaging with a regularized Gauss-Newton method: Theory and experiment. *IEEE Transactions on Antennas and Propagation, 55*(11), 3279–3292. doi:10.1109/TAP.2007.908824

Estatico, C., Pastorino, M., & Randazzo, A. (2005). An inexact-Newton method for short-range microwave imaging within the second-order Born approximation. *IEEE Transactions on Geoscience and Remote Sensing, 43*(11), 2593–2605. doi:10.1109/TGRS.2005.856631

Fang, Q., Meaney, P. M., Geimer, S. D., Streltsov, A. V., & Paulsen, K. D. (2004). Microwave image reconstruction from 3-D fields coupled to 2-D parameter estimation. *IEEE Transactions on Medical Imaging, 23*(4), 475–484. doi:10.1109/TMI.2004.824152

Franchois, A., Joisel, A., Pichot, C., & Bolomey, J.-C. (1998). Quantitative microwave imaging with a 2.45-GHz planar microwave camera. *IEEE Transactions on Medical Imaging, 17*(4), 550–561. doi:10.1109/42.730400

Gilmore, C., Abubakar, A., Hu, W., Habashy, T. M., & van den Berg, P. M. (2009). Microwave biomedical data inversion using the finite-difference contrast source inversion method. *IEEE Transactions on Antennas and Propagation, 57*(5), 1528–1538. doi:10.1109/TAP.2009.2016728

Gilmore, C., Mojabi, P., & LoVetri, J. (2009). Comparison of an enhanced distorted Born iterative method and the multiplicative-regularized contrast source inversion method. *IEEE Transactions on Antennas and Propagation, 57*(8), 2341–2351. doi:10.1109/TAP.2009.2024478

Johnson, J. M., & Rahmat-Samii, V. (1997). Genetic algorithms in engineering electromagnetics. *IEEE Antennas and Propagation Magazine, 39*(4), 7–21. doi:10.1109/74.632992

Massa, A., Pastorino, M., & Randazzo, A. (2006). Optimization of the directivity of a monopulse antenna with a subarray weighting by a hybrid differential evolution method. *Antennas and Wireless Propagation Letters, 5*(1), 155–158. doi:10.1109/LAWP.2006.872435

Michalski, K. A. (2000). Electromagnetic imaging of circular-cylindrical conductors and tunnels using a differential evolution algorithm. *Microwave and Optical Technology Letters, 27*(5), 330–334. doi:10.1002/1098-2760(20001205)27:5<330::AID-MOP13>3.0.CO;2-H

Papadopoulos, T. G., & Rekanos, I. T. (2012). Time-domain microwave imaging of inhomogeneous Debye dispersive scatterers. *IEEE Transactions on Antennas and Propagation, 60*(2), 1197–1202. doi:10.1109/TAP.2011.2173150

Pastorino, M., Caorsi, S., Massa, A., & Randazzo, A. (2004). Reconstruction algorithms for electromagnetic imaging. *IEEE Transactions on Instrumentation and Measurement, 53*(3), 692–699. doi:10.1109/TIM.2004.827093

Qing, A. (2003). Electromagnetic inverse scattering of multiple two-dimensional perfectly conducting objects by the differential evolution strategy. *IEEE Transactions on Antennas and Propagation, 51*(6), 1251–1262. doi:10.1109/TAP.2003.811492

Qing, A., & Lee, C. K. (2010). *Differential evolution in electromagnetics*. Springer. doi:10.1007/978-3-642-12869-1

Rekanos, I. T., & Tsiboukis, T. D. (1999). An iterative numerical method for inverse scattering problems. *Radio Science, 34*(6), 1401. doi:10.1029/1999RS900082

Semnani, A., Rekanos, I. T., Kamyab, M., & Papadopoulos, T. G. (2010). Two-dimensional microwave imaging based on hybrid scatterer representation and differential evolution. *IEEE Transactions on Antennas and Propagation, 58*(10), 3289–3298. doi:10.1109/TAP.2010.2055793

Tijhuis, A. G., Belkebir, K., Litman, A. C. S., & de Hon, B. P. (2001). Theoretical and computational aspects of 2-D inverse profiling. *IEEE Transactions on Geoscience and Remote Sensing, 39*(6), 1316–1330. doi:10.1109/36.927455

Tijhuis, A. G., Belkebir, K., Litman, A. C. S., & de Hon, B. P. (2001). Multiple-frequency distorted-wave Born approach to 2D inverse profiling. *Inverse Problems, 17*(6), 1635–1644. doi:10.1088/0266-5611/17/6/307

Van den Berg, P. M., & Abubakar, A. (2001). Contrast source inversion method: State of art. *Progress in Electromagnetics Research, 34*, 189–218. doi:10.2528/PIER01061103

Van den Berg, P. M., & Kleinman, R. E. (1997). A contrast source inversion method. *Inverse Problems, 13*(6), 1607–1620. doi:10.1088/0266-5611/13/6/013

Weile, D. S., & Michielssen, E. (1997). Genetic algorithm optimization applied to electromagnetics: A review. *IEEE Transactions on Antennas and Propagation, 45*(3), 343–353. doi:10.1109/8.558650Zhou, Y., Li, J., & Ling, H. (2003). Shape inversion of metallic cavities using hybrid genetic algorithm combined with tabu list. *Electronics Letters, 39*(3), 280. doi:10.1049/el:20030207

KEY TERMS AND DEFINITIONS

Ant Colony Optimization: Stochastic optimization method inspired by the way ants find the optimal path from their nest to the food.

Global Optimization Algorithms: Algorithms aimed at finding the globally best solution of a (possibly nonlinear) functional in the presence of multiple local optima.

Inverse Problem: Mathematical problem consisting in finding the unknown causes of known consequences.

Inverse Scattering Problem: Inverse problem aimed at the identification of some characteristics of an object (e.g., its shape, dielectric properties, etc.) starting from the knowledge of the electromagnetic field it scatters.

Microwave Imaging: Techniques for the identification of the internal structure of an object by means of low-power electromagnetic fields at microwave frequencies.

Nondestructive Testing: Set of techniques for evaluating the properties of a material, a component or a system without changing or altering it in any way.

Stochastic Optimization Algorithms: Global optimization algorithms with a random search procedure.

Chapter 16

The Use of Evolutionary Algorithm-Based Methods in EEG Based BCI Systems

Adham Atyabi
Flinders University, Australia

Sean P. Fitzgibbon
Flinders University, Australia

Martin Luerssen
Flinders University, Australia

David M. W. Powers
Flinders University, Australia

ABSTRACT

Electroencephalogram (EEG) based Brain Computer Interface (BCI) is a system that uses human brain-waves recorded from the scalp as a means for providing a new communication channel by which people with limited physical communication capability can effect control over devices such as moving a mouse and typing characters. Evolutionary approaches have the potential to improve the performance of such system through providing a better sub-set of electrodes or features, reducing the required training time of the classifiers, reducing the noise to signal ratio, and so on. This chapter provides a survey on some of the commonly used EA methods in EEG study.

INTRODUCTION

Motor nerves and muscles used in nervous system are traditional communication channel for interaction between brain and computer. A Brain Computer Interface (BCI) is a communication device that bypasses the peripheral nervous system and derives intention directly from brain activity, which it then translates into executable commands. A formal definition for BCI proposed at the first International BCI Meeting (Rensselaerville, New York, 1999) is *"A brain-computer interface is a communication system that does not depend on the brain's normal output pathways of peripheral nerves and muscles"* (Wolpaw et al., 2000).

BCI devices typically incorporate stages such as signal acquisition, feature extraction, and classification in their operation. Electroencephalogram (EEG) is one of the commonly used non-invasive techniques in BCI for signal acquisition. EEG

DOI: 10.4018/978-1-4666-2666-9.ch016

records variations of the surface potential from the scalp using some electrodes. The recorded signal is expected to reflect the functional activity of the underlying brain. The EEG signal is a mixture of signals that includes the desired brain activity (the summated signal of millions of cells in the cortex), as well as the heartbeat, eye movement, voluntary and involuntary muscle activity and some possible noise.

Feature extraction stage is used to provide alternative representations of the raw measured signals that help the classifier to better discriminate the set of BCI operations. It is common to use preprocessing stage containing activities such as re-referencing electrodes, demeaning, normalizing, dimension reduction, artifact removal, and so on prior to feature extraction.

Although BCI systems have proven successful for providing new options and communication channels for people with severe neuromuscular disabilities, the use of such systems for controlling or communicating tasks that require high speed and accurate results is still difficult and far from being realized. This is due to complexities such as the complex nature of brain signals, signal smearing and attenuation due to volume conduction, environmental noise and biological contamination, sensitivity of recording equipment, and the difficulty for subjects in achieving and maintaining a requisite brain states. To date, much BCI research has focused strongly on using signal processing to overcome the poor SNR of the EEG signal measured from the scalp.

EEG can be considered as a set of data-points that follow some pattern with a notable degree of randomness. Considering the high level of dynamism in EEG signal, it would be an appropriate suggestion to address the problem of distinguishing these patterns from each other using theories that address dynamic probability optimization. This reasoning guides us toward the Evolutionary Algorithms (EA) in which a performance func-

tion called "population generator function" is optimized under the influence of some predefined assessment criteria called "fitness function". In EA based methods, complexities such as search dynamism, uncertainty, and multi-objective nature of the targeting optimization can be addressed using different types of "population generator" and "fitness" functions. Various evolutionary based search optimization methods exist such as Evolutionary Programming (EP), Evolution Strategy (ES), Differential Evolution (DE), Genetic Algorithm (GA), Genetic Programming (GP), Particle Swarm Optimization (PSO), Ant Colony Optimization (ACO), etc. Among all, GA, PSO, and ACO methods attracted most attention due to their applicability across different search optimization categories such as self-learning, un-supervised learning, stochastic search, population-based, and behavior-based search. This chapter provides an introduction and some review on these methods and their use in BCI research and discusses their achievements. The chapter is focused on:

- Using EA based methods as decomposition/filtering stage for reducing either the feature or electrode set sizes. This is a multi-objective problem in which it is desirable to reduce the size of feature or electrode sets whilst maintaining the classification accuracy. It is possible to use evolutionary based methods to optimize the classification performance by only selecting a subset of the actual data that favors the classification. The subset can be selected by only including electrodes that most robustly capture the underlying intention of the subject or based on selecting the subset of features that contribute most strongly to correct classification.
- Using EA based methods for classification and pattern recognition as a classifier or to enhance the training mechanism. In some

studies, PSO and GA are used to enhance the training methodology of classifiers like neural network. In addition, modified versions of ACO are used as classifiers for pattern recognition in some studies.

- Using EA based methods for artifact removal as in blind source separation (BSS) problems.
- Using EA based methods for providing a decision fusion method in ensemble classifiers.

A BRIEF INTRODUCTION TO SOME EVOLUTIONARY METHODS

Genetic Algorithm (GA)

The Genetic Algorithm (GA) was first developed by John Holland in 1975 (Hollan, 1975). GA has been successfully applied to numerous problems where the main objective was to optimize or select the best solution out of a number of possible solutions such as the classic traveling salesman problem, flow-shop optimization, and job shop scheduling (Konar, 2005). The basic characteristic that makes GA attractive in developing near-optimal solutions is that it is inherently parallel search technique (Elshamli, et al., 2004).

The GA comprises the three stages of Selection, Crossover, and Mutation (Han, 2007). Selection refers to the operation in which chromosomes (solution candidates) are ranked based on their fitness, with chromosomes of high rank chosen for generating a new generation of chromosomes. Crossover refers to the operation by which new chromosomes are generated from two or more existing chromosomes. Typically, chromosomes are divided into two parts and each chromosome exchanges one part of itself with another chromosome to produce a new chromosome. Mutation involves a random change to one or more bits of the chromosome. This step is useful when the entire population is converging towards a local optimum, trapped between obstacles, or generally not improving for a long time due to a lack of genetic diversity.

It is also common to mutate more than one chromosome and to keep (repeat) the best chromosome from each generation, which is known as Elitism. The main drawback of the classic GA is its unsuitability for dynamic problems, as it operates in a grid map or utilizes a fixed resolution in the search space and does not explicitly control the population diversity.

Ant Colony Optimization (ACO)

Marco Dorigo and his colleagues introduced the first ACO algorithm called the ANT System (Latombe, 1991). The ACO method is inspired by ants' social behaviors. The idea is based on using the collective behavior of ants foraging from a nest toward a food source (Han, 2007). To do so, ants use a chemical trace called a pheromone, which is used by ants to mark routes that they used. Since each ant releases pheromone on the route it took, the quality of the route measures can be based on the quantity of the pheromone on that route as shorter routes exhibit higher pheromone densities. Hence, ants can find shortest routes based on the probability of proportion[1] to the concentration[2] of the pheromone [3](Hoa et al., 2006).

In ACO, a group of agents (ants) would be released from their initial station (in the search space) foraging toward the goal location (optimum). As the deposited quantity of pheromone along the path differs they would be able to optimize the path. In traditional ACO, the pheromone concentrations of all elements are equally initialized. So the solutions are constructed blindly in the beginning and it takes a long time to find a better path toward the optimum among all other paths. The ACO algorithm has an advantage over the genetic algorithm in terms of the algorithm execution time (Han, 2007). It is due to the fact that the method does not devote an inordinate amount of time in an iteration process. However, ACO is inferior to the GA method due to unnecessary steps that ACO takes. These additional steps help algorithm to return the best solution. Furthermore, the use of global attraction term seems necessary since it helps to lead ants toward the optimum point in the search space. Eliminating this term causes ants wandering and becoming stuck at a point. Considering the mentioned drawbacks of ACO, a number of modifications are suggested. A larger population size in problems in which ACO performs poorly due to high number of local optimums can overcome this issue. However, this approach is computationally expensive. It is also possible to improve the method by modifying the pheromone operator and its characteristics by introducing ants with varying pheromone quantities based on other heuristics. In addition, it is possible to define a lifetime for ants' pheromones. These solutions can make the ACO method robust to dynamic problems in which a variety of local optima exist.

A version of ACO called Ant Miner widely used in data-mining field for rule discovery and learning classification rules. The structure of Ant Miner and its process for extracting rules from a subset of data points (features) and classifying them is quite similar to that of an artificial neural network. However, the advantage of Ant Miner is its lower computational requirements. This is due to the fact that unlike a neural network, Ant Miner does not try to reevaluate previously calculated weights each time that a new feature is entered inside the network. Instead, in each step/iteration, it only considers subsets of remaining features that can be classified without any necessity for changing previously set weights. Ji, Zhang, and Liu (2006) introduced an enhanced version of Ant Miner using a mutation operator. In their study, the value of each attribute node is mutated whenever the best rule is discovered. Later, the

validity of the mutation is assessed by examining the quality difference between the new and original rules, and finally the better rule among the new and original rule is approved. Chandrasekar, Suresh, and Ponnambalam (2006) claimed that by applying an obstacle avoidance strategy on ACO, it is possible to achieve better accuracy in comparison to Ant Miner in rule discovery problems. Piatrik and Izquierdo (2006) used a hybrid ACO and k-means algorithm for image classification/clustering. In clustering problems, the similarity is targeted to provide partitioned data points in a way that data points inside each partition have more similarity compared with the data points inside other partitions.

Particle Swarm Optimization (PSO)

Particle Swarm Optimization (PSO) is an evolutionary algorithm, introduced by Kennedy and Eberhart in 1995, that is inspired by the social and cognitive interactions of animals with one another and with the environment (e.g., fish schooling and birds flying). PSO operates by iteratively directing its particles toward the optimum using social and cognitive components. Locations of particles, denoted by $x_{i,j}$, are influenced by their velocity component in the n-dimensional search space, denoted by $v_{i,j}$, where i represents the particle's index and j is the dimension in the search space. In PSO, particles are considered to be possible solutions; they fly through the virtual space with respect to maximum velocity limitations denoted by v_{max}. Particles are generally attracted to the positions that yield the best results (Cichocki et al., 2004, Hinterberger, et. al., 2004). The best positions, for example local best ($p_{i,j}$) and global best (g_i), are stored in each particle's memory. In general, the local best of each particle can be seen as the position in which the particle achieved its highest performance; whereas the global-best of each particle can be seen as the best local-best position achieved by neighboring particles.

Using EA based approaches in EEG and BCI study

EA based methods have been applied in EEG research in a number of contexts including source localization (Qin, Li, and Yao, 2005), artefact removal (Ghanbari, et al., 2009, Fairley, et al., 2010), blind source separation (Zhao and Zheng, 2004), musical mood detection (Ito, et al., 2007), and feature projection (Khushaba, et al., 2008). This chapter describes three main approaches in which EA-based methods have been specifically applied in the BCI domain:

1. EA based feature/electrode reduction/ selection
2. EA based learning phase in neural network classifiers
3. EA based Ensembles and Fusions

Evolutionary Based Feature/ Electrode Reduction/Selection

EEG data has a high dimension that makes it difficult for the classifier to distinguish the underlying patterns. In EEG, the dimension of the dataset can be considered to be the number of channels used for recording multiplied by the number of trials. Given that typically many trials are recorded (with a certain sample rate e.g., 1000Hz) each lasting from some number of seconds to hours, the EEG data's dimension can easily grow to a high value that makes it difficult to assess and analyze. Therefore dimension reduction (also referred to as feature extraction) and decomposition is a common pre-processing stage in EEG signal classification (Majumdar, 2011).

The EEG signal recorded at each scalp electrode is the potential difference between that electrode and another reference electrode. This data contains artifacts, noise and some latent signal. For clarity, in this chapter the general term of the *dimension reduction* is expanded to *Reduction* and *Selection* categories in which reduction is mainly used for situations in which the feature or electrode sets are homogeneous (from the same type) while selection is used for heterogeneous situations where data contains the signal from various sources of different types of electrodes and/or features (e.g. frequency, time-frequency, special-temporal, entropy, etc.). The overall goal of dimension reduction is to reduce the size of the data prior to classification by selecting a subset of electrodes or features that best represent the underlying patterns.

Following this taxonomy, Feature Reduction (FR) and Feature Selection (FS) are the study of selecting a subset of features that best represent the performed task. This can be based on choosing the sub-set from the same feature-set as in FR or from different feature-sets as in FS. It is also possible to improve the classification through the use of a sub-set of electrodes that best record the underlying tasks. This process is called Electrode Selection/Reduction (ES/R). Several studies have investigated the potential of evolutionary methods for ES/R and FR. The advantage of Evolutionary approaches in comparison with conventional methods such as Common Spatial Pattern (CSP) and Single Value Decomposition (SVD) is that they can be used in on-line mode while conventional methods require the complete data for dimension reduction (Hassan, Gan, and Zhang, 2010).

GA is one of the most commonly used EA based methods for feature and electrode reduction. Yom-Tov and Inbar in (Tov and Inbar, 2002) used GA for feature selection in a movement related potential study using EEG data containing 2 and 3 classes. The results indicate 87% and 63% classification accuracy with only 10 subject-specific features for the two and three class cases respectively. Dias et al. (Dias, et al., 2009) compared binary GA with variety of methods such as Across-Group Variance (AGV), recursive feature elimination (RFE), and RELIEF. The study employed EEG data of 5 subjects performing 2 motor imagery tasks. Results indicate an average of 18.5, 22.5, 25.7, and 27.7 classification errors across 5 subjects for AGV, RFE, GA, and RELIEF respectively. Largo et al. (2005) used GA with extra reservoir sub-population mutation for tuning the parameters of the detector and classifier. The study is based on detecting the A phases in one channel sleep EEG data. The results indicate 81% accuracy after tuning for learning set. Zhang and Wang (2008) studied the impact of GA for parameter selection in a BCI study. The GA is used to select parameters such as the overlaps of the FFT windows, the length of the STFT window, and the interpolation intervals for EMG-EEG data. Palaniappan and Raveendran (2002) used a binary GA for selecting optimum features in a study based on distinguishing alcoholics from non-alcoholics based on VEP data and achieved approximately 90% classification accuracy.

PSO is used in several studies for FR and ES/R. Jin et al, (2008) used a version of PSO called Discrete PSO (DPSO) for selecting electrodes from certain regions of the subject's scalp that best describe his/her intention. In their study DPSO outperforms a comparative technique called F-score. DPSO and F-score achieved 77% and 69% average classification accuracy respectively. Hasan et al. (Hasan, Gan, and Zhang, 2010; Hasan and Gan, 2009) used a multi-objective version of PSO (MOPSO) in a similar study. The comparison of MOPSO and sequential floating forward search (SFFS) method indicates that MOPSO is capable of selecting a much smaller number of channels than SFFS. This is done with insignificant loss of classification accuracy (59% and 61% classification accuracy with MOPSO and SFFS respectively). In Hasan, Gan, and Zhang (2010), MOPSO is compared with Sequential Floating Forward Search (SFFS) and Multi-Objective

Evolutionary Algorithm based on Decomposition (MOEA/D) and they report 57%, 60, and 58% averaged cross-validation accuracy with MOPSO, SFFS, and MOEA/D respectively. The average number of used electrodes with MOPSO, SFFS, and MOEA/D are 4, 3, and 10 electrodes across subjects respectively. Moubayed et al. (2010) used a binary version of MOPSO called Binary-SDMOPSO for channel reduction. They compare Binary-SDMOPSO, MOEA/D and OMOPSO and report that Binary-SDMOPSO and MOEA/D are less prone to outliers compared to OMOPSO. In this instance, outlier refers to channels with low contribution in terms of representing motor-imagery movements. Classification accuracy of 58%, 58%, and 57% is reported across subjects with Binary-SDMOPSO, MOEA/D, and OMOPSO respectively. The results are averaged across subjects and cross-validations.

Khushaba et al. (2008) investigate the use of a combination of ACO and differential evolution called ANTDE for feature selection. They compare ANTDE, GA and BPSO, and report that ANTDE's outperforms GA and BPSO due to the use of a mutual information based heuristic measure. Poli et al. (2011) used Genetic Programming (GP) in a BCI study. The study used GP in mouse controlling BCI application and showed better performance compared to SVM.

Table 1 provides a list of EA based methods used for F/ER. Despite the achieved improvements in comparison with other conventional methods, it is noticeable that in most studies the employed paradigm for the use of EA based methods are designed in a way that it allows EA based methods to find the sub-set of features or electrodes that improves the classification of a testing set. That is, the dataset is divided into the training and the testing sets and the selection of final sub-set is achieved based on the results of testing set. Such a paradigm allows the contamination of training

Table 1. A selection of EA based feature/electrode reduction studies and the achieved performance

Reference	EA-based Method	Objective of the Study	EEG data type	Number of Classes	Accuracy
(Ito et al., 2007)	GA for matching music with subjects' mood	subject mood detection for music	brain builder unit EEG signal	up to average 21 moods across 8 subjects	70% hit rate (detection) and 25% mix rate
(Zhiping, et al., 2010)	PSO GA	feature selection	motor imagery	3 classes	average of 88.3% Tp and 4.27% FP with PSO and 89.7 TP and 4.7% FP with GA
(Al-Ani, 2009)	PSO GA DE	feature selection	motor imagery	2 classes	for 30 features, 88% with PSO and GA, and 90% with DE
(Ghanbari et al., 2009)	GA-ICA	Artefact rejection	motor imagery	2 classes	91.7% average across classifiers
(Petyrantonaksi & Hadjileontiadis, 2009)	GA	selection of intrinsic mode functions (IMF) that contain majority of energy content of signal	vP	6 classes of emotions	maximum 84% and minimum 70%
(Sabeti et al., 2007)	GA	feature/electrode selection	Continuous EEG recording from Schizophrenic subjects	2 classes	maximum of 79.4% and minimum of 59.5 with selected channels and frequency bands
(Ravi Palaniappan, 2007)	GA	electrode reduction	vP		average of 75% across 20 fold cross validation
(Nakamura, et al., 2009)	PSO	Feature selection	EEG recording from sense of touch	2 classes	average of 0.19 and 0.22 error rate for false acceptance and rejection rates respectively
(Lv & Liu, 2008)	PSO-CSP	electrode reduction	motor imagery	2 classes	maximum 81%
(Largo, et al., 2005)	ARGA	parameter tunning	Sleep EEG		81%
(Zhang & wang, 2008)	GA	parameter selection	EMG-EEG		84%
(Palaniappan & Raveendran, 2002)	GA	Feature selection	VEP		86%
(Dias, et al., 2009)	Binary-GA	Feature selection	motor imagery	2 classes	74% averaged across 5 subjects
(Jin, Wang, & Zhang, 2008)	DPSO	Electrode Reduction	limb movement	4 classes	77%
(Moubayed, et al., 2010)	Binary SDMOPSO	Electrode Reduction	motor imagery	3 classes	58% averaged across 6 subjects
(Tov & Inbar, 2002)	GA	Feature selection	limb movement	2 and 3 classes	87% and 63%
(Hasan, et al., 2010)	MOPSO	Electrode Reduction	motor imagery	3 classes	57%
(Hasan, et al., 2009)	MOPSO	Electrode Reduction	motor imagery	3 classes	57%
(Atyabi, et al., 2012a)	ga, pso, random search	feature & electrode reduction (50% to 90% reduction)	motor imagery	2 classes	GA ER, GA FR, Random ER, Random FR, and PSO FR, improved informedness from 0.152 (with fullset) to 0.384, 0.342, 0.498, 0.494, and 0.498 respectively

continued on following page

Table 1. Continued

Reference	EA-based Method	Objective of the Study	EEG data type	Number of Classes	Accuracy
(Atyabi, et al.., 2012b)	pso	feature-electrode reduction (99% reduction)	motor imagery	2 classes	averaged informedness across subjects is changed from 0.47 with fullset to 0.412 with combination of most commonly used features/electrodes and polynomial svm with 99% reduction
(Atyabi, et al., 2012c)	pso	feature-electrode reduction (99% reduction)	motor imagery	2 classes	averaged informedness across subjects is changed from 0.47 with fullset to 0.412 with combination of most commonly used features/electrodes and polynomial svm with 99% reduction
(Kolodziej et al., 2011)	GA	feature selection	motor imagery	3 classes	mean classification error of 10.5%
(Dobrea et al., 2010)	GA	selection of optimal band limits and spectrum bands	mental tasks	12 classes	improve average classification by 6% with the peak at 97.1%

and testing sets in the classification. This is not desirable in machine learning studies due to possible lack of generalizability in the found solution.

A solution for this is to use an extra internal cross-validation step which partitions the training set into two sets of learning and evaluation sets and uses the achieved classification results from the learning and evaluation sets as a reference for fitness and finally further evaluate the final solution by assessing the performance through the use of the training and the testing sets. Such a paradigm guarantees the assessment of the results by the use of unseen data (Atyabi et al., 2012a, 2012b, and 2012c).

Evolutionary Based Classifiers

In addition to using EA based methods as a decomposition stage for either reducing the feature or electrode set sizes, a common use of EA based methods is for classification and pattern recognition. Jaganathan et al. (2007) and Ji et al. (2006)

used ACO method for rule discovery and classification. Martens et al, (2007) also used ACO for classification. Piatrik and Izquierdo (2006) used ACO for pattern recognition and image classification. The other possibility is to use EA methods for enhancing the training in the classifiers.

A number of studies have employed such methods to shorten the training phase of classifiers as in the Hema et al. (2008b) study in which Particle Swarm optimization (PSO) is used for training NN. The study demonstrated the sufficiency of PSO for training Neural Network (NN) and it provided comparison between PSO and the Back Propagation (BP) technique. A NN trained with PSO is more time efficient and achieves better performance (Hema, et al., 2008b). Likewise, Hema, et al. (2008a) classified extracted features of EEG signal (using PCA features) with a PSO-based NN classifier. The PSO-based NN classifier was used for classifying five different mental tasks. The study reported 77% and 100% performance. Lin and Hsieh (2009) proposed the use of a PSO-based method for training phase with NN classifiers. In the study, PCA is used as feature extractor and it is reported that PSO-NN classifier achieved faster convergence in comparison to BP-NN classifier. Zhao and Yang (2009) compared BP-NN, conventional PSO-NN and a cooperative random learning version of PSO-NN in a time series prediction problem and reported similar results. BP-NN showed the worst performance compared with variations of PSO-NN. The Extreme Learning Machine (ELM) is considered an alternative to the BP-NN and contains random connections between the input and the first layer. ELM is faster than traditional gradient-based learning methods (Zhu, et al., 2005). Zhu et al. (2005) proposed evolutionary ELM (E-ELM) aiming to fix the necessity of having higher number of hidden nodes and problem with the unseen data in testing samples of ELM. The results indicate the advantage of E-Elm in comparison with other algorithms including ELM in terms of classification accuracy and number of needed hidden neurons.

In addition to training a network, it is possible to use these types of methods in blind source separation (BSS) problems. Zhao and Zheng (2004) proposed the use of PSO for source separation problem. In their study, a PSO-based algorithm is used to iteratively separate source components one by one (deflation-based BSS). They suggested the possibility of using PSO-based algorithms as Symmetric-based BSS method. Implementation simplicity is considered as the contribution of the technique.

Table 2 presents aggregate details about some of the recent EA based EEG classification studies.

Evolutionary Based Ensembles and Fusions

Ensembles of classifiers have been successfully employed in a number of studies (Veeramachaneni, et al., 2006, Polikar, et al., 2006, Zhao and Zhang, 2004, Quinlan, 2006, Rutaq and Gabrys,

2000, Lotte, et al., 2007, Ayache, et al., 2007, Esmaeili, 2007, Parikh, et al., 2005, Ting, et al., 2011). The strategy used in most ensembles involves aggregating the decisions of a population of weak-learners using an aggregation/combination function. Polikar (2006) describes numerous combination functions such as majority voting, weighted majority voting, maximum/minimum/sum/product, fuzzy integral, Dempster-Shafer, and decision templates. An ensemble can be formed in several ways based on the diversity in terms of the used classifiers and the type of training samples. That is, it is possible to form an ensemble by the use of multiple copies of the same classifier and vary the input data among them following some criterion or use a group of different classifiers that all receive the same data for training. Some of the well-known ensembles are Bagging, Random Forest, Boosting, AdaBoost, Stacked Generalization, and Mixture of Experts.

Recently, the combination of ensemble learning and EA based approaches is an area of interest for several researchers (Zhang and Yang, 2008 He, et al., 2009, Gange, et al., 2009, Sylvester and Chawla, 2005, Kim and Cho, 2008, Kim, street and Menczer, 2005, Zhang and Sun, 2010, Tang, Sun and Zhu, 2005, Veeramachaneni, et al., 2006). This combination can be achieved in various ways such as i) applying EA based methods in feature extraction/reduction/selection stage and using ensemble of classifiers to provide final product, ii) using an ensemble of classifiers that utilize EA based approaches for their internal individual learning phase, and iii) using EA based approaches for assessing the expertness of each member of the ensemble aiming to provide more data-sensitive combination rule in the ensemble.

The other possibility is to provide an ensemble of classifiers with a combination of the mentioned methods as in Veeramachaneni et al's study (Veeramachaneni, et al., 2006). The study used PSO in fusion level to combine the classifiers decision and generate the final output of the ensemble. The results indicate marginal advantage of PSO

Table 2. A selection of recent EA based classification studies and the achieved performance

Reference	EA-based Method	Objective of the Study	EEG data type	number of Classes	accuracy	Training time
(Hema et al., 2008)	PSO-NN	Training NN	Mental task	5 classes	92% for 2 classes	101.27s maximum
(Aberg & Wessberg, 2007)	EA-BPNN,EA-MLR	optimization of feature subset and classifier parameters	index finger movement	2 classes	73.5% with EA-MLR 75% with EA-BPNN	not reported
(Hema et al., 2008)	Functional Link PSO-NN	Training NN	Mental task	5 classes	average 86.72% for 2 classes	not reported
(Paulraj, et al., 2007)	PSO-RBFNN	Training radial based NN	Mental task	5 classes	94.87% for 2 classes	7-9s average
(Huang, et al., 2011)	GA-MLD	Training MLD	Mental & motor imagery	4 classes	77% for mental & 57% for motor imagery tasks	7-9s average
(Liyanage, aet al., 2009)	GA-NN PSO-NN	Training NN	self-paced finger tapping	2 classes	77% GA-NN 73% PSO-NN	not reported
(Jiao, Wu, & Guo, 2010)	GA-BPNN GA-SVM	Training NN and SVM	Motor Imagery	2 classes	maximum 84% GA-SVM and 86% with GA-BPNN	not reported
(Li, et al., 2011)	GA-SVM	optimizing alpha & c in SVM	Motor Imagery	2 classes	maximum 86% in training and 82% in testing	not reported
(Harikumar, et al., 2004)	GA-Fuzzy	optimization of fuzzy output	continues EEG data	epilepsy risk level	over 90% classification rate for epilepsy risk level	not reported
(Harikumar, et al., 2005)	Binary coded GA	Ga based epilepsy risk level classifier	continues EEG data	epilepsy risk level	over 90% classification rate for epilepsy risk level	not reported
(Cinar & Sahin, 2010)	PSO-RBF	clustering and pattern classification	Motor Imagery	2 classes	average of 84.8% across 5 subjects	not reported
(Guozheng, et al., 2006)	GA-SVM	feature selection and SVm parameterization	Motor Imagery	3 classes	average of 71.6% across subjects with hybrid of feature and parameter optimization	not reported
(Lin & Hsieh, 2009)	IPSO-NN	training nn	mental tasks	3 classes	average of 68%, 66%, 58%, and 64% for IPso-nn, pso-nn, ga-nn, and bp-nn	not reported

based ensemble in comparison with an ensemble with averaged sum rule for decision combination. The assessment is done using up to 10 single layer feed-forward NN classifiers. Random sampling with replacement is used for selecting sub-set of feature sets for each classifier. The feature set consists of a set of features that are reduced by GA. The study reports an average true positive rate of 98% with PSO.

To the best of our knowledge, a combination of EA based approaches and ensemble learning is not used in EEG study so far.

CONCLUSION

This chapter reviewed some of the more recent studies in which EA based methods are employed to improve the EEG signal analysis. The chapter focused on three trends of i) using EA based methods for dimension reduction through feature or electrode reduction, ii) using EA based methods for facilitating the learning phase of the classifiers and iii) using EA based methods in ensemble learning. Such a hybrid system can take advantage from EA based characteristics to optimize the necessary parameters of ensemble and improve the overall results. To the best of our knowledge, such a system has not been used in EEG and BCI studies yet.

REFERENCES

Aberg, M. C., & Wessberg, J. (2007). Evolutionary optimization of classifiers and features for single-trial EEG discrimination. *Biomedical Engineering Online, 6*, 1–8. doi:10.1186/1475-925X-6-32

Al-Ani, A. (2009). A dependency-based search strategy for feature selection. *Journal of Expert Systems with Applications, 36*, 12392–12398. doi:10.1016/j.eswa.2009.04.057

Atyabi, A., Luerssen, M., Fitzgibbon, S. P., & Powers, D. M. W. (2012a). *Dimension reduction in EEG data using particle swarm optimization.* Congress of Evolutionary Computation CEC12, Brisbane, Australia.

Atyabi, A., Luerssen, M., Fitzgibbon, S. P., & Powers, D. M. W. (2012b). *Adapting subject-independent task-specific EEG feature masks using PSO.* Congress of Evolutionary Computation CEC12, Brisbane, Australia.

Atyabi, A., Luerssen, M., Fitzgibbon, S. P., & Powers, D. M. W. (2012c). *Evolutionary feature selection and electrode reduction for EEG classification.* Congress of Evolutionary Computation CEC12, Brisbane, Australia.

Ayache, S., Quenot, G., & Gensel, J. (2007). Classifier fusion for SVM-based multimedia semantic indexing. *LNCS, 4425*, 494–504.

Chandrasekar, R., Suresh, R. K., & Ponnambalam, S. G. (2006). Evaluating an obstacle avoidance strategy to ant colony optimization algorithm for classification in event logs. In *Proceedings of the 14th International Conference on Advanced Computing and Communications* (ADCOM), (pp. 628-630).

Cichocki, A., & Zurada, J. M. (2004). Blind signal separation and extraction: Recent trends, future perspectives, and applications. *International Conference on Artificial Intelligence and Soft Computing, 3070*, (pp. 30-37).

Cinar, E., & Sahin, F. (2010). *A study of recent classification algorithms and a novel approach for EEG data classification* (pp. 3366–3372). doi:10.1109/ICSMC.2010.5642424

Dias, N. S., Jacinto, L. R., Mendes, P. M., & Correia, J. H. (2009). Feature down selection in brain computer interface. *Proceeding of the 4th International IEEE EMBS Conference on Neural Engineering*, (pp. 323-326).

Dobrea, D. M., & Dobrea, M. C. (2009). Optimization of a BCI system using the GA technique. *2nd International Symposium on Applied Sciences in Biomedical and Communication Technologies, ISABEL 2009*, (pp. 1-6).

Dobrea, D. M., Dobrea, M. C., & Sirbu, A. (2010). GA methods for selecting the proper EEG individual spectral bands limits. *2010 International Workshop on Multimedia Signal Processing*, (pp. 1-5).

Elshamli, A., Abdullah, H. A., & Areibi, S. (2004). Genetic algorithm for dynamic path planning. *Proceeding of IEEE CCECE 2004*, (pp. 677-680).

Esmaeili, M. (2007). Classifier fusion for EEG signals processing in human-computer interface systems. *AAAI'07 Proceedings of the 22nd National Conference on Artificial Intelligence*, Vol. 2, (pp. 1856-1857).

Fairley, J., Johnson, A. N., Georgoulas, G., & Vachtsevanos, G. (2010). Automated polysomnogram artifact compensation using the generalized singular value decomposition algorithm. *32nd Annual International Conference on the IEEE EMBS,* (pp. 5097-5100).

Gange, C., Sebag, M., Schoenauer, M., & Tomassini, M. (2009). *Generic methods for evolutionary ensemble learning.* France: Université Paris Sud, Orsay CEDEX.

Ghanbari, A. A., Kousarrizi, M. R. N., Teshnelab, M., & Aliyari, M. (2009). An evolutionary artifact rejection method for brain computer interface using ICA. *International Journal of Electrical & Computer Sciences, 9*(9), 461–466.

Guozheng, Y., Banghua, Y., Shuo, C., & Rongguo, Y. (2006). Pattern recognition using hybrid optimization for a robot controlled by human thoughts. *Proceeding of the 19th IEEE Symposium on Computer-Based Medical Systems* (CBMS'06), (pp. 1-5).

Han, K. M. (2007). *Collision free path planning algorithms for robot navigation problem.* M.Sc Thesis, Department of Electrical and Computer Engineering, Missouri-Columbia University.

Hao, M., Yantao, T., & Linan, Z. (2006). A hybrid ant colony optimization algorithm for path planning of robot in dynamic environment. *International Journal of Information Technology, 12*(3).

Harikumar, R., Raghavan, S., & Sukanesh, R. (2005). Genetic algorithm for classification of epilepsy risk levels from EEG signals. *TENCON 2005 2005 IEEE Region 10,* (pp. 1-6).

Harikumar, R., Sukanesh, R., & Bharathi, P. A. (2004). Genetic algorithm optimization of fuzzy outputs for classification of epilepsy risk levels from EEG signals. *Conference Record of the Thirty-Eighth Asilomar Conference on Systems and Computers,* Vol. 2, (pp. 1585-1589).

Hasan, B. A. S., & Gan, J. Q. (2009). *Multi-objective particle swarm optimization for channel selection in brain computer interface.* The UK Workshop on Computational Intelligence (UKCI2009), Nottingham, UK.

Hasan, B. A. S., Gan, J. Q., & Zhang, Q. (2010). Multi-objective evolutionary methods for channel selection in brain computer interface: Some preliminary experimental results. *2010 IEEE Congress on Evolutionary Computation* (CEC), (pp. 1-6).

He, D., Wang, Z., Yang, B., & Zhou, C. (2009). Genetic algorithm with ensemble learning for detecting community structure in complex networks. *Fourth International Conference on Computer Sciences and Convergence Information Technology,* (pp. 702-707).

Hema, C. R., Paulraj, M. P., Nagarajan, R., Yaacob, S., & Adom, A. H. (2008). Application of particle swarm optimization for EEG signal classification. *Biomedical Soft Computing and Human Sciences, 13*(1), 79–84.

Hema, C. R., Paulraj, M. P., Yaacob, S., Adom, A. H., & Nagarajan, R. (2008). Functional link PSO neural network based classification of EEG mental task signals. *International Symposium on Information Technology, ITSim 2008,* (pp. 1-6).

Hinterberger, T., Weiskopf, N., Veit, R., Wilhelm, B., Betta, E., & Birbaumer, N. (2004). An EEG - driven brain - computer interface combined with functional magnetic resonance imaging (fMRI). *IEEE Transactions on Bio-Medical Engineering, 51*(6), 971–974. doi:10.1109/TBME.2004.827069

Holland, J. H. (1975). *Adaptation in natural and artificial systems*. Ann Arbor, MI: University of Michigan Press.

Huang, D., Qian, K., Oxenham, S., Fei, D. Y., & Bai, O. (2011). Event-related desynchronization/Synchronization-based brain-computer interface towards volitional cursor control in a 2D center-out paradigm. *IEEE Symposium on Computational Intelligence, Cognitive Algorithms, Mind, and Brain* (CCMB), (pp. 1-8).

Ito, S. I., Mitsukura, Y., Fukumi, M., & Cao, J. (2007). Detecting method of music to match the user's mood in prefrontal cortex EEG activity using GA. *International Conference on Control, Automation and Systems*, (pp. 2142-2145).

Jaganathan, P., Thangavel K., Pethalakshmi, A., & Karnan, M. (2007). Classification rule discovery with ant colony optimization and improved quick reduct algorithm. *IAENG International Journal of Computer Science, 33*(1).

Ji, J., Zhang, N., & Liu, C. (2006). An ant colony optimization algorithm for learning classification rules. *WI '06 Proceedings of the 2006 IEEE/WIC/ACM International Conference on Web Intelligence*, (pp. 1034 -1037).

Jiao, Y., Wu, X., & Guo, X. (2010). Motor imagery classification based on the optimized SVM and BPNN by GA. *International Conference on Intelligent Control and Information Processing*, (pp. 344-347).

Jin, J., Wang, X., & Zhang, J. (2008). *Optimal selection of EEG electrodes via DPSO algorithm. Proceeding of the 7th World Congress on Intelligent Control and Automation*, (pp. 5095-5099).

Khushaba, N. R., Al-Ani, A., Al-Jumaily, A., & Nguyen, H. T. (2008). A hybrid nonlinear-discriminant analysis feature projection technique. In Wobcke, W., & Zhang, M. (Eds.), *AI 2008: Advances in Artificial Intelligence, LNAI 5360* (pp. 544–550). Lecture Notes in Computer Science Berlin, Germany: Springer. doi:10.1007/978-3-540-89378-3_55

Khushaba, R. N., Al-Ani, A., AlSukker, A., & Al-Jumaily, A. (2008). Lecture Notes in Computer Science: *Vol. 5217. A combined ant colony and differential evolution feature selection algorithm. Ant Colony Optimization and Swarm Intelligence* (pp. 1–12).

Kim, K. J., & Cho, S. B. (2008). Evolutionary ensemble of diverse artificial neural networks using speciation. *Neurocomputing, 71,* 1604–1618. doi:10.1016/j.neucom.2007.04.008

Kim, Y. S., Street, W. N., & Menczer, F. (2005). Optimal ensemble construction via meta-evolutionary ensembles. *Expert Systems with Applications: An International Journal, 30*(4), 705–714. doi:10.1016/j.eswa.2005.07.030

Kolodziej, M., Majkowski, A., & Rak, R. J. (2011). A new method of EEG classification for BCI with feature extraction based on higher order statistics of wavelet componenets and selection with genetic algorithms. In A. Dobnikar, U. Lotic, & B. Ster (Eds.), *International Conference on Neural Networks and Genetic Algorithms* (pp. 280-289). Berlin, Germany: Springer-Verlag.

Konar, A. (2005). *Computational intelligence principles, techniques and applications*. Springer Verlag.

Largo, R., Munteanu, C., & Rosa, A. (2005). CAP event detection by wavelets and GA tuning. *WISP, 2005*, 44–48.

Latombe, J. C. (1991). *Robot motion planning*. London, UK: Kluwer International Series in Engineering and Computer Science. doi:10.1007/978-1-4615-4022-9

Lbanez, M. L., Blum, C., Thiruvady, D., Ernst, A. T., & Meyer, B. (2009). Beam-ACO based on stochastic sampling for makespan optimization concerning the TSP with time windows. *Evolutionary Computation in Combinatorial Optimization, Lecture Notes in Computer Science, 2009, 5482/2009*, (pp. 97-108).

Li, X., Wang, Y., Song, J., & Shan, J. (2011). Research on classification method of combining support vector machine and genetic algorithm for motor imager EEG. *Journal of Computer Information Systems, 7*(12), 4351–4358.

Lin, C. J., & Hsieh, M. H. (2009). Classification of mental task from EEG data using neural networks based on particle swarm optimization. *Journal of Neurocomputing, 72*, 1121–1130. doi:10.1016/j.neucom.2008.02.017

Liyanage, S. R., Xu, J. X., Guan, C., Ang, K. K., Zhang, C. S., & Lee, T. H. (2009). Classification of self-paced finger movements with EEG signals using neural network and evolutionary approaches. *International Conference on Control and Automation*, (pp. 1807-1812).

Lotte, F., Congedo, M., Lecuyer, A., Lambarche, F., & Arnaldi, B. (2007). A review of classification algorithms for EEG-based brain computer interfaces. *Journal of Neural Engineering, 4*, 1–24. doi:10.1088/1741-2560/4/2/R01

Lv, J., & Liu, M. (2008). Common spatial pattern and particle swarm optimization for channel selection in BCI. *The 3rd International Conference on Innovative Computing information and Control (ICICIC'08)*, (pp. 1-4).

Majumdar, K. (2011). Human scalp EEG processing: Various soft computing approaches. *Applied Soft Computing, 11*, 4433–4447. doi:10.1016/j.asoc.2011.07.004

Martens, D., De Backer, M., Haesen, R., Vanthienen, J., Snoeck, M., & Baesens, B. (2007). Classification with ant colony optimization. *IEEE Transactions on Evolutionary Computation, 11*(5), 651–665. doi:10.1109/TEVC.2006.890229

Moubayed, N. A., Hasan, B. A. S., Gan, J. Q., Petrovski, A., & McCall, J. (2010). Binary-SD-MOPSO and its application in channel selection for brain computer interfaces. 2010 UK Workshop on Computational Intelligence (UKCI), (pp. 1-6).

Nakamura, T., Ito, S. I., Mitsukura, Y., & Setokawa, H. (2009). A method for evaluating the degree of human's preference based on EEG analysis. *Fifth International Conference on Intelligent Information Hiding and Multimedia Signal Processing, IIH-MSP '09*, (pp. 732–735).

Palaniappan, R., & Raveendran, P. (2002). *Genetic algorithm to select features for fuzzy ART-MAP classification of evoked EEG* (pp. 53–56). doi:10.1109/APCCAS.2002.1115119

Parikh, D., Stepenosky, N., Topalis, A., Green, D., Kounios, J., Clark, C., & Polikar, R. (2005). Ensemble based data fusion for early diagnosis of Alzheimer's disease. *Proceedings of 2005 IEEE Engineering in Medicine and Biology 27th Annual Conference*, (pp. 2479-2482).

Paulraj, M. P., Hema, C. R., Nagarajan, R., Yaacob, S., & Adom, A. H. (2007). EEG classification using radial basis PSO neural network for brain machine interfaces. *5th Student Conference on Research and Development, 2007 SCOReD*, (pp. 1-5).

Petyrantonaksi, P. C., & Hadjileontiadis, L. J. (2009). EEG-based emotion recognition using hybrid filtering and higher order crossing. *3rd International Conference on Affective Computing and Intelligent Interaction and Workshops, ACII 2009*, (pp. 1-6).

Piatrik, T., Chandramouli, K., & Izquierdo, E. (2006). Image classification using biologically inspired systems. In *Proceedings of the 2nd International Mobile Multimedia Communications Conference.*

Poli, R., Salvaris, M., & Cinel, C. (2011). Evolution of a brain computer interface mouse via genetic programming. *EuroGP 2011. LNCS, 6621*, 203–214.

Polikar, R. (2006). Ensemble based systems in decision making. *IEEE Circuits and Systems Magazine,* 21-45.

Polikar, R., Topalis, A., Parikh, D., Green, D., Frymiar, J., Kounios, J., & Clark, C. M. (2006). An ensemble based data fusion approach for early diagnosis of Alzheimer's disease. *Information Fusion, 9*, 83–95. doi:10.1016/j.inffus.2006.09.003

Qin, L., Li, Y., & Yao, D. (2005). A feasibility study of EEG dipole source localization using particle swarm optimization. *The 2005 IEEE Congress on Evolutionary Computation*, Vol. 1, (pp. 720-726).

Quinlan, J. R. (2006). *Boosting first order learning* (pp. 1–14). Sydney, Australia: University of Sydney.

Ravi, K. V. R., & Palaniappan, R. (2007). A minimal channel set for individual identification with EEG biometric using genetic algorithm. *Conference on Computational Intelligence and Multimedia Applications*, (pp. 328-332).

Ruta, D., & Gabrys, B. (2000). An overview of classifier fusion methods. *Computing and Information Systems, 7*, 1–10.

Sabeti, M., Boostani, R., & Katebi, S. D. (2007). A new approach to classify the schizophrenic and normal subjects by finding the best channels and frequency bands. *2007 15th International Conference on Digital Signal Processing*, (pp. 123-126).

Sylvester, J., & Chawla, N. V. (2005). Evolutionary ensembles: Combining learning agents using genetic algorithms. *Proceedings of AAAI Workshop on Multi-agent Systems*, (pp. 46–51).

Tang, J., Sun, Z., & Zhu, J. (2005). Using ensemble information in swarming artificial neural networks. *LNCS, 3496*, 515–519.

Ting, K. M., Wells, J. R., Tan, S. C., Teng, S. W., & Webb, G. I. (2011). Feature-subspace aggregation: Ensembles for stable and unstable learners. *Machine Learning, 82*, 375–397. doi:10.1007/s10994-010-5224-5

Tov, E. Y., & Inbar, G. F. (2002). Feature selection for the classification of movements from single movement-related potentials. *IEEE Transactions on Neural Systems and Rehabilitation Engineering, 10*(3), 170–177. doi:10.1109/TNSRE.2002.802875

Veeramachaneni, K., Yan, W., Goebel, K., & Osadciw, L. (2006). Improving classifier fusion using particle swarm optimization. *IEEE Symposium on Computational Intelligence in Multicriteria Decision Making*, (pp. 128-135).

Wolpaw, J. R., Birbaumer, N., Heetderks, W. J., McFarland, D. J., Peckham, P. H., & Schalk, G. … Vaughan, T. M. (2000). Brain-computer interface technology: A review of the first inter- national meeting. *IEEE Transactions in Rehabilitation Engineering, 8*(2), 164–173.

Zhang, Q., & Sun, M. S. (2010). Evolutionary classifier ensembles for semi-supervised learning. *The 2010 International Joint Conference on Neural Networks* (IJCNN), (pp. 1-6).

Zhang, X., & Wang, X. (2008). A genetic algorithm based time-Frequency approach to a movement prediction task. *Proceeding of the 7th World Congress on Intelligent Control and Automation,* (pp. 1032-1036).

Zhang, Z., & Yang, P. (2008). An ensemble of classifiers with genetic algorithm based feature selection. *IEEE Intelligent Information Bulletin, 9*(1), 18–24.

Zhao, L., & Yang, Y. (2009). PSO-based single multiplicative neuron model fir time series prediction. *Expert Systems with Applications, 36,* 2805–2812. doi:10.1016/j.eswa.2008.01.061

Zhao, Y., & Zheng, J. (2004). Particle swarm optimization algorithm in signal detection and blind extraction. *Proceeding of the 7th International Symposium on Parallel Architectures, Algorithms and Networks* (ISPAN'04), (pp. 1-5).

Zhiping, H., Guanaming, C., Cheng, C., He, X., & Jiacai, Z. (2010). A new feature selection method for self-paced brain computer interface. *2010 10th International Conference on Intelligent Systems Design and Applications* (ISDA), (pp. 845-849).

Zhu, Q. Y., Qin, A. K., Suganthan, P. N., & Huang, G. B. (2005). Evolutionary extreme learning machine. *Pattern Recognition, 38,* 1759–1763. doi:10.1016/j.patcog.2005.03.028

ADDITIONAL READING

Asadi, A. G., B, A., Navidi, H., & A, A. (2012). Brain computer interface with genetic algorithm. *International Journal of Information and Communication Technology Research, 2*(1), 79–86.

Bergqvist, G., & Larsson, E. G. (2009). High-order singular value decomposition: Theory and an application. *IEEE Signal Processing Magazine, 27,* 151–154. doi:10.1109/MSP.2010.936030

Bin, G., Gao, X., Wang, Y., Li, Y., Hong, B., & Gao, S. (2011). A high-speed BCI based on code modulation VEP. *Journal of Neural Engineering, 8*(2), 025015. doi:10.1088/1741-2560/8/2/025015

Brunner, P., Bianchi, L., Guger, C., Cincotti, F., & Schalk, G. (2011). Current trends in hardware and software for brain-computer interfaces (BCIs). *Journal of Neural Engineering, 8*(2), 025001. doi:10.1088/1741-2560/8/2/025001

Chen, Y., Rege, M., Dong, M., & Hua, J. (2008). Non-negative matrix factorization for semi-supervised data clustering. *Knowledge and Information Systems, 17*(3), 355–379. doi:10.1007/s10115-008-0134-6

Collobert, R. (2004). *Links between perceptrons, MLPs and SVMs.* 21st International Conference on Machine Learning.

Fitzgibbon, S., Lewis, T., Powers, D., Whitham, E., Willoughby, J., & Pope, K. (2012). Surface Laplacian of central scalp electrical signals is insensitive to muscle contamination. *IEEE Transactions on Bio-Medical Engineering, 5001*(618), 1–6. doi:10.1109/TBME.2012.2195662

Fitzgibbon, S. P., & Powers, D. M. W. (2007). Removal of EEG noise and artifact using blind. *Journal of Computational Neuroscience, 24*(3), 232–243.

Gaito, S., Greppi, A., & Grossi, G. (2007). Random projections for dimensionality reduction in ICA. *International Journal of Mathematical and Computer Sciences, 3*(4), 211–215.

Garro, B. A., Sossa, H., & Vázquez, R. A. (2011). Back-propagation vs particle swarm optimization algorithm: Which algorithm is better to adjust the synaptic weights of a feed-forward ANN? *International Journal of Artificial Intelligence, 7.*

Guan, Y., & Dy, J. G. (2008). Sparse probabilistic principal component analysis. *12th International Conference on Artificial Intelligence and Statistics (AISTATS),* (pp. 185-192).

Hasan, B. A. S., & Gan, J. Q. (2011). Conditional random fields as classifiers for three-class motor-imagery brain-computer interfaces. *Journal of Neural Engineering, 8*(2), 025013. doi:10.1088/1741-2560/8/2/025013

Hill, N. J., & Schölkopf, B. (2012). An online brain-computer interface based on shifting attention to concurrent streams of auditory stimuli. *Journal of Neural Engineering, 9*(2), 026011. doi:10.1088/1741-2560/9/2/026011

Huang, G.-B., Zhou, H., Ding, X., & Zhang, R. (2012). Extreme learning machine for regression and multiclass classification. *IEEE Transactions on Systems, Man, and Cybernetics. Part B, Cybernetics, 42*(2), 513–529. doi:10.1109/TSMCB.2011.2168604

Kindermans, P.-J., Verstraeten, D., & Schrauwen, B. (2012). A Bayesian model for exploiting application constraints to enable unsupervised training of a P300-based BCI. *PLoS ONE, 7*(4), e33758. doi:10.1371/journal.pone.0033758

Lopez-Gordo, M. a, Fernandez, E., Romero, S., Pelayo, F., & Prieto, A. (2012). An auditory brain-computer interface evoked by natural speech. *Journal of Neural Engineering, 9*(3), 036013. doi:10.1088/1741-2560/9/3/036013

Lu, H., Eng, H.-L., Guan, C., Plataniotis, K. N., & Venetsanopoulos, A. N. (2010). Regularized common spatial pattern with aggregation for EEG classification in small-sample setting. *IEEE Transactions on Bio-Medical Engineering, 57*(12), 2936–2946. doi:10.1109/TBME.2010.2082540

Mak, J. N., Arbel, Y., Minett, J. W., McCane, L. M., Yuksel, B., & Ryan, D. (2011). Optimizing the P300-based brain-computer interface: Current status, limitations and future directions. *Journal of Neural Engineering, 8*(2), 025003. doi:10.1088/1741-2560/8/2/025003

Ng, K. B., Bradley, A. P., & Cunnington, R. (2012). Stimulus specificity of a steady-state visual-evoked potential-based brain-computer interface. *Journal of Neural Engineering, 9*(3), 036008. doi:10.1088/1741-2560/9/3/036008

Nunez, P. L., & Westdorp, A. F. (1994). The surface Laplacian, high resolution EEG and controversies. *Brain Topography, 6*(3), 221–226. doi:10.1007/BF01187712

Onton, J., & Makeig, S. (2006). Information-based modeling of event-related brain dynamics. *Progress in Brain Research, 159*, 99–120. doi:10.1016/S0079-6123(06)59007-7

Paige, C. C., & Saunders, M. A. (2011). Towards a generalized singular value decomposition. *SIAM Journal on Numerical Analysis, 18*(3), 398–405. doi:10.1137/0718026

Powers, D. M. W. (2003). Recall and precision versus the bookmaker. *International Conference on Cognitive Linguistics*, (pp. 529-534).

Powers, D. M. W. (2011). Evaluation: From precision, recall and F-measure to ROC, informedness, markedness & correlation. *Journal of Machine Learning Technology, 2*(1), 37–63.

Reuderink, B., Poel, M., & Nijholt, A. (2011). The impact of loss of control on movement BCIs. *IEEE Transactions on Neural Systems and Rehabilitation Engineering, 19*(6), 628–637. doi:10.1109/TNSRE.2011.2166562

Sannelli, C., Vidaurre, C., Müller, K.-R., & Blankertz, B. (2011). CSP patches: An ensemble of optimized spatial filters: An evaluation study. *Journal of Neural Engineering, 8*(2), 025012. doi:10.1088/1741-2560/8/2/025012

Tam, W.-K., Tong, K.-yu, Meng, F., & Gao, S. (2011). A minimal set of electrodes for motor imagery BCI to control an assistive device in chronic stroke subjects: A multi-session study. *IEEE Transactions on Neural Systems and Rehabilitation Engineering, 19*(6), 617–627. doi:10.1109/TNSRE.2011.2168542

Tong, D. L., & Mintram, R. (2010). Genetic algorithm-neural network (GANN): A study of neural network activation functions and depth of genetic algorithm search applied to feature selection. *International Journal of Machine Learning and Cybernetics, 1*(1-4), 75–87. doi:10.1007/s13042-010-0004-x

van Rijsbergen, N. J., & Schyns, P. G. (2009). Dynamics of trimming the content of face representations for categorization in the brain. *PLoS Computational Biology, 5*(11), e1000561. doi:10.1371/journal.pcbi.1000561

Wang, Y., & Jung, T.-P. (2011). A collaborative brain-computer interface for improving human performance. *PLoS ONE, 6*(5). doi:10.1371/journal.pone.0020422

Wilson, J. J., & Palaniappan, R. (2011). Analogue mouse pointer control via an online steady state visual evoked potential (SSVEP) brain-computer interface. *Journal of Neural Engineering, 8*(2), 025026. doi:10.1088/1741-2560/8/2/025026

Zanchettin, C., Ludermir, T. B., & Almeida, L. M. (2011). Hybrid training method for MLP: Optimization of architecture and training. *IEEE Transactions on Systems, Man, and Cybernetics. Part B, Cybernetics, 41*(4), 1097–1109. doi:10.1109/TSMCB.2011.2107035

Zander, T. O., Ihme, K., Gärtner, M., & Rötting, M. (2011). A public data hub for benchmarking common brain-computer interface algorithms. *Journal of Neural Engineering, 8*(2), 025021. doi:10.1088/1741-2560/8/2/025021

Zhang, Y., Xu, P., Liu, T., Hu, J., Zhang, R., & Yao, D. (2012). Multiple frequencies sequential coding for SSVEP-based brain-computer interface. *PLoS ONE, 7*(3), e29519. doi:10.1371/journal.pone.0029519

Zhu, D., Bieger, J., Garcia Molina, G., & Aarts, R. M. (2010). A survey of stimulation methods used in SSVEP-based BCIs. *Computational Intelligence and Neuroscience, •••*, 702357.

KEY TERMS AND DEFINITIONS

Brain Computer Interface: Is a communication system that does not depend on the brain's normal output pathways of peripheral nerves and muscles.

Electroencephalogram: Electroencephalogram (EEG) is one of the commonly used non-invasive techniques in BCI for signal acquisition. EEG records variations of the surface potential from the scalp using some electrodes.

Feature Reduction and Electrode Reduction: Feature reduction and feature selection are the study of selecting a subset of features that best represent the performed task.

Genetic Algorithm: A technique in artificial intelligence that uses the use specific genetic operator such as mutation, selection, and crossover to solve global optimization problems.

Particle Swarm Optimization: Particle Swarm Optimization (PSO) is an evolutionary algorithm that is inspired by the social and cognitive interactions of animals with one another and with the environment (e.g., fish schooling and birds flying). PSO operates by iteratively directing its particles toward the optimum using social and cognitive components.

ENDNOTES

[1] Proportion of the pheromone shows the length of the route.

[2] Concentration of the pheromone shows the amount of the ants that used the route.

[3] The length of the route and the amount of ants used the route influence the amount of pheromones accumulated on the route.

Compilation of References

Aberg, M. C., & Wessberg, J. (2007). Evolutionary optimization of classifiers and features for single-trial EEG discrimination. *Biomedical Engineering Online*, *6*, 1–8. doi:10.1186/1475-925X-6-32

Adly, A., & Abd-El-Hafiz, S. (2006). Using the particle swarm evolutionary approach in shape optimization and field analysis of devices involving nonlinear magnetic media. *IEEE Transactions on Magnetics*, *42*(10), 3150–3152. doi:10.1109/TMAG.2006.880103

Adra, S. F., Griffin, I. A., & Fleming, P. J. (2006). *An adaptive memetic algorithm for enhanced diversity*. The International Adaptive Computing in Design and Manufacture Conference.

Afkhami, H., Mowla, A., Granpayeh, N., & Hormozi, A. R. (2010). Wideband gain flattened hybrid erbium-doped fiber amplifier/fiber Raman amplifier. *Journal of the Optical Society of Korea*, *14*(4), 342–350. doi:10.3807/JOSK.2010.14.4.342

Afshinmanesh, F., Marandi, A., & Rahimi-Kian, A. (2005, November 21-24). *A novel binary particle swarm optimization method using artificial immune system*. Paper presented at the EUROCON 2005 – The International Conference on "Computer as a Tool", Belgrade, Serbia and Montenegro.

Afshinmanesh, F., Marandi, A., & Shahabadi, M. (2008, July). Design of a single-feed dual-band dual-polarized printed microstrip antenna using a Boolean particle swarm optimization. *IEEE Transactions on Antennas and Propagation*, *56*(7), 1845–1852. doi:10.1109/TAP.2008.924684

Ahn, J. T., & Kim, K. H. (2004). All-optical gain-clamped erbium-doped fiber amplifier with improved noise figure and freedom from relaxation oscillation. *IEEE Photonics Technology Letters*, *16*(1), 84–86. doi:10.1109/LPT.2003.818906

Akbari, T., Rahimikian, A., & Kazemi, A. (2011). A multistage stochastic transmission expansion planning method. *Energy Conversion and Management*, *52*, 2844–2853. doi:10.1016/j.enconman.2011.02.023

Al-Ani, A. (2009). A dependency-based search strategy for feature selection. *Journal of Expert Systems with Applications*, *36*, 12392–12398. doi:10.1016/j.eswa.2009.04.057

Albanese, R., Rubinacci, G., Takagi, T., & Udpa, S. S. (Eds.). (1998). *Electromagnetic nondestructive evaluation (II). Studies in Applied Electromagnetics and Mechanics*. Amsterdam, The Netherlands: IOS Press.

Alguacil, N., Carrión, M., & Arroyo, J. M. (2009). Transmission network expansion planning under deliberate outages. *International Journal of Electrical Power & Energy Systems*, *31*, 553–561. doi:10.1016/j.ijepes.2009.02.001

Alireza, A. (2011). PSO with adaptive mutation and inertia weight and its application in parameter estimation of dynamic systems. *Acta Automatica Sinica*, *37*(5), 541–548.

Arena, P., Caponetto, R., Fortuna, L., & Porto, D. (2000). Nonlinear noninteger order circuits and systems - An introduction. In Chua, L. (Ed.), *World Scientific Series on Nonlinear Science, Series A* (*Vol. 38*). Singapore: World Scientific.

Åström, K. J., & Wittenmark, B. (1997). *Computer-controlled systems - Theory and design,* 3rd ed. New Jersey: Prentice Hall.

Åström, K. J., & Hagglund, T. (1995). *PID controllers: Theory, design, and tuning* (2nd ed.). Research Triangle Park, NC: Instrument Society of America.

Atyabi, A., & Powers, D. M. W. (2010). The use of area extended particle swarm optimization (AEPSO) in swarm robotics. 11th International Conference on Control Automation Robotics & Vision (ICARCV 2010), (pp. 591–596).

Atyabi, A., Luerssen, M., Fitzgibbon, S. P., & Powers, D. M. W. (2012). *Dimension reduction in EEG data using particle swarm optimization.* Congress of Evolutionary Computation CEC12, Brisbane, Australia.

Atyabi, A., Luerssen, M., Fitzgibbon, S. P., & Powers, D. M. W. (2012). *Adapting subject-independent task-specific EEG feature masks using PSO.* Congress of Evolutionary Computation CEC12, Brisbane, Australia.

Atyabi, A., Luerssen, M., Fitzgibbon, S. P., & Powers, D. M. W. (2012). *Evolutionary feature selection and electrode reduction for EEG classification.* Congress of Evolutionary Computation CEC12, Brisbane, Australia.

Audet, C., Savard, G., & Zghal, W. (2008). Multiobjective optimization through a series of single-objective formulations. *SIAM Journal on Optimization, 19*(1), 188–210. doi:10.1137/060677513

Ayache, S., Quenot, G., & Gensel, J. (2007). Classifier fusion for SVM-based multimedia semantic indexing. *LNCS, 4425,* 494–504.

Azadeh, M. (2009). *Fibre Optics Engineering.* New York, NY: Springer. doi:10.1007/978-1-4419-0304-4

Azaro, R., Boato, G., Donelli, M., Massa, A., & Zeni, E. (2006). Design of a prefractal monopolar antenna for 3.4–3.6 GHz wi-max band portable devices. *IEEE Antennas and Wireless Propagation Letters, 5*(1), 116–119. doi:10.1109/LAWP.2006.872427

Azaro, R., De Natale, F., Donelli, M., & Massa, A. (2006). PSO-based optimization of matching loads for lossy transmission lines. *Microwave and Optical Technology Letters, 48*(8), 1485–1487. doi:10.1002/mop.21738

Azaro, R., De Natale, F., Donelli, M., & Zeni, E. (2006). Optimized design of a multi-function/multi-band antenna for automotive rescue systems. *IEEE Transactions on Antennas and Propagation, 54*(2), 897–904. doi:10.1109/TAP.2005.863387

Azaro, R., De Natale, F., Zeni, E., Donelli, M., & Massa, A. (2006). Synthesis of a pre-fractal dual-band monopolar antenna for GPS applications. *IEEE Antennas and Wireless Propagation Letters, 5*(1), 361–364. doi:10.1109/LAWP.2006.880695

Azaro, R., Donelli, M., Benedetti, M., Rocca, P., & Massa, A. (2008). A GSM signals based positioning technique for mobile applications. *Microwave and Optical Technology Letters, 50*(4), 2128–2130. doi:10.1002/mop.23568

Azaro, R., Donelli, M., Franceschini, D., Zeni, E., & Massa, A. (2006). Optimized synthesis of a miniaturized SARSAT band pre-fractal antenna. *Microwave and Optical Technology Letters, 48*(11), 2205–2207. doi:10.1002/mop.21922

Balanis, C. A. (1989). *Advanced engineering electromagnetics.* New York, NY: Wiley.

Balanis, C. A. (2008). *Modern antenna handbook.* Hoboken, NJ: Wiley-Interscience. doi:10.1002/9780470294154

Barbosa, M. R., Solteiro Pires, E. J., & Lopes, A. M. (2010). Optimization of parallel manipulators using evolutionary algorithms. *Soft Computing Models in Industrial and Environmental Applications, 5th International Workshop* (SOCO 2010), Guimarães, Portugal, (pp. 79-86).

Barbosa, R. S., Tenreiro Machado, J. A., & Ferreira, I. M. (2005, 4-8 July). Pole-zero approximations of digital fractional-order integrators and differentiators using signal modelling techniques. In *Proceedings of the 16th IFAC World Congress*, Prague, Czech Republic.

Barbosa, R. S., & Tenreiro Machado, J. A. (2006). Implementation of discrete-time fractional-order controllers based on LS approximations. *Acta Polytechnica Hungarica, 3*(4), 5–22.

Barbosa, R. S., Tenreiro Machado, J. A., & Silva, M. F. (2006, October). Time domain design of fractional differintegrators using least squares. *Signal Processing, 86*(10), 2567–2581. doi:10.1016/j.sigpro.2006.02.005

Barros, M., Guilherme, J., & Horta, N. (2010). Analog circuits optimization based on evolutionary computation techniques. *Integration- The VLSI Journal, 43*, 136–155.

Barros, M. F. M., Guilherme, J. M. C., & Horta, N. C. G. (2010). *Analog circuits and systems optimization based on evolutionary techniques*. Springer-Verlag. doi:10.1007/978-3-642-12346-7

Baskar, S., Alphones, A., Suganthan, P. N., & Liang, J. J. (2005). Design of Yagi-Uda antennas using comprehensive learning particle swarm optimisation. *IEE Proceedings. Microwaves, Antennas and Propagation, 152*(5), 340–346. doi:10.1049/ip-map:20045087

Baussard, A., Miller, E. L., & Prémel, D. (2004). Adaptive B-spline scheme for solving an inverse scattering problem. *Inverse Problems, 20*(2), 347–365. doi:10.1088/0266-5611/20/2/003

Bayraktar, Z., Werner, P. L., & Werner, D. H. (2006). The design of miniature three-element stochastic Yagi-Uda arrays using particle swarm optimization. *IEEE Antennas and Wireless Propagation Letters, 5*(1), 22–26. doi:10.1109/LAWP.2005.863618

Bayraktar, Z., Werner, P. L., & Werner, D. H. (2006, December). The design of miniature three-element stochastic Yagi-Uda arrays using particle swarm optimization. *IEEE Antennas and Wireless Propagation Letters, 5*(1), 22–26. doi:10.1109/LAWP.2005.863618

Benedetti, M., Azaro, R., Franceschini, D., & Massa, A. (2006). PSO-based real-time control of planar uniform circular arrays. *IEEE Antennas and Wireless Propagation Letters, 5*(1), 545–548. doi:10.1109/LAWP.2006.887553

Beni, G., & Wang, J. (1989). Swarm intelligence in cellular robotic systems. *Proceedings of the NATO Advanced Workshop on Robotics and Biological Systems*, Il Ciocco, Tuscany, Italy.

Berman, S., Halasz, A., Kumar, V., & Pratt, S. (2007). Bio-inspired group behaviors for the deployment of a swarm of robots to multiple destinations. In *Proceedings of 2007 IEEE International Conference on Robotics and Automation*, (pp. 2318–2323).

Berman, S., Halasz, A., Hsieh, M. A., & Kumar, V. (2009). Optimized stochastic policies for task allocation in swarms of robots. *IEEE Transactions on Robotics, 25*(4), 927–937. doi:10.1109/TRO.2009.2024997

Bernholtz, B., & Graham, L. J. (1960). Hydrothermal economic scheduling part 1. Solution by incremental dynamic programing. *Transactions in AIEE Power Apparatus and Systems, 141*(5), 497–506.

Bevelacqua, P. J., & Balanis, C. A. (2009). Geometry and weight optimization for minimizing sidelobes in wideband planar arrays. *IEEE Transactions on Antennas and Propagation, 57*(4), 1285–1289. doi:10.1109/TAP.2009.2015853

Bhattacharya, R., Joshi, A., & Bhattacharya, T. K. (2006). PSO-based evolutionary optimization for black-box modeling of arbitrary shaped on-chip RF inductors. In *Proceedings of 2006 Topical Meeting on Silicon Monolithic Integrated Circuits in RF Systems*, 18-20 Jan. 2006.

Bilchev, G., & Parmee, I. C. (1995). The ant colony metaphor for searching continuous design spaces. *Selected Papers from AISB Workshop on Evolutionary Computing* (pp. 25–39). London, UK: Springer-Verlag.

Binato, S., de Oliveira, G. C., & de Araujo, J. L. (2001). A greedy randomized adaptive search procedure for transmission expansion planning. *IEEE Transactions on Power Systems, 16*(2), 247–253. doi:10.1109/59.918294

Bode, H. W. (1945). *Network analysis and feedback amplifier design*. New York, NY: Van Nostrand.

Boeringer, D. W., & Werner, D. H. (2004). Particle swarm optimization versus genetic algorithms for phased array synthesis. *IEEE Transactions on Antennas and Propagation, 52*(3), 771–779. doi:10.1109/TAP.2004.825102

Boeringer, D., & Werner, D. (2004). Particle swarm optimization versus genetic algorithms for phased array synthesis. *IEEE Transactions on Antennas and Propagation, 52*(3), 771–779. doi:10.1109/TAP.2004.825102

Boeringer, D., & Werner, D. (2005). Efficiency-constrained particle swarm optimization of a modified bernstein polynomial for conformal array excitation amplitude synthesis. *IEEE Transactions on Antennas and Propagation, 53*.

Bonabeau, E., Dorigo, M., & Theraulaz, G. (1999, August). *Swarm intelligence: From natural to artificial systems.* New York, NY: Oxford University Press.

Bonabeau, E., Dorigo, M., & Theraulaz, G. (1999). *Swarm intelligence: From natural to artificial systems.* New York, NY: Oxford University Press, Inc.

Bonabeau, E., Theraulaz, G., & Dorigo, M. (1999). *Swarm intelligence: From natural to artificial systems.* Oxford University Press.

Bouallegue, S., Haggege, J., Ayadi, M., & Benrejeb, M. (2012). PID-type fuzzy logic controller tuning based on particle swarm optimization. *Engineering Applications of Artificial Intelligence, 25,* 484–493. doi:10.1016/j.engappai.2011.09.018

Bozza, G., Estatico, C., Pastorino, M., & Randazzo, A. (2006). An inexact Newton method for microwave reconstruction of strong scatterers. *Antennas and Wireless Propagation Letters, 5*(1), 61–64. doi:10.1109/LAWP.2006.870360

Brignone, M., Bozza, G., Randazzo, A., Aramini, R., Piana, M., & Pastorino, M. (2008). Hybrid approach to the inverse scattering problem by using ant colony optimization and no-sampling linear sampling. *Proceedings of the 2008 IEEE Antennas and Propagation Society International Symposium (APS2008)* (pp. 1–4). San Diego, CA: IEEE. doi:10.1109/APS.2008.4619941

Brignone, M., Bozza, G., Randazzo, A., Piana, M., & Pastorino, M. (2008). A hybrid approach to 3D microwave imaging by using linear sampling and ACO. *IEEE Transactions on Antennas and Propagation, 56*(10), 3224–3232. doi:10.1109/TAP.2008.929504

Bromage, J. (2004). Raman amplification for fiber communications systems. *Journal of Lightwave Technology, 22*(1), 79–93. doi:10.1109/JLT.2003.822828

Burgard, W., Moors, M., Stachniss, C., & Schneider, F. E. (2005). Coordinated multi-robot exploration. *IEEE Transactions on Robotics, 21*(3), 376–386. doi:10.1109/TRO.2004.839232

Buygi, M., Balzer, G., Shanechi, H., & Shahidehpour, M. (2004). Market-based transmission expansion planning. *IEEE Transactions on Power Systems, 19*(4), 2060–2067. doi:10.1109/TPWRS.2004.836252

Calderon, L. R., & Galiana, F. D. (1987). Continuous solution simulation in the short-term hydrothermal coordination problem. *IEEE Transactions on Power Systems, 2*(3), 737–743. doi:10.1109/TPWRS.1987.4335203

Camazine, S., Franks, N. R., Sneyd, J., Bonabeau, E., Deneubourg, J.-L., & Theraulaz, G. (2001). *Self-organization in biological systems.* Princeton, NJ: Princeton University Press.

Camci, F. (2008, August). *Analysis of velocity calculation methods in binary PSO on maintenance scheduling.* Paper presented at the First International Conference on the Applications of Digital Information and Web Technologies, Ostrava, Czech Republic.

Campo, A., & Dorigo, M. (2007). Efficient multi-foraging in swarm robotics. In *Proceedings of 9th European Conference on Advances in Artificial Life, Ser. ECAL'07,* (pp. 696–705). Berlin, Germany: Springer-Verlag.

Cao, J.-Y., & Cao, B.-G. (2006, December). Design of fractional order controller based on particle swarm optimization. *International Journal of Control, Automation, and Systems, 4*(6), 775–781.

Caorsi, S., Donelli, M., Lommi, A., & Massa, A. (2004). Location and imaging of two-dimensional scatterers by using a Particle Swarm algorithm. *Journal of Electromagnetic Waves and Applications, 18*(4), 481–494. doi:10.1163/156939304774113089

Caorsi, S., Donelli, M., Massa, A., & Raffetto, M. (2002). A parallel implementation of an evolutionary-based automatic tool for microwave circuit synthesis: Preliminary results. *Microwave and Optical Technology Letters, 35*(3). doi:10.1002/mop.10547

Caorsi, S., Gragnani, G. L., Medicina, S., Pastorino, M., & Zunino, G. (1991). Microwave imaging method using a simulated annealing approach. *IEEE Microwave and Guided Wave Letters, 1*(11), 331–333. doi:10.1109/75.93902

Carena, A., Curri, V., & Poggiolini, P. (2001). On the optimization of hybrid Raman/erbium-doped fiber amplifiers. *IEEE Photonics Technology Letters*, *13*(11), 1170–1172. doi:10.1109/68.959353

Carlisle, A., & Doizier, G. (2001). An off-the-shelf PSO. *Proceedings of the Workshop Particle Swarm Optimization*, Indianapolis, IN.

Carlisle, A., & Dozler, G. (2002, December). *Tracking changing extrema with adaptive particle swarm optimizer*. Paper presented at the 5th Biannual World Automation Congress, Orlando, USA.

Carlson, G. E., & Halijak, C. A. (1964). Approximation of fractional capacitors $(1/s)^{1/n}$ by a regular Newton process. *IEEE Transactions on Circuit Theory*, *11*(2), 210–213.

Castanon, J. D., Nasieva, I. O., Turitsyn, S. K., Brochier, N., & Pincemin, E. (2005). Optimal span length in high-speed transmission systems with hybrid Raman-erbium-doped fiber amplification. *Optics Letters*, *30*(1), 23–25. doi:10.1364/OL.30.000023

Cedeño, E. B., & Arora, S. (2011). Performance comparison of transmission network expansion planning under deterministic and uncertain conditions. *International Journal of Electrical Power & Energy Systems*, *33*, 1288–1295. doi:10.1016/j.ijepes.2011.05.005

Chamaani, S., Abrishamian, M. S., & Mirtaheri, S. A. (2010, May). Time-domain design of UWB Vivaldi antenna array using multiobjective particle swarm optimization. *IEEE Antennas and Wireless Propagation Letters*, *9*, 666–669. doi:10.1109/LAWP.2010.2053691

Chamaani, S., Mirtaheri, S. A., & Abrishamian, M. S. (2011). Improvement of time and frequency domain performance of antipodal Vivaldi antenna using multi-objective particle swarm optimization. *IEEE Transactions on Antennas and Propagation*, *59*(5), 1738–1742. doi:10.1109/TAP.2011.2122290

Chan, E.-J., Chao, C.-H., Jheng, K.-Y., Hsin, H.-K., & Wu, A.-Y. (2010). ACO-based cascaded adaptive routing for traffic balancing in NoC systems. *Proceedings of the 2010 International Conference on Green Circuits and Systems (ICGCS)* (pp. 317–322). Shanghai, China: IEEE. doi:10.1109/ICGCS.2010.5543045

Chandrasekar, R., Suresh, R. K., & Ponnambalam, S. G. (2006). Evaluating an obstacle avoidance strategy to ant colony optimization algorithm for classification in event logs. In *Proceedings of the 14th International Conference on Advanced Computing and Communications* (ADCOM), (pp. 628-630).

Chan, F. T. S., & Tiwari, M. K. (2007). *Swarm Intelligence: Focus on ant and particle swarm optimization*. I-Tech Education and Publishing.

Chang, M., Chou, P., & Lee, H. (1995). Tomographic microwave imaging for nondestructive evaluation and object recognition of civil structures and materials. *Proceedings of the 29th Asilomar Conference on Signals, Systems and Computers*, ASILOMAR '95 (Vol. 2, pp. 1061–1065). Washington, DC: IEEE Computer Society.

Chang, C. L., Wang, L., & Chiang, Y. J. (2006). A dual pumped double-pass L-band EDFA with high gain and low noise. *Optics Communications*, *267*(1), 108–112. doi:10.1016/j.optcom.2006.06.025

Chang, H. C., & Cheng, P. H. (1996). Genetic aided scheduling of hydraulically coupled plants in hydrothermal coordination. *Proceedings of IET, Generation, Transmission, and Distribution*, *11*(2), 975–981.

Charef, A., Sun, H. H., Tsao, Y. Y., & Onaral, B. (1992, September). Fractal system as represented by singularity function. *IEEE Transactions on Automatic Control*, *37*(9), 1465–1470. doi:10.1109/9.159595

Chatterjee, A., Ghoshal, S. P., & Mukherjee, V. (2011). Craziness-based PSO with wavelet mutation for transient performance augmentation of thermal system connected to grid. *Expert Systems with Applications*, *38*, 7784–7794. doi:10.1016/j.eswa.2010.12.128

Chen, C. Y., & Ye, F. (2004). Particle swarm optimization algorithm and its application to clustering analysis. *IEEE International Conference on Networking, Sensing and Control* (pp. 789-794).

Chen, C. T., & Liao, T. T. (2011). A hybrid strategy for the time- and energy-efficient trajectory planning of parallel platform manipulators. *Robotics and Computer-integrated Manufacturing*, *27*(1), 72–81. doi:10.1016/j.rcim.2010.06.012

Chen, C., & Pham, H. (2012). Trajectory planning in parallel kinematic manipulators using a constrained multi-objective evolutionary algorithm. *Nonlinear Dynamics*, *67*, 1669–1681. doi:10.1007/s11071-011-0095-2

Cheng, C. (2004). A global design of an erbium-doped fibre and an erbium-doped fibre amplier. *Optics & Laser Technology*, *36*, 607–612. doi:10.1016/j.optlastec.2004.01.006

Cheng, C., & Xiao, M. (2005). Optimization of an erbium-doped fiber amplifier with radial effects. *Optics Communications*, *254*(4-6), 215–222. doi:10.1016/j.optcom.2005.05.049

Cheng, C., & Xiao, M. (2006). Optimization of a dual pumped L-band erbiumdoped fiber amplifier by genetic algorithm. *Journal of Lightwave Technology*, *24*(10), 3824–3829. doi:10.1109/JLT.2006.881476

Chen, M. R., Li, X., Zhang, X., & Lu, Y. Z. (2010). A novel particle swarm optimizer hybridized with extremal optimization. *Applied Soft Computing*, *10*, 367–373. doi:10.1016/j.asoc.2009.08.014

Chen, W.-N., Zhang, Z., Chung, H. S. H., Zhong, W.-L., Wu, W.-G., & Shi, Y.-H. (2010). A novel set-based particle swarm optimization method for discrete optimization problems. *IEEE Transactions on Evolutionary Computation*, *14*(2), 278–300. doi:10.1109/TEVC.2009.2030331

Chen, Y. Q., & Vinagre, B. M. (2003). A new IIR-type digital fractional order differentiator. *Signal Processing*, *83*(11), 2359–2365. doi:10.1016/S0165-1684(03)00188-9

Chen, Y. Q., Vinagre, B. M., & Podlubny, I. (2004). Continued fraction expansions approaches to discretizing fractional order derivatives. An expository review. *Nonlinear Dynamics*, *38*(1-2), 155–170. doi:10.1007/s11071-004-3752-x

Chew, W. C. (1995). *Waves and fields in inhomogeneous media*. New York, NY: IEEE Press.

Chiang, Y. C., & Chen, C. Y. (2001). Design of a wideband lumped-element 3-dB quadrature coupler. *IEEE Transactions on Microwave Theory and Techniques*, *9*, 476–479. doi:10.1109/22.910551

Choi, B. H., Park, H. H., & Chu, M. J. (2003). New pumped wavelength of 1540-nm band for long-wavelength-band erbium-doped fiber amplifier (L-band EDFA). *IEEE Journal of Quantum Electronics*, *39*(10), 1272–1280. doi:10.1109/JQE.2003.817582

Choi, J., Mount, T., & Thomas, R. (2007). Transmission expansion planning using contingency criteria. *IEEE Transactions on Power Systems*, *22*, 2249–2261. doi:10.1109/TPWRS.2007.908478

Cho, J., Yoon, J., Cho, S., Kwon, K., Lim, S., & Kim, D. (2006, August). In-vivo measurements of the dielectric properties of breast carcinoma xenografted on nude mice. *International Journal of Cancer*, *119*(3), 593–598. doi:10.1002/ijc.21896

Chowdhury, N. M., Baker, S. M., & Choudhury, E. H. (2008). A new adaptive routing approach based on ant colony optimization (ACO) for ad hoc wireless networks. *Proceedings of the 11th International Conference on Computer and Information Technology (ICCIT2008)* (pp. 51–56). Khulna, Bangladesh: IEEE. doi:10.1109/ICCITECHN.2008.4803126

Chu, M., & Allstot, D. J. (2005). An elitist distributed particle swarm algorithm for RF IC optimization. In *Proceedings of IEEE Asia and South Pacific Design Automation Conference 2005*, (pp. 671-674).

Cichocki, A., & Zurada, J. M. (2004). Blind signal separation and extraction: Recent trends, future perspectives, and applications. *International Conference on Artificial Intelligence and Soft Computing, 3070*, (pp. 30-37).

Cinar, E., & Sahin, F. (2010). *A study of recent classification algorithms and a novel approach for EEG data classification* (pp. 3366–3372). doi:10.1109/ICSMC.2010.5642424

Clerc, M. (1999, July 6-9). *The swarm and the queen: Towards a deterministic and adaptive particle swarm optimization*. Paper presented at the 1999 Congress on Evolutionary Computation, Washington, USA.

Clerc, M. (Ed.). (2006). *Particle swarm optimization*. International Scientific and Technical Encyclopaedia. doi:10.1002/9780470612163

Clerc, M., & Kennedy, J. (2002). The particle swarm - Explosion, stability and convergence in a multidimensional complex space. *IEEE Transactions on Evolutionary Computation, 6*(1), 58–73. doi:10.1109/4235.985692

Coello Coello, C. A., Pulido, G. T., & Lechuga, M. S. (2004). Handling multiple objectives with particle swarm optimization. *IEEE Transactions on Evolutionary Computation, 8*(3), 256–279. doi:10.1109/TEVC.2004.826067

Colorni, A., Dorigo, M., & Maniezzo, V. (1991). Distributed optimization by ant colonies. *Proceedings of ECAL91, European Conference on Artificial life* (pp. 134-142). Paris, France: Elsevier Publishing.

Colton, D., Haddar, H., & Piana, M. (2003). The linear sampling method in inverse electromagnetic scattering theory. *Inverse Problems, 19*(6), S105–S137. doi:10.1088/0266-5611/19/6/057

Conca, P., Nicosia, G., Stracquadanio, G., & Timmis, J. (2009). *Nominal-yield-area tradeoff in automatic synthesis of analog circuits: A genetic programming approach using immune-inspired operators.* NASA/ESA Conference on Adaptive Hardware and Systems.

Conn, A. R., Coulman, P. K., Haring, R. A., Morrill, G. L., & Visweswariah, C. (1996). *Optimization of custom MOS circuits by transistor sizing.* The International Conference on Computer Aided Design.

Constantinescu, E. M., Zavala, V. M., Rocklin, M., Lee, S., & Anitescu, M. (2011). A computational framework for uncertainty quantification and stochastic optimization in unit commitment with wind power generation. *IEEE Transactions on Power Systems, 26*, 431–441. doi:10.1109/TPWRS.2010.2048133

Cooren, Y., Fakhfakh, M., Loulou, M., & Siarry, P. (2007). Optimizing second generation current conveyors using particle swarm optimization. In *Proceedings of IEEE International Conference on Microelectronics*, December 2007.

Cooren, Y., Clerc, M., & Siarry, P. (2008). Initialization and displacement of the particles in TRIBES, a parameter-free particle swarm optimization algorithm. *Adaptive and Multilevel Metaheuristics Studies in Computational Intelligence, 136*, 199–219. doi:10.1007/978-3-540-79438-7_10

Cooren, Y., Clerc, M., & Siarry, P. (2009). Performance evaluation of TRIBES, an adaptive particle swarm optimization algorithm. *Swarm Intelligence, 3*(2), 149–178. doi:10.1007/s11721-009-0026-8

Cooren, Y., Clerc, M., & Siarry, P. (2011). MO-TRIBES, an adaptive multiobjective particle swarm optimization algorithm. *Computational Optimization and Applications, 49*(2), 379–400. doi:10.1007/s10589-009-9284-z

Correll, N., Cui, Z., Gao, X.-Z., & Gross, R. (Eds.). (2010). Swarm intelligence and swarm robotics - SISR 2010: Special issue on swarm robotics. *Neural Computing & Applications, 19*(6).

Courat, J. P., Raynaud, G., Mrad, I., & Siarry, P. (1994). Electronic component model minimization based on Log simulated annealing. *IEEE Transactions on Circuits and Systems, 41*, 790–795. doi:10.1109/81.340841

Crespi, V., Galstyan, A., & Lerman, K. (2008). Top-down vs bottom-up methodologies in multi-agent system design. *Autonomous Robots, 24*(3), 303–313. doi:10.1007/s10514-007-9080-5

Cui, S., & Weile, D. (2006). Application of a parallel particle swarm optimization scheme to the design of electromagnetic absorbers. *IEEE Transactions on Antennas and Propagation, 54*(3), 1107–1110.

Cui, S., & Weile, D. S. (2005). Application of a parallel particle swarm optimization scheme to the design of electromagnetic absorbers. *IEEE Transactions on Antennas and Propagation, 53*(11), 3616–3624. doi:10.1109/TAP.2005.858866

Cui, Z., Cai, X., Zeng, J., & Sun, G. (2008). Particle swarm optimization with FUSS and RWS for high dimensional functions. *Applied Mathematics and Computation, 205*, 98–108. doi:10.1016/j.amc.2008.05.147

Curtice, W. R. (1980). A MESFET model for use in the design of GaAs integrated circuits. *IEEE Transactions on Microwave Theory and Techniques, 28*(5), 448–455. doi:10.1109/TMTT.1980.1130099

D'Orazio, A., De Sario, M., Mescia, L., Petruzzelli, V., & Prudenzano, F. (2005). Refinement of Er^{3+}-doped hole-assisted optical fibre amplifier. *Optics Express, 13*, 9970–9981. doi:10.1364/OPEX.13.009970

Da Silva, E. L., Ortiz, J. M. A., De Oliveira, G. C., & Binato, S. (2001). Transmission network expansion planning under a Tabu search approach. *IEEE Transactions on Power Systems, 16*(1), 62–68. doi:10.1109/59.910782

Dambrine, G., Cappy, A., Heliodore, F., & Playez, E. (1988). A new method for determining the FET small-signal equivalent circuit. *IEEE Transactions on Microwave Theory and Techniques, 36*(7), 1151–1159. doi:10.1109/22.3650

Danneville, F., Happy, H., Dambrine, G., Belquin, J.-M., & Cappy, A. (1994). Microscopic noise modeling and macroscopic noise models: How good a connection? *IEEE Transactions on Electron Devices, 41*(5), 779–786. doi:10.1109/16.285031

Das, I., & Dennis, J. (1998). Normal-boundary intersection: A new method for generating the Pareto surface in nonlinear multicriteria optimization problems. *SIAM Journal on Optimization, 8,* 631–657. doi:10.1137/S1052623496307510

Datta, K., Datta, R., Dutta, A., & Bhattacharyya, T. K. (2010). PSO optimized concurrent dual-band LNA with RF Switch for better inter-band isolation. In *Proceedings of 5th European Conference on Circuits and Systems for Communications*, November 23-25, 2010, Belgrade, Serbia.

De la Torre, S., Conejo, A. J., & Contreras, J. (2008). Transmission expansion planning in electricity markets. *IEEE Transactions on Power Systems, 23*(1), 238–248. doi:10.1109/TPWRS.2007.913717

de Moura Oliveira, P. B., Solteiro Pires, E. J., Cunha, J. B., & Vrancic, D. (2009, June). *Multi-objective particle swarm optimization design of PID controllers. 10th International Work-Conference on Artificial Neural Networks, IWANN 2009 Workshops*, Salamanca, Spain, (pp. 1222-1230).

De Salvo, C. A., & Smith, H. L. (1965). Automatic transmission planning with ac load flow and incremental transmission loss evaluation. *IEEE Transactions on Power Apparatus and Systems, 84,* 156–163. doi:10.1109/TPAS.1965.4766166

De Sario, M., Mescia, L., Prudenzano, F., Smektala, F., Deseveday, F., & Nazabal, V. (2009). Feasibility of Er^{3+}-doped, $Ga_5Ge_{20}Sb_{10}S_{65}$ chalcogenide microstructured optical fibre amplifiers. *Journal of Optics & Laser Technology, 41,* 99–106. doi:10.1016/j.optlastec.2008.03.007

Dearn, A., & Devlin, L. (1999). A 40-45 GHz monolithic Gilbert cell mixer. In *MM-Wave Circuits and Technology for Commercial Applications* (Ref. No. 1999/007), (pp. 7/1–7/6). IEE Colloquium.

DeBao, C., & ChunXia, Z. (2009). Particle swarm optimization with adaptive population size and its application. *Applied Soft Computing, 9,* 39–48. doi:10.1016/j.asoc.2008.03.001

Deb, K. (2001). *Multi-objective optimization using evolutionary algorithms*. Wiley-Interscience Series in Systems and Optimization.

Deb, K., Pratap, A., Agarwal, S., & Meyarivan, T. (2002). A fast and elitist multiobjective genetic algorithm: NSGA-II. *IEEE Transactions on Evolutionary Computation, 6*(2), 182–197. doi:10.1109/4235.996017

Del Valle, Y. (2008). Particle swarm optimization: Basic concepts, variants and applications in power systems. *IEEE Transactions on Evolutionary Computation, 12*(2). doi:10.1109/TEVC.2007.896686

Deligkaris, K. V., Zaharis, Z. D., Kampitaki, D. G., Goudos, S. K., Rekanos, I. T., & Spasos, M. N. (2009). Thinned planar array design using Boolean PSO with velocity mutation. *IEEE Transactions on Magnetics, 45*(3), 1490–1493. doi:10.1109/TMAG.2009.2012687

Delille, J. P., Slanetz, P. J., Yeh, E. D., Kopans, D. B., & Garrido, E. (2002, May). Breast cancer: Regional blood flow and blood volume measured with magnetic susceptibility-based MR imaging-initial results. *Radiology, 223*(2), 558–565. doi:10.1148/radiol.2232010428

Desurvire, E. (2009). *Erbium-doped fibre amplifiers: principles and applications*. Wiley Interscience.

Desurvire, E., Simpson, J. R., & Becker, P. C. (1987). High-gain erbium-doped traveling-wave fiber amplifier. *Optics Letters, 12*(11), 888–890. doi:10.1364/OL.12.000888

Devireddy, V., & Reed, P. (2004). *Efficient and reliable evolutionary multiobjective optimization using epsilon-dominance archiving and adaptive population sizing*. The Genetic and Evolutionary Computation Conference.

Dias, N. S., Jacinto, L. R., Mendes, P. M., & Correia, J. H. (2009). Feature down selection in brain computer interface. *Proceeding of the 4th International IEEE EMBS Conference on Neural Engineering*, (pp. 323-326).

Dias, M. B., Zlot, R., Kalra, N., & Stentz, A. (2006). Market-based multirobot coordination: A survey and analysis. *Proceedings of the IEEE, 94*(7), 1257–1270. doi:10.1109/JPROC.2006.876939

Dierckx, P. (1996). *Curve and surface fitting splines.* Oxford, UK: Clarendon Press.

Dimopoulos, G. G. (2007). Mixed-variable engineering optimization based on evolutionary and social metaphors. *Computer Methods in Applied Mathematics and Engineering, 196,* 803–817. doi:10.1016/j.cma.2006.06.010

Dinger, R. H. (1998). *Engineering design optimization with genetic algorithm.* The IEEE Northcon Conference.

Dobrea, D. M., & Dobrea, M. C. (2009). Optimization of a BCI system using the GA technique. *2nd International Symposium on Applied Sciences in Biomedical and Communication Technologies,* ISABEL 2009, (pp. 1-6).

Dobrea, D. M., Dobrea, M. C., & Sirbu, A. (2010). GA methods for selecting the proper EEG individual spectral bands limits. *2010 International Workshop on Multimedia Signal Processing,* (pp. 1-5).

Donelli, M., Azaro, A., De Natale, F., & Massa, A. (2006). An innovative computational approach based on a particle swarm strategy for adaptive phased-arrays control. *IEEE Transactions on Antennas and Propagation, 54*(3), 888–898. doi:10.1109/TAP.2006.869912

Donelli, M., Azaro, R., Massa, A., & Raffetto, M. (2006). Unsupervised synthesis of microwave components by means of an evolutionary-based tool exploiting distributed computing resources. *Progress in Electromagnetics Research, 56,* 93–108. doi:10.2528/PIER05010901

Donelli, M., Martini, A., & Massa, A. (2009). A hybrid approach based on PSO and Hadamard difference sets for the synthesis of square thinned arrays. *IEEE Transaction on Antennas and Propagation Letters, 57*(8), 2491–2495. doi:10.1109/TAP.2009.2024570

Donelli, M., & Massa, A. (2005). Computational approach based on a particle swarm optimizer for microwave imaging of two-dimensional dielectric scatterers. *IEEE Transactions on Microwave Theory and Techniques, 53*(5), 1761–1776. doi:10.1109/TMTT.2005.847068

Dong-Xiao, N., Yun-Peng, L., Zhao, Q., & Qing-Ying, Z. (2006). An improved particle swarm optimization method based on borderline search strategy for transmission network expansion planning. *Proceedings of the Fifth International Conference on Machine Learning and Cybernetics,* (pp. 2846-2850). Dalian, 13-16 August.

Dong, Y. (2005). An application of swarm optimization to nonlinear programming. *Computers & Mathematics with Applications (Oxford, England), 49,* 1655–1668. doi:10.1016/j.camwa.2005.02.006

Dorell, D. G., Thomson, W. D., & Roach, S. (1997). Analysis of air gap flux, current, and vibration signals as a function of the combination of static and dynamic air gap eccentricity in 3-phase induction motors. *IEEE Transactions on Industry Applications, 33*(1), 24–34. doi:10.1109/28.567073

Dorica, M., & Giannacopoulos, D. D. (2007). Evolution of wire antennas in three dimensions using a novel growth process. *IEEE Transactions on Magnetics, 43*(4), 1581–1584. doi:10.1109/TMAG.2006.892105

Dorica, M., & Giannacopoulos, D. D. (2007). Evolution of two-dimensional electromagnetic devices using a novel genome structure. *IEEE Transactions on Magnetics, 43*(4), 1585–1588. doi:10.1109/TMAG.2006.892106

Dorigo, M. (1992). *Optimization, learning and natural algorithms.* PhD thesis, Politecnico di Milano.

Dorigo, M., & Stützle, T. (2004, July). *Ant colony optimization.* Cambridge, MA: MIT Press/Bradford Books.

Dorigo, M., & Blum, C. (2005). Ant colony optimization theory: A survey. *Theoretical Computer Science, 344,* 243–278. doi:10.1016/j.tcs.2005.05.020

Dorigo, M., & Colombetti, M. (1998). *Robot shaping: An experiment in behavior engineering.* Cambridge, MA: MIT Press/Bradford Books. doi:10.1177/105971239700500308

Dorigo, M., & Gambardella, L. M. (1997). Ant colony system: A cooperative learning approach to the traveling salesman problem. *IEEE Transactions on Evolutionary Computation, 1*(1), 53–66. doi:10.1109/4235.585892

Dorigo, M., Maniezzo, V., & Colorni, A. (1996). Ant system: optimization by a colony of cooperating agents. *IEEE Transactions on Systems, Man, and Cybernetics. Part B, Cybernetics*, *26*(1), 29–41. doi:10.1109/3477.484436

Dorigo, M., & Sahin, E. (2004). Swarm robotics - Special Issue. *Autonomous Robots*, *17*, 111–113. doi:10.1023/B:AURO.0000034008.48988.2b

Dorigo, M., & Stützle, T. (2004). *Ant colony optimization*. The MIT Press. doi:10.1007/b99492

Dréo, J., & Siarry, P. (2002). A new ant colony algorithm using the heterarchical concept aimed at optimization of multiminima continuous functions. *Proceedings of the Third International Workshop on Ant Algorithms (ANTS'02)* (pp. 216–221). London, UK: Springer-Verlag.

Du, Y., Zhang, Q., & Chen, Q. (2008). ACO-IH: An improved ant colony optimization algorithm for airport ground service scheduling. *Proceedings of the 2008 IEEE International Conference on Industrial Technology (ICIT2008)* (pp. 1–6). Chengdu, China: IEEE. doi:10.1109/ICIT.2008.4608674

Dudek, G., Jenkin, M., & Milios, E. (2002). A taxonomy of multirobot systems. In Balch, T., & Parker, L. (Eds.), *Robot teams: From diversity to polymorphism* (pp. 3–22). Natick, MA: A.K. Peters.

Durbin, F., Haussy, J., Berthiau, G., & Siarry, P. (1992). Circuit performance optimization and model fitting based on simulated annealing. *International Journal of Electronics*, *73*, 1267–1271. doi:10.1080/00207219208925797

Dusonchet, Y. P., & El-Abiad, A. H. (1973). Transmission planning using discrete dynamic optimization. *IEEE Transactions on Power Apparatus and Systems*, *92*, 1358–1371. doi:10.1109/TPAS.1973.293543

Dutta Roy, S. C. (1982). Rational approximation of some irrational functions through a flexible continued fraction expansion. *Proceedings of the IEEE*, *70*(1), 84–85. doi:10.1109/PROC.1982.12233

Eberhart, R. C., & Kennedy, J. (1995). A new optimizer using particle swarm theory. *Proceedings 6th International Symposium on Micro Machine and Human Science*, (pp. 39-43). Nagoya, Japan.

Eberhart, R. C., & Shi, Y. (2000). Comparing inertia weights and constriction factors in particle swarm optimization. *Proceedings of Congress on Evolutionary Computing* (pp. 84–89).

Eberhart, R., & Shi, Y. (2001, May). *Particle swarm optimization: Developments, applications, and resources.* Paper presented at the Congress on Evolutionary Computation, Seoul, Korea.

Eichhorn, M. (2008). Quasi-three-level solid-state lasers in the near and mid infrared based on trivalent rare earth ions. *Applied Physics. B, Lasers and Optics*, *93*, 269–316. doi:10.1007/s00340-008-3214-0

Elshamli, A., Abdullah, H. A., & Areibi, S. (2004). Genetic algorithm for dynamic path planning. *Proceeding of IEEE CCECE 2004*, (pp. 677-680).

Enamorado, J. C., Gómez, T., & Ramos, A. (1999). Multi-area regional interconnection planning under uncertainty. *Proceedings of the 13th Power Systems Computer Conference* Trondheim, 1999, (pp. 599–606).

Engelbrecht, A. P. (2005). *Fundamentals of computational swarm intelligence*. New York, NY: John Wiley & Sons.

Escobar, A. H., Gallego, R. A., & Romero, R. (2004). Multistage and coordinated planning of the expansion of transmission systems. *IEEE Transactions on Power Systems*, *19*(2), 735–744. doi:10.1109/TPWRS.2004.825920

Esmaeili, M. (2007). Classifier fusion for EEG signals processing in human-computer interface systems. *AAAI'07 Proceedings of the 22nd National Conference on Artificial Intelligence*, Vol. 2, (pp. 1856-1857).

Fairley, J., Johnson, A. N., Georgoulas, G., & Vachtsevanos, G. (2010). Automated polysomnogram artifact compensation using the generalized singular value decomposition algorithm. *32nd Annual International Conference on the IEEE EMBS,* (pp. 5097-5100).

Fakhfakh, M., Cooren, Y., Sallem, A., Loulou, M., & Siarry, P. (2009). Analog circuit design optimization through the particle swarm optimization technique. *Analog Integrated Circuits and Signal Processing*, *63*(1), 71–82. doi:10.1007/s10470-009-9361-3

Fakhfakh, M., Loulou, M., & Masmoudi, N. (2009). A novel heuristic for multi-objective optimization of analog circuit performances. *Journal of Analog Integrated Circuits and Signal Processing, 61*(1), 47–64. doi:10.1007/s10470-008-9275-5

Fakhfakh, M., Tlelo-Cuautle, M., & Fernandez, F. V. (2011). *Design of analog circuits through symbolic analysis*. Bentham Scientific Publisher.

Fang, H., Chen, L., & Shen, Z. (2011). Application of an improved PSO algorithm to optimal tuning of PID gains for water turbine governor. *Energy Conversion and Management, 52*, 1763–1770. doi:10.1016/j.enconman.2010.11.005

Farinelli, A., Iocchi, L., Nardi, D., & Ziparo, V. A. (2006). Assignment of dynamically perceived tasks by token passing in multirobot systems. *Proceedings of the IEEE, 94*(7), 1271–1288. doi:10.1109/JPROC.2006.876937

Fernandez, F. V., & Fakhfakh, M. (2009). *Applications of evolutionary computation techniques to analog, mixed-signal and RF circuit design*. The IEEE International Conference on Electronics, Circuits, and Systems.

Fimognari, L., Donelli, M., Massa, A., & Azaro, R. (2007). A planar electronically reconfigurable wi-fi band antenna based on a parasitic microstrip structure. *IEEE Antennas and Wireless Propagation Letters, 6*, 623–626. doi:10.1109/LAWP.2007.913274

Fornarelli, G., Mescia, L., Prudenzano, F., De Sario, M., & Vacca, F. (2009). A neural network model of erbium-doped photonic crystal fibre amplifiers. *Journal of Optics & Laser Technology, 41*, 580–585. doi:10.1016/j.optlastec.2008.10.010

Franchi, A., Freda, L., Oriolo, G., & Vendittelli, M. (2009). The sensor-based random graph method for cooperative robot exploration. *IEEE/ASME Transactions on Mechatronics, 14*(2), 163–175. doi:10.1109/TMECH.2009.2013617

Gallego, L. A., Rider, M. J., Romero, R., & Garcia, A. V. (2009). *A specialized genetic algorithm to solve the short term transmission expansion planning*. IEEE Bucharest Power Tech Conference, June 28th – July 2nd 2009, Bucharest, Romania.

Gamagami, P., Silverstein, M. J., & Waisman, J. R. (1997, November). *Infra-red imaging in breast cancer*. Paper presented at the proceedings of the 19th Annual International Conference of the IEEE Engineering in Medicine and Biology Society, Chicago, USA.

Gandhi, K. R., Karnan, M., & Kannan, S. (2010, February). *Classification rule construction using particle swarm optimization algorithm for breast cancer data sets*. Paper presented at the International Conference on Signal Acquisition and Processing, Singapore, Singapore.

Gange, C., Sebag, M., Schoenauer, M., & Tomassini, M. (2009). *Generic methods for evolutionary ensemble learning*. France: Université Paris Sud, Orsay CEDEX.

Gao, H., & Xu, W. (2011, October). A new particle swarm algorithm and its globally convergent modifications. *IEEE Transactions on Systems, Man, and Cybernetics. Part B, Cybernetics, 41*(5), 1334–1351. doi:10.1109/TSMCB.2011.2144582

Gao, P. C., Tao, Y. B., Bai, Z. H., & Lin, H. (2012). Mapping the SBR and TS-ILDCs to heterogeneous CPU-GPU architecture for fast computation of electromagnetic scattering. *Progress in Electromagnetics Research, 122*, 137–154. doi:10.2528/PIER11092303

Garnero, L., Franchois, A., Hugonin, J.-P., Pichot, C., & Joachimowicz, N. (1991). Microwave imaging-complex permittivity reconstruction-by simulated annealing. *IEEE Transactions on Microwave Theory and Techniques, 39*(11), 1801–1807. doi:10.1109/22.97480

Gazi, P., Jevtić, A., Andina, D., & Jamshidi, M. (2010). A mechatronic system design case study: Control of a robotic swarm using networked control algorithms. In *Proceedings of 4th Annual IEEE Systems Conference (SysCon 2010)*, (pp. 169-173).

Genovesi, S., Monorchio, A., Mittra, R., & Manara, G. (2007). A subboundary approach for enhanced particle swarm optimization and its application to the design of artificial magnetic conductors. *IEEE Transactions on Antennas and Propagation, 55*(3), 766–770. doi:10.1109/TAP.2007.891559

Genovesi, S., Salerno, E., Monorchio, A., & Manara, G. (2009). Permittivity range profile reconstruction of multilayered structures from microwave backscattering data by using particle swarm optimization. *Microwave and Optical Technology Letters*, *51*(10), 2390–2394. doi:10.1002/mop.24642

Georgilakis, P. S., Korres, G. N., & Hatziargyriou, N. D. (2011). Transmission expansion planning by enhanced differential evolution. *16ᵗʰ International Conference on Intelligent System Application to Power Systems (ISAP)*, 25 – 28 Sept.

Gerkey, B. P., & Matarić, M. J. (2004). A formal analysis and taxonomy of task allocation in multi-robot systems. *The International Journal of Robotics Research*, *23*(9), 939–954. doi:10.1177/0278364904045564

Ghanbari, A. A., Kousarrizi, M. R. N., Teshnelab, M., & Aliyari, M. (2009). An evolutionary artifact rejection method for brain computer interface using ICA. *International Journal of Electrical & Computer Sciences*, *9*(9), 461–466.

Giakos, G. C., Pastorino, M., Russo, F., Chowdhury, S., Shah, N., & Davros, W. (1999). Noninvasive imaging for the new century. *IEEE Instrumentation & Measurement Magazine*, *2*(2), 32–35. doi:10.1109/5289.765967

Gibson, W. C. (2008). *The method of moments in electromagnetics*. Boca Raton, FL: CRC Press.

Gies, D., & Rahmat-Samii, Y. (2003, August). Particle swarm optimization for reconfigurable phase-differentiated array design. *Microwave and Optical Technology Letters*, *38*(3), 172–175. doi:10.1002/mop.11005

Gil, E., Bustos, J., & Rudnick, H. (1983). Short-term hydrothermal generation scheduling model using a genetic algorithm. *IEEE Transactions on Power Systems*, *18*(4), 1256–1264. doi:10.1109/TPWRS.2003.819877

Giles, C. R., & Desurvire, E. (1991). Modeling erbium-doped fiber amplifiers. *Journal of Lightwave Technology*, *9*(2), 271–283. doi:10.1109/50.65886

Girard, S., Ouerdane, Y., Tortech, B., Marcandella, C., Robin, T., & Cadier, B. (2009). Radiation effects on Ytterbium and Ytterbium/Erbium-doped double-clad optical fibres. *IEEE Transactions on Nuclear Science*, *56*, 3293–3299. doi:10.1109/TNS.2009.2033999

Golberg, D. E. (1989). *Genetic algorithms in search, optimization and machine learning*. IN, Lebanon: Addison Wesley USA.

Golnabi, A. H., Meaney, P. M., Epstein, N. R., & Paulsen, K. D. (2011, August). *Microwave imaging for breast cancer detection: Advances in three- dimensional image reconstruction.* Paper presented at the Annual International Conference of the IEEE on Engineering in Medicine and Biology (EMBC), Boston, MA.

Gosselin, C., & Angeles, J. (1991). A global performance index for the kinematic optimization of robotic manipulators. *ASME Journal of Mechanical Design*, *13*, 220–226. doi:10.1115/1.2912772

Goudos, S. K., Moysiadou, V., Samaras, T., Siakavara, K., & Sahalos, J. N. (2010). Application of a comprehensive learning particle swarm optimizer to unequally spaced linear array synthesis with sidelobe level suppression and null control. *IEEE Antennas and Wireless Propagation Letters*, *9*, 125–129. doi:10.1109/LAWP.2010.2044552

Goudos, S. K., Rekanos, I. T., & Sahalos, J. N. (2008). EMI reduction and ICs optimal arrangement inside high-speed networking equipment using particle swarm optimization. *IEEE Transactions on Electromagnetic Compatibility*, *50*(3), 586–596. doi:10.1109/TEMC.2008.924389

Goudos, S. K., & Sahalos, J. N. (2006). Microwave absorber optimal design using multi-objective particle swarm optimization. *Microwave and Optical Technology Letters*, *48*(8), 1553–1558. doi:10.1002/mop.21727

Goudos, S. K., Zaharis, Z. D., Kampitaki, D. G., Rekanos, I. T., & Hilas, C. S. (2009). Pareto optimal design of dual-band base station antenna arrays using multi-objective particle swarm optimization with fitness sharing. *IEEE Transactions on Magnetics*, *45*(3), 1522–1525. doi:10.1109/TMAG.2009.2012695

Graeb, H., Zizala, S., Eckmueller, J., & Antreich, K. (2001). *The sizing rules method for analog integrated circuit design.* The IEEE/ACM International Conference on Computer-Aided Design.

Gragnani, G. L. (2002). *Two-dimensional imaging of dielectric scatterers based on Markov random field models: A short review. Microwave Nondestructive Evaluation and Imaging* (pp. 121–145). Trivandrum, India: Research Signpost.

Grimaccia, F., Mussetta, M., & Zich, R. E. (2007). Genetical swarm optimization: self-adaptive hybrid evolutionary algorithm for electromagnetics. *IEEE Transactions on Antennas and Propagation, 55*(3), 781–785. doi:10.1109/TAP.2007.891561

Grimbleby, J. B. (2000). Automatic analogue circuit synthesis using genetic algorithms. *IEE Proceedings. Circuits, Devices and Systems, 147*(6), 319–323. doi:10.1049/ip-cds:20000770

Groß, R., Nouyan, S., Bonani, M., Mondada, F., & Dorigo, M. (2008). Division of labour in self-organised groups. In *Proceedings of 10th international conference on Simulation of Adaptive Behavior: From Animals to Animats, SAB '08*, (pp. 426–436). Berlin, Germany: Springer-Verlag.

Groß, R., Bonani, M., Mondada, F., & Dorigo, M. (2006). Autonomous self-assembly in swarm-bots. *IEEE Transactions on Robotics, 22*(6), 1115–1130. doi:10.1109/TRO.2006.882919

Guerra-Gomez, I., Tlelo-Cuautle, E., McConaghy, T., & Gielen, G. (2009). Optimizing current conveyors by evolutionary algorithms including differential evolution. *The IEEE International Conference on Electronics, Circuits, and Systems.*

Guozheng, Y., Banghua, Y., Shuo, C., & Rongguo, Y. (2006). Pattern recognition using hybrid optimization for a robot controlled by human thoughts. *Proceeding of the 19th IEEE Symposium on Computer-Based Medical Systems* (CBMS'06), (pp. 1-5).

Gutiérrez, A., Campo, A., Dorigo, M., Donate, J., Monasterio-Huelin, F., & Magdalena, L. (2009). Open e-puck range and bearing miniaturized board for local communication in swarm robotics. In *Proceedings of the IEEE International Conference on Robotics and Automation*, (pp. 3111–3116).

Gutiérrez, A., Campo, A., Dorigo, M., Amor, D., Magdalena, L., & Monasterio-Huelin, F. (2008). An Open localization and local communication embodied sensor. *Sensors (Basel, Switzerland), 8*(8), 7545–7563. doi:10.3390/s8117545

Gutiérrez, A., Campo, A., Monasterio-Huelin, F., Magdalena, L., & Dorigo, M. (2010). Collective decision-making based on social odometry. *Neural Computing & Applications, 19*(6), 807–823. doi:10.1007/s00521-010-0380-x

Hall, P., & Hao, Y. (2006). *Antennas and propagation for body-centric wireless communications*. Norwood, MA: Artec House, INC.

Han, D., & Chatterjee, A. (2004). Simulation-in-the-loop analog circuit sizing method using adaptive model-based simulated annealing. *The IEEE International Workshop on System-on-Chip for Real-Time Applications.*

Han, K. M. (2007). *Collision free path planning algorithms for robot navigation problem*. M.Sc Thesis, Department of Electrical and Computer Engineering, Missouri-Columbia University.

Hao, M., Yantao, T., & Linan, Z. (2006). A hybrid ant colony optimization algorithm for path planning of robot in dynamic environment. *International Journal of Information Technology, 12*(3).

Harikumar, R., Raghavan, S., & Sukanesh, R. (2005). Genetic algorithm for classification of epilepsy risk levels from EEG signals. *TENCON 2005 2005 IEEE Region 10*, (pp. 1-6).

Harikumar, R., Sukanesh, R., & Bharathi, P. A. (2004). Genetic algorithm optimization of fuzzy outputs for classification of epilepsy risk levels from EEG signals. *Conference Record of the Thirty-Eighth Asilomar Conference on Systems and Computers*, Vol. 2, (pp. 1585-1589).

Harrington, R. (1993). *Field computation by moment methods*. Piscataway, NJ: IEEE Press. doi:10.1109/9780470544631

Hartley, T. T., Lorenzo, C. F., & Qammar, H. K. (1995). Chaos in fractional order Chua's system. *IEEE Transactions on Circuits & Systems – Part I, 42*(8), 485–490. doi:10.1109/81.404062

Hasan, B. A. S., & Gan, J. Q. (2009). *Multi-objective particle swarm optimization for channel selection in brain computer interface*. The UK Workshop on Computational Intelligence (UKCI2009), Nottingham, UK.

Hasan, B. A. S., Gan, J. Q., & Zhang, Q. (2010). Multi-objective evolutionary methods for channel selection in brain computer interface: Some preliminary experimental results. *2010 IEEE Congress on Evolutionary Computation* (CEC), (pp. 1-6).

Hasegawa, A. (1983). Amplification and reshaping of optical solitons in a glass fiber-IV: Use of the stimulated Raman process. *Optics Letters*, *8*(12), 650–652. doi:10.1364/OL.8.000650

Hassan, A., & El-Shenawee, M. (2011, September). Review of electromagnetic techniques for breast cancer detection. *IEEE Reviews in Biomedical Engineering*, *4*, 103–114. doi:10.1109/RBME.2011.2169780

Haupt, R. L. (1995). An introduction to genetic algorithms for electromagnetics. *IEEE Antennas and Propagation Magazine*, *37*(2), 7–15. doi:10.1109/74.382334

He, D., Wang, Z., Yang, B., & Zhou, C. (2009). Genetic algorithm with ensemble learning for detecting community structure in complex networks. *Fourth International Conference on Computer Sciences and Convergence Information Technology*, (pp. 702-707).

Headley, C., & Agrawal, G. P. (2005). *Raman amplification in fiber optical communication systems* (pp. 33–163). Burlington, MA: Elsevier.

Hedar, A.-R., & Fukushima, M. (2006). Tabu search directed by direct search methods for nonlinear global optimization. *European Journal of Operational Research*, *170*(2), 329–349. doi:10.1016/j.ejor.2004.05.033

Hema, C. R., Paulraj, M. P., Yaacob, S., Adom, A. H., & Nagarajan, R. (2008). Functional link PSO neural network based classification of EEG mental task signals. *International Symposium on Information Technology, ITSim 2008*, (pp. 1-6).

Hema, C. R., Paulraj, M. P., Nagarajan, R., Yaacob, S., & Adom, A. H. (2008). Application of particle swarm optimization for EEG signal classification. *Biomedical Soft Computing and Human Sciences*, *13*(1), 79–84.

He, N., Xu, D., & Huang, L. (2009, August). The application of particle swarm optimization to passive and hybrid active power filter design. *IEEE Transactions on Industrial Electronics*, *56*(8), 2841–2851. doi:10.1109/TIE.2009.2020739

He, Q., & Wang, L. (2007). An effective co-evolutionary particle swarm optimization for constrained engineering design problems. *Engineering Applications of Artificial Intelligence*, *20*, 89–99. doi:10.1016/j.engappai.2006.03.003

He, Q., Yan, R., Kong, F., & Du, R. (2009). Machine condition monitoring using principal component representations. *Mechanical Systems and Signal Processing*, *23*, 446–466. doi:10.1016/j.ymssp.2008.03.010

Hinterberger, T., Weiskopf, N., Veit, R., Wilhelm, B., Betta, E., & Birbaumer, N. (2004). An EEG - driven brain - computer interface combined with functional magnetic resonance imaging (fMRI). *IEEE Transactions on Bio-Medical Engineering*, *51*(6), 971–974. doi:10.1109/TBME.2004.827069

Holland, J. H. (1975). *Adaptation in natural and artificial systems*. Ann Arbor, MI: University of Michigan Press.

Hoorfar, A. (2007, March). Evolutionary programming in electromagnetic optimization: A review. *IEEE Transactions on Antennas and Propagation*, *55*(3), 523–537. doi:10.1109/TAP.2007.891306

Ho, S. Y., Lin, H. S., Liauh, W. H., & Ho, S. J. (2008, March). OPSO: Orthogonal particle swarm optimization and its application to task assignment problems. *IEEE Transactions on Systems, Man, and Cybernetics. Part A, Systems and Humans*, *38*(2), 288–298. doi:10.1109/TSMCA.2007.914796

Ho, S., Yang, S., Ni, G., & Wong, H. (2006). A particle swarm optimization method with enhanced global search ability for design optimizations of electromagnetic devices. *IEEE Transactions on Magnetics*, *42*(4), 1107–1110. doi:10.1109/TMAG.2006.871426

Hosseini, H. S. (2007). *Problem solving by intelligent water drops*. IEEE Congress on Evolutionary Computation, CEC. Singapore.

Hota, P. K., Barisal, A. K., & Chakrabarti, R. (2009). An improved PSO technique for short-term optimal hydrothermal scheduling. *Electric Power Systems Research*, *79*(7), 1047–1053. doi:10.1016/j.epsr.2009.01.001

Hota, P. K., Chakrabarti, R., & Chattopadhyay, P. K. (1999). Short-term hydrothermal scheduling through evolutionary programming technique. *Electric Power Systems Research*, *52*, 189–196. doi:10.1016/S0378-7796(99)00021-8

Huang, D., Qian, K., Oxenham, S., Fei, D. Y., & Bai, O. (2011). Event-related desynchronization/Synchronization-based brain-computer interface towards volitional cursor control in a 2D center-out paradigm. *IEEE Symposium on Computational Intelligence, Cognitive Algorithms, Mind, and Brain* (CCMB), (pp. 1-8).

Huang, R., Zhu, J., & Xu, G. (2007). An optimization coverage mechanism in sensor networks based on ACO. *Proceedings of the 2007 International Conference on Microwave and Millimeter Wave Technology (ICMMT'07)* (pp. 1–4). Gulin, China: IEEE. doi:10.1109/ICMMT.2007.381508

Huang, C.-H., Chen, C.-H., Chiu, C.-C., & Li, C. L. (2010). Reconstruction of the buried homogenous dielectric cylinder by FDTD and asynchronous particle swarm optimization. *Applied Computational Electromagnetics Society Journal*, *25*(8), 672–681.

Huang, H., Wu, C.-G., & Hao, Z.-F. (2009). A pheromone-rate-based analysis on the convergence time of ACO algorithm. *IEEE Transactions on Systems, Man, and Cybernetics. Part B, Cybernetics*, *39*(4), 910–923. doi:10.1109/TSMCB.2009.2012867

Huang, T., & Mohan, A. S. (2007). A microparticle swarm optimizer for the reconstruction of microwave images. *IEEE Transactions on Antennas and Propagation*, *55*(3), 568–576. doi:10.1109/TAP.2007.891545

Hung, C. M., Chen, N. K., Lai, Y., & Chi, S. (2007). Double-pass high-gain low-noise EDFA over S- and C+L-bands by tunable fundamental-mode leakage loss. *Optics Express*, *15*(4), 1454–1460. doi:10.1364/OE.15.001454

Hwang, K. C. (2009, September). Design and optimization of a broadband waveguide magic-T using a stepped conducting cone. *IEEE Microwave and Wireless Components Letters*, *19*(9), 539–541. doi:10.1109/LMWC.2009.2027052

Ingber, L. (1996). Adaptive simulated annealing (ASA): Lessons learned. *Control and Cybernetics*, *25*(1), 33–54.

Inostroza, J. C., & Hinojosa, V. H. (2011). Short-term scheduling solved with a particle swarm optimizer. *IET Generation, Transmission, and Distribution*, *5*(11), 1091–1104. doi:10.1049/iet-gtd.2011.0117

Intelligence, S. (n.d.). *Codes*. Retrieved from http://www.swarmintelligence.org/codes.php

Islam, M. N. (2004). *Raman amplifiers for telecommunications I and II*. New York, NY: Springer.

Ismail, T. H., & Hamici, Z. M. (2010). Array pattern synthesis using digital phase control by quantized particle swarm optimization. *IEEE Transactions on Antennas and Propagation*, *58*(6), 2142–2145. doi:10.1109/TAP.2010.2046853

Ito, S. I., Mitsukura, Y., Fukumi, M., & Cao, J. (2007). Detecting method of music to match the user's mood in prefrontal cortex EEG activity using GA. *International Conference on Control, Automation and Systems*, (pp. 2142-2145).

Jaganathan, P., Thangavel K., Pethalakshmi, A., & Karnan, M. (2007). Classification rule discovery with ant colony optimization and improved quick reduct algorithm. *IAENG International Journal of Computer Science*, *33*(1).

Jamshidi, M. (Ed.). (2009). *System of systems engineering - Innovations for the 21st century*. New York, NY: John Wiley & Sons.

Jamwal, P. K., Xie, S., & Aw, K. C. (2009). Kinematic design optimization of a parallel ankle rehabilitation robot using modified genetic algorithm. *Robotics and Autonomous Systems*, *57*(10), 11018–11027. doi:10.1016/j.robot.2009.07.017

Janson, S., & Middendorf, M. (2005). A hierarchical particle swarm optimizer and its adaptive variant. *IEEE Transactions in Systems, Man, and Cybernetics - Part B*, *35*, 1272–1282. doi:10.1109/TSMCB.2005.850530

Jespers, P. G. (Ed.). (2009). *The gm/ID Methodology, a sizing tool for low-voltage analog CMOS Circuits: The semi-empirical and compact model approaches.* Springer.

Jevtić, A., Gazi, P., Andina, D., & Jamshidi, M. (2010). Building a swarm of robotic bees. *World Automation Congress (WAC 2010),* (pp. 1–6).

Jevtić, A., & Gutiérrez, Á. (2011). Distributed bees algorithm parameters optimization for a cost efficient target allocation in swarms of robots. *Sensors (Basel, Switzerland), 11*(11), 10880–10893. doi:10.3390/s111110880

Jevtić, A., Gutiérrez, A., Andina, D., & Jamshidi, M. (2011). Distributed bees algorithm for task allocation in swarm of robots. *IEEE Systems Journal, 5*(3), 1–9.

Ji, J., Zhang, N., & Liu, C. (2006). An ant colony optimization algorithm for learning classification rules. *WI '06 Proceedings of the 2006 IEEE/WIC/ACM International Conference on Web Intelligence,* (pp. 1034 -1037).

Jiang, M., Luo, Y. P., & Yang, S. Y. (2007). Stochastic convergence analysis and parameter selection of standard particle swarm optimization algorithm. *Information Processing Letters, 102,* 8–16. doi:10.1016/j.ipl.2006.10.005

Jiang, Y., Hu, T., Huang, C. C., & Wu, X. (2007). An improved particle swarm optimization algorithm. *Applied Mathematics and Computation, 193,* 231–239. doi:10.1016/j.amc.2007.03.047

Jiang, Y., Liu, C., Huang, C., & Wu, X. (2010). Improved particle swarm algorithm for hydrological parameter optimization. *Applied Mathematics and Computation, 217,* 3207–3215. doi:10.1016/j.amc.2010.08.053

Jiao, Y., Wu, X., & Guo, X. (2010). Motor imagery classification based on the optimized SVM and BPNN by GA. *International Conference on Intelligent Control and Information Processing,* (pp. 344-347).

Jie, J., Zeng, J., Han, C., & Wang, Q. (2008). Knowledge-based cooperative particle swarm optimization. *Applied Mathematics and Computation, 205,* 861–873. doi:10.1016/j.amc.2008.05.100

Jin, J., Wang, X., & Zhang, J. (2008). *Optimal selection of EEG electrodes via DPSO algorithm. Proceeding of the 7th World Congress on Intelligent Control and Automation,* (pp. 5095-5099).

Jin, N., & Rahmat-Samii, Y. (2008, July). *Particle swarm optimization for multi-band handset antenna designs: A hybrid real-binary implementation.* Paper presented at IEEE International Symposium of Antennas and Propagation, Taipei, Taiwan.

Jin, N., & Rahmat-Samii, Y. (2005). Parallel particle swarm optimization and finite-difference time-domain (PSO/FDTD) algorithm for multiband and wide-band patch antenna designs. *IEEE Transactions on Antennas and Propagation, 53*(11), 3459–3468. doi:10.1109/TAP.2005.858842

Jin, N., & Rahmat-Samii, Y. (2005, November). Parallel particle swarm optimization and finite difference time-domain (PSO/FDTD) algorithm for multiband and wide-band patch antenna designs. *IEEE Transactions on Antennas and Propagation, 53*(11), 3459–3468. doi:10.1109/TAP.2005.858842

Jin, N., & Rahmat-Samii, Y. (2007). Advances in particle swarm optimization for antenna designs: real-number, binary, single-objective and multi-objective implementations. *IEEE Transactions on Antennas and Propagation, 55*(3), 556–567. doi:10.1109/TAP.2007.891552

Jin, N., & Rahmat-Samii, Y. (2007, March). Advances in particle swarm optimization for antenna designs: real-number, binary, single-objective and multiobjective implementations. *IEEE Transactions on Antennas and Propagation, 55*(3), 556–567. doi:10.1109/TAP.2007.891552

Jones, G. C., & Houde-Walter, S. N. (2006). Determination of the macroscopic upconversion parameter in Er^{3+}-doped transparent glass ceramics. *Journal of the Optical Society of America. B, Optical Physics, 23,* 1600–1608. doi:10.1364/JOSAB.23.001600

Joordens, M. A., & Jamshidi, M. (2010). Consensus control for a system of underwater swarm robots. *IEEE Systems Journal, 4*(1), 65–73. doi:10.1109/JSYST.2010.2040225

Junkin, G. (2011). Conformal FDTD modeling of imperfect conductors at millimeter wave bands. *IEEE Transactions on Antennas and Propagation, 59*(1), 199–205. doi:10.1109/TAP.2010.2090490

Kankar, P. K., Sharma, S. C., & Harsha, S. P. (2011). Fault diagnosis of ball bearings using machine learning methods. *Expert Systems with Applications, 33,* 1876–1886. doi:10.1016/j.eswa.2010.07.119

Kanović, Ž., Rapaić, M., & Erdeljan, A. (2008). Generalized PSO algorithm in optimization of water distribution. *Proceedings of Planning and Management of Water Resources Systems*, (pp. 203-210). Novi Sad.

Kanović, Ž., Rapaić, M., & Jeličić, Z. (2011). Generalized particle swarm optimization algorithm - Theoretical and empirical analysis with application in fault detection. *Applied Mathematics and Computation*, *217*, 10175–10186. doi:10.1016/j.amc.2011.05.013

Kao, I. W., Tsai, C. Y., & Wang, Y. C. (2007). An effective particle swarm optimization method for data clustering. *Proceedings of the 2007 IEEE Int. Conf. on Industrial Engineering and Engineering Management (IEEM)* (pp. 548–552).

Karaboga, D., & Basturk, B. (2008). On the performance of artificial bee colony (ABC) algorithm. *Applied Soft Computing*, *8*(1), 687–697. doi:10.1016/j.asoc.2007.05.007

Kaveh, A., & Talatahari, S. (2010). A novel heuristic optimization method: charged system search. *Acta Mechanica*, *213*(3-4), 267–289. doi:10.1007/s00707-009-0270-4

Kavitha, D., & Swarup, K. S. (2006). *Transmission expansion planning using LP-based particle swarm optimization*. Power Indian Conference, IEEE, India.

Kayali, S., Ponchak, G., & Shaw, R. (1996). *GaAs MMIC reliability assurance guideline for space applications*. Technical report, NASA, Jet Propulsion Laboratory. Retrieved from http://parts.jpl.nasa.gov/mmic/contents.htm

Kennedy, J., & Eberhart, R. (1995). Particle swarm optimization. *Proceedings of IEEE International Conference on Neural Networks* (pp. 1942-1948).

Kennedy, J., & Eberhart, R. (1997, October 12-15). *A discrete binary version of the particle swarm algorithm*. Paper presented at the 1997 IEEE International Conference on Systems, Man and Cybernetics, Orlando, USA.

Kennedy, J., & Eberhart, R. C. (2001). *Swarm intelligence*. San Francisco, CA: Morgan Kaufmann Publishers.

Khan, U. A., Al-Moayed, N., Nguyen, N., Korolev, K. A., Afsar, M. N., & Naber, S. P. (2007, December). Broadband dielectric characterization of tumorous and nontumorous breast tissues. *IEEE Transactions on Microwave Theory and Techniques*, *55*(12), 2887–2893. doi:10.1109/TMTT.2007.909621

Kharkovsky, S., & Zoughi, R. (2007). Microwave and millimeter wave nondestructive testing and evaluation - Overview and recent advances. *IEEE Instrumentation & Measurement Magazine*, *10*(2), 26–38. doi:10.1109/MIM.2007.364985

Khodier, M. M., & Christodoulou, C. G. (2005). Linear array geometry synthesis with minimum sidelobe level and null control using particle swarm optimization. *IEEE Transactions on Antennas and Propagation*, *53*(8), 2674–2679. doi:10.1109/TAP.2005.851762

Khovanskii, A. N. (1965). Continued fractions. In L. A. Lyusternik & A. R. Yanpol'skii (Eds.), *Mathematical analysis: Functions, limits, series, continued fractions* (D. E. Brown, Trans.). International Series Monographs in Pure and Applied Mathematics. Oxford, UK: Pergamon Press.

Khovanskii, A. N. (1963). *The application of continued fractions and their generalizations to problems in approximation theory*. Groningen, The Netherlands: P. Noordhoff N. V.

Khushaba, N. R., Al-Ani, A., Al-Jumaily, A., & Nguyen, H. T. (2008). A hybrid nonlinear-discriminant analysis feature projection technique. In Wobcke, W., & Zhang, M. (Eds.), *AI 2008: Advances in Artificial Intelligence, LNAI 5360* (pp. 544–550). Lecture Notes in Computer ScienceBerlin, Germany: Springer. doi:10.1007/978-3-540-89378-3_55

Khushaba, R. N., Al-Ani, A., AlSukker, A., & Al-Jumaily, A. (2008). Lecture Notes in Computer Science: *Vol. 5217. A combined ant colony and differential evolution feature selection algorithm. Ant Colony Optimization and Swarm Intelligence* (pp. 1–12).

Kim, H., Bae, J., & Chun, J. (2009). Synthesis method based on genetic algorithm for designing EDFA gain flattening LPFGs having phase-shifted effect. *Optical Fiber Technology*, *15*, 320–323. doi:10.1016/j.yofte.2009.02.001

Kim, K. J., & Cho, S. B. (2008). Evolutionary ensemble of diverse artificial neural networks using speciation. *Neurocomputing*, *71*, 1604–1618. doi:10.1016/j.neucom.2007.04.008

Kim, Y. S., Street, W. N., & Menczer, F. (2005). Optimal ensemble construction via meta-evolutionary ensembles. *Expert Systems with Applications: An International Journal, 30*(4), 705–714. doi:10.1016/j.eswa.2005.07.030

Kirkpatrick, S., Gellat, C. D., & Vecchi, M. P. (1983). Optimization by simulated annealing. *Science, 220*(4598), 671–680. doi:10.1126/science.220.4598.671

Knowles, J., & Corne, D. (2003). Properties of an adaptive archiving algorithm for storing nondominated vectors. *IEEE Transactions on Evolutionary Computation, 7*(2), 100–116. doi:10.1109/TEVC.2003.810755

Ko, C. N., Chang, Y. P., & Wu, C. J. (2009, February). A PSO method with nonlinear time-varying evolution for optimal design of harmonic filters. *IEEE Transactions on Power Systems, 24*(1), 437–444. doi:10.1109/TPWRS.2008.2004845

Kolodziej, M., Majkowski, A., & Rak, R. J. (2011). A new method of EEG classification for BCI with feature extraction based on higher order statistics of wavelet componenets and selection with genetic algorithms. In A. Dobnikar, U. Lotic, & B. Ster (Eds.), *International Conference on Neural Networks and Genetic Algorithms* (pp. 280-289). Berlin, Germany: Springer-Verlag.

Konar, A. (2005). *Computational intelligence principles, techniques and applications.* Springer Verlag.

Kumar, K. S., Bhaskar, U. P., Chattopadhyay, S., & Mandal, P. (2009). Circuit partitioning using particle swarm optimization for pseudo-exhaustive testing. In *Proceedings of International Conference on Advances in Recent Technologies in Communication and Computing 2009*, ARTCom '09, (pp. 346-350).

Kyziol, P., & Rutkowski, J. (2010). Searching groups and layouts in n-terminal based test method using heuristic PSO algorithm. In *Proceedings of 2010 International Conference on Signals and Electronic Systems (ICSES)*, (pp. 217-220).

Kyziol, P., Rutkowski, J., & Grzechca, D. (2010). Testing analog electronic circuits using N-terminal network. In *Proceedings of 2010 IEEE 13th International Symposium on Design and Diagnostics of Electronic Circuits and Systems (DDECS).*

Laakso, T. I., Välimäki, V., Karjalainen, M., & Laine, U. K. (1996). Splitting the unit delay: Tool for fractional delay filter design. *IEEE Signal Processing Magazine*, 30–60. doi:10.1109/79.482137

Labella, T. H., Dorigo, M., & Deneubourg, J.-L. (2006). Division of labor in a group of robots inspired by ants' foraging behavior. *ACM Transactions on Autonomous and Adaptive Systems, 1*(1), 4–25. doi:10.1145/1152934.1152936

Lakshminarasimman, L., & Subramanian, S. (2006). Short-term scheduling of hydrothermal power system with cascaded reservoirs by using modified differential evolution. *IEE Proceedings. Generation, Transmission and Distribution, 153*(6), 693–700. doi:10.1049/ip-gtd:20050407

Largo, R., Munteanu, C., & Rosa, A. (2005). CAP event detection by wavelets and GA tuning. *WISP, 2005*, 44–48.

Latombe, J. C. (1991). *Robot motion planning.* London, UK: Kluwer International Series in Engineering and Computer Science. doi:10.1007/978-1-4615-4022-9

Latorre, G., Cruz, R., Areiza, J., & Villegas, A. (2003). Classification of publications and models on transmission expansion planning. *IEEE Transactions on Power Systems, 18*(2). doi:10.1109/TPWRS.2003.811168

Lazebnik, M., Changfang, Z., Palmer, G. M., Harter, J., Sewall, S., Ramanujam, N., & Hagness, S. C. (2008, October). Electromagnetic spectroscopy of normal breast tissue specimens obtained from reduction surgeries: Comparison of optical and microwave properties. *IEEE Transactions on Bio-Medical Engineering, 55*(10), 2444–2451. doi:10.1109/TBME.2008.925700

Lazebnik, M., Popovic, D., McCartney, L., Watkins, C. B., Lindstorm, M. J., & Harter, J. (2007, April). A large-scale study of the ultrawideband microwave dielectric properties of normal, benign and malignant breast tissues obtained from cancer surgeries. *Physics in Medicine and Biology Journal, 52*(20), 6093–6115. doi:10.1088/0031-9155/52/20/002

Lbanez, M. L., Blum, C., Thiruvady, D., Ernst, A. T., & Meyer, B. (2009). Beam-ACO based on stochastic sampling for makespan optimization concerning the TSP with time windows. *Evolutionary Computation in Combinatorial Optimization, Lecture Notes in Computer Science, 2009, 5482/2009,* (pp. 97-108).

Lee, C. H., Shin, S. W., & Chung, J. W. (2006). Network intrusion detection through genetic feature selection. *7th ACIS International Conference on Software Engineering, Artificial Intelligence, Networking and Parallel/Distributed Computing* (pp. 109–114).

Lee, K. H., Ahmed, I., Goh, R. S. M., Khoo, E. H., Li, E. P., & Hung, T. G. G. (2011). Implementation of the FDTD method based on Lorentz-Drude dispersive model on GPU for plasmonics applications. *Progress in Electromagnetics Research, 116,* 441–456.

Lee, K. Y., & El-Sharkawi, M. A. (2008). *Modern heuristic optimization techniques.* IEEE Press. doi:10.1002/9780470225868

Lee, T. H. (2004). *Planar microwave engineering* (2nd ed.). Cambridge, UK: Cambridge University Press.

Lerman, K., Jones, C., Galstyan, A., & Matarić, M. J. (2006). Analysis of dynamic task allocation in multi-robot systems. *The International Journal of Robotics Research, 25*(3), 225–241. doi:10.1177/0278364906063426

Lesselier, D., & Razek, A. (Eds.). (1999). *Electromagnetic nondestructive evaluation (III). Studies in Applied Electromagnetics and Mechanics.* Amsterdam, The Netherlands: IOS Press.

Liang, J. J., & Suganthan, P. N. (2005). Dynamic multi-swarm particle swarm optimizer. *Proceedings of IEEE Swarm Intelligence Symposium* (pp. 124-129). Pasadena, California.

Liang, J. J., Qin, A. K., Suganthan, P. N., & Baskar, S. (2006). Comprehensive learning particle swarm optimizer for global optimization of multimodal functions. *IEEE Transactions on Evolutionary Computation, 10*(3), 281–295. doi:10.1109/TEVC.2005.857610

Liang, T. C., & Hsu, S. (2007). The L-band EDFA of high clamped gain and low noise figure implemented using fiber Bragg grating and double-pass method. *Optics Communications, 281*(5), 1134–1139. doi:10.1016/j.optcom.2007.11.020

Lin, C. J., & Hsieh, M. H. (2009). Classification of mental task from EEG data using neural networks based on particle swarm optimization. *Journal of Neurocomputing, 72,* 1121–1130. doi:10.1016/j.neucom.2008.02.017

Lin, C., Zhang, F.-S., Zhao, G., Zhang, F., & Jiao, Y.-C. (2010). Broadband low-profile microstrip antenna design using GPSO based on mom. *Microwave and Optical Technology Letters, 52*(4), 975–979. doi:10.1002/mop.25069

Lin, F., & Kompa, G. (1994). FET model parameter extraction based on optimization with multiplane data-fitting and bidirectional search–A new concept. *IEEE Transactions on Microwave Theory and Techniques, 42*(7), 1114–1121. doi:10.1109/22.299745

Ling, S. H., Iu, H. H. C., Chan, K. Y., Lam, H. K., Yeung, B. C. W., & Leung, F. H. (2008). Hybrid particle swarm optimization with wavelet mutation and its industrial applications. *IEEE Transactions on Systems, Man, and Cybernetics. Part B, Cybernetics, 38*(3), 743–763. doi:10.1109/TSMCB.2008.921005

Liu, W. (2005, October). Design of multiband CPW-fed monopole antenna using a particle swarm optimization approach. *IEEE Transactions on Antennas and Propagation, 53*(10), 3273–3279. doi:10.1109/TAP.2005.856339

Liu, X., Chen, J., Lu, C., & Zhou, X. (2004). Optimizing gain profile and noise performance for distributed fiber Raman amplifiers. *Optics Express, 12*(24), 6053–6066. doi:10.1364/OPEX.12.006053

Liu, X., & Lee, B. (2003). A fast and stable method for Raman amplifier propagation equations. *Optics Express, 11*(18), 2163–2176. doi:10.1364/OE.11.002163

Liu, X., & Lee, B. (2003). Effective shooting algorithm and its application to fiber amplifiers. *Optics Express*, *11*(12), 1452–1461. doi:10.1364/OE.11.001452

Li, X., Wang, Y., Song, J., & Shan, J. (2011). Research on classification method of combining support vector machine and genetic algorithm for motor imager EEG. *Journal of Computer Information Systems*, *7*(12), 4351–4358.

Li, Y. (2009). A simulation-based evolutionary approach to LNA circuit design optimization. *Applied Mathematics and Computation*, *209*(1), 57–67. doi:10.1016/j.amc.2008.06.015

Li, Y., Yu, S., & Li, Y. (2008). Electronic design automation using a unified optimization framework. *Mathematics and Computers in Simulation*, *79*, 1137–1152. doi:10.1016/j.matcom.2007.11.001

Liyanage, S. R., Xu, J. X., Guan, C., Ang, K. K., Zhang, C. S., & Lee, T. H. (2009). Classification of self-paced finger movements with EEG signals using neural network and evolutionary approaches. *International Conference on Control and Automation*, (pp. 1807-1812).

Li, Z., Zhao, C. L., Wen, Y. J., Lu, C., Wang, Y., & Chen, J. (2006). Optimization of Raman/EDFA hybrid amplifier based on dual-order stimulated Raman scattering using a single-pump. *Optics Communications*, *265*(2), 655–658. doi:10.1016/j.optcom.2006.03.049

Lopes, A. M., Freire, H., de Moura Oliveira, P. B., & Solteiro Pires, E. J. (2011). *Multi-objective optimization of a parallel robot using particle swarm algorithm*. Paper presented at the meeting of the 7th European Nonlinear Dynamics Conference (ENOC 2011), Rome, Italy, 2011.

Lopes, A. M. (2009). Dynamic modeling of a Stewart platform using the generalized momentum approach. *Communications in Nonlinear Science and Numerical Simulation*, *14*, 3389–3401. doi:10.1016/j.cnsns.2009.01.001

Lopes, A. M., & Solteiro Pires, E. J. (2011). Optimization of the workpiece location in a machining robotic cell. *International Journal of Advanced Robotic Systems*, *8*, 37–46.

Lopez, L., Paez, G., & Strojnik, M. (2006). Characterization of up-conversion coefficient in erbium-doped materials. *Optics Letters*, *31*, 1660–1662. doi:10.1364/OL.31.001660

Lotte, F., Congedo, M., Lecuyer, A., Lambarche, F., & Arnaldi, B. (2007). A review of classification algorithms for EEG-based brain computer interfaces. *Journal of Neural Engineering*, *4*, 1–24. doi:10.1088/1741-2560/4/2/R01

Lu, J., Wang, N., Chen, J., & Su, F. (2011). Cooperative path planning for multiple UCAVs using an AIS-ACO hybrid approach. *Proceedings of the 2011 International Conference on Electronic and Mechanical Engineering and Information Technology (EMEIT)* (pp. 4301–4305). Harbin, China: IEEE. doi:10.1109/EMEIT.2011.6023113

Luan, F., Choi, J. H., & Jung, H. K. (2012, February). A particle swarm optimization algorithm with novel expected fitness evaluation for robust optimization problems. *IEEE Transactions on Magnetics*, *48*(2), 331–334. doi:10.1109/TMAG.2011.2173753

Lurie, B. J. (1994). Three-parameter tunable tilt-integral-derivative (TID) controller. US Patent US5371670. Alexandria, VA: United States Patent and Trademark Office (USPTO).

Lurie, B. J., & Enright, P. J. (2000). *Classical feedback control with Matlab. Control Engineering Series.* New York, NY: Marcel Dekker, Inc.

Lu, Y. B., & Chu, P. L. (2000). Gain flattening by using dual-core fiber in erbium-doped fiber amplifier. *IEEE Photonics Technology Letters*, *12*(12), 1616–1617. doi:10.1109/68.896325

Lu, Y. B., Chu, P. L., Alphones, A., & Shum, P. (2004). A 105-nm ultrawideband gain-flattened amplifier combining C- and L-band dual-core EDFAs in a parellel configuration. *IEEE Photonics Technology Letters*, *16*(7), 1640–1642. doi:10.1109/LPT.2004.827964

Lv, J., & Liu, M. (2008). Common spatial pattern and particle swarm optimization for channel selection in BCI. *The 3rd International Conference on Innovative Computing information and Control* (ICICIC'08), (pp. 1-4).

Maas, S. A. (1993). *Microwave mixers*. Norwood, MA: Artech House Publishers.

Maas, S. A. (2003). *Nonlinear microwave and RF circuits* (2nd ed.). Norwood, MA: Artech House Publishers.

Maione, G. (2011, 28 August - 2 September). Conditions for a class of rational approximants of fractional differentiators/integrators to enjoy the interlacing property. In S. Bittanti, A. Cenedese, & S. Zampieri (Eds.), *Preprints of the 18th IFAC World Congress* (pp. 13984-13989). Milan, Italy.

Maione, G. (2006). Concerning continued fractions representation of noninteger order digital differentiators. *IEEE Signal Processing Letters, 13*(12), 725–728. doi:10.1109/LSP.2006.879866

Maione, G. (2008). Continued fractions approximation of the impulse response of fractional order dynamic systems. *IET Control Theory & Applications, 2*(7), 564–572. doi:10.1049/iet-cta:20070205

Maione, G. (2011, March). High-speed digital realizations of fractional operators in the delta domain. *IEEE Transactions on Automatic Control, 56*(3), 697–702. doi:10.1109/TAC.2010.2101134

Majhi, B., & Panda, G. (2011). Robust identification of nonlinear complex systems using low complexity ANN and particle swarm optimization technique. *Expert Systems with Applications, 38*, 321–333. doi:10.1016/j.eswa.2010.06.070

Majumdar, K. (2011). Human scalp EEG processing: Various soft computing approaches. *Applied Soft Computing, 11*, 4433–4447. doi:10.1016/j.asoc.2011.07.004

Mallipeddi, R., Jeyadevi, S., Suganthan, P. N., & Baskar, S. (2012). Efficient constraint handling for optimal reactive power dispatch problems. *Swarm and Evolutionary Computation, 5*, 28–36. doi:10.1016/j.swevo.2012.03.001

Manabe, S. (1961). The non-integer integral and its application to control systems. *Japanese Institute of Electrical Engineers Journal, 6*(3/4), 83–87.

Mandal, K. K., & Chakraborty, N. (2008). Differential evolution technique-based short-term economic generation scheduling of hydrothermal systems. *Electric Power Systems Research, 78*, 1972–1979. doi:10.1016/j.epsr.2008.04.006

Marandi, A., Afshinmanesh, F., Shahabadi, M., & Bahrami, F. (2006, July 16-21). *Boolean particle swarm optimization and its application to the design of a dual-band dual-polarized planar antenna*. Paper presented at the 2006 IEEE Congress on Evolutionary Computation, Vancouver, Canada.

Marhic, M. E., & Nikonov, D. E. (2002). Low third-order glass-host nonlinearities in erbium-doped waveguide amplifiers. *Proceedings of SPIE, 4645*, (p. 193).

Marseguerra, M., & Zio, E. (2000). *System design optimization by genetic algorithms*. The IEEE Annual Reliability and Maintainability Symposium.

Martens, D., De Backer, M., Haesen, R., Vanthienen, J., Snoeck, M., & Baesens, B. (2007). Classification with ant colony optimization. *IEEE Transactions on Evolutionary Computation, 11*(5), 651–665. doi:10.1109/TEVC.2006.890229

Martini, A., Donelli, M., Franceschetti, M., & Massa, A. (2008). Particle density retrieval in random media using a percolation model and a particle swarm optimizer. *IEEE Antennas and Wireless Propagation Letters, 921140*, 213–216. doi:10.1109/LAWP.2008.921140

Martin, J. C. (2001). Erbium transversal distribution influence on the effectiveness of a doped fiber: Optimization of its performance. *Optics Communications, 194*(4-6), 331–339. doi:10.1016/S0030-4018(01)01312-8

Massa, A., Franceschini, D., Franceschini, G., Pastorino, M., Raffetto, M., & Donelli, M. (2005). Parallel GA-based approach for microwave imaging applications. *IEEE Transactions on Antennas and Propagation, 53*(10), 3118–3127. doi:10.1109/TAP.2005.856311

Masuda, H., & Kawai, S. (1999). Wide-band and gain-flattened hybrid fiber amplifier consisting of an EDFA and a multi-wavelength pumped Raman amplifier. *IEEE Photonics Technology Letters, 11*(6), 647–649. doi:10.1109/68.766772

Matarić, M. J., Sukhatme, G. S., & Østergaard, E. H. (2003). Multi-robot task allocation in uncertain environments. *Autonomous Robots, 14*(2-3), 255–263. doi:10.1023/A:1022291921717

Matekovits, L., Mussetta, M., Pirinoli, P., Selleri, S., & Zich, R. (2005, July). *Improved PSO algorithms for electromagnetic optimization*. Paper presented at IEEE Antennas and Propagation Society International Symposium, Washington, DC.

Matsuda, K., & Fujii, H. (1993). H$_\infty$ optimized wave-absorbing control: Analytical and experimental results. *Journal of Guidance, Control, and Dynamics, 16*(6), 1146–1153. doi:10.2514/3.21139

McIsaac, P. R. (1975). Symmetry-induced modal characteristic of uniform waveguide-II: Theory. *IEEE Transactions on Microwave Theory and Techniques, 23*, 429–433. doi:10.1109/TMTT.1975.1128585

Mears, R., Reekie, L., Jauncey, I., & Payne, D. (1987). Low-noise erbium-doped fibre amplifier operating at 1.54μm. *Electronics Letters, 23*(19), 1026–1028. doi:10.1049/el:19870719

Mears, R., Reekie, L., Poole, S., & Payne, D. (1986). Low-threshold tunable CW and Q-switched fibre laser operating at 1.55 μm. *Electronics Letters, 22*(3), 159–160. doi:10.1049/el:19860111

Medeiro, F., Rodríguez-Macías, R., Fernández, F. V., Domínguez-Astro, R., Huertas, J. L., & Rodríguez-Vázquez, A. (1994). Global design of analog cells using statistical optimization techniques. *Analog Integrated Circuits and Signal Processing, 6*(3), 179–195. doi:10.1007/BF01238887

Mendes, R., Cortez, O., Rocha, M., & Neves, J. (2002). Particle swarms for feedforward neural network training. *Proceedings of International Joint Conference on Neural Networks* (pp. 1895-1899).

Meng, L., & Xue, D. (2009, 9-11 December). Automatic loop shaping in fractional-order QFT controllers using particle swarm optimization. In *Proceedings of the IEEE International Conference on Control and Automation* (pp. 2182-2187). Christchurch, New Zealand.

Menhas, M. I., Wang, L., Fei, M., & Pan, H. (2012). Comparative performance analysis of various binary coded PSO algorithms in multivariable PID controller design. *Expert Systems with Applications, 39*, 4390–4401. doi:10.1016/j.eswa.2011.09.152

Menozzi, R., Piazzi, A., & Contini, F. (1996). Small-signal modeling for a microwave FET linear circuits based on a genetic algorithm. *IEEE Transactions on Circuits and Systems, 43*(10), 839–847. doi:10.1109/81.538990

Merlet, J.-P., & Gosselin, C. (1991). Nouvelle architecture pour un manipulateur parallele a six degres de liberte. *Mechanism and Machine Theory, 26*, 77–90. doi:10.1016/0094-114X(91)90023-W

Mescia, L., Fornarelli, G., Magarielli, D., Prudenzano, F., De Sario, M., & Vacca, F. (2011). Refinement and design of rare earth doped photonic crystal fibre amplifier using an ANN approach. *Journal of Optics & Laser Technology, 43*, 1096–1103. doi:10.1016/j.optlastec.2011.02.005

Meyer, K. D., Nasuto, S. J., & Bishop, M. (2006). Stochastic diffusion search: Partial function evaluation in swarm intelligence dynamic optimisation. *Studies in Computational Intelligence. Stigmergic Optimization, 31*, 185–207. doi:10.1007/978-3-540-34690-6_8

Michael, N., Zavlanos, M. M., Kumar, V., & Pappas, G. J. (2008). Distributed multi-robot task assignment and formation control. In *Proceedings of IEEE International Conference on Robotics and Automation (ICRA 2008)*, (pp. 128–133).

Michalewicz, Z. (1999). *Genetic algorithms + data structures = Evolution programming* (3rd ed.). Berlin, Germany: Springer.

Michielssen, E., Sajer, J.-M., Ranjithan, S., & Mittra, R. (1993). Design of lightweight, broad-band microwave absorbers using genetic algorithms. *IEEE Transactions on Microwave Theory and Techniques, 41*(6), 1024–1031. doi:10.1109/22.238519

Migliore, M., Pinchera, D., & Schettino, F. (2005). A simple and robust adaptive parasitic antenna. *IEEE Transactions on Antennas and Propagation*, *53*(10), 3262–3272. doi:10.1109/TAP.2005.856361

Mikki, S., & Kishk, A. (2005, July). *Investigation of the quantum particle swarm optimization technique for electromagnetic applications.* Paper presented at IEEE Antennas and Propagation Society International Symposium, Washington, DC.

Mikki, S., & Kishk, A. (2006). Quantum particle swarm optimization for electromagnetics. *IEEE Transactions on Antennas and Propagation*, *54*(10), 2764–2775. doi:10.1109/TAP.2006.882165

Miniscalco, W. J. (1991). Erbium-doped glasses for fiber amplifiers at 1500 nm. *Journal of Lightwave Technology*, *9*(2), 234–250. doi:10.1109/50.65882

Miranda, V. (2005). *Evolutionary algorithms with particle swarm movements.* Paper presented at 13th International Conference on Intelligent Systems Application to Power Systems.

Miranda, V. (2005). Evolutionary algorithms with particle swarm movements. *Proceedings of the 13th International Conference on Intelligent Systems Application to Power Systems*, (pp. 6-21).

Miranda, V., & Fonseca, N. (2002). *EPSO best of two worlds meta-heuristic applied to power system problems.* Paper presented at IEEE Congress on Evolutionary Computing.

Miranda, V., De Magalhaes, L., da Rosa, M. A., Da Silva, A. M. L., & Singh, C. (2009). Improving power system reliability calculation efficiency with EPSO variants. *IEEE Transactions on Power Systems*, *24*, 1772–1779. doi:10.1109/TPWRS.2009.2030397

Miranda, V., de Magalhaes, L., daRosa, M. A., daSilv, A. M. L., & Singh, C. (2009). Improving power system reliability calculation efficiency with EPSO variants. *IEEE Transactions on Power Systems*, *24*, 1772–1779. doi:10.1109/TPWRS.2009.2030397

Miranda, V., Keko, H., & Jaramillo, A. (2007). EPSO: Evolutionary particle swarms. *Evolutionary Computing for System Design. Series* (Springer). *Studies in Computational Intelligence*, *66*, 139–168. doi:10.1007/978-3-540-72377-6_6

Mitsi, S., Bouzakis, K. D., Sagris, D., & Mansour, G. (2008). Determination of optimum robot base location considering discrete end-effector positions by means of hybrid genetic algorithm. *Robotics and Computer-integrated Manufacturing*, *24*, 50–59. doi:10.1016/j.rcim.2006.08.003

Modiri, A., & Kiasaleh, K. (2010, May). *Efficient design of microstrip antennas using modified PSO algorithm.* Paper presented at the 14th Biennial IEEE Conference on Electromagnetic Field Computation (CEFC), Chicago, IL.

Modiri, A., & Kiasaleh, K. (2011, January). *Real time reconfiguration of wearable antennas.* Paper presented at IEEE Topical Conference on Biomedical Wireless Technologies, Networks & Sensing Systems, Phoenix, AZ, USA.

Modiri, A., & Kiasaleh, K. (2011, September). *Permittivity estimation for breast cancer detection using particle swarm optimization algorithm.* Paper presented at the 33rd Annual International Conference of the IEEE Engineering in Medicine and Biology Society (EMBC), Boston, MA.

Modiri, A., & Kiasaleh, K. (2011a, August 30 - September 3). *Permittivity estimation for breast cancer detection using particle swarm optimization algorithm.* Paper presented at the 2011 Annual International Conference of the IEEE Engineering in Medicine and Biology Society, Boston, USA.

Modiri, A., & Kiasaleh, K. (2011, May). Efficient design of microstrip antennas for SDR applications using modified PSO algorithm. *IEEE Transactions on Magnetics*, *47*(5), 1278–1281. doi:10.1109/TMAG.2010.2087316

Modiri, A., & Kiasaleh, K. (2011). Modification of real-number and binary PSO algorithms for accelerated convergence. *IEEE Transactions on Antennas and Propagation*, *59*(1), 214–224. doi:10.1109/TAP.2010.2090460

Mollenauer, L. F., Stolen, R. H., & Islam, M. N. (1985). Experimental demonstration of soliton propagation in long fibers: Loss compensated by Raman gain. *Optics Letters, 10*(5), 229–231. doi:10.1364/OL.10.000229

Monmarché, N., Venturini, G., & Slimane, M. (2000). On how Pachycondyla apicalis ants suggest a new search algorithm. *Future Generation Computer Systems, 16,* 937–946. doi:10.1016/S0167-739X(00)00047-9

Montes de Oca, M. A., Stutzle, T., Birattari, M., & Dorigo, M. (2009, October). Frankenstein's PSO: A composite particle swarm optimization algorithm. *IEEE Transactions on Evolutionary Computation, 13*(5), 1120–1132. doi:10.1109/TEVC.2009.2021465

Moubayed, N. A., Hasan, B. A. S., Gan, J. Q., Petrovski, A., & McCall, J. (2010). Binary-SDMOPSO and its application in channel selection for brain computer interfaces. 2010 UK Workshop on Computational Intelligence (UKCI), (pp. 1-6).

Movassaghi, M., & Jackson, M. K. (2001). Design and compact modeling of saturated Erbium-doped fibre amplifiers with nonconfined doping. *Optical Fiber Technology, 7,* 312–323. doi:10.1006/ofte.2000.0354

Mowla, A., & Granpayeh, N. (2008). A novel design approach for erbium doped fiber amplifiers by particle swarm optimization. *Progress in Electromagnetic Research M, 3,* 103–118. doi:10.2528/PIERM08061003

Mowla, A., & Granpayeh, N. (2008). Design of a flat gain multi-pumped distributed fiber Raman amplifier by particle swarm optimization. *Journal of the Optical Society of America. A, Optics, Image Science, and Vision, 25*(5), 3059–3066. doi:10.1364/JOSAA.25.003059

Mowla, A., & Granpayeh, N. (2009). Optimum design of a hybrid erbium doped fiber amplifier/fiber Raman amplifier using particle swarm optimization. *Applied Optics, 48*(5), 979–984. doi:10.1364/AO.48.000979

Mudanyal, O., Yldz, S., Semerci, O., Yapar, A., & Akduman, I. (2008). A microwave tomographic approach for nondestructive testing of dielectric coated metallic surfaces. *IEEE Geoscience and Remote Sensing Letters, 5*(2), 180–184. doi:10.1109/LGRS.2008.915602

Murata, Y., Shibata, N., Yasumoto, K., & Ito, M. (2002). *Agent oriented self adaptive genetic algorithm* (pp. 348–353). The IASTED Communications and Computer Networks.

Mussetta, M., Selleri, S., Pirinoli, P., Zich, R. E., & Matekovits, L. (2008). Improved particle swarm optimization algorithms for electromagnetic optimization. *Journal of Intelligent and Fuzzy Systems, 19*(1), 75–84.

Nakamura, T., Ito, S. I., Mitsukura, Y., & Setokawa, H. (2009). A method for evaluating the degree of human's preference based on EEG analysis. *Fifth International Conference on Intelligent Information Hiding and Multimedia Signal Processing, IIH-MSP '09,* (pp. 732–735).

Nakano, S., Ishigame, A., & Yasuda, K. (2007, September). *Particle swarm optimization based on the concept of Tabu search.* Paper presented at IEEE Congress on Evolutionary Computation (CEC), Singapore.

Namiki, S., & Emori, Y. (2001). Ultrabroad-band Raman amplifiers pumped and gain-equalized by wavelength-division-multiplexed high power laser diodes. *IEEE Journal on Selected Topics in Quantum Electronics, 7*(1), 3–16. doi:10.1109/2944.924003

Nanbo, J., & Rahmat-Samii, Y. (2010). Hybrid real-binary particle swarm optimization (HPSO) in engineering electromagnetics. *IEEE Transactions on Antennas and Propagation, 58*(12), 3786–3794. doi:10.1109/TAP.2010.2078477

Nanda, J., & Bijwe, P. R. (1981). Optimal hydrothermal scheduling with cascaded plants using progressive optimality algorithm. *IEEE Transactions on Power Apparatus and Systems, 100*(4), 2093–2099. doi:10.1109/TPAS.1981.316486

Naresh, R., & Sharma, J. (1999). Two-phase neural network based solution technique for short term hydrothermal scheduling. *IEE Proceedings. Generation, Transmission and Distribution, 146*(6), 657–663. doi:10.1049/ip-gtd:19990855

Navan, R. R., Thakker, R. A., Tiwari, S. P., Baghini, M. S., Patil, M. B., Mhaisalkar, S. G., & Rao, V. R. (2009). DC transient circuit simulation methodologies for organic electronics. In *Proceedings of 2ⁿᵈ International Workshop on Electron Devices and Semiconductor Technology 2009*, India.

Nikolova, N. K. (2011, December). Microwave imaging for breast cancer. *IEEE Microwave Magazine, 12*(7), 78–94. doi:10.1109/MMM.2011.942702

Nikonorov, N., Przhevuskii, A., Prassas, M., & Jacob, D. (1999). Experimental determination of the upconversion rate in Erbium-doped silicate glasses. *Applied Optics, 38*, 6284–6291. doi:10.1364/AO.38.006284

Niu, B., Zhu, Y., He, X., & Wu, H. (2006). MCPSO: A multi-swarm cooperative particle swarm optimizer. *Applied Mathematics and Computation, 185*, 1050–1062. doi:10.1016/j.amc.2006.07.026

O'Connor, I., & Kaiser, A. (2000). Automated synthesis of current memory cells. *IEEE Transactions on Computer-Aided Design of Integrated Circuits and Systems, 19*(4), 413–424. doi:10.1109/43.838991

Olamaei, Y., Niknam, T., & Gharehpetian, G. (2008). Application of particle swarm optimization for distribution feeder reconfiguration considering distributed generators. *Applied Mathematics and Computation, 201*, 575–586. doi:10.1016/j.amc.2007.12.053

Olamei, J., Niknam, T., Arefi, A., & Mazinan, A. H. (2011). A novel hybrid evolutionary algorithm based on ACO and SA for distribution feeder reconfiguration with regard to DGs. *Proceedings of the 2011 IEEE GCC Conference and Exhibition (GCC)* (pp. 259–262). Dubai, UAE: IEEE. doi:10.1109/IEEEGCC.2011.5752495

Oliveira, G., Binato, S., & Pereira, M. (2007). Value-based transmission expansion planning of hydrothermal systems under uncertainty. *IEEE Transactions on Power Systems, 22*, 1429–1435. doi:10.1109/TPWRS.2007.907161

Olsson, T., Haage, M., Kihlman, H., Johansson, R., Nilsson, K., & Robertsson, A. (2010). Cost-efficient drilling using industrial robots with high-bandwidth force feedback. *Robotics and Computer-integrated Manufacturing, 26*, 24–38. doi:10.1016/j.rcim.2009.01.002

Oñate, P., Ramirez, J., & Coello, C. (2009). Optimal power flow subject to security constraints solved with a particle swarm optimizer. *IEEE Transactions on Power Systems, 23*, 96–104.

Orero, S. O., & Irving, M. R. (1998). A genetic algorithm modelling framework and solution technique for short term optimal hydrothermal scheduling. *IEEE Transactions on Power Systems, 13*(2), 501–518. doi:10.1109/59.667375

Orfanidis, S. J. (2008). *Electromagnetic waves and antennas*. eBook, Retrieved January 3, 2012, from http://www.ece.rutgers.edu/~orfanidi/ewa/

Oustaloup, A., Cois, O., Lanusse, P., Melchior, P., Moreau, X., & Sabatier, J. (2006, 19-21 July). The CRONE approach: Theoretical developments and major applications. In *Proceedings of the 2nd IFAC Workshop on Fractional Differentiation and its Applications*, Porto, Portugal.

Oustaloup, A., Sabatier, J., Lanusse, P., Malti, R., Melchior, P., Moreau, X., & Moze, M. (2008, 6-11 July). An overview of the CRONE approach in system analysis, modeling and identification, observation and control. In *Proceedings of the 17ᵗʰ IFAC World Congress* (pp. 14254-14265). Seoul, Korea.

Oustaloup, A., Levron, F., Mathieu, B., & Nanot, F. M. (2000). Frequency band complex noninteger differentiator: Characterization and synthesis. *IEEE Transactions on Circuits and Systems. I, Fundamental Theory and Applications, 47*(1), 25–39. doi:10.1109/81.817385

Oustaloup, A., Mathieu, B., & Lanusse, P. (1995). The CRONE control of resonant plants: Application to a flexible transmission. *European Journal of Control, 1*(2), 113–121.

Oustaloup, A., Moreau, X., & Nouillant, M. (1996). The CRONE suspension. *Control Engineering Practice, 4*(8), 1101–1108. doi:10.1016/0967-0661(96)00109-8

Ozcan, E., & Mohan, C. K. (1998). Analysis of simple particle swarm optimization system. *Intelligent Engineering Systems through Artificial. Neural Networks, 8*, 253–258.

Palaniappan, R., & Raveendran, P. (2002). *Genetic algorithm to select features for fuzzy ARTMAP classification of evoked EEG* (pp. 53–56). doi:10.1109/APC-CAS.2002.1115119

Pal, M., Paul, M. C., Dhar, A., Pal, A., Sen, R., & Dasgupta, K. (2007). Investigation of the optical gain and noise figure for multichannel amplification in EDFA under optimized pump condition. *Optics Communications, 273*(2), 407–412. doi:10.1016/j.optcom.2007.01.039

Pam, S., Bhattacharya, A. K., & Mukhopadhyay, S. (2010). An efficient method for bottom-up extraction of analog behavioral model parameters. In *Proceedings of 23rd International Conference on VLSI Design*, (pp. 363-368).

Papoulis, A. (1965). *Probability, random variables and stochastic processes*. New York, NY: McGraw-Hill.

Parikh, D., Stepenosky, N., Topalis, A., Green, D., Kounios, J., Clark, C., & Polikar, R. (2005). Ensemble based data fusion for early diagnosis of Alzheimer's disease. *Proceedings of 2005 IEEE Engineering in Medicine and Biology 27th Annual Conference*, (pp. 2479-2482).

Park, J., Choi, K., & Allsot, D. J. (2004). Parasitic-aware RF circuit design and optimization. *IEEE Transactions on Circuits and Systems, 51*(10).

Parmee, I. C. (Ed.). (2001). *Evolutionary and adaptive computing in engineering design*. Springer. doi:10.1007/978-1-4471-0273-1

Parsopoulos, K., & Vrahatis, M. (2010). *Particle swarm optimization and intelligence: Advances and applications*. Information Science Reference. doi:10.4018/978-1-61520-666-7

Particle Swarm. (n.d.). Retrieved from http://www.particleswarm.info/

Pastorino, M. (2007). Stochastic optimization methods applied to microwave imaging: A review. *IEEE Transactions on Antennas and Propagation, 55*(3), 538–548. doi:10.1109/TAP.2007.891568

Pastorino, M. (2010). *Microwave imaging*. Hoboken, NJ: John Wiley. doi:10.1002/9780470602492

Paulraj, M. P., Hema, C. R., Nagarajan, R., Yaacob, S., & Adom, A. H. (2007). EEG classification using radial basis PSO neural network for brain machine interfaces. *5th Student Conference on Research and Development, 2007 SCOReD*, (pp. 1-5).

Peng, S., & Nie, Z. (2008). Acceleration of the method of moments calculations by using graphics processing units. *IEEE Transactions on Antennas and Propagation, 56*(7), 2130–2133. doi:10.1109/TAP.2008.924768

Pereira, M. V. F., & Pinto, L. M. V. G. (1983). Application of decomposition techniques to the mid-and short-term scheduling of hydrothermal systems. *IEEE Transactions on Power Apparatus and Systems, 102*, 3611–3618. doi:10.1109/TPAS.1983.317709

Perlin, V. E., & Winful, H. G. (2002). Optimal design of flat-gain wide-band fiber Raman amplifiers. *Journal of Lightwave Technology, 20*(2), 250–254. doi:10.1109/50.983239

Petyrantonaksi, P. C., & Hadjileontiadis, L. J. (2009). EEG-based emotion recognition using hybrid filtering and higher order crossing. *3rd International Conference on Affective Computing and Intelligent Interaction and Workshops, ACII 2009*, (pp. 1-6).

Piatrik, T., Chandramouli, K., & Izquierdo, E. (2006). Image classification using biologically inspired systems. In *Proceedings of the 2nd International Mobile Multimedia Communications Conference*.

Pittens, K., & Podhorodeski, R. (1993). A family of stewart platforms with optimal dexterity. *Journal of Robotic Systems, 10*, 463–479. doi:10.1002/rob.4620100405

Podell, A. F. (1981). A functional GaAs FET noise model. *IEEE Transactions on Electron Devices, 28*(5), 511–517. doi:10.1109/T-ED.1981.20375

Podlubny, I. (1999, January). Fractional-order systems and PI$^\lambda$D$^\mu$-controllers. *IEEE Transactions on Automatic Control, 44*(1), 208–214. doi:10.1109/9.739144

Podlubny, I., Petráš, I., Vinagre, B. M., O'Leary, P., & Dorčák, L. (2002). Analogue realizations of fractional-order controllers. *Nonlinear Dynamics, 29*(1-4), 281–296. doi:10.1023/A:1016556604320

Polikar, R. (2006). Ensemble based systems in decision making. *IEEE Circuits and Systems Magazine,* 21-45.

Polikar, R., Topalis, A., Parikh, D., Green, D., Frymiar, J., Kounios, J., & Clark, C. M. (2006). An ensemble based data fusion approach for early diagnosis of Alzheimer's disease. *Information Fusion, 9*, 83–95. doi:10.1016/j.inffus.2006.09.003

Poli, R. (2009, August). Mean and variance of the sampling distribution of particle swarm optimizers during stagnation. *IEEE Transactions on Evolutionary Computation, 13*(4), 712–721. doi:10.1109/TEVC.2008.2011744

Poli, R., Salvaris, M., & Cinel, C. (2011). Evolution of a brain computer interface mouse via genetic programming. *EuroGP 2011. LNCS, 6621*, 203–214.

Poole, S., Payne, D., Mears, R., Fermann, M., & Laming, R. (1986). Fabrication and characterization of low-loss optical fibers containing rare-earth ions. *Journal of Lightwave Technology, 4*(7), 870–876. doi:10.1109/JLT.1986.1074811

Pospiezalski, M. W. (1989). Modeling of noise parameters of MES-FET's and MODFET's and their frequency and temperature dependence. *IEEE Transactions on Microwave Theory and Techniques, 37*(9), 1340–1350. doi:10.1109/22.32217

Pozar, D. (1998). *Microwave engineering.* New York, NY: John Wiley & Sons.

Pozar, D. M. (2005). *Microwave engineering* (3rd ed.). Hoboken, NJ: John Wiley & Sons.

Price, K. V. (1999). *An introduction to differential evolution* (pp. 79–108). Maidenhead, UK: McGraw-Hill.

Pringles, R., Miranda, V., & Garcés, F. (2007). Expansión optima del sistema de transmisión utilizando EPSO. *VII Latin American Congress on Electricity Generation & Transmission*, October 24-27.

Prudenzano, F., Mescia, L., Allegretti, L., De Sario, M., Smektala, F., & Moizan, V. (2009). Simulation of MID-IR amplification in Er^{3+} doped chalcogenide microstructured optical fibre. *Optical Materials, 31*, 1292–1295. doi:10.1016/j.optmat.2008.10.004

Prudenzano, F., Mescia, L., D'Orazio, A., De Sario, M., Petruzzelli, V., Chiasera, A., & Ferrari, M. (2007). Optimization and characterization of rare earth doped photonic crystal fibre amplifier using genetic algorithm. *Journal of Lightwave Technology, 25*, 2135–2142. doi:10.1109/JLT.2007.901331

Qin, L., Li, Y., & Yao, D. (2005). A feasibility study of EEG dipole source localization using particle swarm optimization. *The 2005 IEEE Congress on Evolutionary Computation*, Vol. 1, (pp. 720-726).

Quinlan, J. R. (2006). *Boosting first order learning* (pp. 1–14). Sydney, Australia: University of Sydney.

Rabanal, P., Rodríguez, I., & Rubio, F. (2007). Using river formation dynamics to design heuristic algorithms. *Unconventional Computation. Lecture Notes in Computer Science, 4618*, 163–177. doi:10.1007/978-3-540-73554-0_16

Rahmani, M., Rashididejad, M., Carreno, E. M., & Romero, R. (2010). Efficient method for AC transmission network expansion planning. *Electric Power Systems Research*, 1056–1064. doi:10.1016/j.epsr.2010.01.012

Rajanaronk, P., Namahoot, A., & Akkareakthalin, P. (2006). A single diode frequency doubler using a feed-forward technique. *Proceeding of Asia-Pacific Microwave Conference*.

Ramesh, L., Chakraborthy, N., Chowdhury, S. P., & Chowdhury, S. (2012). Intelligent DE algorithm for measurement location and PSO for bus voltage estimation in power distribution system. *Electrical Power and Energy Systems, 39*, 1–8. doi:10.1016/j.ijepes.2011.10.009

Rana, O. F., & Stout, K. (2000). What is scalability in multi-agent systems? *Proceedings of the 4th International Conference on Autonomous Agents,* AGENTS '00, (pp. 56–63). New York, NY: ACM.

Rapaić, M. R., & Kanović, Ž. (2009). Time-varying PSO - Convergence analysis, convergence related parameterization and new parameter adjustment schemes. *Information Processing Letters, 109*, 548–552. doi:10.1016/j.ipl.2009.01.021

Rashedi, E., Nezamabadi-Pour, H., & Saryazdi, S. (2009). GSA: A gravitational search algorithm. *Information Sciences, 179*(13), 2232–2248. doi:10.1016/j.ins.2009.03.004

Ratnaweera, A., Halgamuge, S. K., & Watson, H. C. (2004, June). Self-organizing hierarchical particle swarm optimizer with time-varying acceleration coefficients. *IEEE Transactions on Evolutionary Computation, 8*(3), 240–255. doi:10.1109/TEVC.2004.826071

Ratnaweera, A., Saman, K. H., & Watson, H. C. (2004). Self-organizing hierarchical particle swarm optimizer with time-varying acceleration coefficients. *IEEE Transactions on Evolutionary Computation, 8*(3), 240–255. doi:10.1109/TEVC.2004.826071

Ravi, K. V. R., & Palaniappan, R. (2007). A minimal channel set for individual identification with EEG biometric using genetic algorithm. *Conference on Computational Intelligence and Multimedia Applications,* (pp. 328-332).

Ray, A. K., Benavidez, P., Behera, L., & Jamshidi, M. (2009). Decentralized motion coordination for a formation of rovers. *IEEE Systems Journal, 3*(3), 369–381. doi:10.1109/JSYST.2009.2031012

Ren, P., Gao, L., Li, N., Li, Y., & Lin, Z. (2005). Transmission network optimal planning using the particle swarm optimization method. *Proceedings of the Fourth International Conference on Machine Learning and Cybernetics, Guangzhou,* 16-21 August.

Reyes-Sierra, M., & Coello, C. A. C. (2006). Multi-objective particle swarm optimizers: A survey of the state-of-the-art. *International Journal of Computer Intelligence Research, 2*, 287–30.

Reyes-Sierra, M., & Coello-Coello, C. A. (2006). Multi-objective particle swarm optimizers: A survey of the state-of-the-art. *International Journal of Computational Intelligence Research, 2*(3), 287–308.

Rider, M. (2006). *Planejamento da expansão de sistemas de transmissão usando os modelos CC – CA e técnicas de programação não – Linear.* Ph.D Thesis in Portuguese, University of Campinas, Brazil.

Rider, M. J., Gallego, L. A., Romero, R., & Garcia, A. V. (2007). Heuristic algorithm to solve the short term transmission expansion planning. *Proceedings of the IEEE General Meeting.*

Rider, M. J., Garcia, A. V., & Romero, R. (2007). Power system transmission network expansion planning using AC model. *Proceedings of IET Generation, Transmission, and Distribution, 1*(5), 731–742. doi:10.1049/iet-gtd:20060465

Robinson, J., Sinton, S., & Rahmat-Samii, Y. (2002). Particle swarm, genetic algorithm, and their hybrids: Optimization of a profiled corrugated horn antenna. *IEEE Antennas Propagation Society International Symposium Digest, 1,* 314–317.

Robinson, J., & Rahmat-Samii, Y. (2004). Particle swarm optimization in electromagnetics. *IEEE Transactions on Antennas and Propagation, 52*(2), 397–407. doi:10.1109/TAP.2004.823969

Rochat, E., Dändliker, R., Haroud, K., Czichy, R. H., Roth, U., Costantini, D., & Holzner, R. (2001). Fibre amplifiers for coherent space communication. *IEEE Journal on Selected Topics in Quantum Electronics, 7*(1), 74–81. doi:10.1109/2944.924012

Rodriguez, J. I., Falcão, D. M., & Taranto, G. N. (2008). *Short-term transmission expansion planning with AC network model and security constraints.* 16th PSCC, Glasgow, Scotland, July 14-18.

Rodriguez, J. I., Falcão, D. M., Taranto, G. N., & Almeida, H. L. S. (2009). *Short-term transmission expansion planning by a combined genetic algorithm and hill-climbing technique*. 15th International Conference on Intelligent System Applications to Power Systems, ISAP '09.

Romero, R., & Monticelli, A. (1994). A hierarchical decomposition approach for transmission network expansion planning. *IEEE Transactions on Power Systems, 9*, 373–380. doi:10.1109/59.317588

Romero, R., Monticelli, A., García, A., & Haffner, S. (2002). Test systems and mathematical models for transmission network expansion planning. *IEE Proceedings. Generation, Transmission and Distribution, 149*(1), 27–36. doi:10.1049/ip-gtd:20020026

Romero, R., Rocha, C., Mantovani, J. R. S., & Sanchez, I. G. (2005). Constructive heuristic algorithm for the DC model in network transmission expansion planning. *IEE Proceedings. Generation, Transmission and Distribution, 152*(2), 277–282. doi:10.1049/ip-gtd:20041196

Rorsman, N., Garcia, M., Karlsson, C., & Zirath, H. (1996). Accurate small-signal modeling of HFETs for millimeter-wave applications. *IEEE Transactions on Microwave Theory and Techniques, 44*(3), 432–437. doi:10.1109/22.486152

Ross, T., Cormier, G., Hettak, K., & Amaya, R. E. (2011). Particle swarm optimization in the determination of the optimal bias current for noise performance of gallium nitride HEMTs. *Microwave and Optical Technology Letters, 53*(3), 652–656. doi:10.1002/mop.25758

Roudas, I., Richards, D. H., Antoniades, N., Jackel, J. L., & Wagner, R. E. (1999). An efficient simulation model of the erbium-doped fibre for the study of multiwavelength optical networks. *Optical Fiber Technology, 5*, 363–389. doi:10.1006/ofte.1999.0306

Ruta, D., & Gabrys, B. (2000). An overview of classifier fusion methods. *Computing and Information Systems, 7*, 1–10.

Ryu, J., Kim, S., & Wan, H. (2009). *Pareto front approximation with adaptive weighted sum method in multiobjective simulation optimization*. The Winter Simulation Conference.

Sabat, S. L., Raju, V., & Ali, L. (2008). MESFET small signal model parameter extraction using particle swarm optimization. In *International Conference on Microelectronics* (pp. 208-211).

Sabeti, M., Boostani, R., & Katebi, S. D. (2007). A new approach to classify the schizophrenic and normal subjects by finding the best channels and frequency bands. *2007 15th International Conference on Digital Signal Processing*, (pp. 123-126).

Sabouni, A., Noghanian, S., & Pistorius, S. (2010, Winter). A global optimization technique for microwave imaging of the inhomogeneous and dispersive breast. *Canadian Journal of Electrical and Computer Engineering, 35*(1), 15–24. doi:10.1109/CJECE.2010.5783380

Sahin, C., Zuyi, L., Shahidehpour, M., & Erkmen, I. (2011). Impact of natural gas system on risk-constrained midterm hydrothermal scheduling. *IEEE Transactions on Power Systems, 26*, 520–531. doi:10.1109/TP-WRS.2010.2052838

Salazar-Lechuga, M., & Rowe, J. E. (2005, September 2-5). *Particle swarm optimization and fitness sharing to solve multi-objective optimization problems*. Paper presented at the 2005 IEEE Congress on Evolutionary Computation, Edinburgh, UK.

Salvade, A., Pastorino, M., Monleone, R., Bozza, G., & Randazzo, A. (2009). A new microwave axial tomograph for the inspection of dielectric materials. *IEEE Transactions on Instrumentation and Measurement, 58*(7), 2072–2079. doi:10.1109/TIM.2009.2015521

Samanta, B., & Nataraj, C. (2009). Use of particle swarm optimization for machinery fault detection. *Engineering Applications of Artificial Intelligence, 22*, 308–316. doi:10.1016/j.engappai.2008.07.006

Sanabria, C., Chakraborty, A., Xu, H., Rodwell, M. J., Mishra, U. K., & York, R. A. (2006). The effect of gate leakage on the noise figure of AlGaN/GaN HEMTs. *IEEE Electron Device Letters, 27*(1), 19–21. doi:10.1109/LED.2005.860889

Sanks, R. L. (Ed.). (1998). *Pumping station design*. Woburn, MA: Butterworth – Heinemann.

Satapathy, S. C., Naga, B., Murthy, J. V. R., & Reddy, P. (2007). A comparative analysis of unsupervised k-means, PSO and self-organizing PSO for image clustering. *International Conference on Computational Intelligence and Multimedia Applications (ICCIMA)* (pp. 229–237).

Sawai, H., & Adachi, S. (1999). *Genetic algorithm inspired by gene duplication.* The Congress on Evolutionary Computing.

Schnecke, V., & Vornberger, O. (1996). *An adaptive parallel genetic algorithm for VLSI-layout optimization.* The International Conference on Parallel Problem Solving from Nature.

Schutte, J. C., & Groenwold, A. A. (2005). A study of global optimization using particle swarms. *Journal of Global Optimization*, *31*, 93–108. doi:10.1007/s10898-003-6454-x

Schwager, M., McLurkin, J., Slotine, J.-J., & Rus, D. (2009). From theory to practice: Distributed coverage control experiments with groups of robots. In, O. Khatib, V. Kumar, & G. Pappas (Eds.), *Experimental robotics, Springer tracts in advanced robotics, 54*, 127–136. Berlin, Germany: Springer.

Seifi, H., Sepasian, M. S., Haghighat, H., Foroud, A. A., Yousefi, G. R., & Rae, S. (2007). Multi-voltage approach to long-term network expansion planning. *IET Generation. Transmission & Distribution*, *1*(5), 826–835. doi:10.1049/iet-gtd:20070092

Sensarma, P. S., Rahmani, M., & Carvalho, M. (2002). A comprehensive method for optimal expansion planning using particle swarm optimization. *IEEE Power Engineering Society Winter Meeting*, Vol. 2, (pp. 1317–1322).

Senthil Arumugam, M., & Rao, M. V. C. (2007). On the improved performances of the particle swarm optimization algorithms with adaptive parameters, cross-over operators and root mean square (RMS) variants for computing optimal control of a class of hybrid systems. *Applied Soft Computing*, *8*, 324–336. doi:10.1016/j.asoc.2007.01.010

Senthil Arumugam, M., Rao, M. V. C., & Tan, A. W. C. (2009). A novel and effective particle swarm optimization like algorithm with extrapolation technique. *Applied Soft Computing*, *9*(1), 308–320. doi:10.1016/j.asoc.2008.04.016

Shahzad, F., Rauf Baig, A., Masood, S., Kamran, M., & Naveed, N. (2009). Opposition-based particle swarm optimization wit velocity clamping (OVCPSO). In Yu, W., & Sanchez, E. N. (Eds.), *Advances in computational intelligence* (pp. 339–348). Berlin, Germany: Springer. doi:10.1007/978-3-642-03156-4_34

Sharaf, A. M., & El-Gammal, A. A. A. (2009, November). *A discrete particle swarm optimization technique (DPSO) for power filter design.* Paper presented at 4th International Design and Test Workshop (IDT).

Sharkey, A. J. C. (2009). Swarm robotics. In Rabunal, J. R., Dorado, J., & Pazos, A. (Eds.), *Encyclopedia of artificial intelligence* (pp. 1537–1542).

Shayegui, H., Mahdavi, M., & Bagheri, A. (2009). Discrete PSO algorithm based optimization of transmission lines loading in TNEP problem. (Elsevier.). *Energy Conversion and Management*, 112–121.

Shi, Y., & Eberhart, R. (1999, July). *Empirical study of particle swarm optimization.* Paper presented at Congress on Evolutionary Computation, Washington, DC.

Shi, Y., & Eberhart, R. C. (1998, 25-27 March). Parameter selection in particle swarm optimization. In V. W. Porto, N. Saravanan, D. E. Waagen, & A. E. Eiben (Eds.), *Proceedings of the 7th International Conference on Evolutionary Programming, Lecture Notes in Computer Science, Vol. 1447*, San Diego, CA, (pp. 591-600). Berlin, Germany: Springer-Verlag.

Shi, Y., & Eberhart, R. C. (1999). Empirical study of particle swarm optimization. *Proceedings of IEEE International Congress on Evolutionary Computation*, Vol. 3, (pp. 101-106). Washington, DC.

Shi, Y., & Eberhart, R. C. (1998). Parameter selection in particle swarm optimization. In Porto, V. W., Saravanan, N., Waagen, D., & Eiben, A. E. (Eds.), *Evolutionary Programming VII (Vol. 1447*, pp. 591–600). Lecture Notes in Computer Science Berlin, Germany: Springer. doi:10.1007/BFb0040810

Shi, Y., & Eberhart, R. C. (2001). *Fuzzy adaptive particle swarm optimization.* Congress on Evolutionary Computation.

Sifuentes, W. S., & Vargas, A. (2007). Hydrothermal scheduling using benders decomposition: Accelerating techniques. *IEEE Transactions on Power Systems*, *22*(3), 1351–1359. doi:10.1109/TPWRS.2007.901751

Silva, I. J., Rider, M. J., Romero, R., & Murari, C. A. F. (2006). Transmission network expansion planning considering uncertainty in demand. *IEEE Transactions on Power Systems*, *21*(4), 1565–1573. doi:10.1109/TP-WRS.2006.881159

Singh, R., Sunanda, & Sharma, E. K. (2004). Gain flattening by long period gratings in erbium doped fibers. *Optics Communications*, *240*(1-3), 123–132. doi:10.1016/j.optcom.2004.06.023

Singh, S. V., Maheshwari, S., & Chauhan, D. S. (2011). Single MO-CCCCTA-based electronically tunable current/trans-impedance-mode biquad universal filter. *Circuits and Systems*, *2*, 1–6. doi:10.4236/cs.2011.21001

Singh, S. V., Maheshwari, S., Mohan, J., & Chauhan, D. S. (2009). An electronically tunable SIMO biquad filter using CCCCTA. *Contemporary Computing Communications in Computer and Information Science*, *40*(11), 544–555. doi:10.1007/978-3-642-03547-0_52

Sinha, N., Chakrabarti, R., & Chattopadhyay, P. K. (2003). Fast evolutionary programming techniques for short-term hydrothermal scheduling. *IEEE Transactions on Power Systems*, *18*(1), 214–220. doi:10.1109/TPWRS.2002.807053

Smith, K., & Mollenauer, L. F. (1989). Experimental observation of soliton interaction over long fiber paths: Discovery of a long-range interaction. *Optics Letters*, *14*(22), 1284–1286. doi:10.1364/OL.14.001284

Soares, S., Lyra, C., & Tavares, H. (1980). Optimal generation scheduling of hydro-thermal power system. *IEEE Transactions on Power Apparatus and Systems*, *PAS-99*(3), 1107–1115. doi:10.1109/TPAS.1980.319741

Socha, K., & Dorigo, M. (2008). Ant colony optimization for continuous domains. *European Journal of Operational Research*, *185*(3), 1155–1173. doi:10.1016/j.ejor.2006.06.046

Solteiro Pires, E. J., de Moura Oliveira, P. B., & Tenreiro Machado, J. A. (2005, March). Multi objective MaxiMin sorting scheme. *Evolutionary Multi-Criterion Optimization*, (pp. 165-175). Guanajuato, Mexico.

Solteiro Pires, E. J., de Moura Oliveira, P. B., & Tenreiro Machado, J. A. (2007). Manipulator trajectory planning using a MOEA original research article. *Applied Soft Computing*, *7*(3), 659–667. doi:10.1016/j.asoc.2005.06.009

Spanier, J., & Oldham, K. B. (1987). *An atlas of functions*. New York, NY: Hemisphere Publishing Co.

Staicu, S., & Zhang, D. (2008). A novel dynamic modelling approach for parallel mechanisms analysis. *Robotics and Computer-integrated Manufacturing*, *24*, 167–172. doi:10.1016/j.rcim.2006.09.001

Stepanić, P., Latinović, I., & Đurović, Ž. (2009). A new approach to detection of defects in rolling element bearings based on statistical pattern recognition. *International Journal of Advanced Manufacturing Technology*, *45*, 91–100. doi:10.1007/s00170-009-1953-7

Su, H., Michael, C., & Ismail, M. (1994). *Statistical constrained optimization of analog MOS circuits using empirical performance models*. The IEEE International Symposium on Circuits and Systems.

Suganthan, P. N. (1999). *Particle swarm optimisation with a neighbourhood operator*. Congress on Evolutionary Computation. doi:10.1109/CEC.1999.785514

Sun, J., Fang, W., & Xu, W. (2010, February). A quantum-behaved particle swarm optimization with diversity-guided mutation for the design of two-dimensional IIR digital filters. *IEEE Transactions on Circuits and Wystems. II, Express Briefs*, *57*(2), 141–145. doi:10.1109/TCSII.2009.2038514

Sylvester, J., & Chawla, N. V. (2005). Evolutionary ensembles: Combining learning agents using genetic algorithms. *Proceedings of AAAI Workshop on Multi-agent Systems*, (pp. 46–51).

Tang, J., & Luh, P. B. (1995). Hydrothermal scheduling via extended differential dynamic programming and mixed coordination. *IEEE Transactions on Power Systems*, *10*(4), 2021–2028. doi:10.1109/59.476071

Tang, J., Sun, Z., & Zhu, J. (2005). Using ensemble information in swarming artificial neural networks. *LNCS*, *3496*, 515–519.

Tao, Q., Chang, H. Y., Yi, Y., Gu, C. Q., & Li, W. J. (2010, July). *An analysis for particle trajectories of a discrete particle swarm optimization*. 3rd IEEE International Conference on Computer Science and Information Technology (ICCSIT).

Tao, Y., Lin, H., & Bao, H. (2010). GPU-based shooting and bouncing ray method for fast RCS prediction. *IEEE Transactions on Antennas and Propagation, 58*(2), 494–502. doi:10.1109/TAP.2009.2037694

Tenglong, K., Xiaoying, Z., Jian, W., & Yihan, D. (2011). A modified ACO algorithm for the optimization of antenna layout. *Proceedings of the 2011 International Conference on Electrical and Control Engineering (ICECE)* (pp. 4269–4272). Yichang, China: IEEE. doi:10.1109/ICECENG.2011.6057613

Tenreiro Machado, J. A. (1997). Analysis and design of fractional-order digital control systems. *Systems Analysis Modeling and Simulation, 27*(2-3), 107–122.

Tenreiro Machado, J. A. (2001). Discrete-time fractional-order controllers. *Fractional Calculus & Applied Analysis, 4*(1), 47–66.

Tenreiro Machado, J. A. (2010). Optimal tuning of fractional controllers using genetic algorithms. *Nonlinear Dynamics, 62*, 447–452. doi:10.1007/s11071-010-9731-5

Tenreiro Machado, J. A., & Galhano, A. M. (2009). Approximating fractional derivatives in the perspective of system control. *Nonlinear Dynamics, 56*(4), 401–407. doi:10.1007/s11071-008-9409-4

Thakker, R. A., Baghini, M. S., & Patil, M. B. (2009). Low-power low-voltage analog circuit design using HPSO. In *Proceedings of IEEE International Conference on VLSI Design 2009*, India.

Thakker, R. A., Sathe, C., Sachid, A. B., Baghini, M. S., Rao, V. R., & Patil, M. B. (2009). Automated design and optimization of circuits in emerging technologies. In *Proceedings of IEEE Asia and South Pacific Design Automation Conference, 2009*, Japan.

Thakker, R. A., Sathe, C., Baghini, M. S., & Patil, M. B. (2010). A table-based approach to study the impact of process variations on FinFET circuit performance. *IEEE Transactions on Computer-Aided Design of Integrated Circuits and Systems, 29*(4), 627–631. doi:10.1109/TCAD.2010.2042899

Thakker, R. A., Sathe, C., Sachid, A. B., Baghini, M. S., Rao, V. R., & Patil, M. B. (2009). A novel table based approach for design of FinFET circuits. *IEEE Transactions on Computer-Aided Design of Integrated Circuits and Systems, 28*(7), 1061–1070. doi:10.1109/TCAD.2009.2017431

Thakur, K. P., Chan, K., Holmes, W. S., & Carter, G. (2002, June). *An inverse technique to evaluate thickness and permittivity using reflection of plane wave from inhomogeneous dielectrics*. Paper presented at 59th Automatic RF Techniques Group (ARFTG) Conference, Seattle, WA.

Thangaraj, R., Pant, M., Abraham, A., & Bouvry, P. (2011). Particle swarm optimization: Hybridization perspectives and experimental illustrations. *Applied Mathematics and Computation, 217*, 5208–5226. doi:10.1016/j.amc.2010.12.053

Timmis, J., Neal, M., & Hunt, J. (2000). An artificial immune system for data analysis. *Bio Systems, 55*(1-3), 143–150. doi:10.1016/S0303-2647(99)00092-1

Ting, K. M., Wells, J. R., Tan, S. C., Teng, S. W., & Webb, G. I. (2011). Feature-subspace aggregation: Ensembles for stable and unstable learners. *Machine Learning, 82*, 375–397. doi:10.1007/s10994-010-5224-5

Tlelo-Cuautle, E., Guerra-Gómez, I., de la Fraga, L. G., Flores-Becerra, G., Polanco-Martagón, S., & Fakhfakh, M. ... Reyes-Salgado, G. (2010). Evolutionary algorithms in the optimal sizing of analog circuits. In M. Köppen, G. Schaefer, & A. Abraham (Eds.), *Intelligent computational optimization in engineering: Techniques & applications*. Springer.

Torres, S. P., Castro, C. A., Pringles, R., & Guaman, W. (2011). Comparison of particle swarm based metaheuristics for the electric transmission network expansion planning problem. *Proceedings of the IEEE General Meeting*, Detroit, 24-29 July 2011, USA.

Toumazou, C., Moschytz, G., & Gilbert, B. (2010). *Trade-offs in analog circuit design: The designer's companion.* Kluwer Academic Publishers.

Tov, E. Y., & Inbar, G. F. (2002). Feature selection for the classification of movements from single movement-related potentials. *IEEE Transactions on Neural Systems and Rehabilitation Engineering, 10*(3), 170–177. doi:10.1109/TNSRE.2002.802875

Trianni, V., Nolfi, S., & Dorigo, M. (2006). Cooperative hole avoidance in a swarm-bot. *Robotics and Autonomous Systems, 54*(2), 97–103. doi:10.1016/j.robot.2005.09.018

Trianni, V., Nolfi, S., & Dorigo, M. (2008). Swarm Robotics. *Design, 4433*(31), 163–191.

Tricaud, C., & Chen, Y. Q. (2009, 10-12 June). Solution of fractional order optimal control problems using SVD-based rational approximations. In *Proceedings of the 2009 American Control Conference,* Hyatt Regency Riverfront, St. Louis, MO, USA (pp. 1430-1435).

Tripathi, J. N. (2011). Designing, optimization and modeling of analog/RF circuits by design of experiments. In *Proceedings of Ph.D. Forum, IEEE/IFIP 19th International Conference on VLSI System- On-Chip,* (pp. 457-460). Oct. 2-5, 2011, Hong Kong, China.

Tripathi, J. N., Nagpal, R. K., Chhabra, N. K., Malik, R., & Mukherjee, J. (2012). Maintaining power integrity by damping the cavity-mode anti-resonances peaks on a power plane by particle swarm optimization. In *Proceedings of 13th International Symposium on Quality Electronic Design,* March 2012, Santa Clara, USA.

Tripathi, J. N., Mukherjee, J., & Apte, P. R. (in press). Nonlinear modeling and optimization by design of experiments: A 2 GHz RF oscillator case study. *International Journal of Design* (in press). *Analysis and Tools for Integrated Circuits and Systems.*

Tseng, C. C. (2001). Design of fractional order digital FIR differentiators. *IEEE Signal Processing Letters, 8*(3), 77–79. doi:10.1109/97.905945

Tustin, A. (1958). The design of systems for automatic control of the position of massive objects. *The Proceedings of the Institution of Electrical Engineers, 105*(Part C, Suppl. No. 1), 1-57.

Uma, K., Palanisamy, P. G., & Poornachandran, P. G. (2011). Comparison of image compression using GA, ACO and PSO techniques. *Proceedings of the 2011 International Conference on Recent Trends in Information Technology (ICRTIT)* (pp. 815–820). Chennai, India: IEEE. doi:10.1109/ICRTIT.2011.5972298

Vaitkus, R. L. (1983). Uncertainty in the values of GaAs MESFET equivalent circuit elements extracted from measured two-port scattering parameters. In *IEEE/Cornell Conference on High-Speed Semiconductor Devices & Circuits,* (pp. 301–308).

Välimäki, V., & Laakso, T. I. (2000, 5-9 June). Principles of fractional delay filters. In *Proceedings of the IEEE International Conference on Acoustics, Speech, and Signal Processing (ICASSP '00)* (Vol. 6, pp. 3870-3873). Istanbul, Turkey.

van Ast, J., Babuška, R., & De Schutter, B. (2008, 6-11 July). Particle swarms in optimization and control. In *Proceedings of the 17th IFAC World Congress* (pp. 5131-5136). Seoul, Korea.

van den Bergh, F., & Engelbrecht, A. P. (2004). A cooperative approach to particle swarm optimization. *IEEE Transactions on Evolutionary Computation, 8*(3), 225–239. doi:10.1109/TEVC.2004.826069

van den Bergh, F., & Engelbrecht, A. P. (2006). A study of particle swarm optimization particle trajectories. *Information Sciences, 176,* 937–971. doi:10.1016/j.ins.2005.02.003

Van der Bergh, F. (2001). *An analysis of particle swarm optimizers.* PhD Thesis, University of Pretoria, Pretoria.

van Niekerk, C., Meyer, P., Schreurs, D. M. M.-P., & Winson, P. B. (2000). A robust integrated multibias parameter-extraction method for MESFET and HEMT models. *IEEE Transactions on Microwave Theory and Techniques, 48*(5), 777–786. doi:10.1109/22.841871

Veeramachaneni, K., Yan, W., Goebel, K., & Osadciw, L. (2006). Improving classifier fusion using particle swarm optimization. *IEEE Symposium on Computational Intelligence in Multicriteria Decision Making*, (pp. 128-135).

Verma, A., Panigrahi, B. K., & Bijwe, P. R. (2009). Transmission network expansion planning with adaptive particle swarm optimization. *World Congress on Nature & Biologically Inspired Computing (NaBIC 2009)*, (pp. 1099–1104).

Verma, A., Panigrahi, B. K., & Bijwe, P. R. (2010). Harmony search algorithm for transmission expansion planning. *IET Generation. Transmission & Distribution*, *4*(6), 663–673. doi:10.1049/iet-gtd.2009.0611

Villanova, R., & Visioli, A. (Eds.). (2012). *PID control in the third millennium – Lessons learned and new approaches*. London, UK: Springer. doi:10.1007/978-1-4471-2425-2

Vilovic, I., Burum, N., Sipus, Z., & Nad, R. (2007). PSO and ACO algorithms applied to location optimization of the WLAN base station. *Proceedings of the 19th International Conference on Applied Electromagnetics and Communications (ICECom2007)* (pp. 1–5). Dubrovnik, Croatia: IEEE. doi:10.1109/ICECOM.2007.4544491

Vinagre, B. M., & Chen, Y. Q. (2002, 10-13 December). Fractional calculus applications in automatic control and robotics. In *Lecture Notes for Tutorial Workshop #2, 41ˢᵗ IEEE International Conference on Decision and Control* (pp. 1-310). Las Vegas, NV, USA.

Vinagre, B. M., Podlubny, I., Hernandez, A., & Feliu, V. (2000). Some approximations of fractional order operators used in control theory and applications. *Fractional Calculus and Applied Analysis*, *3*(3), 231–248.

Vosniakos, G., & Matsas, E. (2010). Improving feasibility of robotic milling through robot placement optimization. *Robotics and Computer-integrated Manufacturing*, *26*, 517–525. doi:10.1016/j.rcim.2010.04.001

Vural, R. A., Yildirim, T., Kadioglu, T., & Basargan, A. (2012, February). Performance evaluation of evolutionary algorithms for optimal filter design. *IEEE Transactions on Evolutionary Computation*, *16*(1), 135–147. doi:10.1109/TEVC.2011.2112664

Wang, H., Wang, N., & Wang, D. (2011, May 23-25). *A memetic particle swarm optimization algorithm for multimodal optimization problems*. Paper presented at the 2011 Chinese Control and Decision Conference, Mianyang, China.

Wang, J., Gao, X.-G., Zhu, Y.-W., & Wang, H. (2010). A solving algorithm for target assignment optimization model based on ACO. *Proceedings of the 2010 Sixth International Conference on Natural Computation (ICNC)* (pp. 3753–3757). Yantai, China: IEEE. doi:10.1109/ICNC.2010.5583099

Wang, L. (Ed.). (2005). *Support vector machines: Theory and applications*. Berlin, Germany: Springer.

Wang, W., Lu, Y., Fu, J. S., & Xiong, Y. Z. (2005, May). Particle swarm optimization and finite-element based approach for microwave filter design. *IEEE Transactions on Magnetics*, *41*(5), 1800–1803. doi:10.1109/TMAG.2005.846467

Wei, J., & Zhang, M. (2011, June 5-8). *A memetic particle swarm optimization for constrained multi-objective optimization problems*. Paper presented at the 2011 IEEE Congress on Evolutionary Computation, New Orleans, USA.

Weis, G., Lewis, A., Randall, M., & Thiel, D. (2010). Pheromone pre-seeding for the construction of RFID antenna structures using ACO. *Proceedings of the 2010 IEEE Sixth International Conference on e-Science (e-Science)* (pp. 161–167). Brisbane, Australia: IEEE. doi:10.1109/eScience.2010.39

Werner, T. G., & Verstege, J. F. (1999). An evolution strategy for short-term operation planning of hydrothermal power systems. *IEEE Transactions on Power Systems*, *14*(4), 1362–1368. doi:10.1109/59.801897

Westerlund, S., & Ekstam, L. (1994, October). Capacitor theory. *IEEE Transactions on Dielectrics and Electrical Insulation*, *1*(5), 826–839. doi:10.1109/94.326654

Widodo, A., Yang, B., & Han, T. (2007). Combination of independent component analysis and support vector machines for intelligent faults diagnosis of induction motors. *Expert Systems with Applications*, *32*, 299–312. doi:10.1016/j.eswa.2005.11.031

Wikipedia. (n.d.). *Swarm intelligence*. Retrieved from http://en.wikipedia.org/wiki/Swarm_intelligence

Wolpaw, J. R., Birbaumer, N., Heetderks, W. J., McFarland, D. J., Peckham, P. H., & Schalk, G. … Vaughan, T. M. (2000). Brain-computer interface technology: A review of the first inter- national meeting. *IEEE Transactions in Rehabilitation Engineering, 8*(2), 164–173.

Wong, K. P., & Wong, Y. W. (1994). Short-term hydrothermal scheduling - Part-I and Part-II: Simulated annealing approach. *IEE Proc. Gen. Transactions Dist., 141 (5)*, 497-506.

Wood, A. J., & Wollenberg, B. F. (1996). *Power generation, operation and control*. John Wiley & Sons.

Wright, M. W., & Valley, G. C. (2005). Yb-doped fibre amplifier for deep-space optical communications. *Journal of Lightwave Technology, 23*(3), 1369–1374. doi:10.1109/JLT.2005.843532

Xie, H., Li, L.-L., Bo, H., & Zhang, Y.-N. (2009). A novel method for ship detection based on NSCT and ACO. *Proceedings of the 2nd International Congress on Image and Signal Processing (CISP '09)* (pp. 1–4). Tianjin, China: IEEE. doi:10.1109/CISP.2009.5304472

Xu, C., El-Hajjar, M., Maunder, R. G., Yang, L.-L., & Hanzo, L. (2010). Performance of the space-time block coded DS-CDMA uplink employing soft-output ACO-aided multiuser space-time detection and iterative decoding. *Proceedings of the 2010 IEEE 71st Vehicular Technology Conference (VTC 2010-Spring)* (pp. 1–5). Taipei, Taiwan: IEEE. doi:10.1109/VETECS.2010.5493768

Xu, S., Bing, Z., Lina, Y., Shanshan, L., & Lianru, G. (2010). Hyperspectral image clustering using ant colony optimization (ACO) improved by K-means algorithm. *Proceedings of the 2010 3rd International Conference on Advanced Computer Theory and Engineering (ICACTE)* (Vol. 2, pp. V2–474–V2–478). IEEE. doi:10.1109/ICACTE.2010.5579337

Xu, X., Liu, Q. H., & Zhang, Z. W. (2002). The stabilized biconjugate gradient fast Fourier transform method for electromagnetic scattering. *Proceedings of the 2002 IEEE Antennas and Propagation Society International Symposium* (Vol. 2, pp. 614–617). San Antonio, TX: IEEE. doi:10.1109/APS.2002.1016722

Yamada, M., Mori, A., Kobayashi, K., Ono, H., Kanamori, T., & Oikawa, K. (1998). Gain-flattened tellurite-based EDFA with a flat amplification bandwidth of 76 nm. *IEEE Photonics Technology Letters, 10*(9), 1244–1246. doi:10.1109/68.705604

Yang, X. S., & Deb, S. (2009). Cuckoo search via Lévy flights. *World Congress on Nature & Biologically Inspired Computing*, (pp. 210-214). Coimbatore.

Yang, J. S., & Chen, N. (1989). Short-term hydrothermal generation scheduling model using a genetic algorithm. *IEEE Transactions on Power Systems, 4*(3), 1050–1056.

Yang, P. C., Yang, H. T., & Huang, C. L. (1996). Scheduling short-term hydrothermal generation using evolutionary programming techniques. *Proceedings of IET, Generation, Transmission, and Distribution, 143*(4), 371–376. doi:10.1049/ip-gtd:19960463

Yang, X. S. (2010). *Nature-inspired metaheuristic algorithms* (2nd ed.). Luniver Press.

Yeh, C. H., Lee, C. C., & Chi, S. (2004). S- plus C-band erbium-doped fiber amplifier in parallel structure. *Optics Communications, 241*(4-6), 443–447. doi:10.1016/j.optcom.2004.07.018

Yeung, C. W., Leung, F. H., Chan, K. Y., & Ling, S. H. (2009, June). *An integrated approach of particle swarm optimization and support vector machine for gene signature selection and cancer prediction*. Paper presented at the International Joint Conference on Neural Network, Atlanta, Georgia.

Yi, L. L., Zhan, L., Hu, W. S., Tang, Q., & Xia, Y. X. (2006). Tunable gainclamped double-pass erbium-doped fiber amplifier. *Optics Express, 14*(2), 570–574. doi:10.1364/OPEX.14.000570

Yi, L. L., Zhan, L., Ji, J. H., Ye, Q. H., & Xia, Y. X. (2004). Improvement of gain and noise figure in double-pass L-band EDFA by incorporating a fiber Bragg grating. *IEEE Photonics Technology Letters, 16*(4), 1005–1007. doi:10.1109/LPT.2004.823697

Yi, L., Zhan, L., Taung, C. S., Luo, S. Y., Hu, W. S., & Su, Y. K. (2005). Low noise figure all-optical gainclamped parallel C+L band erbium-doped fiber amplifier using an interleaver. *Optics Express, 13*(12), 4519–4524. doi:10.1364/OPEX.13.004519

Yilmaz, A. E. (2010). Swarm behavior of the electromagnetics community as regards using swarm intelligence in their research studies. *Acta Polytechnica Hungarica*, *7*(2), 81–93.

Yi-Xiong, J., Hao-Zhong, G., Jian-Yong, Y., & Li, Z. (2005). *Local optimum embranchment based convergence guarantee particle swarm optimization and its application in transmission network planning.* IEEE/PES Transmission and Distribution Conference & Exhibition: Asia and Pacific, Dalian, China.

Yi-Xiong, J., Hao-Zhong, G., Jian-Yong, Y., & Li, Z. (2006). New discrete method for particle swarm optimization and its application in transmission network expansion planning. (Elsevier). *Electric Power Systems Research*, 227–233.

Youssef, H. K., & Hackam, R. (1989). New transmission planning model. *IEEE Transactions on Power Systems*, *4*, 9–18. doi:10.1109/59.32451

Zaghbani, I., Lamraoui, M., Songmene, V., Thomas, M., & El Badaoui, M. (2011). Robotic high speed machining of aluminium alloys. *Advanced Materials Research*, *188*, 584–589. doi:10.4028/www.scientific.net/AMR.188.584

Zaharis, Z. D. (2008). Radiation pattern shaping of a mobile base station antenna array using a particle swarm optimization based technique. *Electrical Engineering*, *90*(4), 301–311. doi:10.1007/s00202-007-0078-y

Zaharis, Z. D., Kampitaki, D. G., Lazaridis, P. I., Papastergiou, A. I., & Gallion, P. B. (2007). On the design of multifrequency dividers suitable for GSM/DCS/PCS/UMTS applications by using a particle swarm optimization-based technique. *Microwave and Optical Technology Letters*, *49*(9), 2138–2144. doi:10.1002/mop.22658

Zamani, M., Karimi-Ghartemani, M., Sadati, N., & Parniani, M. (2009). Design of a fractional order PID controller for an AVR using particle swarm optimization. *Control Engineering Practice*, *17*(12), 1380–1387. doi:10.1016/j.conengprac.2009.07.005

Zhang, Q., & Sun, M. S. (2010). Evolutionary classifier ensembles for semi-supervised learning. *The 2010 International Joint Conference on Neural Networks (IJCNN)*, (pp. 1-6).

Zhang, X., & Wang, X. (2008). A genetic algorithm based time-Frequency approach to a movement prediction task. *Proceeding of the 7th World Congress on Intelligent Control and Automation*, (pp. 1032-1036).

Zhang, B., Sun, X., Gao, L., & Yang, L. (2011). Endmember extraction of hyperspectral remote sensing images based on the ant colony optimization (ACO) algorithm. *IEEE Transactions on Geoscience and Remote Sensing*, *49*(7), 2635–2646. doi:10.1109/TGRS.2011.2108305

Zhang, Z., & Yang, P. (2008). An ensemble of classifiers with genetic algorithm based feature selection. *IEEE Intelligent Information Bulletin*, *9*(1), 18–24.

Zhan, Z. H., Zhang, J., Li, Y., & Chung, H. S. (2009). Adaptive particle swarm optimization. *IEEE Transactions on Systems, Man, and Cybernetics. Part B, Cybernetics*, *39*(6).

Zhao, Y., & Zheng, J. (2004). Particle swarm optimization algorithm in signal detection and blind extraction. *Proceeding of the 7th International Symposium on Parallel Architectures, Algorithms and Networks* (ISPAN'04), (pp. 1-5).

Zhao, L., & Yang, Y. (2009). PSO-based single multiplicative neuron model fir time series prediction. *Expert Systems with Applications*, *36*, 2805–2812. doi:10.1016/j.eswa.2008.01.061

Zhao, S.-Z., & Suganthan, P. N. (2011). Two-*lbests* based multi-objective particle swarm optimizer. *Engineering Optimization*, *43*(1), 1–17. doi:10.1080/03052151003686716

Zhiping, H., Guanaming, C., Cheng, C., He, X., & Jiacai, Z. (2010). A new feature selection method for self-paced brain computer interface. *2010 10th International Conference on Intelligent Systems Design and Applications (ISDA)*, (pp. 845-849).

Zhou, J., Ji, Z., Shen, L., Zhu, Z., & Chen, S. (2011, April 11-15). *PSO based memetic algorithm for face recognition Gabor filters selection*. Paper presented at the 2011 IEEE Workshop on Memetic Computing, Paris, France.

Zhou, X., & Lu, C., Liu, Shum, P., & Cheng, T. H. (2001). A simplified model and optimal design of a multiwavelength backward-pumped fiber Raman amplifier. *IEEE Photonics Technology Letters, 13*(9), 945–947. doi:10.1109/68.942655

Zhu, Q. Y., Qin, A. K., Suganthan, P. N., & Huang, G. B. (2005). Evolutionary extreme learning machine. *Pattern Recognition, 38*, 1759–1763. doi:10.1016/j.patcog.2005.03.028

Zhu, Z., Zhou, J., Ji, Z., & Shi, Y.-S. (2011). DNA sequence compression using adaptive particle swarm optimization-based memetic algorithm. *IEEE Transactions on Evolutionary Computation, 15*(5), 643–658. doi:10.1109/TEVC.2011.2160399

Zi, B., Duan, B. Y., Du, J. L., & Bao, H. (2008). Dynamic modeling and active control of a cable-suspended parallel robot. *Mechatronics, 18*(1), 1–12. doi:10.1016/j.mechatronics.2007.09.004

Zimmerman, R. D., Murillo-Sánchez, C. E., & Thomas, R. J. (2011). MATPOWER: Steady-state operations, planning, and analysis tools for power systems research and education. *IEEE Transactions on Power Systems, 26*(1). doi:10.1109/TPWRS.2010.2051168

About the Contributors

Girolamo Fornarelli received his Master's degree in Electronic Engineer and the Ph.D. degree in Electrical Engineering from the Politecnico di Bari where at present, he is Assistant Professor in the Circuit Theory group. He taught "Circuit Simulation" and "Fundamentals of Electric Circuits" at the I and II Faculty of Engineering, and concurrently, he teaches "Electrotechnics" at the I Faculty of Engineering. He serves as reviewer for many international conferences and journals, as well as chairman, organizer of the special sessions, and a member of the international program committee at international conferences. His most research interests deal with analysis of non-linear circuits, theoretical aspects for the development of neural networks, application of artificial neural networks and soft computing methods to clustering and non-destructive evaluation of industrial products and installations. Moreover, he works in the field of evolutionary computational, in detail such studies are related the characterization, optimization, and design of neural circuits and optical fibres.

Luciano Mescia joined the Department of Electrical and Electronic Engineering at the Politecnico of Bari as Assistant Professor. His research activity focus on fibre and integrated optic technologies, with emphasis on rare-earth doped active devices, optical sensors for the environmental monitoring, optical fibres (conventional and microstructured) for industrial, biological and biomedical applications. Recently, he works in the field of the swarm intelligence for the design, optimization and characterization of high power fibre amplifiers for space applications and harsh environments. Moreover, his research activity is devoted to the design of lasing action in rare earth doped microspheres and to the design of novel optical antennas for solar energy harvesting. During his activity research he has cooperated with many national and international research institutions and he has been involved in several collaborative projects with academic and industrial partners. Moreover, he serves as reviewer for many international conferences and journals, as well as chairman at international conferences.

* * *

Diego Andina (Prof. Dr.-Eng, IEEE Senior Member SMC&IES), was born in Madrid, Spain, were he received simultaneously two Masters' degrees, in Computer Science and in Electronics & Communications from the Technical University of Madrid (UPM), Spain, in 1990. He achieved the PhD degree in 1995, also from UPM, with a thesis on Artificial Neural Networks applications in Signal Processing. He presently works for the UPM where he heads the Group for Automation in Signals and Communications (GASC/UPM), a research group in-terested in Multidisciplinary Applications of Signal Processing and Computational Intelligence: Man-Machine Systems and Cybernetics. He is author or co-author of about 250 national and international publications, having being director of more than 50 R+D projects financed by National and Local Governments, the European Commission or private Institutions and Firms. He is also Associate Editorial Member of several international research journals and transactions, and has participated in the organization of more than 60 international research, innovation, or technology transfer events.

Prakash R. Apte is currently a Professor at Indian Institute of Technology Bombay. He completed his B.E. (EE) from Indore University, M. Tech (EE) from IIT Kanpur and Ph.D. (Physics) from University of Bombay. From 1970-2001, he was involved in research at TIFR Bombay. From 2001- till date, he is with Indian Institute of Technology Bombay. His research interests confine Silicon Semiconductor Devices & IC Technology and Thin Film Devices & Technology of High T_c Superconductors. His other interests are Design and Analysis of experiments for Research & Development, Taguchi Method for Industrial process/product optimization, technology innovation using TRIZ methods and sensors using MicroElectromechanical Systems (MEMS & NEMS).

Adham Atyabi received his B.Sc in Computer-Engineering from Azad University of Mashad-Iran in 2002. His B.Sc thesis title was "Imitating human speech." He received his M.Sc by research from the faculty of Computer Science and Information Technology at Multimedia University in Malaysia in 2009. His M.Sc thesis title was "Navigating agents in uncertain environment using Particle Swarm Optimization." Currently, he is studying his PhD under Prof. David M. W. Powers' supervision at CSEM, Flinders University. In the course of his study, he was PhD Student representative of Flinders at Australian Computer Society (ACS) in 2010-2011 and University Relation Manager of Young IT SA in 2011. He was also a member of Magician team (one of the top 5 teams in the world in the Magic2010 competition) in Flinders University in 2009-2010.

Konstantinos B. Baltzis was born in Thessaloniki, Greece in 1973. He received his B.Sc. degree in Physics in 1996, his M.Sc. degree in Communications and Electronics in 1999, and his PhD degree in Communication Engineering in 2005, all from the Aristotle University of Thessaloniki (AUTh), Greece. He is the author or co-author of more than fifty scientific articles in international journals and conferences. He is a research associate in the RadioCommunications Laboratory of AUTh and a teaching staff member at the Program of Postgraduate Studies in Electronic Physics of AUTh. His current research interests are focused in the areas of wireless communication systems and networks, propagation, antennas, and evolutionary optimization. Dr. Baltzis is a member of IEEE, FITCE, Hellenic Physical Society, and Greek Association of Physicists in Electronics and Communications.

Gabriel Cormier obtained a B.A.Sc (Electrical Engineering) and an M.A.Sc (Electrical Engineering) from the Université de Moncton, in Moncton, New Brunswick, Canada, in 1998 and 2000, respectively. He received his Ph.D. in Electrical Engineering from Carleton University in Ottawa, Canada, in 2007. He has been an Assistant Professor at the Université de Moncton since 2006. His research interests include the design of millimeter-wave integrated circuits and devices, using GaAs and GaN substrates, as well as the use of evolutionary algorithms, such as genetic algorithms and particle swarm optimization, applied to various engineering optimization problems.

Massimo Donelli received the Electronic Engineering degree and Ph.D. degree in space science and engineering from the University of Genoa, Italy, in 1998 and 2003, respectively. Actually he is an Assistant Professor of Electromagnetic Field at the Department of Information and Communication Technology, University of Trento, Italy. His main interests are electromagnetic inverse scattering, adaptive antennas synthesis, optimization techniques for microwave imaging and microwave systems and devices design, wave propagation in superconducting materials and urban environment. Massimo Donelli is member of the European Microwave Association (EuMa), of the Italian Society of Electromagnetism (Siem) and he is Senior Member of the IEEE.

Mourad Fakhfakh was born in Sfax-Tunisia in 1969. He received the Engineering, the PhD and the Habilitation degrees from the National Engineering School of Sfax Tunisia in 1996, 2006, and 2011, respectively. From 1998 to 2004 he worked in the Tunisian National Society of Electricity and Gas (STEG) as a department head. In September 2004, he joined the higher institute of electronics and communications (ISECS)/University of Sfax, Tunisia, where he is working as an Associate Professor. His research interests include symbolic analysis techniques, analog design automation, and optimization techniques.

Antonio Giaquinto received his Master's degree in Electronic Engineer in 2001 and the Ph.D. degree in Electrical Engineer from the Politecnico di Bari in 2005. His main research interests include cellular neural networks working as associative memories particularly in the field of the signal processing applications, application of evolutionary computation to the synthesis of circuits, linear and nonlinear electric circuits. Moreover, his research topics include the development of artificial neural networks and soft computing methods in clustering and comparison of multi-dimensional data sets in the field of industrial non-destructive diagnosis.

Sotirios K. Goudos was born in Thessaloniki, Greece in 1968. He received the B.Sc. degree in Physics in 1991 and the M.Sc. degree in Electronics in 1994 both from the Aristotle University of Thessaloniki. In 2001 he received the Ph.D. degree in Physics from the Aristotle University of Thessaloniki and in 2005 the Master in Information Systems degree from the University of Macedonia, Greece. In 2011 he obtained the Diploma degree in Electrical and Computer Engineering from the Aristotle University of Thessaloniki. Since 1996 he has been working in the Telecommunications Center of the Aristotle University of Thessaloniki, Greece. His research interests include antenna and microwave structures design, electromagnetic compatibility of communication systems, evolutionary computation algorithms, mobile communications and semantic web technologies. Dr. Goudos is a member of the IEEE, the Greek Physics Society, the Technical Chamber of Greece, and the Greek Computer Society.

Nosrat Granpayeh has received his B.Sc., M.Sc., and Ph.D. degrees in Telecom. Eng. from Telecommunication College of Iran, Radio and Television College, Tehran, Iran and University of NSW, Sydney Australia, in 1975, 1980, and 1996, respectively. In 1975, as an honor graduate of the Faculty of Electrical and Computer Engineering of K. N. Toosi University of Technology, Tehran, Iran, he was employed as an instructor there, where he was later promoted to Lecturer, Assistant, and Associate Professor in 1980, 1996, and 2007, respectively. His research interests are in optical devices, equipments and materials, optical fibers, and optical fiber nonlinear effects. He is the author or co-author of more than 40 journal- and 60 conference- papers. He is a member of OPSI, IAEEE, IEEE, IEEE Photonic Society, and OSA.

V. H. Hinojosa was born in Quito, Ecuador, on January 30, 1975. He received his B.Sc. in Electrical Engineer from the Escuela Politécnica Nacional (EPN), Quito, Ecuador, in 2000. In 2001, he was awarded with a scholarship by German Academy Exchange Program (DAAD) to carry out his Ph.D. studies at the Instituto de Ingeniería Eléctrica from the Universidad Nacional de San Juan, San Juan, Argentina. He received his Ph.D. in Electrical Engineering on December 2007. Since 2008, he works as Professor in Department of Electrical Engineering at Universidad Técnica Federico Santa Maria (UT-FSM), Valparaíso, Chile. His special fields of interest include optimization models, operation research, and artificial intelligence applied to planning and operation of power systems, and renewable energies. He is a member of IEEE-Chile and the Computer Intelligence Society from Ecuador.

Azadeh Rastegari Hormozi was born in Shiraz, Iran. She entered the Islamic Azad University of Arsanjan to study physics where she has received her BSc. in 2004. During her B.Sc. she has done his thesis on simulation of the noise of mode-locked lasers. At 2005 she has entered the research institute of Fereshtegan where she conducts research on the optical fiber amplifiers. She is also active in student scientific educations and has cooperation with some of the student educational institutions.

Mo M. Jamshidi (Fellow IEEE, Fellow ASME, A. Fellow-AIAA, Fellow AAAS, Fellow TWAS, Fellow NYAS, Foreign Fellow HAE) received BS in EE, Oregon State University, Corvallis, OR, USA in 1967, the MS and Ph.D. degrees in EE from the University of Illinois at Urbana-Champaign, IL, USA in June 1969 and February 1971, respectively. He holds honorary Doctorate degrees from University of Waterloo, Canada, 2004 and Technical University of Crete, Greece, 2004. Currently, he is the Lutcher Brown Endowed Chaired Professor and Leader of Sustainable Energy Research Group at the University of Texas, San Antonio, TX, USA. He has also been the founding Director of Center for Autonomous Control Engineering (ACE – ace.utsa.edu) at the University of New Mexico in 1995, and has moved the Center to University of Texas, San Antonio in 2006. He was a Senior Research Advisor at US Air Force Research Laboratory, KAFB, NM from 2002-2005 and 1984-1990. He was also a consultant with US Department of Energy Office of Industrial Technologies and DOE Laboratories Oak Ridge, Sandia and Los Alamos. He was also an advisor for the NASA Headquarters from 1998-2004 and on NASA JPL's Pathfinder Project mission and Surface Systems Track Review Board. He has worked in various academic and industrial positions at various national and international locations including with IBM and GM Corporations. In summer 1999, he was a NATO Distinguished Professor in Portugal conducting lectures on intelligent systems and control.

Zoran D. Jeličić received the Electronic Engineer degree from the University of Novi Sad, Serbia in 1995 and the PhD degree from same University in 2003. During 2001, he held research position in Germany at TU Berlin. Since 2003 he has held various positions at the Department of Computing and Control, Faculty of Technical Sciences, University of Novi Sad, Serbia, where he is currently an Associate Professor. His research interests are optimization, optimal control, and fractional control systems.

Aleksandar Jevtić received his B.Sc. and M.Sc. degrees in Electrical Engineering from the University of Belgrade, Serbia in 2005, and the M.Sc. and the Ph.D. degree in Computer Science from the Technical University of Madrid (UPM), Spain in 2007 and 2011, respectively. He currently works as a Marie Curie postdoctoral researcher in Robosoft, France. On many occasions he was invited to universities in the United States, the United Kingdom, Israel, Germany, and Serbia where he worked as a visiting researcher or lecturer. He was an international committee member and a special-session chair or co-chair on several international conferences. He is the author of two book chapters and more than 20 referee journal papers and conference proceeding papers. His research interests include human-robot interaction, robot control, machine learning, swarm intelligence and multi-agent systems, and image processing and analysis.

Željko S. Kanović was born on July 18th, 1976 in Sombor, Serbia. He enrolled at Mechanical Engineering Department of Faculty of Technical Sciences, University of Novi Sad in 1995, where he graduated in October 2000. In 2007 he received Master of Technical Sciences degree and in 2012 PhD degree in the field of Automatic Control at Computing and Control Department at the same university. His main scientific interest is in the area of optimal control and global optimization techniques, particularly swarm optimization, with practical application in industrial process control and monitoring, water distribution systems control and optimization and energy efficiency. He is employed at the Faculty of Technical Sciences in Novi Sad as research and teaching assistant.

Kamran Kiasaleh received his B.S. (Cum Laude), M.S., and Ph.D. degrees all in Electrical Engineering from the Communications Sciences Institute at the University of Southern California in 1981, 1982, and1986, respectively. He has been with the University of Texas at Dallas, where he is currently a Full Professor of Electrical Engineering. Dr. Kiasaleh is the recipient of the Research Initiation Award from the National Science Foundation. He was also the recipient of the NASA/ASEE faculty fellowship award at the Jet Propulsion Laboratory (JPL) in 1992 where he participated in the Galileo Optical Experiment (GOPEX) demonstration, the first successful demonstration of an optical communications link involving a deep-space vehicle. He received the NASA Group Achievement Award in 1992. In 1993, he was the recipient of NASA/ASEE faculty fellowship award at JPL where he participated in Compensated Earth-Moon-Earth Laser Link (CEMERLL) demonstration. He was an Associate Editor for *IEEE Communications Letters*.

Kang Li received PhD degree on Control Theory and Applications from Shanghai Jiaotong University, China, in 1995. He is currently a Professor of Intelligent Systems and Control at the School of Electronics, Electrical Engineering and Computer Science, Queen's University Belfast, UK, where he teaches and conducts researches in control engineering. His research interests cover nonlinear system modelling, identification and control, and bio-inspired computational intelligence, with recent applications to power systems and polymer extrusion. He has also extended his research to bioinformatics and systems biology with applications on food safety and healthcare. He has published over 160 papers in the above areas. Dr Li serves in the editorial board as an associate editor or member of the editorial board for *Neurocomputing*, the *Transactions of the Institute of Measurement & Control, Cognitive Computation*, and *Int. J. of Modelling, Identification and Control*. Dr Li is a senior member of the IEEE, a Fellow of the Higher Education Academy, UK, a member of the IFAC Technical Committee on Computational Intelligence in Control, and a member of the Executive Committee of UK Automatic Control Council.

António M. Lopes received a PhD degree in Mechanical Engineering from the University of Porto, Portugal, in March 2000. He is currently Assistant Professor at the School of Engineering, University of

Porto, where he started teaching in 1991. Besides the research interests in robotics, control, and complex systems, he has also interests in distance learning and remote labs.

Martin Luerssen's research forte is in the field of nature-inspired computing, specifically the application of biological principles to machine learning. His dissertation, which was nominated for the CORE Australasian Distinguished Doctoral Dissertation award, innovated the use of graph grammars in evolutionary design and optimisation. With a comprehensive background in computational intelligence, Dr Luerssen has served as a program committee member for several international conferences and has delivered talks on many diverse domains, including cognitive science, artificial life, and user interfaces. At present, he is a key participant of the Thinking Head project, an ARC/NH&MRC Special Research Initiative that targets the development of intelligent Embodied Conversational Agents (ECAs). Dr Luerssen is responsible for the successful audiovisual embodiment and evolution of the agent, which encompasses the agent's adaptive capacity to comprehend its social context from audiovisual input and synthesise realistic speech and animation patterns in response.

Guido Maione received the *Laurea* degree with honors in Electronic Engineering in 1992, and the PhD degree in Electrical Engineering (curriculum "Industrial Automation") in 1997, both from Politecnico di Bari. In 1996 he joined the University of Lecce as Assistant Professor of Control Systems Engineering. He was also Visiting Scholar at the Electrical, Computer and Systems Engineering Department of Rensselaer Polytechnic Institute (Troy, New York, USA) in 1997. In December 2002, he moved back to Technical University of Bari. His research interests cover: non-integer (fractional) order systems and controllers; modelling, simulation, and control of discrete event systems (e.g. automated manufacturing systems, intermodal container terminals); soft computing techniques applied to multi-agent systems; Petri nets and digraph modelling tools for manufacturing control. He is currently member of: IFAC; IEEE; IEEE Control Systems Society; IEEE Robotics and Automation Society Technical Committee on Marine Robotics; Italian SIDRA society of researchers and professors in control systems engineering; COMES (COntrol and Models of Event Systems) group. In 1996 he was also founder and *ad interim* president of the IEEE Student Branch at Politecnico di Bari.

Arezoo Modiri received her B.S. and M.S. degrees in Electrical Engineering from University of Tehran and Iran University of Science and Technology, Tehran, Iran, respectively. She was with Iran Telecommunication Research Center as an RF Engineer for more than three years and collaborated with Islamic Azad University as an instructor for three semesters. She has joined University of Texas at Dallas in 2009 where she is currently a Ph.D. candidate in Electrical Engineering department working on microwave interaction with live tissue. She is the winner of the 2012 best teacher assistant award in Electrical Engineering department. Her research interests include microwave imaging, electromagnetics, radio frequency systems and components, antennas, optimization algorithms, and reflectometry.

Alireza Mowla was born in 1981 in Shiraz, Iran. He has entered Shiraz University to study Electrical Engineering and has received his B.Sc. in 2005 and his M.Sc. in Electrical Engineering in 2008 from K. N. Toosi University of Technology. Since then he has research affiliations with K. N. Toosi University of Technology and work as a lecturer at Azad University of Sepidan and also at Shiraz University of Applied Science and Technology. Mr. Mowla has done researches in the area of optical fibers and amplifiers, lasers and optical components. At the moment he is going to strat his Ph.D in Electrical Engineering at the University of Queensland, Brisbane, Australia. He is also a member of the Optical Society of America.

Jayanta Mukherjee is currently an Associate Professor at Indian Institute of Technology Bombay. His research interests include RFIC design and testing, antennas, and biomedical VLSI circuits. Previously he was a Texas Instruments Fellow during his Ph.D. 2001-2004. He was also awarded the gold medal at Birla Institute of Technology during his Bachelor of Engineering. He has interned at Thomson Multimedia, Princeton, New Jersey during 2002-2003. He is author of two books.

Paulo Moura Oliveira received the Electrical Engineering degree in 1991, from the UTAD University, Portugal, MSc in Industrial Control Systems in 1994 and PhD in Control Engineering in 1998, both from Salford University, Manchester in the UK. He is an Assistant Professor with Aggregation at the Engineering Department of UTAD University. His research interests are focused on the fields of intelligent control, PID control, control education, and evolutionary and natural inspired algorithms, with applications in several domains.

Matteo Pastorino received the *laurea* degree in Electronic Engineering from the University of Genoa, Italy, in 1987 and the Ph.D. degree in Electronics and Computer Science in 1992. Since 2008, he has been a Full Professor of Electromagnetic Fields at the University of Genoa. He has been the past Director of the Department of Biophysical and Electronic Engineering (2008-2011). Currently, he is with the Department of Naval, Electrical, Electronic, and Telecommunication Engineering and teaches the courses of "Acoustic and electromagnetic emission of ships" and "Yacht navigation support systems." His main research interests are in the field of nondestructive testing and evaluation, antennas, biolectromagnetics, and applied electromagnetics. He is the author of the book Microwave Imaging (Wiley, 2010) and the coauthor of more than 350 papers in international journals and congress proceedings. Prof. Pastorino has been an Associate Editor of the *IEEE Transactions on Antennas and Propagation* (2004-2010). At present, he is an Associate Editor of the *IEEE Trans. on Instrumentation and Measurement*. He is also a member of the editorial boards and technical program committees of several other international journals and conferences in the field of microwaves and antennas.

E. J. Solteiro Pires graduated in 1993 in Electrical Engineering at the University of Coimbra, in 1993. He pursued post graduate studies and obtained, in 1999, an MSc degree in Electrical and Computer Engineering at the University of Oporto. In 2006, he graduated with a PhD degree at UTAD University. Since 2006 he works as an Assistant Professor at the Engineering Department of UTAD University. His main research interests are in evolutionary computation, multi-objective problems, and fractional calculus.

David Powers is a Professor in the School of Computer Science, Engineering and Mathematics at Flinders University, where he is Director of the Centre for Knowledge and Interaction Technologies. He has been awarded a Diploma in Technical Analysis, Dip. TA (ATAA), Securities Institute Australia in 2001, a Ph.D. (Electrical Engineering and Computer Science, University of NSW) in 1989, a Certificate in Theology (Hons; Moore Theological College) in 1984, a Certificate in Linguistics (Introductory; Summer Institute of Linguistics) in 1984, and a B.Sc. (Hons in Computer Science and major in Pure Mathematics; U. Sydney) in 1979. David's major research interests lie in the general area of artificial intelligence and cognitive science, taking language, learning and logic as the cornerstones for a broad cognitive science perspective on artificial intelligence and its practical applications, with a particular focus on human factors, medical devices, and assistive technologies.

Antonio Punzi was born in Cisternino (BR), Italy, on 20 September 1984. He earned a first-level undergraduate degree in Electronic and Information Engineering (27 July 2007) and a second-level graduate degree in Electronic and Information Engineering (22 July 2011), both from the II School of Engineering (in Taranto) of the Politecnico di Bari, Italy. The first-level graduation project and the second-level graduation thesis were both supervised by Dr. Guido Maione. Part of the last work was developed at the Intelligent Systems and Control (ISAC) research cluster, School of Electronics, Electrical Engineering and Computer Science, Queen's University Belfast, where Punzi was a visiting student (March-June 2010). There he studied advanced heuristic optimization methods supervised by Dr. Kang Li. In November 2011 Punzi moved to Stuttgart, Germany, where he is currently a system developer for a small software company in the automotive industry.

Milan R. Rapaić was born on November 17, 1982 in Ruma, Serbia. In 2001 he enrolled at Computing and Control Department of Faculty of Technical Sciences, University of Novi Sad, where he received Master of Sciences degree in 2006. He received PhD degree in the field of Automatic Control Systems from the same university in 2011. His primary research interests include control theory, with particular emphasis on fractional order systems and optimal control, and global optimization, particularly particle swarm optimization (PSO). Currently he is employed as an Assistant Professor (Docent) at Faculty of Technical Sciences in Novi Sad.

Andrea Randazzo received the *laurea* degree in Telecommunication Engineering from the University of Genoa, Genoa, Italy, in 2001 and the Ph.D. degree in Information and Communication Technologies from the same university in 2006. Currently, he is an Assistant Professor at the Department of Naval, Electrical, Electronic, and Telecommunication Engineering of the University of Genoa. His primary research interests are in the field of electromagnetic scattering (both direct and inverse), smart antennas, and numerical methods for microwave nondestructive evaluations and imaging.

Marcos J. Rider was born in Lima, Peru; in 1975. He received his BSc (with honors) and the Professional Engineer (PE) degrees in 1999 and 2000, respectively, from National University of Engineering (UNI), Faculty of Electrical and Electronics Engineering (FIEE), Lima, Peru; his MSc degree in 2002 from the Federal University of the Maranhao (UFMA), Brazil, and his PhD degree in 2006 from University of Campinas (UNICAMP), Brazil, all in Electrical Engineering. Currently he is a Professor in the Electrical Engineering Department at the Universidade Estadual Paulista (UNESP), Ilha Solteira, Brazil. His professional experiences include several research projects, international consultings and technical papers publications. His areas of research are the development of methodologies for the optimization, planning, and control of electrical power systems and applications of artificial intelligence in power systems.

Cristhoper Leyton Rojas was born in La Serena, Chile, on June 03, 1986. He received his B.Sc. in Electrical Engineer from the Universidad Técnica Federico Santa Maria (UTFSM), Valparaíso, Chile, in 2011. Since 2011, he works as Planning Engineer in the Operation Area from the Centre for Economic Load Dispatch of the Large Northern Interconnected System (CDEC-SING), Chile. His interest areas are operation and planning of electrical power systems using optimization models and techniques of artificial intelligence.

Tyler Ross was born in Dieppe, New Brunswick, Canada in 1985. He received the B.Eng and M.A.Sc. degrees in Electrical Engineering from the Université de Moncton in Moncton, New Brunswick, Canada, in 2008 and 2010, respectively. He is currently pursuing the Ph.D. degree in Electrical and computer engineering at Carleton University in Ottawa, Ontario, Canada. His current research interests include the design of microwave and millimetre-wave integrated circuits and the modeling of microwave transistors, particularly gallium nitride (GaN) circuits and devices. He has been the recipient of several Natural Sciences and Engineering Research Council of Canada (NSERC) scholarships and awards.

Patrick Siarry was born in France in 1952. He received the PhD degree from the University Paris 6, in 1986 and the Doctorate of Sciences (Habilitation) from the University Paris 11, in 1994. He was first involved in the development of analog and digital models of nuclear power plants at Electricité de France (E.D.F.). Since 1995 he is a Professor in Automatics and Informatics. His main research interests are computer-aided design of electronic circuits, and the applications of new stochastic global optimization heuristics to various engineering fields. He is also interested in the fitting of process models to experimental data, the learning of fuzzy rule bases, and of neural networks.

J. A. Tenreiro Machado graduated with Licenciatura (1980), PhD. (1989), and Habilitation(1995) degrees in Electrical and Computer Engineering at the University of Porto. During 1980-1998 he worked as Professor at the Department of Electrical and Computer Engineering of the University of Porto. Since 1998 he is Coordinator Professor at the Institute of Engineering of the Polytechnic Institute of Porto, Department of Electrical Engineering. His main research work is on the areas of nonlinear dynamics, modeling, control, fractional calculus, evolutionary computing, intelligent transportation systems, and robotics.

Santiago P. Torres (S'05, M'07, SM'10) received the B.S. from the University of Cuenca, in Ecuador, in 1998, and the Ph.D. degree from the Institute of Electrical Energy of the National University of San Juan in Argentina, in 2007. In 2011, Dr. Torres was a Post-Doctoral Fellow at the School of Electrical and Computing Engineering at Cornell University. He is a Post-Doctoral Fellow at the Power Systems Department, University of Campinas, Brazil. His research interests are operation and planning of smart transmission and distribution systems, computational intelligence, and optimization applications in smart power systems.

Jai Narayan Tripathi was born in Gangapur (Bhilwara), a town in Rajasthan, India. He received his Bachelor of Engineering (ECE) in 2007 from University of Rajasthan, Jaipur, and M.Tech (ICT) from

DA- IICT, Gandhinagar in 2009. He has worked as a Research Intern at TR&D, STMicroelectronics Pvt. Ltd., Greater Noida, India for one year. He worked on Signal Integrity and Power Integrity issues for high speed serial links. At present he is a Ph.D. Scholar at IIT Bombay, Mumbai. His areas of interest are Signal Integrity and RF Circuits Optimization. He is author of one monograph "Analysis of Signal Integrity and Power Integrity at System Level." He has served as reviewer for some conferences and various journals.

Zaharias D. Zaharis received the B.Sc. degree in Physics in 1987, the M.Sc. degree in Electronics in 1994 and the Ph.D. degree in 2000 from Aristotle University of Thessaloniki. Also, in 2011 he obtained the Diploma degree in Electrical and Computer Engineering from the same university. Since 2002 he has been working in the administration of the telecommunications network at the Aristotle University of Thessaloniki. His research interests include design and optimization of antenna arrays, design and optimization of microwave circuits and systems, mobile communications systems, RF measurements, wave propagation, electromagnetic scattering, and computational electromagnetics. Dr. Zaharis is a member of the Greek Physics Society and the Technical Chamber of Greece.

Index

A

AC model 260-263, 270-271, 274-275, 277-278, 280, 283
across-group variance (AGV) 330
Adaptive Particle Swarm Optimization (APSO) 262, 267, 283
agent 54, 102, 193
amplified spontaneous emission (ASE) 130-131, 153
Analog 40, 42, 47, 50-55, 57-58, 62-65, 67-70, 177
analog behavioral modeling 62
ant colony optimization (ACO) 238, 309, 320, 323, 327-328
approximation 17, 27, 32-33, 54, 155, 194, 198-203, 205-210, 212-217, 219, 222, 312, 318, 323-324
autonomous robot 193
average final value (AFV) 79
average iteration number (AIN) 79

B

binary PSO (BPSO) 73
brain computer interface (BCI) 326

C

cage fault frequency 250
circuit optimization 39
circuit under test (CUT) 64
common spatial pattern (CSP) 330
complex number PSO (CPSO) 74, 91
computational swarm intelligence 146, 191, 193
constriction factor PSO (CFPSO) 100-101
control theory 174, 217, 219, 237-238, 240-242, 253, 258
CRONE controller 201
crossover 4, 101, 109, 205, 225, 245, 253, 328, 343
Current Conveyor Transconductance Amplifier (CCTA) 40, 47

D

DC model 260-261, 269-270, 273-277, 281, 283
design automation 57-58, 68-70
differential evolution (DE) 327
discrete PSO (DPSO) 74, 91, 331
dispersion-compensating fiber (DCF) 150
dispersion-shifted fiber (DSF) 150
Distributed Bees Algorithm (DBA) 170-171, 189
distributed generator (DG) 255, 267, 309, 322
distributed robotic system 193
distributed system 170

E

earliest due date (EDD) 310
economic dispatch problem (EDP) 286, 288
electrode selection/reduction (ES 330
electroencephalogram (EEG) 326, 343
electronic design automation (EDA) 57
elitist distributed PSO (ED-PSO) 66
Elitist Learning Strategy (ELS) 267
equivalent series resistance (ESR) 64
erbium-doped fiber amplifier (EDFA) 132, 144, 147-166, 168
evolutionary algorithm (EA) 4, 14, 16-17, 53-56, 75, 79, 83, 89, 92, 97, 101, 119-120, 130, 224, 233, 235, 238, 253, 265, 280, 282, 293, 295, 301, 304, 322, 326-336, 344
evolutionary computation 13, 17, 52-55, 69-70, 93-94, 96-98, 101, 121-124, 144-146, 164, 167-168, 216-218, 229, 235-236, 254-256, 258, 272, 279, 282, 321, 336-337, 339-340
Evolutionary Particle Swarm Optimization (EPSO) 262, 265, 283
evolutionary programming (EP) 75, 327
Evolutionary State Estimation (ESE) 267
evolution strategy (ES) 327